Advances in Geographical and Environmental Sciences

Advances in Geographical and Environmental Sciences synthesizes series diagnostigation and prognostication of earth environment, incorporating challenging interactive areas within ecological envelope of geosphere, biosphere, hydrosphere, atmosphere and cryosphere. It deals with land use land cover change (LUCC), urbanization, energy flux, land-ocean fluxes, climate, food security, ecohydrology, biodiversity, natural hazards and disasters, human health and their mutual interaction and feedback mechanism in order to contribute towards sustainable future. The geosciences methods range from traditional field techniques and conventional data collection, use of remote sensing and geographical information system, computer aided technique to advance geostatistical and dynamic modeling.

The series integrate past, present and future of geospheric attributes incorporating biophysical and human dimensions in spatio-temporal perspectives. The geosciences, encompassing land-ocean-atmosphere interaction is considered as a vital component in the context of environmental issues, especially in observation and prediction of air and water pollution, global warming and urban heat islands. It is important to communicate the advances in geosciences to increase resilience of society through capacity building for mitigating the impact of natural hazards and disasters. Sustainability of human society depends strongly on the earth environment, and thus the development of geosciences is critical for a better understanding of our living environment, and its sustainable development.

Geoscience also has the responsibility to not confine itself to addressing current problems but it is also developing a framework to address future issues. In order to build a 'Future Earth Model' for understanding and predicting the functioning of the whole climatic system, collaboration of experts in the traditional earth disciplines as well as in ecology, information technology, instrumentation and complex system is essential, through initiatives from human geoscientists. Thus human geoscience is emerging as key policy science for contributing towards sustainability/survivality science together with future earth initiative.

Advances in Geographical and Environmental Sciences series publishes books that contain novel approaches in tackling issues of human geoscience in its broadest sense-books in the series should focus on true progress in a particular area or region. The series includes monographs and edited volumes without any limitations in the page numbers.

Subhash Anand · Rituparna Bhattacharyya ·
Madhushree Das · Tulshi Kumar Das ·
Pushkar K. Pradhan
Editors

Sustainability in South Asian Cities

Springer

Editors
Subhash Anand
Department of Geography
University of Delhi
New Delhi, Delhi, India

Madhushree Das
Department of Geography
Gauhati University
Guwahati, Assam, India

Pushkar K. Pradhan
Central Department of Geography
Tribhuvan University
Kathmandu, Nepal

Rituparna Bhattacharyya
Space and Culture, India
North Shields, UK

Centre for Indian Knowledge Systems
Indian Institute of Technology
Guwahati, India

Department of Earth Science
University of Science and Technology
Meghalaya, India

Tulshi Kumar Das
Department of Social Work
Shahjalal University of Science
and Technology
Sylhet, Bangladesh

ISSN 2198-3542 ISSN 2198-3550 (electronic)
Advances in Geographical and Environmental Sciences
ISBN 978-981-97-7454-8 ISBN 978-981-97-7455-5 (eBook)
https://doi.org/10.1007/978-981-97-7455-5

This Springer imprint is published by the registered company Springer Nature Singapore Pte Ltd.
The registered company address is: 152 Beach Road, #21-01/04 Gateway East, Singapore 189721, Singapore

Foreword: Understanding the Essence of Sustainability

This present anthology, consisting of 24 chapters, takes various facets and social, economic, and environmental sustainability issues, illustrated with cities of South Asia. These essays are arranged into thematic parts—Part I—Towards Blue-Green Infrastructure of South Asian Cities; Part II—Socio-Economic Challenges Towards Sustainable Cities and Part II—Environmental Challenges. Chapter 1, narrates a critical appraisal of the selected literature on sustainability issues in South Asian cities, emphasising the making and paving the path of a sustainable future while keeping in mind that despite the successes, challenges always remain and constantly one should widen and make the sustainable path stronger and humane. In these challenges, population pressure is an alarming issue as around half of the world's population lives in Asia.

Part I of this volume focuses on the concept of 'Blue-Green Infrastructure' aimed at making the cities of South Asia more sustainable and livable.

Part II focuses on the diverse issues of socioeconomic challenges towards sustainable cities, capturing themes like urbanization and migration, drinking water, food culture, the sustainable relationship between nature and society, economic sustainability, the social environment in slums (cases of Delhi, Bangladesh, and Nepal), urban vulnerabilities during COVID-19, fictional story of the making of and coping mechanism in Dhaka, transportation plans and policies in Dhaka, status and role of the informal sector in Kathmandu, urban informal workforce during COVID-19, making of Bolpur as a 'sustainable city' over time, interfaces between urbanization and sustainability in Uttar Pradesh, and urban sprawl and future land uses projections. Finally, Part III presents glimpses of environmental challenges, illustrated with case studies mapping changes in water surface area in Afghanistan, land use/land cover changes in Pakistan, impact of bare lands in the cities of Kathmandu valley, urban growth and land use dynamics in Nepal, and agriculture practice for sustainable livelihood in Kathmandu valley.

The editors' ambition and claim that "This volume would serve as a valuable companion guide for those engaged in sustainability—whether as scholars, practitioners or policymakers" is rationally justified to an optimal level and, to an extent, successfully made the horizon widened and the paving of path stronger. I expect this compendium will throw fresh light on understanding, envisioning, and action orientation, taking the niches and gaps examined in different case studies. Of course, there appears to be a diversity of topics and contents; however, they are woven together by a commonality—to foreground the challenges and achieve sustainability in South Asian cities.

"I emphasise that" through the practice and use of sacred ecology, a strategy for sustainable development considering heritage conservation, preservation, and reverential development should be accepted in the service to human civilisation and its symbolic identity. Let us come to one stop on the cosmic path, at least at this stage, with the words of the African ecologist Babu Diou (as cited in Singh 1995, p. 213):

> In the end
> We will conserve only what we love.
> We will love only what we understand.
> We will understand only what we learn.

Rana P. B. Singh
(Ex)-Professor and Head
Department of Geography
Banaras Hindu University
and President
ACLA—Asian Cultural Landscape
Association

Reference

Singh RPB (1995) Heritage ecology and caring for the Earth: a search for preserving harmony and ethical values. Natl Geogr J India 41(2):191–218

Acknowledgements

The conceptual idea of this volume entitled *Sustainability in South Asian Cities* arose from a discussion with the editors of this volume over an online meeting in early 2020. This volume is made up of 25 diverse chapters (including Chap. 1) pertaining to issues and challenges of sustainability in the cities of South Asia and fits the series *Advances in Geographical and Environmental Sciences*.

Each reviewed and copy-edited chapter was revised by the respective authors. The editors thank each of the authors for re-reading and revising several drafts and the final manuscript. This book project would not have been possible without their continued support.

A big thank you to Dr. Rana P. B. Singh, (ex)-Professor and Head, Department of Geography, Banaras Hindu University; and President, ACLA—Asian Cultural Landscape Association for writing the Foreword of this volume.

The editors are equally thankful to Dr. Avaya Mohapatra, Former Head and Professor (now retired), North Eastern Hill University, Meghalaya, 'Professor Vidhya Bir Singh Kansakar, a veteran geographer of Nepal and Professor Suman Singh, Department of Geography, Banaras Hindu University, Varanasi, India' for writing the blurbs of the book.

Equally, we thank Yosuke Nishida, the publishing editor, for her invaluable editorial assistance in preparing this volume. Thanks also go to the production team Niraja Deshmukh, Deepa Varadharajan (known popularly as Deepa V), Vikesha Thilakkumar, Nobuko Kamikawa and others for rendering their editorial support with patience.

While we pay our humble obeisance to our parents for their unconditional blessings, we also take the opportunity to thank our families and close-knit relatives for their support in various spectrums of life.

Prof. Subhash Anand
Dr. Rituparna Bhattacharyya
Prof. Madhushree Das
Prof. Tulshi Kumar Das
Prof. Pushkar K. Pradhan

Contents

About the Editors

Dr. Subhash Anand is currently a professor of geography at the Department of Geography, Delhi School of Economics, University of Delhi, India. He has more than 25 years of experience in teaching and research and has published seven books and more than 70 research papers, articles and chapters in journals and books. He specialises in urban environmental issues, sustainability, solid waste management and planning and geotourism. Dr. Anand is the vice president of the Indian Geographical Society (IGS) and the vice chair of the Commission on Geoheritage, International Geographical Union (IGU). He is a series editor of the *Advances in Geographical and Environmental Sciences* and an associate editor-in-chief of the *International Journal of Geoheritage and Parks* as well as an editorial board member of many journals. He has held an advanced training workshop, Urban Spatial Planning in Response to Climate Change in Asia, in Taipei and a Commonwealth Geographical Bureau (CGB) workshop, Human Consequences of Climate Change, at the University of Colombo. He has travelled widely to participate in international conferences, seminars and workshops and has delivered lectures in several countries.

Dr. Rituparna Bhattacharyya is an adjunct professor at the Indian Institute of Technology, Guwahati; a visiting professor at the University of Science and Technology, Meghalaya; and an independent research consultant and Editor-in-Chief (Joint) of the SCOPUS indexed international journal Space and Culture, India published from the UK. She accomplished her senior fellow position at Advance HE (formerly Higher Education Academy, UK) in 2021. Rituparna is currently the Series Editor for Gender and Violence, Routledge, Taylor and Francis. She also works as an Associate Editor for two international journals. They are Humanities and Social Sciences Communications (nature.com), Springer Nature and Diversity & Inclusion Research published by Wiley. Rituparna was appointed Executive Council member of Cotton University, India, in November 2023.

Earlier in 2009, she obtained her Ph.D. from the School of Geography, Politics and Sociology, University of Newcastle, UK. This research was funded by an Overseas Research Scholarship (ORS) and an International Research Scholarship (IRS). In 1994, she received the Jawaharlal Nehru Memorial Award for standing First Class

First (Geography Major) from Cotton College (now University). And in 1996, she secured First Class First (in MA, Geography) from Banaras Hindu University. She was also the recipient of the National Merit Scholarship from the Government of India.

She has more than 25 years of teaching and research experience. She bears expertise in geohumanities linked to gendered politics, inequalities, poverty, violence, sustainable goals, and development in the global south. She serves as an anonymous journal reviewer of 26 high-quality peer-reviewed journals. Dr. Bhattacharyya volunteers at the Prag Foundation for Capacity Building, a public charitable trust in India, and the Alliance for Community Capacity Building for Northeast India, a UK-registered charity. She has more than 89 publications to her credit. e-mail: rituparna.bhattacharyya@accb.org.uk

Dr. Madhushree Das is currently a professor at the Department of Geography, Guwahati University, Assam, India. She has more than 24 years of experience in teaching and research. Dr. Das is a social and cultural geographer and works in the field of social, cultural, gender and tribal geography. She has authored three books and published more than 50 research papers in national and international journals. She received the International Geographical travel grant award thrice to present her research in Durban (South Africa), Cologne (Germany), Szeged (Hungary), and Kyoto (Japan). She was the former general secretary (elect) of North East India Geographical Society (2016–2018). Besides, she is the court member of Guwahati University and Cotton University. She has received several awards and honours including Academic Gold Medal for securing 1st class 1st position in M.A. Geography; Bhuvaneswari Memorial Gold Medal for standing 1st in M.A. degree in all subjects in Arts in 1997; Dipali Borthakur Memorial Award from North East India Geographical Society (NEIGS); National Talent Search Examination, Government of Assam, India in 1989; and Common wealth Award in 2005.

Tulshi Kumar Das Ph.D., is a professor of social work at the Shahjalal University of Science and Technology, Bangladesh. He was the head of the Department of Social Work at the university for 10 years and was the dean of the School of Social Sciences for two years. He has widely published articles in national and international journals and has participated conferences around the world. He has completed research projects relating to slum culture, domestic violence, natural disasters, street children, micro-credit, food security, ethnicity, human rights, social work education and practice, nationalism, genocide, and psycho-physical health of Bangladesh police, among others. Professor Das is an editorial board member and a regular reviewer of many international journals. He is involved in 46 social welfare organizations working for marginalized people throughout the country.

Pushkar K. Pradhan is a professor at the Central Department of Geography, Tribhuvan University. He received his Ph.D. degree from the Institute of Geography and Regional Science, Klagenfurt University, Austria, M.Sc. in regional development planning from the Asian Institute of Technology, Thailand, and M.A. in geography from the Tribhuvan University, Nepal. Affiliating to the Tribhuvan University, Dr. Pradhan has four decades of professional experience in teaching and research. During this period, he held several positions at Tribhuvan University, such as the head of the Central Department of Geography, a member of the Academic Council and a faculty board member of Humanities and Social Sciences, and the chairperson of Geography Subject Steering Committee. He served as an executive member of the Steering Committees of the International Geographical Union Commissions on Marginality, Public Policy and Local Governance and convened successfully four IGU conferences in Nepal in 2004, 2007, 2011 and 2019. Dr. Pradhan has also served as a member of the editorial boards of several journals published by the university and academic societies. Dr. Pradhan has published over five dozen research papers and books, participated in over four dozen international and national seminars, and involved as a lead position in over 40 national and international research works.

Chapter 1
Towards Socio-Economic and Environmentally Sustainable Cities of South Asia

Rituparna Bhattacharyya, Madhushree Das, Subhash Anand, Tulshi Kumar Das, and Pushkar K. Pradhan

Abstract It is well established that the countries of South Asia suffer from multi-dimensional poverty. This volume examines the issues and the challenges linked to socioeconomic and environmental indicators and the various practices entailed in evolving the notion of sustainable cities. The volume is indeed an attempt to answer the 'big question'—the factors and obstacles affecting the building of sustainable cities in South Asia.

Keywords Multi-dimensional poverty · Millennium development goals · Sustainable development goals · Sustainable cities · Socio-economic challenges · Environmental challenges · South Asia

R. Bhattacharyya (✉)
Senior Fellow, Advance HE (Formerly Higher Education Academy), London, UK
e-mail: rituparna.bhattacharyya@accb.org.uk

Centre for Indian Knowledge Systems, Indian Institute of Technology, Guwahati, India

Research Consultant and Editor-in-Chief (Joint), Space and Culture, India, North Shields, UK

Department of Earth Science, University of Science and Technology, Meghalaya, India

Series Editor, Gender and Violence, Routledge, Taylor and Francis, London, UK

M. Das
Department of Geography, Gauhati University, Guwahati, Assam, India
e-mail: madhushreedas@gauhati.ac.in

S. Anand
Department of Geography, Delhi School of Economics, University of Delhi, Delhi, India

T. K. Das
Department of Social Work, Shahjalal University of Science and Technology, Sylhet, Bangladesh

P. K. Pradhan
Central Department of Geography, Tribhuvan University, Kathmandu, Nepal

S. Anand et al. (eds.), *Sustainability in South Asian Cities*, Advances in Geographical and Environmental Sciences, https://doi.org/10.1007/978-981-97-7455-5_1

1.1 Introduction

There is an emerging publication on Sustainable Cities and Communications (SDG 11) by Springer Nature. Some of its publications include—greening cities by growing food; sustainable energy constructions, artificial intelligence and Cy-Phy future and many others. These publications aim to build both sustainable cities and sustainable future for communities worldwide in this era of staggering climate change (Breuste et al. 2022; Hammelman 2022; Littlewood et al. 2022; Mishra et al. 2022; Scardovi 2021). Seemingly, the findings of the chapters of this volume aim to benefit the residents of the cities of South Asia. Notwithstanding, Henri Lefebvre, the French Marxist philosopher and sociologist, introduced the notion of the 'right to city' in his 1968 book entitled *Le Droit à la ville.* This 'the right to the city' notion has emerged as one of the robust academic, activist, and policy endeavours (Castro 2021; Harvey 2008, 2012; Purcell 2002, 2014). The civil unrest and outrageous rebels that sprouted in the streets of France in May 1968, led mainly by the students, followed by the workers against the so-called hierarchical, autocratic, and tradition-bound French society of the time (Wolin 2017), perhaps, compelled Henri Lefebvre (1996, [1968], 174) to think about the phrase the 'right to the city' displaying 'superior form of rights' in understanding the development process of urbanization, urban life vis-à-vis capitalist growth (Castro 2021) as Lefebvre argued: "right to freedom, to individualization in socialization, to habitat and to inhabit" (1996, [1968], 174). Lefebvre's arguments perhaps sought normative urban spaces seeking social justice (Harvey 2000). However, Lefebvre's urged for multi-dimensional aspects of the right to the city—"to urban life, to renewed centrality, to places of encounter and exchange, to life rhythms and time uses, enabling the full and complete usage of these moments and places" (Lefebvre 1996 [1968], 179; see also, Castro 2021). Nevertheless, both Lefebvre and Harvey (2008, 2012) called for collective rights rather than individualized ones. As Harvey stated:

> The right to the city is, therefore, far more than a right of individual access to the resources that the city embodies: it is a right to change ourselves by changing the city more after our heart's desire. It is, moreover, a collective rather than an individual right since changing the city inevitably depends upon the exercise of a collective power over the processes of urbanization (Harvey 2008, p. 23).

In a simple sense, the right to the city is a complex play of some of the fundamental rights that a city should affirm and comply with— access to clean and affordable drinking water, food, cooking, electricity, sanitation, affordable housing, an environment free from pollution, eco-friendly transportation, safety and security, employment and decent jobs, ease of establishing the business, crime and violence-free urban space, sound sewage system and other associated markers and innovative techniques helping the cities to progress towards socio-economically and environmentally sustainable and vibrant cities (Das 2022) via which residents can collectively seek social justice (Harvey 2000). However, this project is not directly linked to the right to the city but brings to the fore grotesque inequalities and challenges engulfing some of the cities of South Asia. The case studies demonstrate how we can make the cities better. However, it is worth noting that all the countries in South Asia reel on the brink of multi-dimensional poverty.

1.2 Multi-dimensional Poverty and South Asia

South Asia is a region made up of eight countries—Afghanistan, Bangladesh, Bhutan, India, Maldives, Nepal, Pakistan and Sri Lanka. Of course, there is no rigid geographical demarcation, for example, an ocean, strait, etc., locating as to where the geographical boundary of South Asia starts in the north; however, the Himalayas, Karakorum, and Pamir mountain ranges are considered as the boundaries. The Indian Ocean is regarded as the boundary to the south (World Atlas n.d.). The area, population, population density, and Gross Domestic Product (GDP) per capita of the eight countries of South Asia are illustrated in Table 1.1. Home to approximately 1.86 billion people (Table 1.1), South Asia as a region has over ten cities with a population of over 10 million. Most of these cities (Delhi, Mumbai, Kolkata, Bengaluru, Chennai, Hyderabad, Ahmedabad), except Karachi and Lahore (Pakistan), and Dhaka (Bangladesh) are located in India. The challenges are colossal for the South Asian cities mired in multi-dimensional poverty.

Table 1.1 Overview of the countries in south Asia

Country	Capital	Area (Km^2)	Population	Population density (per person Km^2)	GDP per capita (2019)
India	New Delhi	3,287,263 (64.02%)	1,366,417,750	415.6	$2172.00
Afghanistan	Kabul	652,230 (12.7%)	38,041,750	8.3	$513.00
Bangladesh	Dhaka	148,460 (2.87%)	163,046,160	1,104.90	$1906.00
Bhutan	Thimpu	38,394 (0.75%)	763,090	19.8	$3423.00
Maldives	Male	298 (0.006%)	530,950	1781.8	$15,563.00
Nepal	Kathmandu	147,181 (2.87%)	28,608,710	194.3	$1048.00
Pakistan	Islamabad	796,095 (15.5%)	216,565,320	245.8	$1388.00
Sri Lanka	Colombo	65,610 (1.28%)	21,803,000	325	$3947.00

Source World Atlas (n.d.). *The Countries of South Asia.* https://www.worldatlas.com/articles/the-population-and-economy-of-the-south-asian-countries.html

Oxford Poverty and Human Development Initiative (OPHI) developed the concept of multi-dimensional poverty in 2010. This notion embraces a range of deprivations that the poorer sections of societies encounter in everyday life. Some of these deprivations *inter alia* are food insecurity or deprivation from good quality food, lack of access to poor health, poor living standards, poor access to education, indecent work, living in hazardous environments, fear of violence and its associated connotations (like children and women facing the risk of going missing, getting kidnapped or trafficked). Taking into account three key dimensions of poverty (**health**—nutrition, and child mortality; **education**—years of schooling and school attendance and **standard of living**—cooking fuel, sanitation, drinking water, electricity, housing, assets), multi-dimensional Poverty Index (MPI) has been developed by OPHI and UNDP. This means "people are counted as multidimensionally poor if they are deprived in one-third or more of 10 indicators…, where each indicator is equally weighted within its dimension, so the health and education indicators are weighted 1/6 each, and the standard of living indicators are weighted 1/18 each" (UNDP and OPHI 2021, p. 2). Table 1.2 presents the MPI of the countries in South Asia region.

Therefore, sustainable development remains paramount to alleviating the countries of South Asia from the vagaries of multi-dimensional poverty. But, of course, the COVID-19 Pandemic and the ongoing Russia-Ukraine war and the Israel–Hamas war since 07 October 2023 have pushed the countries further into the shackles of poverty.

The Brundtland Report (1987) defines "[s]ustainable development [as the] development that meets the needs of the present without compromising the ability of future generations to meet their own needs" (Brundtland Report 1987). This report, also known as *World Commission on Environment and Development: Our Common Future,* was produced more than 30 years ago by Gro Harlem Brundtland, the Commission's chairperson, which remains outrageously prevalent even today. The Conference of the Parties (COP 26, COP 27, and COP 28) climate change summits held respectively in Glasgow (in 2021), the Arab Republic of Egypt (in 2022), and the Dubai, UAE (in 2023) are witness to it.

In the wake of the publication of the Brundtland Report, the 1992 Earth Summit held in Rio de Janeiro, Brazil, the notion of sustainable development gained momentum when the stakeholders of different governments, countries and local organizations of the United Nations System started taking the pledge of accomplishing sustainable development. Accordingly, the Earth Summit launched an international environmental treaty to tackle climate change which was signed by 154 states and ratified by 50 states, which became effective on 21 March 1994 (United Nations Framework Convention on Climate Change 1992). Simultaneously, on 9 May 1992, the Earth Summit launched the idea for a multilateral environmental treaty known as the Convention on Biological Diversity (or Biodiversity Convention). It became effective on 29 December 1993 after 30 states ratified it. Following the momentum of maintaining sustainable development vis-à-vis accomplishing a carbon–neutral global economy, the 2000 Millennium Summit of the United Nations and the United Nations Millennium Declaration made way for 191 United Nations member states and 22 international organisations to take the pledge to achieve eight Millennium

Table 1.2 Multidimensional poverty index of the countries in South Asia Region

Country	Area	MPI	H* (%)	A** (%)	Vulnerable (%)	Severe poverty (%)	Populationshare (%)
India	National	0.123	27.90	43.90	19.30	8.80	100
	Urban	0.039	9.20	42.60	14	2.40	32.30
	Rural	0.163	36.80	44.10	21.80	11.80	67.70
Afghanistan	National	0.272	55.90	48.60	18.10	24.90	100
	Urban	0.094	21.20	44.40	21.20	7.10	24
	Rural	0.328	66.90	49	17.20	30.50	76
Bangladesh	National	0.104	24.60	42.20	18.20	6.50	100
	Urban	0.06	14.50	41.60	14.70	3.40	21.60
	Rural	0.116	27.40	42.30	19.20	7.30	78.4
Bhutan	National	0.175	37.30	46.80	17.70	14.70	100
	Urban	0.037	9.30	39.50	10.50	1.50	28.30
	Rural	0.229	48.40	47.40	20.50	19.90	71.70
Maldives	National	0.003	0.80	34.40	4.80	0	100
	Urban	0.002	0.50	33.30	2.30	0	36.60
	Rural	0.003	0.90	34.70	6.30	0	63.40
Nepal	National	0.074	17.50	42.55	17.80	4.90	100
	Urban	0.053	12.40	42.60	15.40	3.30	67.30
	Rural	0.119	28.10	42.40	22.90	8.00	32.70
Pakistan	National	0.198	38.30	51.70	12.90	21.50	100
	Urban	0.079	18.00	43.80	8.60	5.70	36.10
	Rural	0.266	49.80	53.30	15.40	30.40	63.90
Sri Lanka	National	0.011	2.90	38.30	14.30	0.30%	100
	Urban	0.004	1.00	36.70	6.60	0.00%	16.50
	Rural	0.013	3.30	38.40	15.90	0.30	83.50

NB: H—* incidence of poverty and A—** the average intensity of poverty
Source Country Briefings (2021). Oxford Poverty and Human Development Initiative. https://ophi.org.uk/multidimensional-poverty-index/mpi-country-briefings/

Development Goals (MDGs) by 2015. However, achieving MDGs by 2015 appeared to be over-ambitious. Hence, on 25 September 2015, the United Nations General Assembly adopted another resolution to achieve sustainability by 2030 through 17 interconnected global goals of sustainable development (Transforming our world: the 2030 Agenda for Sustainable Development 2015), which is, of course, ostentatious and the global crisis provoked by COVID-19 has already demonstrated the acceleration of multi-dimensional poverty "as income losses are expected to exceed $220 billion in developing countries, and an estimated 55 per cent of the global population have no access to social protection. These losses will reverberate across societies; impacting education, human rights and, in the most severe cases, basic food security and nutrition" (United Nations n.d.a). Taking these backdrops into account, this endeavour aims to examine the challenges, achievements, and socio-economic sustainability of the various Sustainable Development Goals (SDGs) intertwined to attain SDG-11 (Sustainable Cities and Communities) in some of the cities of South Asia so that its citizens collectively have their spaces as 'normative' liveable urban life —their basic rights to their cities enabling them social justice (Lefebvre 1996 [1968], Harvey 2000, 2008, 2012).

This volume presents 24 chapters on various issues of sustainable cities in South Asia contributed by social science, humanities, and pure science scholars. The chapters of this volume are highly diverse but are woven together by a commonality—to foreground the challenges and achieve sustainability in South Asian cities. Most chapters in this volume are driven by a local approach to understanding the complex challenges of attaining sustainable urban development. The volume is divided into three parts: **Part I—Towards Blue-Green Infrastructure of South Asian Cities**; **Part II—Socio-Economic Challenges towards Sustainable Cities** and **Part III— Environmental Challenges**

Towards Socio-Economic and Environmental Sustainability

One must look at various sustainability indicators to build a socio-economic and environmentally sustainable city embedded in basic collective rights for its citizens, as 50% of the world's population dwell in cities.

Part I: Towards Blue-Green Infrastructure of South Asian Cities

SDG 11 is inherently intertwined with urbanization, migration, and development. By 2050, people living in urban areas will stand at 6.5 billion (United Nations, n.d.b). This rapid growth of cities stems from rising population and internal and international migration. However, unplanned urbanization has posed a number of interconnected environmental challenges in the cities of South Asia— increased urban flooding, groundwater depletion, drinking water scarcity, food scarcity, traffic congestion, rise in air temperature, change in climate, and increased air and water pollution plus loss in biodiversity. To attain an integrated road towards sustainable cities and build urban resilience, scholars suggest a structured approach to gaining solutions to environmental and urban challenges by enhancing urban ecoservices, reducing the risk of disaster and building a nature-friendly urban ecosystem (Dhyani et al. 2022; Lamond and Everett 2019). In the context of the concept of Blue-Green Infrastructure, Chap. 2 by Praschaya Kaushik and Sanjeeb Kakoty is an attempt to probe the scopes of Blue-Green Infrastructure for a sustainable urban future in a few crowded South Asian cities—Karachi, Dhaka, Bengaluru, and Guwahati.

Part II: Socio-Economic Challenges towards Sustainable Cities

Access to clean and affordable drinking water for the city and town residents is one of the crucial indicators of the sustainable development of a city. Estimates suggest that 40% of the world's population has been affected by water scarcity, with 733 million people living in water-stress areas (Bhattacharyya and Prasad 2020; Bhattacharyya et al. 2023; Pulla et al. 2018). Besides, worldwide, water-linked ecosystems have plummeted at a staggering rate. Evidently, over the past 300 years, the world has lost over 85% of its wetlands. Moreover, due to a lack of appropriate monitoring machinery techniques, it is unknown for approximately 3 billion people the quality of water they depend on.

Further statistics suggest that out of every ten people worldwide, three lack access to the services of 'safely managed drinking water.' In addition, six out of ten people do not have access to adequate sanitation facilities. As a consequence, approximately three billion people lack basic facilities for handwashing at their homes/premises, which is a vital requirement to fend off the spread of COVID-19 and other infections (United Nations n.d.b). Given the current state of crisis, it is reckoned that by 2030, 1.6 billion people will lack access to safely managed drinking water (United Nations n.d.c). Taking similar arguments into context, Mohammed Mahmud Khan, Sang Hyup Shin, Tapos Kormoker, and Tanmoy Roy Tusher in Chap. 3 examine the challenges of accessing safe and affordable drinking water in developing countries taking a case study from Bangladesh.

Worryingly, the increased use of the antimicrobial agent triclosan (TCS) in pharmaceutical and personal care products (PPCPs) during Public Health Crises such as the COVID-19 Pandemic has been a growing public health concern worldwide, including the South Asian countries. This is because TCS not only has side effects, but its presence in water sources, especially in drinking water, further poses challenges to the already existing crises of managing safe drinking water. To mitigate this, adequate investment in infrastructure remains vital. In the said context, in their study titled Chap. 4: *Modelling of Antimicrobial Agent Triclosan (TCS) Transport in Nanofiltration System for Sustainable Design of Drinking Water Treatment Systems for South Asian Countries*, Dina Dzafira Ramlan, Antonios Parasyris, Marco Discacciati, and Diganta Bhusan Das have designed a sustainable nanofiltration (NF) model of Drinking Water Treatment Systems for countries of South Asia as well as other countries of the world aimed at separating TCS and other contamination from drinking water.

SDGs1 and 2 are linked to ending all forms of poverty and ending Hunger. But worldwide multi-dimensional poverty and hunger catastrophes loom large due to the culmination of ongoing conflicts, especially the Russia-Ukraine War, the COVID-19 Pandemic, and climate change (World Food Programme n.d.a) cascading into the rising cost of living. It is worth noting that the Russian Federation and Ukraine contribute 30% of wheat, 20% of maize, and 80% of sunflower seed products—global exports. While 1 in 10 people is suffering from Hunger worldwide, approximately 1 in 3 people lack access to adequate food (United Nations, Goals 2 n.d.d). According to World Food Programme (n.d.b), in 82 countries, 345 million suffer from acute food insecurity. As the World Food Programme (n.d.a) argue:

> From the Central American Dry Corridor and Haiti, through the Sahel, Central African Republic, South Sudan and then eastwards to the Horn of Africa, Syria, Yemen and all the way to Afghanistan, there is a ring of fire stretching around the world where conflict and climate shocks are driving millions of people to the brink of starvation.
>
> In countries like Nigeria, South Sudan and Yemen, WFP is already faced with hard decisions, including cutting rations to be able to reach more people. This is tantamount to taking from the hungry to feed the starving (World Food Programme n.d.a).

Needless to say, the countries of South Asia, too, have been suffering from food insecurity and Hunger. The 2022 Global Hunger Index is evidence of this (Global Hunger Index 2022). With the ongoing Russia-Ukraine War, there remains a dismay that food insecurity and hunger might trigger the second global Holodomor in the 21st Century.[1]

Nevertheless, despite suffering from food crises and Hunger, the countries of South Asia take pride in their rich and diverse culinary heritages. In Chap. 5, titled *Culinary Landscapes- Understanding the Intersections of Food Cultures in South Asia*, Jyotishma Bordoloi and Madhushree Das seek to understand how culinary skills and foodscapes of South Asia intersect, evolving a number of questions— how people of different countries of South Asia eat, what they eat, with whom they eat as well as why they eat, and how the foodscapes of these countries shape their respective identities, culture and everyday life. In a slight deviation from the food rationale but stressing on teleconnection between air pollution and agriculture, Haruhisa Asada, in Chap. 6, titled *Towards Solving Environmental, Societal, and Economic Teleconnections between Delhi, Punjab, and Assam,* discusses the sustainable relationship between nature and society while trying to put forward a sustainable solution. He argues that the problems caused by the invisible ingrained chains of nature, society, and economy of the study regions— causes of air pollution in Delhi, reduction of groundwater in Punjab, and socio-environmental problems in Assam should not be examined as discrete problems. Instead, he calls for teleconnection of the ecological issues of the study regions so that SDGs can be achieved.

Alongside poverty, hunger, and lack of access to proper sanitation and drinking water in urban areas of South Asian cities (see Table 1.2), another offshoot is the unprecedented growth of slums due to the population migration from the rural and semi-rural areas to the urban spaces. Surprisingly, slum dwellers constitute 1 billion worldwide (United Nations n.d. e). These slums are characterised by overcrowding and multi-occupancy with very little or poor access to the needs of everyday life— poor sanitation, electricity, water supply, poor housing conditions etc. (McFarlane 2008; Pulla and Khan 2022). Nevertheless, the slums are usually but not necessarily home to low-wage/migrant labourers, but most of them are engaged in the informal economy. Unsurprisingly, the two largest slums in the world are found in South Asia (World Economic Forum 2016). Orangi Town, home to 2.4 million people and geospatially located in Karachi, Pakistan, is the largest slum in the world, followed

[1] Holodomor is not a natural but human-made famine in Ukraine that occurred between 1932-1933 in former Bolshevik Russia. Ukraine, along with the Volga Valley, the North Caucasus region and Kazakhstan was the bread basket of the world as it produced the largest grains in the former USSR. The etymon, Holodomor, a Ukrainian term is broken as 'holod' meaning hunger or starvation, 'dom' means 'home', and 'mor' signals 'death', indicating the intention of mass killing by starvation and hunger. There remains a consensus that Joseph Stalin's regime planned Holodomor to politically suppress the Ukrainian independence movement. Overall, the famine exterminated approximately 5 million people, out of which 3.9 million are believed to be Ukrainians. The Ukrainian parliament officially declared the famine as Genocide in May 2003, and later on, endorsed by many other countries (for details, please see, Applebaum 2022; Graziosi 2005; Holodomor—Famine in Soviet Ukraine, 1932–1933, 2008; International Recognition of the Holodomor n.d.).

by Ciudad Neza, located in Mexico City, Mexico accommodating 1.2 million population. With one million people, India's Dharavi is the third-largest slum in the world (World Economic Forum 2016 see also, Pulla and Khan 2022). Taking these rationales forward and with the aim of building economically sustainable, inclusive and safe cities in urban India, Krittika Roy, in Chap. 7 entitled *Making Cities Sustainable: Tracking Economic Sustainability in Urban India* critically analyses the living conditions of the migrants and the Urban dwellers' changes in the urban housing between 2012 and 2018. Seemingly, in their joint Chap. 8, *Living by the Edges: Existential Vulnerabilities Among the Slum-dwelling Women of Delhi, India*, Shriya Thakkar and Rituparna Bhattacharyya examine the squalid conditions of the migrant women mainly working as domestic workers in the middle-class households and dwelling in the slums of Delhi. Apace to their sordid living conditions, these women were often subjected to violence against women and psychosocial stresses. The COVID-19 pandemic has further exacerbated the incidences of violence against women, transforming violence against women into a shadow pandemic (Mlambo-Ngcuka 2020; Singh and Bhattacharyya 2020), thereby signalling the prevalence of cultures of 'hegemonic masculinity' and 'misogyny' across nations (Singh and Bhattacharyya 2020) but in varying scales.[2] Statistics suggest that globally, about 736 million women or one in three women, amounting to approximately 30% of women aged 15 years and older, face "physical and/or sexual intimate partner violence, non-partner sexual violence, or both at least once in their life" (UN Women, n.d.; World Health Organization 2021). Arguably, a violence-free environment and women's safety are keys to the development and maintenance of a sustainable society. Against these backdrops, deploying qualitative research, Atia Binta Abdul Wahid and Tulshi Kumar Das, in Chap. 9, entitled *COVID-19 Induced Family Violence in an Urban Setting in Bangladesh: Prospects for Social Work Interventions*, explore family violence in urban areas of Sylhet, Bangladesh, during the COVID-19 lockdown and suggests social work intervention measures.

In another qualitative research in the forms of in-depth interviews, focus groups, and key informants' interviews, Md. Mahmud Hasan and Tulshi Kumar Das, in Chap. 10, titled *Going beyond Microcredit Approach for Socioeconomic Development: A Case of a Slum from Bangladesh*, explore how Pathgudam Duldul Camp Landless Cooperative Society has been functioning the micro-credit program of community development among the dwellers of a slum located in the Mymensingh district of Bangladesh. The authors argue that the cooperative society has successfully recovered almost 100 per cent of the credit without coercive tactics in collecting instalments. Their findings also suggest that the dwellers have benefited from the micro-credit program of the cooperative society.

Notwithstanding, the characteristics of slums and squatter settlements in urban spaces are similar, bearing subtle but often blurry differences between them. The

[2] COVID-19 and Violence against Women: The Evidence Behind the Talk.
Insights from big data analysis in Asian countries. https://data.unwomen.org/sites/default/files/documents/Publications/COVID-19%20and%20VAW_Insights%20from%20big%20data%20analysis_final.pdf.

residents of squatter settlements, usually inhabited by poor and low-income migrants, possess no legal rights over the land they occupy. The names of squatter settlements are different in different geographical localities. For example, Favelas in Brazil, Ranchos in Venezuela and Bustee/Jhuggi and Jhopri in India. Similar to the slums, these settlements' quality of living conditions is poor; houses are dilapidated, and the residents lack access to basic amenities and, therefore, are always subjected to precarity and vulnerability. Uttam Sagar Shrestha, in Chap. 11, entitled *Growth and Development of Squatter Settlements: A Challenge to Urban Sustainability in Nepal-The Case of Kathmandu City,* addresses the problems of squatter settlements of Kathmandu, urging for effective strategies and pragmatic policy matters.

In a slightly different context, reinforcing the lack of access to basic facilities in the urban spaces of India and critically investigating the urban policies and programmes, Anuradha Banerjee, in Chap. 12, titled *The COVID-19 Crisis and Indian Cities: Managing Urban Contours in Times of Pandemic,* charts out the vulnerabilities of India's urban areas during the outbreak of the COVID-19 Pandemic. Turning to socio-economic challenges and vulnerability, in Chap. 13: *Representation of Dhaka in the Book of Dhaka: A City in Short Fiction*, Mohammad Shafiqul Islam discusses the representation of Dhaka as discussed in the *Book of Dhaka*. Dhaka is the capital city of Bangladesh, a country that gained independence from Pakistan in 1971. In other words, East Pakistan became Bangladesh (see, Das et al. 2022). Taking the discussion of vulnerability forward in Dhaka, in their joint endeavour, Chap. 14: *Vulnerability and Coping Mechanisms of the City Dwellers: A Study of Old Dhaka*, Md. Anwar Hossain and Tulshi Kumar Das examine the coping mechanisms of the residents of old Dhaka. The authors argue that the residents of old Dhaka face many perennial challenges in their everyday lives—income insecurity, drainage problems, gas supply disruptions, water logging, fire outbreak, mobility problems, traffic congestion, alongside water, air and noise pollution. However, the residents cope with these everyday challenges by adopting specific indigenous mechanisms such as engaging in alternative jobs, early morning cooking or late night cooking for the next day, abstaining from going outside during the period of waterlogging and being alert for fire outbreaks.

Chapter 15 engages the attention towards transportation policies and plans of the city of Dhaka in their chapter, *Integrating Accessibility: A Critical Review of the Transportation Policies and Plans in Dhaka*. Paromita Nakshi and Anindya Kishore Debnath identify the accessibility measures and critically assesses how and to what extent the notion of accessibility has been integrated into the transport policy and planning framework of Dhaka. The critical review of this chapter provides a better understanding of the practices entailed in planning policies, which might help planners and policymakers to enhance the issue of transportation accessibility in a meaningful manner.

One of the critical characteristics of the cities of South Asia and other emerging and developing economies is the prevalence of informal economies, which possess market value, but its value is often difficult to measure and remains unregistered formally but in(directly) contributes to the process of economic development (Bhattacharyya 2009; IMF Country Focus 2021; Informal Economy n.d.). Approximately 60% of the world's population remains engaged in the informal sector, which comprises more than 90% of micro and small enterprises (MSEs) (Informal Economy n.d.). United Nations statistics suggest that during the first phase of COVID-19 situation, 1.6 billion workers in the informal economy lost their jobs/livelihoods (United Nations n.d.f).

In South Asia, more than 90% of the region's business operates informally, where 80% of its workers remain engaged in informal activities (Bussolo et al. 2020). The Economic Survey 2018–2019 statistics reveal that 93% of workers in India are engaged in the economy's informal sector (Chap. 10: Social Infrastructure, Employment and Human Development; Bhattacharyya et al. 2020). In the aftermath of the first few days of India's first lockdown, the world witnessed one of the most significant migrant labourer crises following the mass exodus of India's internal migrant labourers (Bhattacharyya et al. 2020, 2023; Dandekar and Ghai 2020; Suresh et al. 2020). This exodus brought unprecedented disgrace to the country, foregrounding the precarity and vulnerability of the migrant workers attributed to sheer human rights violations (Kumar and Choudhury 2021). With no formal contracts, no wages and job loss, the migrant labourers of India's informal economy were compelled to opt for reverse migration which the renowned economist, Jean Drèze address as "one of the "harshest in the world"—was almost like a "death sentence" for the poor who have been living hand-to-mouth" (Scroll Staff 2020).

This volume contains two chapters on the challenges entailed in the informal economy in achieving urban sustainability. Puspa Sharma, in Chap. 16, *titled Informal Sector towards Urban Sustainability in Kathmandu Valley, Nepal,* argues that the informal activities/livelihood strategies in the valley generate employment and income among the poor, migrants and women through which they access their basic everyday needs and conserve the environment. She discusses that Kathmandu Valley's informal activities demonstrate unique marketing, spatial location, and transformation features. Therefore, she puts forward policy measures for better management of informal marketplaces, marketing systems, and environmental conservation to achieve urban sustainability in the valley. And in Chapter 17, entitled *Covid-19 Pandemic and Urban Informal Workforce: Challenges Faced by the Small Entrepreneurs in Silchar Town of Assam, India*, Subhabrata Dutta discusses the obstacles encountered by the small-scale entrepreneurs during the COVID-19 pandemic at Silchar, a small town located in the Barak Valley, Assam, India. Chap. 18: *Bolpur—A Historical Eco-Friendly Townscape Promoting Sustainable City* has been contributed by Ankabehari Saha, Syeda Fahima Shahnaz Sultana, Payel Saha, and Madhushree Das. Embracing Santiniketan, Sriniketan and Prantik, Bolpur municipal area, as well as a city, is located in the Birbhum district of West Bengal, India. Deploying multidimensional indicators of sustainability— economic (occupation, taxpayers); environmental (water quality, waste collection, availability of greenery); social (education: school dropouts, household ownership and slum proportion, toilet facility: types of toilets, community toilets, water supply, street light, public transportation, services stemming from government policies), the authors examine how Bolpur is developing gradually towards becoming a 'Sustainable City'. The authors argue that the study findings would help developing cities, popularly known as third-world cities, realise the intrinsic possibilities of gaining sustainability without compromising the inherent qualities of its cityscapes. In Chap. 19, titled *The Pattern of Urbanisation, its Sustainability and Spatial Set-Up of Urban Centres in Uttar Pradesh (India)*, using Census of India data (1981–2011), S.C. Bansal links sustainability with the patterns of urbanization centres in the state of Uttar Pradesh and demonstrates the growth and spatial development of urban centres within the state. Similarly, the final chapter in this section is Chap. 20, entitled *Urban Sprawl and Future Land Resource Utilization Projections for Ghaziabad City, Uttar Pradesh, India, Using High-Resolution Remote Sensing Data and GIS Techniques*. Here, the authors—Virendra Kumar, Kamlesh Bhalavi, A.K.S. Rana, V.K.S. Rathore and Harish Kumar deploy high-resolution Remote Sensing Data and GIS Techniques to examine the dynamics of urban sprawl and land consumption rate for Ghaziabad city. In so doing, the authors reckon that the findings of their study would be beneficial to the urban planners/ decision makers for better planning and management of Ghaziabad's land use and land resources in attaining sustainability.

Part III— Environmental Challenges

Tellingly, the climate crisis remains one of the most intense catastrophes of the current times. On the closing day of the 27th Conference of the Parties to the United Nations Framework Convention on Climate Change (COP27) held at Sharm el-Sheik, the Egyptian coastal city from 6 to 20 November 2022, UN Secretary-General António Guterres, while welcoming the decision of more fund in tackling climate change said:

> The world still needs a giant leap on climate ambition.
>
> The red line we must not cross is the line that takes our planet over the 1.5 degree temperature limit," he stressed, urging the world not to relent "in the fight for climate justice and climate ambition.
>
> We can and must win this battle for our lives, (Delivering for people and the planet n.d).

This section of the volume presents five chapters demonstrating the environmental challenges linked to attaining sustainability. One of the central problems of environmental challenges as already stated above is access to water. Band-e Amir National Park, located in Afghanistan, has been studied by Dilek Eren Akyuza, Filiz Bektas Balcikb, Ahmad Frahmandc, Fusun Balik Sanlic in Chap. 21 titled *Analyzing and Mapping of Change in Water Surface Area Using Remote Sensing Data: Case Study of Band-e Amir National Park, Afghanistan*. The study applied remote sensing data to critically examine and map the change in the water surface of the park. The authors argue that the study results would help gain sustainability.

There are three chapters on Land Use/ Land Cover. In Chap. 22, titled *Land Use/Land Cover Change Dynamics in Rapidly Growing City: A Potential Threat to Sustainability: A Case Study of the Peshawar District of Pakistan*, Iffat Tabassum, Moazma Sultan, Ihsanullah, and Naeemudin, taking the case study of Peshawar district of Pakistan, sheds light as to how the changes in the dynamics of land use/ land cover in the fast-growing Peshawar district has developed a potential threat to environmental sustainability. The authors argue that there are four classifications of land use/ land cover area in the study area— build-up areas, water bodies, open areas (barren lands) and vegetation cover (agricultural lands). To cater to the growing population demands, built-up areas have been increased while proportionately decreasing the areas covered by water bodies, open areas (barren lands) and vegetation cover (agricultural lands). Similarly, Chhabi Lal Chidi, in Chap. 23, titled *Impact of Bare Lands in the Cities of Kathmandu Valley, Nepal: Question for Urban Environmental Sustainability,* appraises the question of urban environmental sustainability in the Kathmandu Valley. She argues that unplanned and rapid urbanization has increased the demand for constructing buildings and other associated infrastructures, which has increased the valley's bare lands. For over one and a half decades, the barren land area has risen because the building materials have been extracted from the surrounding hill slopes. As a consequence, there has been an uncontrolled increase of heavy sediments in the rivers and rivulets, which causes frequent flash floods during the rainy seasons. She further argues that haphazard sand mining from the riverbed as well as dredging and deepening of the riverbed additionally causes infrastructure damage and increased water pollution in the valley. In a similar context, in the penultimate chapter, in their study, Chap. 24, *Urban Growth and Urban Land Use Dynamics of Pokhara city, Nepal: Challenges to Urban Sustainability?* Khagendra Raj Poudel, Rajesh Bahadur Thapa, Motilal Ghimire, and Ramesh Hamal revisit the question of urban sustainability in Nepal's Pokhara city, which was a simple trading and administrative centre. Over time, it became a tourism hub and a provincial capital. The study demonstrates the changes in its land use patterns, along with changes in the environment, demographic and other socio-economic indicators, which continue to produce challenges to its urban and environmental sustainability. In the final chapter of this volume, Shobha Shrestha in Chap. 25, *Agriculture Practice in Kathmandu Valley, Nepal: Adaptive Strategy or Sustainable Livelihood?* provides a critical geographical analysis of the spatial proponents and attributes of urban agricultural practices and examine processes and decision-making of urban farmers in the context of change in land use in the Kathmandu valley. The findings of her study unravel that considering 'risk perception' and 'uncertainty' the farmers of the urban spaces of the valley adopt tactical and strategic decisions to sustain their livelihood.

1.3 Conclusion

This volume brings to the fore 24 diverse chapters demonstrating the multidimensional factors and challenges influencing sustainability in South Asia's cities and urban spaces. Drawing upon exploratory research and case studies, the volume is concerned with contemporary issues linked to sustainability. The connective thesis that weaves the chapters together (in)directly reinforces the right to liveable cities—as geographical spaces for socio-economic and environmentally sustainable liveability sites. This volume should serve as a valuable companion guide for those engaged in sustainability whether as scholars, practitioners or policymakers.

References

Applebaum A (2022) Holodomor: Ukranian history. Britannica. https://www.britannica.com/event/Holodomor

Bhattacharyya R, Prasad S (2020) Water scarcity in Delhi: mapping for solutions and the way forward. In: Singh RB, Srinagesh B, Anand S (eds) Urban health risk and resilience in Asian cities, pp 421–444. https://doi.org/10.1007/978-981-15-1205-6_24

Bhattacharyya R, Sarma PK, Das TK (2023) Mass exodus of India's internal migrant labourers during the first phase of COVID-19: a critical analysis. SN Soc Sci 3:108. https://doi.org/10.1007/s43545-023-00691-x

Bhattacharyya R, Tan B, Das DB (2023) On understanding the drinking water treatment requirements in Assam, India, during emergencies. In: Anand S, Das M, Bhattacharyya R, Singh RB (eds) Sustainable development goals in Northeast India. Advances in geographical and environmental sciences. Springer, Singapore. https://doi.org/10.1007/978-981-19-6478-7_19

Bhattacharyya R (2009) Examining the changing status and role of middle class assamese women: lessons from the lives of university students. PhD thesis, Newcastle University, UK

Brundtland Report (1987) Report of the World Commission on environment and development: our common future. https://sustainabledevelopment.un.org/content/documents/5987our-common-future.pdf

Breuste J, Artmann M, Ioja C, Qureshi S (2022) Making green cities: concepts, challenges and practice. Springer Nature

Bussolo M, Sharma S, Timmer H (2020) COVID-19 has worsened the woes of South Asia's informal sector. World Bank Blogs. https://tinyurl.com/weeund9t

Castro SE (2021) Urban (digital) play and right to the city: a critical perspective. Front Psychol. https://doi.org/10.3389/fpsyg.2021.636111

Chapter 10: Social infrastructure, employment and human development. Economic Survey 2018–2019. https://www.indiabudget.gov.in/budget2019-20/economicsurvey/doc/vol2chapter/echap10_vol2.pdf

Kumar S, Choudhury S (2021) Migrant workers and human rights: a critical study on India's COVID-19 lockdown policy. Soc Sci Human Open 3(1). https://doi.org/10.1016/j.ssaho.2021.100130

Das TK (2022) Towards sustainable social development in Bangladesh. Space and Culture, India 10(2):1–6. https://doi.org/10.20896/saci.v10i2.1298

Das TK, Bhattacharyya R, Sarma PK (2022) Revisiting geographies of nationalism and national identity in Bangladesh. GeoJournal 87:1099–1120. https://doi.org/10.1007/s10708-020-10305-1

Dandekar A, Ghai R (2020) Reverse migration in the age of COVID-19. Econ Pol Wkly 55(19):28–31

Delivering for people and the planet (n.d) United Nations, Climate Action. https://www.un.org/en/climatechange/cop27

Dhyani S, Singh S, Basu M, Dasgupta R, Santhanam H (2022) Blue-green infrastructure for addressing urban resilience and sustainability in the warming world. In: Dhyani S, Basu M, Santhanam H, Dasgupta R (eds) Blue-green infrastructure across asian countries. Springer, Singapore. https://doi.org/10.1007/978-981-16-7128-9_1

Global Hunger Index (2022) Peer-reviewed annual publication designed to comprehensively measure and track hunger at the global, regional, and country levels. https://www.globalhungerindex.org/

Graziosi A (2005) Soviet famines of 1931–1933 and the Ukrainian Holodomor. Foreigners in Russia, Russians abroad. Cahiers du Monde Russe 46 (3):453–472. https://doi.org/10.4000/monderusse.8817

Hammelman C (2022) Greening cities by growing food: a political ecology analysis of urban agriculture in the Americas. Springer Nature

Harvey D (2000) Spaces of hope. University of California Press

Harvey D (2008) The right to the city. New Left Rev 53:23–40
Harvey D (2012) Rebel cities: from the right to the city to the urban revolution. Verso.
Holodomor—Famine in Soviet Ukraine, 1932–1933 (2008) Internet archive: wayback machine. https://web.archive.org/web/20081209085548/, http://kiev.usembassy.gov/files/famine.pdf
IMF Country Focus (2021) Five things to know about the informal economy. https://www.imf.org/en/News/Articles/2021/07/28/na-072821-five-things-to-know-about-the-informal-economy
Informal Economy (n.d.) International Labour Organization. https://www.ilo.org/global/topics/dw4sd/themes/informal-economy/lang--en/index.htm
International Recognition of the Holodomor (n.d.) Holodomor Genocide by Famine. http://www.holodomoreducation.org/news.php/news/4
Lamond J, Everett G (2019) Sustainable blue-green infrastructure: a social practice approach to understanding community preferences and stewardship. Landsc Urban Plan 191:1–10. https://doi.org/10.1016/j.landurbplan.2019.103639
Levebvre H (1996[1968]) Writings on cities. Edited and translated by Elenore Kofman and Elizabeth Lebas. Blackwell
Littlewood JR, Howlett RJ, Jain LC (2022) Conference proceedings: sustainability in energy and buildings 2021. Springer Nature
McFarlane C (2008) Sanitation in Mumbai's informal settlements: state, 'slum' and infrastructure. Environ Planning A 40(1):88–107. https://doi.org/10.1068/a39221
Mishra RK, Kumari CL, Chachra S, Janaki Krishna PS, Dubey A, Singh RB (2022) Smart cities for sustainable development. Springer Nature
Mlambo-Ngcuka P (2020) Violence against women and girls: the shadow pandemic. UN Women. https://www.unwomen.org/en/news/stories/2020/4/statement-ed-phumzile-violence-against-women-during-pandemic
Purcell M (2014) Possible worlds: Henri Lefebvre and the right to the city. J Urban Aff 36(1):141–154. https://doi.org/10.1111/juaf.12034
Purcell M (2002) Excavating Lefebvre: the right to the city and its urban politics of the inhabitant. Geo J 58:99–108. https://doi.org/10.1023/B:GEJO.0000010829.62237.8f
Pulla VR, Bhattarai S, Thapa BJ, Rai S (2022) The pandemic in the himalayan country: Nepal. Space and Culture, India 10(3):110–124. https://doi.org/10.20896/saci.v10i3.1191
Pulla VR, Khan B (2022) Urban housing for the poor still a contested right: case reflections from Mumbai. Space and Culture, India 10(1):16–29. https://doi.org/10.20896/saci.v10i1.1219
Pulla V, Ahmed ZS, Pawar M (2018) Water and communities in south Asia: a case for regional cooperation. Space and Culture, India 6(3):23–40. https://doi.org/10.20896/saci.v6i3.392
Ryabchikov A, Maximovich S, Calambur Y, Yury K, Alexeeva NN (2024, October 2) South Asia. Encyclopedia Britannica. https://www.britannica.com/place/South-Asia
Scardovi C (2021) Sustainable cities big data, artificial intelligence and the rise of green, "Cy-phy" cities. Springer Nature
Scroll Staff (2020) Covid-19 alert: coronavirus: host states don't want to lose 'cheap labour', claims Jean Dreze on migrant crisis. Scroll.in. https://scroll.in/latest/960758/covid-19-host-states-dont-want-to-lose-cheap, https://scroll.in/latest/960758/covid-19-host-states-dont-want-to-lose-cheap-labour-claims-jean-dreze-on-migrant-crisislabour-claims-jean-dreze-on-migrant-crisis
Singh S, Bhattacharyya R (2020) A review of domestic violence against women in India during lockdown. Int J Innov, Creativity and Change. Special Edition: COVID-19 Life Beyond 14(6):230–242. https://www.ijicc.net/images/Vol_14/Iss_6/PUL015_Singh_2020_R.pdf
Suresh R, James J, Balraju SR (2020) Migrant workers at crossroads–the covid-19 pandemic and the migrant experience in India. Social Work in Public Health 35(7):633–643. https://doi.org/10.1080/19371918.2020.1808552
Transforming our world: the 2030 Agenda for sustainable development (2015) United Nations General Assembly. https://undocs.org/A/RES/70/1
United Nations Framework Convention on Climate Change (1992) https://unfccc.int/files/essential_background/background_publications_htmlpdf/application/pdf/conveng.pdf

United Nations (n.d.a.) Goals 1: End poverty in all its forms everywhere. Department of Economic and Social Affairs Sustainable Development. https://sdgs.un.org/goals/goal1
United Nations (n.d.b) Sustainable development goals: 17 goals to transform our world. https://www.un.org/en/exhibits/page/sdgs-17-goals-transform-world
United Nations (n.d.c.) Goals 6: ensure availability and sustainable management of water and sanitation for all. Department of Economic and Social Affairs Sustainable Development. https://sdgs.un.org/goals/goal6
United Nations (n.d. d) Goals 2: end hunger, achieve food security and improved nutrition and promote sustainable agriculture. Department of Economic and Social Affairs Sustainable Development. https://sdgs.un.org/goals/goal2
United Nations (n.d. e) Goals 11: make cities and human settlements inclusive. Department of Economic and Social Affairs Sustainable Development. https://sdgs.un.org/goals/goal11
United Nations (n.d.f) Goals 8: promote sustained, inclusive and sustainable economic growth, full and productive employment and decent work for all, Department of Economic and Social Affairs, Sustainable Development. https://sdgs.un.org/goals/goal8
UNDP and OPHI (2021). *Global Multidimensional Poverty Index 2021 – Unmasking disparities by ethnicity, caste and gender.* United Nations Development Programme and Oxford Poverty and Human Development Initiative.
UN Women (n.d.) Facts and figures: ending violence against women. https://www.unwomen.org/en/what-we-do/ending-violence-against-women/facts-and-figures#notes
World Health Organization (2021) Devastatingly pervasive: 1 in 3 women globally experience violence. New York
Wolin R (2017) Events of May 1968. Encyclopedia Britannica. https://www.britannica.com/event/events-of-May-1968
World Atlas (n.d.) The Countries of South Asia. https://www.worldatlas.com/articles/the-population-and-economy-of-the-south-asian-countries.html
World Economic Forum (2016) Cities and Urbanization: these are the world's five biggest slums. https://www.weforum.org/agenda/2016/10/these-are-the-worlds-five-biggest-slums/
World Food Programme (n.d.a) A global food crisis. Conflict, COVID, the climate crisis and rising costs have combined in 2022 to create jeopardy for up to 828 million hungry people across the world. https://www.wfp.org/global-hunger-crisis
World Food Programme (n.d.b) Emergency: a global food crisis. https://www.wfp.org/emergencies/global-food-crisis

Part I
Towards Blue-Green Infrastructure of South Asian Cities

Chapter 2
Is the Blue-Green Infrastructure (BGI) Key to Sustainable Development of South Asian Cities?

Praschaya Kaushik and Sanjeeb Kakoty

Abstract The global urban population is increasing at an alarming rate, and is estimated to be 66% of the world's total population by 2050. As per the World Bank, the urbanisation pattern in the cities of South Asia is largely unplanned and messy, with expected growth in urban population up to 250 million by 2030. It has been observed that massive land use change that occurred due to unplanned urban growth has already destroyed the natural resources in and around the cities of South Asia, resulting in many environmental problems like urban flood, drinking water scarcity, groundwater depletion, traffic congestion etc. All these are inextricably intertwined with global environmental issues that is, climate change, air and water pollution, and biodiversity loss. Moreover, the cities are going to face more challenges in sustainable development with an increasing urban population, as projected by the United Nations (UN). To overcome such challenges, it is the need of the hour to understand the fragility of the natural system and to value and invest in ecosystem goods and services. The concept of Blue- Green Infrastructure (BGI), a network of designed and natural landscapes covering the blue elements, viz. ponds, rivers, wetlands etc. and the green elements, viz. trees, gardens, playgrounds, forests etc., is considered as a nature-friendly measure to manage urban flood, improve water and air quality, preserve biodiversity etc. In this study, the scopes of Blue-Green Infrastructure for a sustainable urban future are explored based on the analyses of the environmental issues of South Asian cities like Karachi, Dhaka, Bengaluru, Guwahati and so on.

Keywords Urbanisation · Land use land cover (LULC) change · Blue green infrastructure (BGI) · Sustainable development · South Asian cities

P. Kaushik (✉)
Department of Housing and Urban Affairs, Government of Assam, Dispur, India
e-mail: praschaya.kaushik@assam.gov.in

P. Kaushik · S. Kakoty
Indian Institute of Management, Shillong, India

S. Anand et al. (eds.), *Sustainability in South Asian Cities*, Advances in Geographical and Environmental Sciences, https://doi.org/10.1007/978-981-97-7455-5_2

2.1 Introduction

Urbanisation is a process of shifting of human population from rural areas to urban areas for availing opportunities such as employment and education. As urbanisation has a positive impact on economic development and poverty reduction (United Nations and Department of Economic and Social Affairs 2015), it has become one of the important emerging phenomena of most of the developing countries of the world.

Since 2007, it is reported that the urban population has been growing faster than the rural population; hence, in 2018, 55% of the world's population is recorded to reside in urban areas, projected to reach up to 68% by 2050 (United Nations 2019). However, the rate of urbanisation is highest in Asia and Africa. The level of urbanisation of Asia is presently 50%. Although Asia is less urbanised compared to the rest of the world, it is the home to 54% of total urban population of the world.

The South Asia region comprises of eight countries viz. Afghanistan, Bangladesh, Bhutan, India, the Maldives, Nepal, Pakistan and Sri Lanka. It is the home of 1.9 billion people, i.e., almost one-fourth of the world's population. Figure 2.1, shows that the urban population growth is more than the region's overall population.

South Asia's 36.6% population resides in urban areas. The urban population of South Asia expanded by 130 million between 2000 and 2011 (Sivakumar and Stefanski 2010). As projected, the region's urban population is poised to grow by 250 million by 2030 (World Bank Group 2015). Among the South Asian sub-regions, Bhutan is recorded the highest urban population that is, 43% followed by the Maldives 41.1%. On the other hand, Sri Lanka, Nepal and Afghanistan have comparatively lower urban populations than the region's other countries (Table 2.1).

India, the largest country in the region, has shown a rapid urbanisation pattern (Fig. 2.2), with a present urban population of 35.4%. As per UN, India is projected

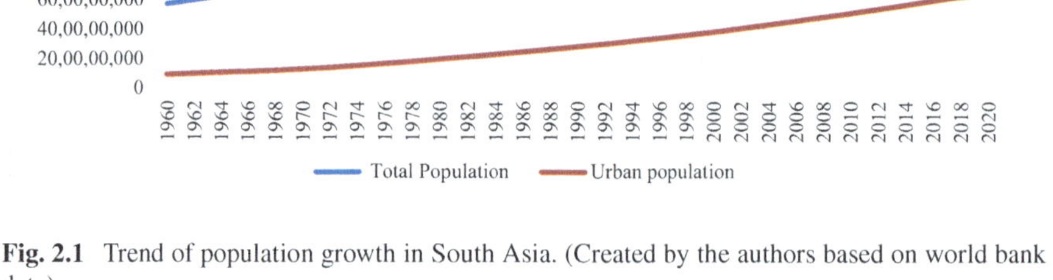

Fig. 2.1 Trend of population growth in South Asia. (Created by the authors based on world bank data)

Table 2.1 Total population and urban population of south Asian countries

Country name	Total population (2021)	Urban population	Urban population (%age)
Afghanistan	4,00,99,462	1,05,51,772	26.3
Bangladesh	169,356,251	6,59,57,486	38.9
Bhutan	777,486	3,34,381	43.0
India	1,407,563,842	49,81,79,071	35.4
Maldives	521,457	2,14,329	41.1
Nepal	30,034,989	63,09,750	21.0
Pakistan	231,402,117	8,66,36,953	37.4
Sri Lanka	22,156,000	41,78,622	18.9

Source World Bank

to add 416 million urban dwellers by 2050, followed by China (255 million) and Nigeria (189 million).

The present trend shows that world's most of the urban population lives in urban settlements with fewer than one million inhabitants, which is close to 58% of the world's population. Moreover, almost 21% of the world's urban population live in medium-sized cities, 8% in large cities, and around 13% in mega cities (United Nations and Department of Economic and Social Affairs 2019). However, the urban spaces of South Asian countries, especially the cities, are mired in myriad environmental challenges—climate change, air and water pollution, and biodiversity loss and so on. Considering these backdrops, this chapter probes the scopes of Blue-Green

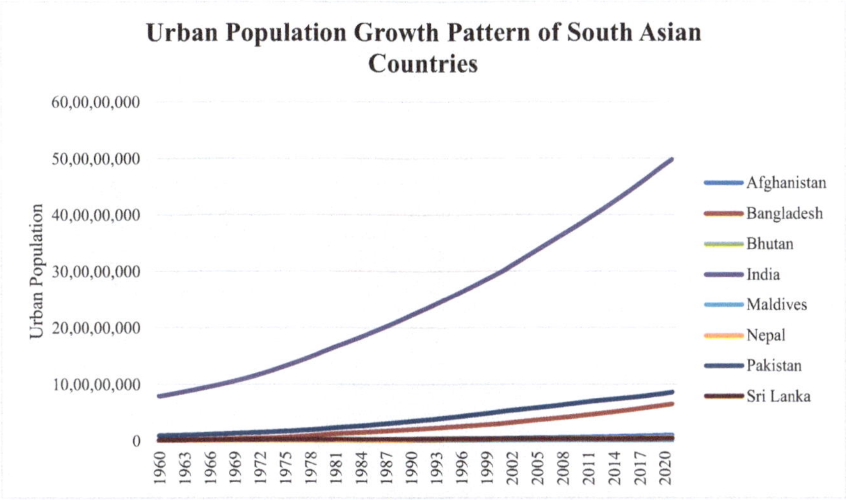

Fig. 2.2 Total and urban population of South Asian countries. (Created by the authors based on world bank data)

Infrastructure for a sustainable urban future in South Asian cities like Karachi, Dhaka, Bengaluru, Guwahati, etc.

The chapter begins with a discussion of the environmental challenges amid the urbanisation patterns in South Asia. Following this, we try to illustrate the scope of Blue-Green Infrastructure, which we reckon would be a nature-based solution for a sustainable future of the South Asian cities.

2.2 Urbanisation Pattern of South Asian Cities & Environmental Challenges

Urbanisation transforms the landscape of cities, and generates agglomeration of economies by creating favourable conditions for the market increasing productivity and economic growth (Bhagat 2011). At the same time, it increases the demand for infrastructure, viz. energy, water, and sanitation as well as for public services, education and health care etc. (United Nations. Department of Economic and Social Affairs n.d.). For sustainable urban development, cities must meet the requisite living standards of the urban population. It is evident that unplanned urban sprawl engulfs natural sites covering water bodies, forest lands, agriculture lands and sub urban areas as shown in Fig. 2.3, leading to environmental problems.

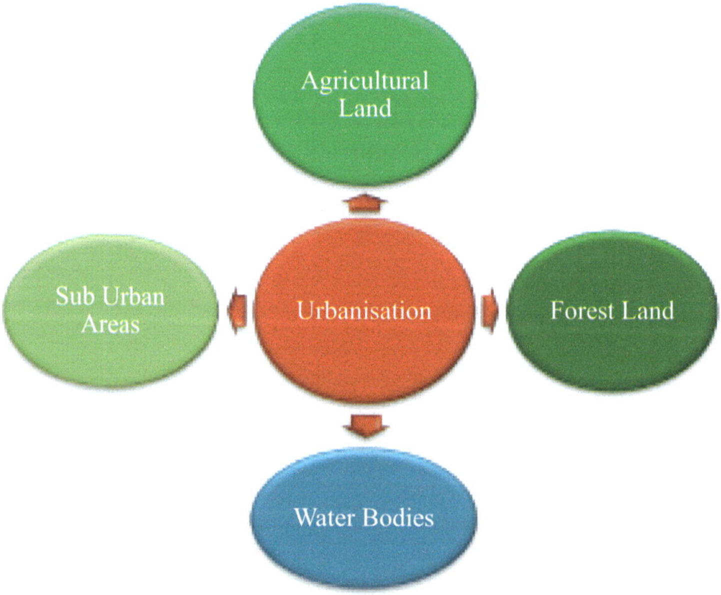

Fig. 2.3 Effect of urban sprawl. (Created by the authors)

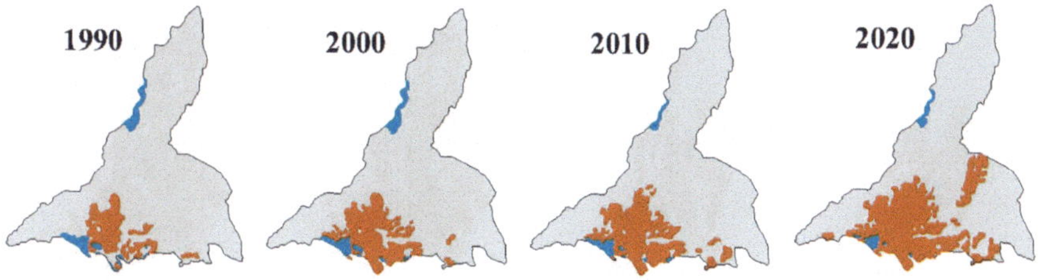

Fig. 2.4 Urban sprawl of Karachi city from 1990 to 2020. (Created by the authors based on studies done by Baqa et al. 2021)

It remains well documented that South Asia, the most densely populated region of the world (Kankesu et al. 2020), have inadequate infrastructure and services at present to cater the urban population (World Bank Group 2015). Most of the cities of the region are having ecological and environmental threats due to haphazard urban development causing massive change in the Land Use and Land Cover (LULC). A few examples of urbanisation in South Asian cities are discussed below.

2.3 Karachi (Pakistan)

Karachi is the largest city in Pakistan and a major commercial and industrial centre of the country. It is located on the Arabian coast and the capital of the Sindh province with an area of 3530 km^2. Reportedly, Karachi is Pakistan's fastest urbanizing city, whose population has increased from 9.8 million in 2000 to 15.4 million in 2018 (Idress et al. 2021).

The Land Use Land Cover (LULC) has drastically changed in Karachi over the last three decades. In the Fig. 2.4 the pattern of urban sprawl of Karachi city is shown from 1990 to 2020. The land covered by the Built-up area has expanded drastically. In contrast, the area covered by agricultural land, mangroves, and open bare ground has declined (Baqa et al. 2021).

The LULC change has created the disturbances in the natural habitation affecting the surface runoff. Evidently in recent years flood have severely hit urban localities; significantly, the area affected by urban flood doubled in the year of 2013 (Zafar and Zaidi 2019).

2.4 Dhaka, Bangladesh

Dhaka is the capital city of Bangladesh and was reported as the ninth most populous city in the world in 2018. The area covered is 306.4 km^2. The city's population has increased from 10.2 million in 2000 to 19.6 million in 2013 with a rate of.6% per

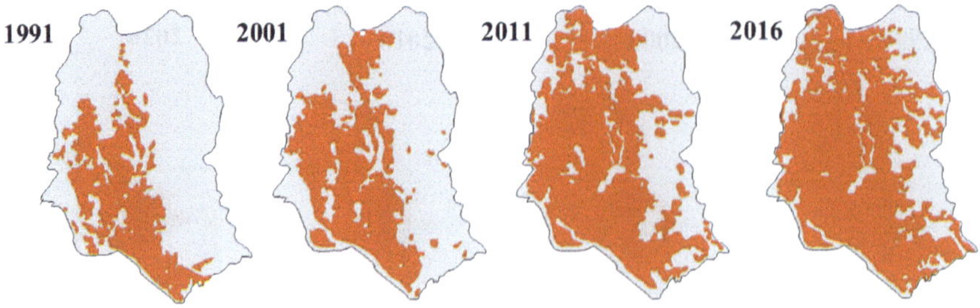

Fig. 2.5 Urban sprawl of Dhaka city from 1991 to 2016. (Created by the authors based on studies done by Ullah and Esmat Enan 2016)

annum (Ranagalage et al. 2021). It is one of the fastest-growing megacities in the world, however, UN predicts that by 2025, Dhaka will be the home of 20 million people. Such rapid urban population growth has already impacted the city's landscape (Hossain et al. 2021).

Figure 2.5, is showing the urban sprawl of Dhaka city where the expansion of the built-up area is clearly visible. As mentioned by (Ullah and Esmat Enan 2016) in 1991 the land cover was mostly dominated by 40% water bodies which has been reduced to 14% by 2016. On the other hand, the land covered by the built-up area was only 33% which has expanded to 72% in 2016. The built-up areas are concentrated around the city core and gradually spread towards the northern and southern part. Similarly, drastic change in green coverage has also occurred in the city. Such changes in the landscape have already disturbed the natural vegetation, and the lowland areas of Dhaka have become more vulnerable. Moreover, flood has become a severe issue in the city day by day due to the disturbances in the natural drainage system (Ahmed and Khakhar 2022).

2.5 Bengaluru, India

Bengaluru, the 'Silicon Valley of India' is one of the largest cities of the southern part of the country. It is also known as the 'Garden city' and the 'City of Lakes' because of the availability of many tree-covered areas and wetlands. The capital of Karnataka state covers an area of 741 km^2 of land with a population of more than 10 million, as recorded (Brinkmann et al. 2020). The city is experiencing rapid urbanisation and sprawl, which has led to significant change in the Land Use and Land Cover.

The urban sprawl as shown in Fig. 2.6, has highlighted the unprecedented increase in built-up area in Bangaluru. As per studies, built up area of the garden city has increased from 7.97% in 1973 to 93.26% in 2020. At the same time, the area covered by vegetation was recorded as 68.27% in 1973, which has reduced to 2.96% in 2020. Similarly, the area covered by water bodies has decreased to 2.09% in 2020 from 3.40% in 1973. As reported, presently, the city has only 17 wetlands, which declined

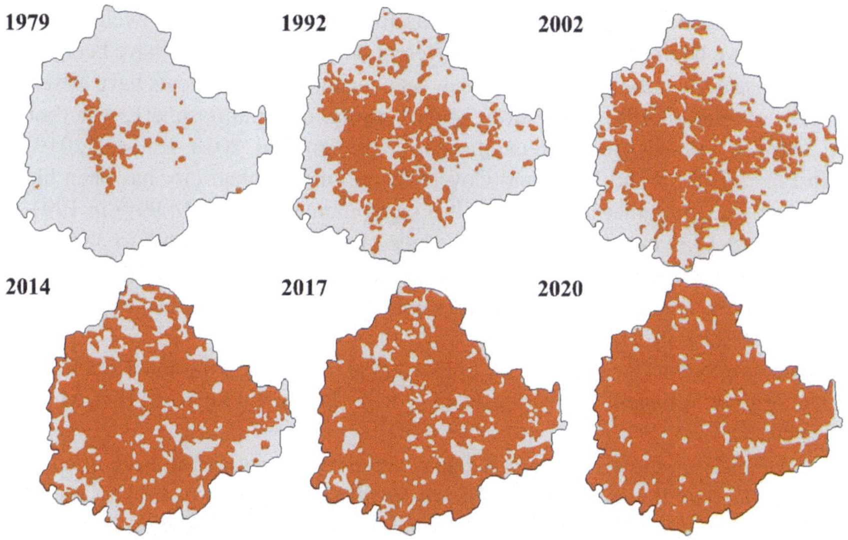

Fig. 2.6 Urban sprawl of Bangaluru city from 1973 to 2020. (Created by the authors based on studies done by Ramachandra et al. 2017)

from 51 in 1973 (Kumar Gupta 2011). Moreover, there is a 1028% rise in the paved surface from 1973 to 2017 (Ramachandra et al. 2017).

Such land use change has already reduced the size of stormwater drainage (Ramachandra and Ahalya 2001), increased the surface runoff, and also has increased the silt deposition in the floodplains. As a result, the Bengaluru city is facing frequent flooding. In September 2022, the city experienced the highest rainfall in the last three decades, leading to severe floods in the city. As reported during that flood, 7700 houses were affected, 394 km of roads were damaged, and a loss of 41.4 million USD had occurred (Times of India 2022).

2.6 Guwahati, India

Guwahati, the Gateway to Northeast India and the region's largest city, covers an area of 314 km^2. The city is located in the southern part of the river Brahmaputra and consists of three broad ecological units, viz., the hills, plains and wetlands. The city has witnessed a high urban population growth rate since 1971 after shifting capital from Shillong. The population had increased from 293,219 in 1971 to 968,549 in 2011. It is estimated that the population of Guwahati is around 1,319,000 in 2023.

Despite rapid urban growth, the necessary infrastructures are not increased in the city for which all three ecological units are affected in different ways. The plain areas, which were the places of early settlers, became over crowded. Hill slopes are extremely disturbed mainly due to terracing for the construction of dwelling units. Wetlands have been encroached on for different purposes, and many have been filled up for construction and garbage disposal. As a result, many problems have arisen in the city, of which traffic congestion, water logging, dusty atmosphere, waterborne and airborne diseases are worth mentioning (Kashyap et al. 2018; Kaushik 2019).

In Fig. 2.7, the Land Use Land Cover change of Guwahati city has been highlighted. It is evident that the city's built-up area increased from 14.99% in 1991 to 49.11% in 2015, whereas forest cover reduced from 55.36% in 1991 to 38.25% in 2015. Guwahati has a historically unique terrain with a number of wetlands within the city. But the area covered by wetlands and waterbodies has reduced from 12.51% in 1991 to 9.22% in 2015.

For example, the Deepor Beel, the only wetland in Assam, designated as a Ramsar site, is severely affected by the haphazard growth (Dutta and Sharma 2020; Monjoy Karigapsa 2013).

As per analysis shown do Fig. 2.8 area covered by the wetland was recorded 6.97 km^2 in 1991, which was reduced to 4.8 km^2 in 2002 and further to 4.2 km^2 in 2015.

A similar situation has also been observed in the case of the Silsako beel, which lies in the south eastern part of the Hengarabari reserve forest. This wetland has

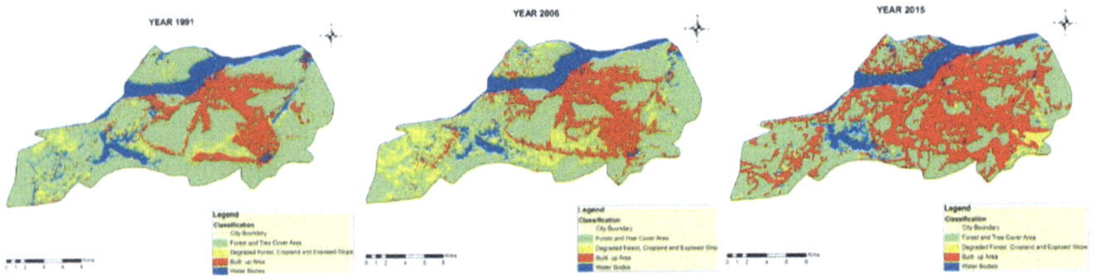

Fig. 2.7 Land use land cover (LULC) change of Guwahati city from 1991 to 2015. (Created by the authors)

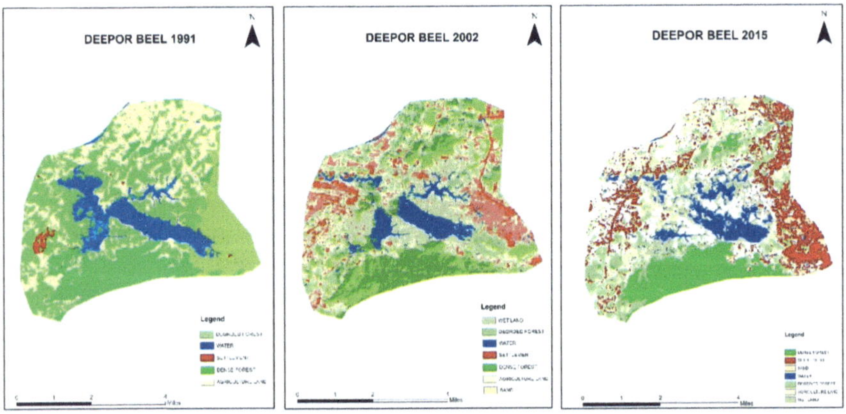

Fig. 2.8 Land use land cover change in deepar beel. (Created by the authors)

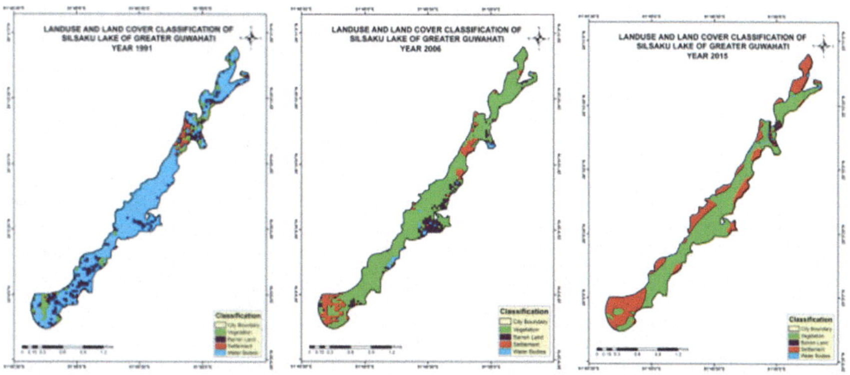

Fig. 2.9 LULC change of Silsaku beel in 1991, 2006, 2015. (Created by the authors)

severely affected by encroachment. In Fig. 2.9, the Land Use Land Cover change is analysed. The area covered by water body has immensely reduced in 2015 in comparison to 1991.

The wetlands have a significant biological value as they are the home of different kinds of aquatic fauna and flora in a region (Dutta and Sharma 2020). The large-scale land use change in reserve forest areas causes rapid degradation of wetlands leading to the existence of fragmented form which lies in the foothills of reserve forests area. A few small wetlands lie in fragmented form inside the city (Gogoi 2012). As the natural drainage system has already been disturbed and the water retention capacity of the city has been reduced, many major environmental problems are being faced by its citizens—Flash floods, Landslides, Groundwater depletion, Air pollution, etc.

From the above discussion related to the LULC change pattern of different South Asian cities, the significant finding highlighted is the urban areas' unplanned growth. It has been observed that in the case of unplanned cities, the urban population growth influences the settlements to sprawl towards the available land, irrespective of its ecological importance. In many cases, for example, in Karachi, Dhaka, and Bengaluru the urban settlements are growing towards the low-lying areas, that is, Water bodies or wetlands. On the other hand, cities like Guwahati sprawls towards the hills and forest lands.

It is evident that rapid urbanisation creates a colossal pressure on land, resulting in encroachment on floodplains, drainage channels and low-lying areas. As cities grow, impervious surfaces like buildings, roads, bridges, pavements etc., expand (Rohilla et al. 2017). The loss of green cover and water bodies deteriorates the watershed area, leading to reduced water retention capacity of the city, increasing the water logging and flash flood issues. Moreover, the disturbances in hilly terrain by deforestation and hill cutting cause landslides. It also increases the urban runoff, leading to groundwater depletion, silt deposition, air and water pollution etc. In Fig. 2.10, such correlation of unplanned urban sprawl and environmental problems are showcased.

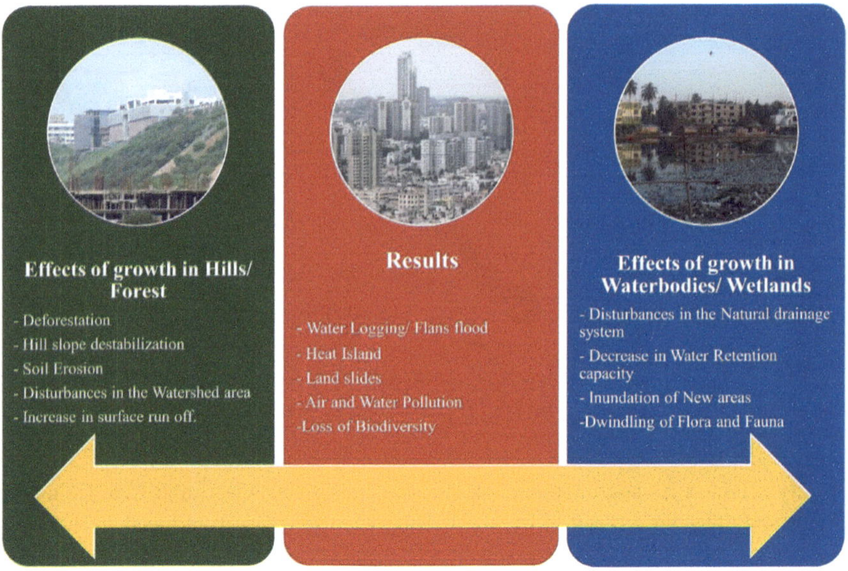

Fig. 2.10 Diagram showing impact of urbanisation on environment. (Created by the authors)

2.7 Blue Green Infrastructure (BGI) as Nature-Based Solution (NBS)

Nature-based solution (NBS) is an umbrella concept that covers different approaches. According to IUCN, NBS can be defined as the actions to protect, sustainably manage and restore natural or modified ecosystems, addressing societal challenges (that is, climate change, natural disaster, food and water security etc.) effectively or adaptively, while simultaneously providing human well-being and biodiversity benefits (Langemeyer and Baró 2021).

Nature Based Solutions (NBS) can intervene as follows:

- Restoration and sustainable management of wetlands and rivers to reduce the risk of flood, maintain fish stock and fisheries-based livelihoods, and provide tourism and recreational benefits.
- Conservation of forests for climate change adaptation and mitigation, to preserve biodiversity and to support food and energy security.
- Restoration of dryland for resilience to climate change impact, to strengthen water security and local livelihoods.
- Developing Blue-Green infrastructure in urban areas to improve air quality, and reduce stormwater runoff, flash floods and water pollution, eventually enhancing quality of life.

– Protection of shoreline and communities by using natural coastal infrastructure from coastal flooding and reduce the impacts of sea level rise (Cohen-Shacham et al. 2016).

This (NBS) concept overlaps with many other ideas like ecosystem-based adaptation, ecological engineering, green infrastructure etc. (Toxopeus and Polzin 2021). This concept develops its major strengths through the capacity to explicitly address concrete societal challenges, most notably climate change (Langemeyer and Baró 2021) As discussed earlier, rapid urbanisation and dependence on grey infrastructure have already created many environmental, economic and social issues in South Asian cities. Major environmental problems these cities face are water, viz, urban floods, groundwater depletion, water pollution, etc. Hence, this study will explore how South Asian cities can be guided by BGI for sustainability.

Traditionally urban infrastructures are comprised of grey elements like roads, stormwater drains, reservoirs, barrages/ dams etc. In recent years it has been realised that 'Grey Infrastructure' is not an adequate solution for urbanisation-induced hydro-climatic problems. Blue-Green Infrastructure is a globally recognised approach to managing urban flood risk (Lamond and Everett 2019). It significantly depends on the urban green cover and natural water systems (Dhyani and Thummarukuddy 2016). BGI is a network system connecting the natural infrastructure that is, the vegetation system or green areas (green infrastructure) and urban hydrological system or water bodies (blue infrastructure) (Xu and Zhao 2021). Basically, BGI integrates the natural system with the built environment. Combining natural (that is, Blue-Green) and built (grey) infrastructure creates the potential to mitigate climate change issues and as well as to provide water, energy, and food to growing urban populations and the quality of life of the citizens (Browder et al. n.d.). For example, to provide water supply in urban areas, different built forms such as reservoirs, water treatment plants, and network of pipeline areas are constructed; similarly, to manage urban floods, stormwater drains are designed, and pumps are fitted. On the other hand, if we see the functions of wetlands, it serves as the stormwater storage. Therefore, integrating wetlands in urban flood management or water supply systems is more efficient than depending only on the grey infrastructure.

BGI entails urban water management and urban green infrastructure to regulate the natural water cycles to maintain urban ecology (Drosou et al. 2019). There are a number of natural components which are considered to be the Blue-Breen Infrastructure, viz, wetlands, urban agriculture, sponge gardens, bioswales, biopores, green roofs, permeable pavement, parks, green belts etc. (Shalini et al. 2022).

Different BGIs can be divided into 4 different categories (Hamel and Tan 2022), viz. Wetlands, Urban open green spaces, Engineered stormwater devices, and Watershed-scale features (Fig. 2.11).

1. Wetlands: It can be defined as submerged or water-saturated lands, natural or human-made, inland or coastal, permanent or temporary, static or dynamic, or vegetated or non-vegetated, which necessarily have a land–water interface (Pradip 2017). Wetlands can be natural or human-made. Peatlands, Mangroves,

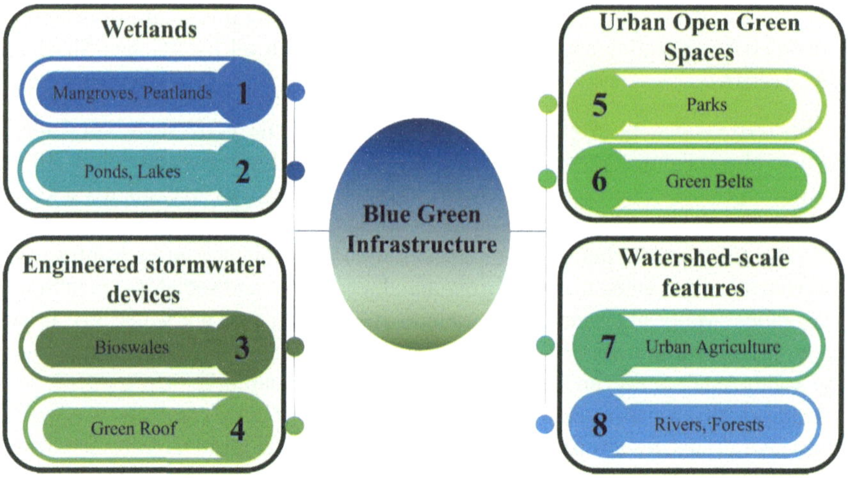

Fig. 2.11 Diagram showing different category and types of blue-green infrastructure. (Created by the authors)

ponds, lakes etc. fall under this category. Wetlands help in reducing peak flow, storing stormwater, and remove wastewater effluent.

2. Urban open green spaces: This includes gardens, parks, green belts etc. Green cover helps increase the water retention capacity by reducing surface runoff, resulting reduce in water and soil pollution and also enhancing groundwater level.
3. Engineered Stormwater devices: These are basically the integration of natural and man-made elements. It includes:

    Bioretention systems or Bioswales: It is also known as rain gardens comprising multiple layers of soil and vegetation designed to reduce surface runoff by storing and infiltering stormwater.
    Green Roofs: These are vegetation cover in buildings similar to Bioswales, including impervious membrane. This helps in reducing surface runoff and is also used for rainwater harvesting
    Sedimentation basins or Retention ponds: These are used for capturing litter, suspended solids and course sediments of wastewater as well as stormwater. It helps reduce pollution in urban water bodies by reducing siltation and acts as storage for access runoff (Hamel and Tan 2022).

4. Watershed scale features: This category includes urban agriculture land, urban forests, rivulets and rivers covering the watershed. These components are responsible for maintaining the natural water flow of the terrain. In addition, it helps reduce the velocity of water flowing from upstream, resulting in reduced soil erosion.

Hence, the effective use of BGI is associated with several benefits like recharge of groundwater, reduction in storm water runoff, reduction in soil erosion, decrease in water and air pollution, controlling riparian ecological system etc. These together help in reducing the risk of urban flood and heat island effect and enhance the resilience of urban infrastructure (Dreiseitl and Wanschura 2016)As a result, there are many countries worldwide practising BGI. A few examples are discussed below:

Copenhagen, Denmark: Copenhagen, the most populous city of Denmark, is a coastal urban area. The rising sea level s increased the flood risk in the city. As a mitigation measure, integration of BGI was adopted in the Copenhagen Climate Adaptation Plan of October 2011 (Dreiseitl and Wanschura 2016). In this initiative, Engineered Stormwater devices like Bioswales, rain gardens, paved recreational areas etc. were designed for stormwater management and to provide multiple benefits. Water treatments often involve bio-filtration with biofilters and sometimes UV treatment. The project focused mainly on flood control combined with city dwellers' liability (Liu et al. 2019).

Singapore: The island state of Singapore is one of the densest urban areas in the world. Earlier, the country faced certain issues like scarcity of drinking water, fluctuating water level in the coastal line, high demand for a quality living environment. Therefore, to provide their citizens with the highest quality of life, the country took certain initiatives. In the 1960s only Green space planning began in Singapore. Presently, Singapore has 56% vegetation cover out of which 27% is actively managed parks, lawns and gardens, etc. (Lim and Xenarios 2021). Such urban green space not only has improved the quality of life but ensures the sustenance of water catchment area.

In 2006, the ABC Waters Program was launched in Singapore, which aimed to improve recreational value, water quality and physical appearance of all the water bodies in the country (Liao 2019).The project involved different categories of BGI, viz. open fields, water bodies, urban plazas, sediment basins, vegetation swales, retention ponds, building components etc. (Driver and Mankikar 2021).

Sponge City, China: Many urban areas of China were experiencing flooding due to rapid urbanisation. As recorded, more than 150 cities were affected by flash floods and water logging from 2006 to 2010 (Guo et al. 2018), damaging property of around 160 billion RMB (Yang 2014) To mitigate this issue, Republic of China adopted the 'Sponge City' concept in 2014. Moreover, this concept is found to be similar to Water Sensitive Urban Design in Australia (Morison and Brown 2011), Low Impact Development approach in the USA (Pyke et al. 2011), Blue-Green cities in the UK (Thorne et al. 2018) etc.

Aims of the Sponge City concept

- To adopt and develop LID (Low impact developments) approach for effective control of surface runoff and create temporary storage of stormwater.
- To improvise the traditional drainage system.

- To integrate all the natural water bodies like wetlands and lakes and drainage systems to create additional green space and water bodies to provide ecosystem services (Chan et al. 2018).

In brief, the idea of Sponge City is to absorb water during rain and storm like a sponge and release it after processing back to urban runoff (Qi et al. 2020). In the first phase of this programme, 30 cities were selected as pilot cities depending on various climatic and socio-economic conditions. The Government invested 652 million USD in Sponge City projects in China from 2015 to 2020. The chosen cities had to provide lands for sponge features, viz. rain gardens, bioswales, bio-retention basins, green roofs, etc.

2.8 Blue Green Infrastructure and Sustainability of South Asian Cities

South Asian countries are already facing the impact of climate change in different forms, viz., variable precipitation, temperature rise, extreme weather events, etc. It is predicted that the challenges would intensify in the future (Nazrul Islam and Sultan 2009). The messy urbanisation of South Asian cities has increased the risks. It has been observed that the current trend of urbanisation in several South Asian cities has disconnected the natural environment, which has increased different environmental issues (Goswami et al. 2022). As the cities grow, Built up areas grow, and green spaces and water bodies get reduced. The term urbanisation is intrinsically intertwined with infrastructure. In South Asia, infrastructure dominantly contains grey components. Adopting the BGI concept may play an important role in ensuring the sustainability and resilience of the urban areas (Andersson et al. 2019).

Amongst South Asian countries, India is the biggest country having the highest population in the world. Many Indian cities have high-density built-up areas. Two such cases, viz. Bengaluru, and Guwahati have already been discussed above. However, India has already started integrating the BGI concept in urban planning. Delhi Masterplan 2041 is the first move by the Government of India to include Blue-Green policies (Driver and Mankikar 2021). But cities like Bengaluru do not have a particular mandate for such resiliency planning. However, in 2014, a blue-green action plan was conceptualised by Indian Institute of Human Settlements and local stakeholders for the future-proof city of Bengaluru (Ravindra 2014). Similarly, the sponge city concept has been taken up in four cities under the scheme Atal Mission for Rejuvenation and Urban Transformation (AMRUT) in the State of Assam. The Silsakoo Beel (Wetland) rejuvenation has been taken as a pilot project, by which it is expected to store more than 3 million litres of water (Town and Country Planning et al. 2023).

2.9 Conclusion

The BGI is a new concept for South Asian countries, but many countries across the world have already initiated the transition to combat against climate change. Unlike typical infrastructure projects, BGI needs a long-term approach by combining human-made (hard) and natural (soft) infrastructure. In South Asian urban areas, scattered attempts of these concept are being made, but it has to be implemented cautiously to ensure the resilience of the region's urban areas. Appropriate institutionalisation and policy framework for the adoption of this concept is the need of the hour for the sustainability of the South Asian cities.

References

Ahmed S, Khakhar S (2022) Adaptive flood resilient city design strategies: an exploratory analysis of the impact of flooding on street network of Dhaka city. Urbanie & Urbanus—Resilient Cities 7:40–54. https://doi.org/10.55412/07.03

Andersson E, Langemeyer J, Borgström S, McPhearson T, Haase D, Kronenberg J, Barton DN, Davis M, Naumann S, Röschel L, Baró F (2019) Enabling green and blue infrastructure to improve contributions to human well-being and equity in urban systems. In: BioScience, vol 69, no 7, pp 566–574. Oxford University Press. https://doi.org/10.1093/biosci/biz058

Baqa MF, Chen F, Lu L, Qureshi S, Tariq A, Wang S, Jing L, Hamza S, Li Q (2021) Monitoring and modeling the patterns and trends of urban growth using urban sprawl matrix and CA-Markov model: a case study of Karachi, Pakistan. Land 10(7). https://doi.org/10.3390/land10070700

Bhagat RB (2011) Emerging pattern of urbanisation in India. Econ Political Weekly 46(34):10–12. https://www.jstor.org/stable/23017782

Brinkmann K, Hoffmann E, Buerkert A (2020) Spatial and temporal dynamics of urban wetlands in an Indian Megacity over the past 50 years. Remote Sens 12(4). https://doi.org/10.3390/rs12040662

Browder G, Ozment S, Bescos IR, Gartner T, Lange G-M (n.d.) Integrating green and gray: creating next generation infrastructure

Chan FKS, Griffiths JA, Higgitt D, Xu S, Zhu F, Tang YT, Xu Y, Thorne CR (2018) "Sponge City" in China—a breakthrough of planning and flood risk management in the urban context. Land Use Policy 76:772–778. https://doi.org/10.1016/j.landusepol.2018.03.005

Cohen-Shacham E, Walters G, Janzen C, Maginnis S (2016) Nature-based solutions to address global societal challenges. In: Nature-based solutions to address global societal challenges. IUCN International Union for Conservation of Nature. https://doi.org/10.2305/iucn.ch.2016.13.en

Dhyani S, Thummarukuddy M (2016) Ecological engineering for disaster risk reduction and climate change adaptation. Environ Sci Pollut Res 23(19):20049–20052. https://doi.org/10.1007/s11356-016-7517-0

Dhyani S, Basu M, Santhanam H, Dasgupta R (2022) Blue-green infrastructure across Asian countries. In: Blue-green infrastructure across Asian countries. Springer Singapore. https://doi.org/10.1007/978-981-16-7128-9

Dreiseitl Prof. H, Wanschura B (2016) Strengthening blue-green in our cities. www.ramboll.com/LCL

Driver B, Mankikar SU (2021) Blue-green infrastructure: an opportunity for Indian cities. Observer Res Found 317:2–38. www.orfonline.org Highlight

Dutta J, Archana S (2020) Valuing fishing activity of the deepor beel. Space and Culture, India 7(4):122–132. https://doi.org/10.20896/saci.v7i4.607

Drosou N, Soetanto R, Hermawan F, Chmutina K, Bosher L, Hatmoko JUD (2019) Key factors influencing wider adoption of blue-green infrastructure in developing cities. Water (Switzerland) 11(6):1–20. https://doi.org/10.3390/w11061234

Goswami S, Basak S, Malik A, Bhagat Palanichamy R (2022) Urban blue-green conundrum: a 10-city study on the impacts of urbanization on natural infrastructure in India. World Resources Institute. https://doi.org/10.46830/wriwp.19.00062

Guo L, He B, Ma M, Chang Q, Li Q, Zhang K, Hong Y (2018) A comprehensive flash flood defense system in China: overview, achievements, and outlook. Nat Hazards 92(2):727–740. https://doi.org/10.1007/s11069-018-3221-3

Hamel P, Tan L (2022) Blue-green infrastructure for flood and water quality management in southeast Asia: evidence and knowledge gaps. Environ Manage 69(4):699–718. https://doi.org/10.1007/s00267-021-01467-w

Hossain MdA, Sultana S, Nazem NI (2021) Spatial distribution of strategic service infrastructures that favour quality of life (QOL) in the Greater Dhaka Region (GDR), Bangladesh. SN Soc Sci 1(8):1–25. https://doi.org/10.1007/s43545-021-00219-1

Idress M, Nergis Y, Butt JA, Sharif M (2021) Ambient air quality assessment in Karachi, Sindh Pakistan. Int J Econ Environ Geol 12(3):60–64. www.econ-environ-geol.org

Indrani G (2012) Changing landuse pattern and environmental problems in greater Guwahati Area, ASSAM [PhD Thesis]. Gauhati University

Kankesu, J. K., Verma Guanghua, R., & Wilson, W. E. (2020). *Internal Migration, Urbanization, and Poverty in Asia: Dynamics and Interrelationships*. Springer Open. https://doi.org/10.1007/978-981-13-1537-4

Kashyap K, Sahariah D, Ghosh S (2018) Monitoring landuse/land cover change and consequent effects on wetland environment—a case study of deepor beel, Assam, India. ESRI India user conference

Kumar Gupta A (2011) Urban floods in Bangalore and Chennai: risk management challenges and lessons for sustainable urban ecology. Curr Sci 100:1638–1645

Lamond J, Everett G (2019) Sustainable blue-green infrastructure: a social practice approach to understanding community preferences and stewardship. Landsc Urban Plan 191:1–10. https://doi.org/10.1016/j.landurbplan.2019.103639

Langemeyer J, Baró F (2021) Nature-based solutions as nodes of green-blue infrastructure networks: a cross-scale, co-creation approach. Nature-Based Solutions 1:1–11. https://doi.org/10.1016/j.nbsj.2021.100006

Liao K (2019) The socio-ecological practice of building blue-green infrastructure in high-density cities: what does the ABC Waters Program in Singapore tell us? Socio-Ecol Pract Res 1:67–81. https://doi.org/10.1007/s42532-019-00009-3

Lim M, Xenarios S (2021) Economic assessment of urban space and blue-green infrastructure in Singapore. J Urban Ecol 7(1):1–11. https://doi.org/10.1093/jue/juab020

Liu L, Fryd O, Zhang S (2019) Blue-green infrastructure for sustainable urban stormwater management-lessons from six municipality-led pilot projects in Beijing and Copenhagen. https://doi.org/10.3390/w11102024

Monjoy K (2013) Change detection of deepor beel (wetland) using remote sensing data & GIS. In: Management of water resources in north east region with special reference to flood & erosion management, pp 56–68

Morison PJ, Brown RR (2011) Understanding the nature of publics and local policy commitment to water sensitive urban design. Landsc Urban Plan 99(2):83–92. https://doi.org/10.1016/j.landurbplan.2010.08.019

Nazrul Islam A, Sultan S (2009) Climate change and south Asia: what makes the region most vulnerable? Asian J Agric Dev 17(2):1–18. https://mpra.ub.uni-muenchen.de/21875/1/MPRA_paper_21875.pdf

Rohilla SK, Jainer S, Matto M (2017) Green infrastructure—a practitioner's guide. www.cseindia.org

Praschaya K (2019) Change of wetlands: a threat to the living conditions of people in Guwahati city, Assam, India. In: Harnessing wetlands for sustainable livelihood. Notion Press, pp 111–124

Pradip S (2017) Land use change in the peripheries of wetlands and its impact on the water bodies: a comparative study in the deepor and urpad beels of Assam, India. IMPACT: Int J Res Appl, Nat Soc Sci 5(2):65–72

Pyke C, Warren MP, Johnson T, Lagro Jr, Scharfenberg J, Groth P, Freed R, Schroeer W, Main E (2011) Assessment of low impact development for managing stormwater with changing precipitation due to climate change. Lands Urban Plan 103:166–173. https://doi.org/10.1016/j.landurbplan.2011.07.006

Qi Y, Chan FKS, Thorne C, O'donnell E, Quagliolo C, Comino E, Pezzoli A, Li L, Griffiths J, Sang Y, Feng M (2020) Addressing challenges of urban water management in Chinese sponge cities via nature-based solutions. In: Water (Switzerland), vol 12, no 10, pp 1–24. MDPI AG. https://doi.org/10.3390/w12102788

Ramachandra TV, Aithal Bharath H, Shivamurthy V (2017) Frequent floods in Bangalore: causes and remedial measures. https://doi.org/10.13140/RG.2.2.17517.90088

Ramachandra TV, Ahalya N (2001) Wetlands restoration and conservation. In: Control of industrial pollution and environmental degradation, pp 262–275

Ranagalage M, Morimoto T, Simwanda M, Murayama Y (2021) Spatial analysis of urbanization patterns in four rapidly growing south Asian cities using sentinel-2 data. Remote Sens 13(8). https://doi.org/10.3390/rs13081531

Ravindra PN (2014) Future proofing Indian cities: Bangalore action plan for water and sanitation infrastructure

Sivakumar MVK, Stefanski R (2010) Climate change in south Asia. In: Climate change and food security in south Asia. Springer Netherlands, pp 13–30. https://doi.org/10.1007/978-90-481-9516-9_2

Town & Country Planning, Department of Housing & Urban Affairs, & Govt. of Assam (2023) Development of systematic, vibrant and inclusive cities across Assam

Thorne CR, Lawson EC, Ozawa C, Hamlin SL, Smith LA (2018) Overcoming uncertainty and barriers to adoption of blue-green infrastructure for urban flood risk management. J Flood Risk Manage 11:S960–S972. https://doi.org/10.1111/jfr3.12218

Times of India (2022) 7,700 homes flooded. https://Timesofindia.Indiatimes.Com/City/Bengaluru/7700 Homes-Flooded-Bruhat-Bengaluru-Mahanagara-Palike-Puts-Overall-Loss-at-Rs-340-Crore/Articleshow/94441510.Cms

Toxopeus H, Polzin F (2021) Reviewing financing barriers and strategies for urban nature-based solutions. J Environ Manage 289:1–11. Academic Press. https://doi.org/10.1016/j.jenvman.2021.112371

Ullah S, Esmat Enan M (2016) Landuse and land cover change in Dhaka metropolitan area during 1991–2022. Oriental Geographer 60(2):1–18

United Nations, & Department of Economic and Social Affairs, P. D (2015) World Urbanization Prospects—The 2014 Revision

United Nations, & Department of Economic and Social Affairs (2019) World urbanization prospects 2018 highlights.

United Nations. Department of Economic and Social Affairs (n.d.) World economic and social survey 2013: sustainable development challenges

World Bank Group (2015) South Asian development matter: overview—leveraging urbanization in south asia; managing spatial transformation for prosperity and liveability

Xu H, Zhao G (2021) Assessing the value of urban green infrastructure ecosystem services for high-density urban management and development: case from the capital core area of Beijing, China. Sustain (Switzerland) 13(21):1–19. https://doi.org/10.3390/su132112115

Yang L (2014) Climate change, water risks and urban responses in the pearl river Delta, China [Dissertation]. University of Hamburg
Zafar S, Zaidi A (2019) Impact of urbanization on basin hydrology: a case study of the Malir Basin, Karachi, Pakistan. Regional Environ Change 19(6):1815–1827. https://doi.org/10.1007/s10113-019-01512-9

Part II
Socio-Economic Challenges towards Sustainable Cities

Chapter 3
Challenges for Achieving Sustainable Development Goal 6.1 'Safe and Affordable Drinking Water' in Developing Countries: A Case Study of Patharghata in Bangladesh

Mohammed Mahmud Khan, Sang Hyup Shin, Tapos Kormoker, and Tanmoy Roy Tusher

Abstract Ensuring safe drinking water supply becomes major development challenges in Bangladesh. The indicator 6.1 of UN Sustainable Development Goals (SDGs) focusses on achieving safe and affordable drinking water with collective and impartial access for all by 2030. In developing countries, contamination of water with harmful microorganisms, hazardous metal(loid)s, or salts puts a substantial challenge to ensure safe drinking water supply. This study aimed to evaluate the existing conditions of drinking water supply in Patharghata upazila in Bangladesh, and to ascertain the challenges to achieve SDG 6.1. In the area, deep and shallow aquifers contain excessive salinity; discouraging people to drink deep or shallow tube well water. Thus, majority of the local inhabitants largely consume pond water without any prior treatment, which indicates that drinking water supply remains in poor condition in Patharghata upazila. Although the government and NGOs have constructed many pond sand filters (PSFs), most of them do not sustain for a longer period and even some of PSFs become useless within a year. Moreover, contamination with fecal coliform (e.g. *E. coli*) was observed 83% of samples in sources and

M. M. Khan (✉)
Department of Public Health Engineering, Dhaka, Bangladesh
e-mail: mmk7795@gmail.com

M. M. Khan · S. H. Shin
College of International Studies, Kyung Hee University, Seoul, South Korea
e-mail: shshin@khu.ac.kr

T. Kormoker
Department of Emergency Management, Faculty of Environmental Science and Disaster Management, Patuakhali Science and Technology University, Patuakhali, Bangladesh

T. R. Tusher (✉)
Department of Environmental Science and Resource Management, Mawlana Bhashani Science and Technology University, Tangail, Bangladesh
e-mail: trtusher.esrm@gmail.com

S. Anand et al. (eds.), *Sustainability in South Asian Cities*, Advances in Geographical and Environmental Sciences, https://doi.org/10.1007/978-981-97-7455-5_3

73% of samples in households; indicating a significant barrier to achieve SDG 6.1. It is, therefore, urgently needed to strengthen government support and to take specific strategies and guidelines to achieve SDG 6.1 within 2030 in developing countries like Bangladesh.

Keywords Sustainable development goals · Water supply · Drinking water · Groundwater contamination · Bangladesh

3.1 Introduction

Drinking water is considered as one of the universal human rights and is also included in the Human Development Index (HDI)[1] and directly reflects a civilized standard of living. All United Nations (UN) member countries have agreed to ensure safe water in all spheres of society under the Sustainable Development Goals (SDGs) agenda by 2030 (United Nations 2018). SDG 6.1, the drinking water indicator, focuses on three criteria: accessibility, availability and quality of the water. This indicator represents the percentage of people using an improved water source that is easily reachable on their premises, available throughout the year, and free from harmful microbes and chemical pollutants.

One of the major focuses of the SDGs is the bottom-up planning that ensures understanding of deep-rooted problems, while the Millennium Development Goals (MDGs) had a top down, reductionist and fragmented approach that did not effectively address the grass roots improvement of obstacles to ensure equitable and sustainable development. The SDGs universally claim the agenda to comprise 'all' people 'everywhere' across 'all ages' (United Nations 2018). Besides, access to drinking water is a human right according to the UN Charter (Scanlon et al. 2004), but the Bangladesh Constitution has not included water as a particular right like most other developing countries. World Bank (2018) shows the contamination of water with dangerous microbes, heavy metals or salts in many parts of Bangladesh places a substantial challenge on providing fresh drinking water across the country. In addition, 97% of people use technologically improved drinking water sources and the rest (nearly 4 million people) depend on polluted water sources (Hoque et al. 2019).

As Bangladesh is a low-lying country, there is an abundance of surface and groundwater, but the quality of its water resources is poor in a few areas due to effects of climate change, haphazard urban development, and population growth (World Bank 2018). Most of the people primarily depend on groundwater sources, which are comparatively safer than surface water. However, many shallow and a few deep aquifers in different places of the country have excessively been affected by toxic metal(liod)s like arsenic (As) that are seriously harmful for human health. Bangladesh's coastal aquifers are vulnerable due to climate change and salinity intrusion that happens along the coastline (Rabbani et al. 2013). In addition, the

[1] http://hdr.undp.org/en/content/human-development-index-hdi.

groundwater-based water supply in the coastal areas of Bangladesh suffers from a number of major troubles: As contamination, lowering of the freshwater table, and the non-availability of proper aquifers (Shamsuzzoha et al. 2018).

Patharghata upazila under the Barguna district is situated in the southern coastal region of Bangladesh. Safe drinking water has become one of the major concerning issues in that area where people mostly depend on ponds for fulfilling their domestic needs including drinking, cooking, bathing and washing. This study aimed to: (i) evaluate the existing conditions of drinking water supply in the Patharghata upazila in Bangladesh focusing the SDG clean water target and indicators; and (ii) point out the challenges to achieving the SDG 6.1 'safe and affordable drinking water' in the study area as a representative area in Bangladesh.

3.2 Literature Review

This section summarizes the overall water supply situation throughout the country based on the data collected by Multiple Indicator Cluster Survey (MICS) accomplished by United Nation Children's Fund (UNICEF) and Joint Monitoring Program (JMP) prepared by both UNICEF and World Health Organization (WHO). It also depicts the water supply status of Bangladesh in light of the Bangladesh Census and Household Survey (2011).

Water supply in Bangladesh based on SDG indicator

Water in Bangladesh is highly reliant on external sources and vulnerable due to its low-lying nature and geographic impacts on the river flow (Azam and Sarkar 2012). Since the surface water is contaminated in some places, the majority of the people depend on groundwater for both drinking and agricultural uses. A number of household drinks water that contains *E. coli*, a marker of fecal contamination (World Bank 2018). The Department of Public Health Engineering (DPHE), Water Supply Authorities (WASA), and municipalities under the Ministry of Local Government and Rural Development (LGRD), and cooperatives operate the drinking water supply across the country. Although, the people use different sources for drinking water, the majority of the people depend on tube wells, a low-cost hand pump-based technology. Figure 3.1 shows that 89.05% of the population uses tube well sources, while others depend on different sources such as tap, wells, ponds, or rivers as ranged from 8.08, 0.71, 1.29, and 0.38%, respectively.

Drinking water coverage of Bangladesh

According to JMP (2013),[2] drinking water services can be categorized into different groups like safely managed, basic, limited, and unimproved drinking water services. Figure 3.2 depicts that safely managed drinking water coverage was found to be 55.44% based on the SDG indicators. Besides, 41.57% was found as for basic water

---

[2] https://www.who.int/water_sanitation_health/publications/2013/jmp_report/en/.

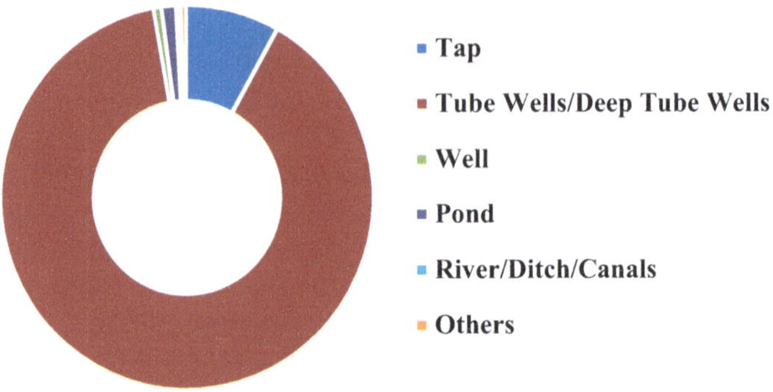

Fig. 3.1 Different sources of drinking water in Bangladesh. *Source* Bangladesh Bureau of Statistics (BBS 2015)

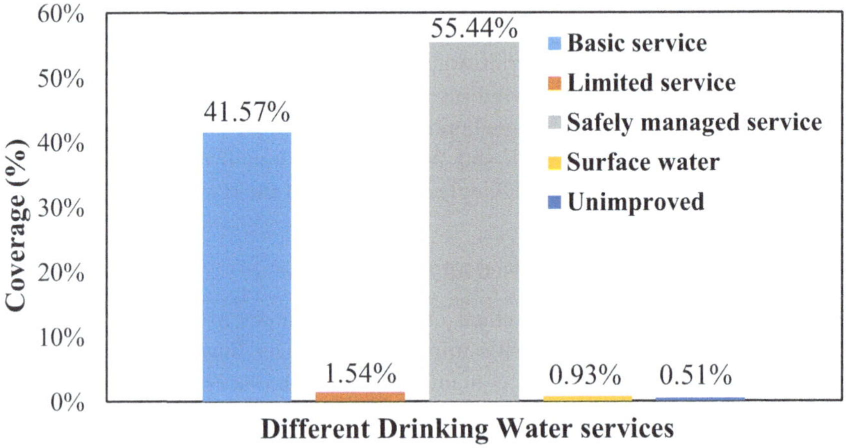

Fig. 3.2 Drinking water supply status in Bangladesh based on the SDG indicators. *Source* JMP (2017)[3]

service that are fixed by collection time. This data represents the country's drinking water status which is slightly more than 40%, showing challenges to achieve the target within 2030. Moreover, limited drinking water services (collection time >30 min for round trip) and unimproved drinking water services (collection from unprotected spring or dug wells) is about 1.54 and 0.51%, respectively.

On the other hand, Fig. 3.3 describes the water supply coverage in Bangladesh based on different indicators (MICS 2019).[4] It shows that 98.50% of people use a technologically improved water source. This level slightly shifted to 98% when

[3] https://www.who.int/water_sanitation_health/publications/jmp-2017/en/.

[4] https://www.unicef.org/bangladesh/media/3281/file/Bangladesh%202019%20MICS%20Report_English.pdf.

adding round trip water collection time within 30 min. Figure 3.3 also depicts that 96.9% people have access to water when considering the availability indicator, which can be defined as anyone who can collect water when needed. The coverage sharply falls to 40.3% when considering the *E. coli* in user drinking water sources. In contrast, the coverage is found 81.9%, counting *E. coli* at the household level. The coverage with *E. coli* at the household and source shows a great difference, which means that people have adopted a few precautionary processes at the household before drinking. In most cases, urban households boil water before using the contaminated source water. This figure also shows that the safely managed drinking water coverage is 47.9%. According to Fig. 3.3, drinking water coverage is 39.1% in the WHO standard and 42.6% in the Bangladesh standard while considering the As.

Status of the drinking water supply in coastal areas of Bangladesh

Groundwater is one of the main sources for drinking water in Bangladesh and the proportion of the population in the coastal areas of Bangladesh has rare access to safe drinking water (Azam and Sarkar 2012). The coastal area covers nearly 32% of the total landmass of the country with approximately 48 million people living across the coast (Bangladesh Bureau of Statistics 2015). In addition, nearly 74% of the population in the coastal zones of Bangladesh have little or no access to safe drinking water (Sarker and Ahmed 2015). The shallow and deep aquifers are heavily contaminated with salt, which exceeds the permissible limit. The pattern

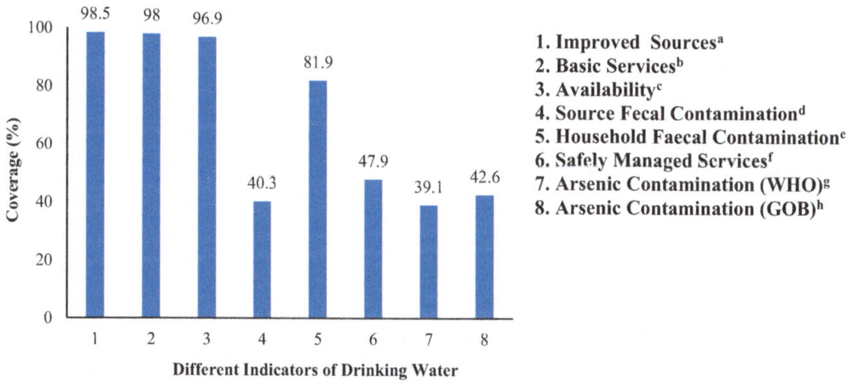

Fig. 3.3 Water coverage in Bangladesh based on different indicators. *Source* Multiple Indicator Cluster Survey (MICS) 2019. *Note*: a. Proportion of people using advanced source water; b. Proportion of people using advanced sources water whose collection time within 30 min round trip; c. Proportion of people using a water source which is available during the year; d. A number of people whose source water was examined and found bacterial contamination; e. The number of people whose household water was examined and found bacterial contamination; f. Percentage of household population with drinking water free from fecal contamination, available when needed, and accessible on premises; g. Proportion of people using advanced source water situated on premises, safe from microbes, available during the year and <= 10 ppb As; h. Proportion of people using advanced source water situated on premises, safe from microbes, available during the year and <= 50 ppb As

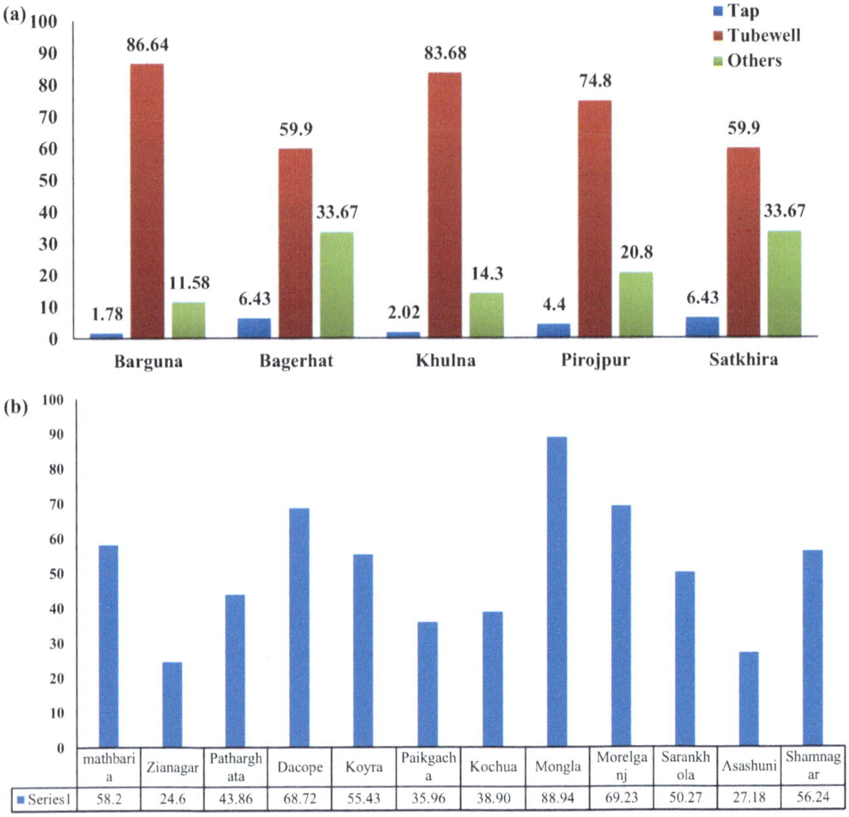

Fig. 3.4 Drinking water status in the coastal districts of Bangladesh: **a** in different costal districts, and **b** in different coastal upazilas including ponds, rivers, and rainwater. *Source* Bangladesh Bureau of Statistics (BBS 2013)

of salinity intrusion is acute due to climate change, which causes an increase in the sea surface temperature that results in frequency of cyclonic events and storm surges, and eventually salinity intrusion occurs along the main land due to sea level rise (IPCC 2014)[5] and the reduction of upstream freshwater flow (Azam and Sarkar 2012). Shamsuzzoha et al. (2018) noted that technologies, which are being used for different water sources, have a few deficiencies like deep tube wells and shallow tube wells, which are not useful due to the excessive range of salinity in both deep and shallow aquifers. Figure 3.4 shows the drinking water sources used by the local inhabitants in the coastal areas of Bangladesh.

[5] https://www.ipcc.ch/site/assets/uploads/2018/05/SYR_AR5_FINAL_full_wcover.pdf.

About 97% of the people of Bangladesh dependent on hand-pump based tube wells as the main sources of drinking water (Azam and Sarkar 2012). Ponds can be a safe surface water source, but *E. coli* affects most of it, especially during the dry season. In Bangladesh, the governmental and non-governmental organizations (NGOs) are responsible to ensure safe water for vulnerable communities (Haq and Rabbani 2014). The local government institutions are particularly responsible for water supply, but they are unable to ensure safe water in the rural coastal areas. In addition, rainwater is suitable for drinking, but it is available only in the rainy season (Bashar 2012). Pond sand filters (PSF) constructed to ensure safe drinking water supply do not work properly due to lack of operation and maintenance. Very shrouded aquifers are also found in some specific areas, but it cannot be sustained in long run due to a shortage of layers.

3.3 Methodology

3.3.1 Study Area

Patharghata upazila is one of the southernmost coastal areas in Bangladesh. Two major national rivers, namely the Baleshwar and Bishkhali, drain into the sea through this region. This upazila covers an area of 387.36 km^2 and is situated at 22°14′ and 22°58 N latitude and at 89°53′ and 90°05′ E longitude (Fig. 3.5). This upazila is also divided into seven unions (every union is divided into nine wards), which are the lowest administrative units of the local government system of Bangladesh.

3.3.2 Data Collection

Data was collected from both primary and secondary sources. The primary data was gathered through questionnaire surveys, while laboratory analysis of the water samples was also performed. The survey questionnaire was prepared based on the research questions and SDG drinking water target and indicators. Besides, secondary data was collected from both the Departments of Statistics and the Department of Public Health Engineering of Bangladesh. The demographic features of the population surveyed are presented in Table 3.1 and the major themes of the questionnaire survey are described in Table 3.2.

The cluster sampling method was used that is mainly of two types: one stage and two stage (Galway et al. 2012). Among the 18 wards, 3 wards (Patharghata ward no. 2 and 9, and Chardoani ward no. 6) from two unions (Patharghata and Chardoani) were randomly selected to collect the necessary data. The sample size using the Taro Yamani formula (Eq. 3.1) is as follows:

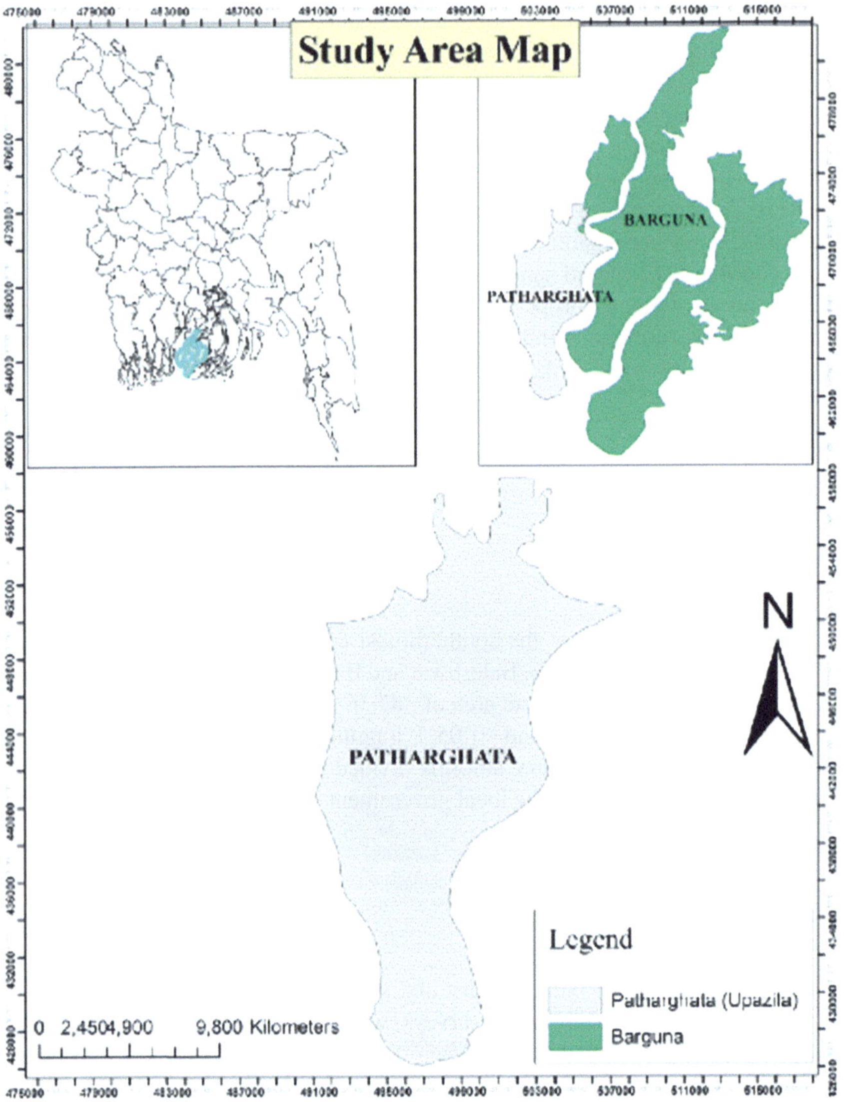

Fig. 3.5 Map showing the Patharghata upazila in the southern region of Bangladesh. *Source* Banglapedia (2012)

$$n = N/1 + N\,(e)^2 \tag{3.1}$$

where, n indicates the sample size, N represents the population size, and e represents the level of precision or sampling error which is ±10%. Therefore, the sample size was determined as 270 (87 from Patharghata ward no. 2, 93 from Patharghata ward

Table 3.1 Demographic features of the population surveyed

Upazila	Name of union	Population	Male	Female
Patharghata	Patharghata	28,284	14,534	13,750
	Kalmegha	26,080	13,021	13,059
	Kakchira	19,888	10,006	9882
	Raihanpur	15,496	7724	7772
	Chardoani	24,049	12,385	11,664
	Kathaltali	19,519	9799	9720
	Nachnapara	14,434	7415	7019

Source Banglapedia (2012)

Table 3.2 Major themes of the questionnaire survey

Variables	Indicators
1. Demographic and livelihood information	Name, age, gender, mobile number, marital status, educational level, monthly income
2. Perception about the drinking water	Type and status of the drinking water sources, location and distance from the drinking water source, availability of the drinking water, ideas about safely managed drinking water
3. Health issues due to drinking water	Frequently affected diseases, most common affected diseases in last five years, average yearly costs of health hazards for drinking water

no. 9, and 90 from Chardoani ward no. 6) based on the sampling error and population size.

3.3.3 *Water Sample Collection and Analysis*

Water samples were collected from both the water source and household level for laboratory analysis. Different water quality parameters including turbidity, pH, fecal coliforms, dissolved oxygen (DO), arsenic (As), iron (Fe), and chloride (Cl^-), and total suspended solids (TDS) were investigated. Total 60 samples were collected from both households and water sources, while ponds as the mostly used water source were given priority for source water sample collection. This study followed the water quality analysis procedures as described by Baird et al. (2017). However, for measuring the turbidity, the Multi-Parameter Water Quality Checker was used and was measured by Jackson turbidity units (JTU).

The concentrations of As and Fe were measured using the atomic absorption spectrometer (AAS), whereas the concentration of Cl^- was measured by argent metric method and titration procedure using the Eq. 3.2 as follows:

$$MgCl/L = \frac{(A - B) * N * 35450}{mLsample} \quad (3.2)$$

where, A and B represents the mL titration for the sample and blank, respectively, and N indicates the normality of $AgNO_3$.

The allowable limit of As, Fe, and Cl^- in drinking water is considered as 0.05, 0.3–1.0, and up to 600 mg/L, respectively, in Bangladesh. The following formula is useful in the argent metric method to calculate the chloride concentration.

3.4 Results and Discussions

Drinking water sources

Table 3.3 shows that 164 respondents collect their drinking water from a single source, and 106 participants depend on mixed sources. Among the single-source users, 154 households depend on pond water and 1.11, 1.11, and 2.96% depend on very shrouded shallow tube wells (VSST), pond sand filters (PSF), and rainwater harvesting systems (RWHS), respectively. Within the three study clusters, all participants from Patharghata ward no. 2 use pond water, all respondents from Charduani ward no. 6 depend on mixed sources and the inhabitants of Patharghata ward no. 9 collect drinking water through mixed sources. A typical PSF and RWHS used in the study area are shown in Figs. 3.6 and 3.7, respectively.

The result shows that 57% of the households collect drinking water from ponds throughout the year, while 39.3% inhabitants manage it from mixed sources that include two, three or even four different sources. It indicates that residents have to waste excess time for collecting water, which does not meet the SDG coverage. Women are mostly involved in collecting the drinking water. In some cases, girls are engaged to manage the water even though it is very far from their home. As a result, girls lose working hours for their academic study. Sultana et al. (2014) showed that pond water become contaminated mostly by *E. coli* during the summer season due to a shortage of water, and many people have suffered from water-borne diseases due to

Table 3.3 Uses of drinking water sources in the study area

Sources	Patharghata union				Charduani union		Total	
	Ward 9		Ward 2		Ward 6			
	No	% within	No	% within	No	% within	No	%
VSST	3	3.57	0	0	0	0	3	1.11
PSF	3	3.57	0	0	0	0	3	1.11
Pond water	48	57.14	106	100	0	0	154	57.0
RWHS	4	4.76	0	0	0	0	4	2.96
Mixed	26	30.95	0	0	80	100	106	39.3

Fig. 3.6 A typical pond sand filter (PSF) used in the study area

consuming of polluted water. Ensuring alum to purify water at household level can give better solution for removing turbidity and other physical properties. In addition, mixing bleaching powder into the pond water can remove disease-causing harmful microorganisms.

Moreover, Patharghata upazila has shortage of freshwater aquifers that are affected by a higher range of salinity (Shamsuzzoha et al. 2018), compelling the local people to be dependent on surface and rainwater. However, PSF are found to be unusable due to lack of operation and maintenance, while rainwater becomes useful only during the rainy season. In some cases, a number of people collect rainwater in a large tank and partly use it in the dry season. However, owing to poor management, the rainwater storage tank is often affected by harmful microorganisms. In some pocket areas, very shallow aquifers are found at 20 m depth from the ground level. People collect this water through much shrouded hand pumping systems, which can be effective for few years.

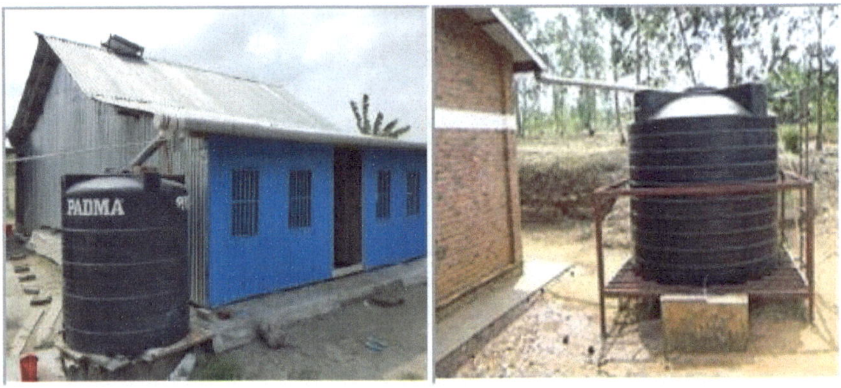

Fig. 3.7 Rainwater harvesting system (RWHS) in the study area

Figure 3.8 illustrates that around 53% of the respondents depend on pond water and RWHS throughout the year, and nearly 23% of the participants use PSF and RWHS. On the other hand, 12.26% of people collect water from both a PSF and pond water, while 3.77% participants depend on rivers osmosis (RO), pond water and RWHS, and 4.71% of respondents manage water from combined PSF, pond water and RWHS. Moreover, nearly 1% of respondents accumulate water from four different categories such as pond water and others; VSST and pond water; RO and pond water; and RO, pond water, PSF and RWHS. Besides, a few participants collect water from pond and river. The study found that pond water is mostly used as both the single and mixed categories. Since rainwater is safer than other sources although it can only be available during the rainy season, people suffer during the summer season and, therefore, mostly manage water from ponds, which become contaminated by bacteria.

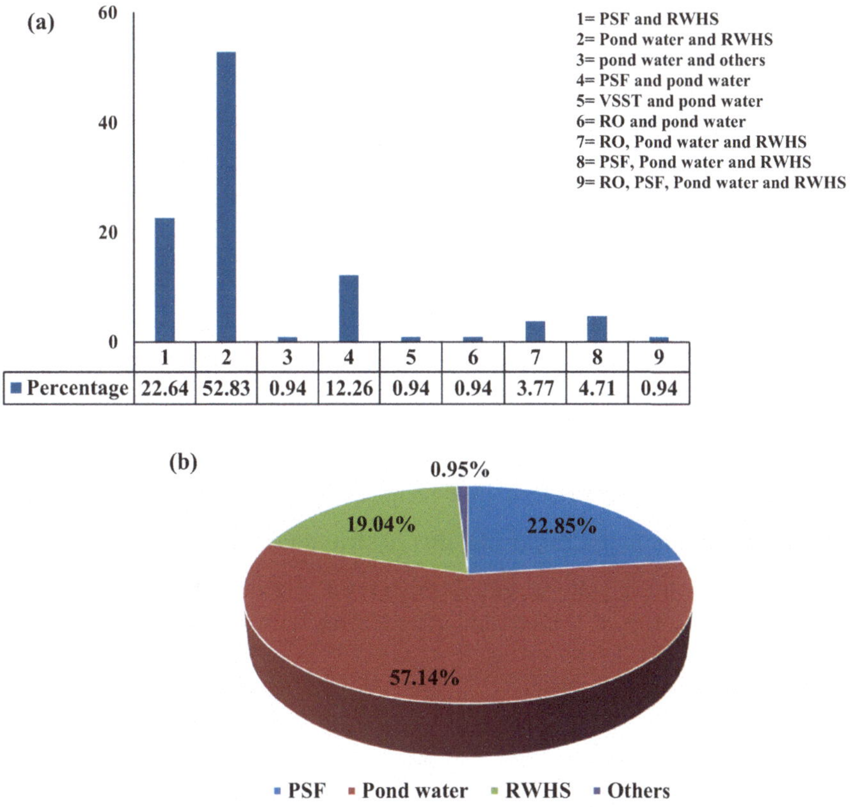

Fig. 3.8 **a** Uses of multiple water sources in percentage in the study area, and **b** Maximum used water source among the mixed sources

Peoples' perception about safely managed drinking water

The study found that 77.40% people have no idea about safely managed drinking water, while 22.60% participants knew about the concept and believed that their water is out of safely managed category. Table 3.4 shows the respondent's perception about the quality of drinking water. It also describes about the quality of water that 70.37% of participants certified it as moderate and 19.25% as good, while 7.77% believed it bad. The peoples' perception about the quality of water was different among the surveyed wards in the study area.

Accessibility of the drinking water

Figure 3.9a shows the accessibility-indicator where 20.74% of respondents ensured about the safe water sources on their premises, while 78.86% households did not have any freshwater source on their premises. They collect drinking water outside the terrace. In some cases, vulnerable people lose their working hour for collecting water and many children sacrifice the school-lesson. Figure 3.9b indicates 79% of

Table 3.4 Respondent's perception about the quality of the drinking water

Quality	Patharghata union				Charduani union		Total	
	Ward 9		Ward 2		Ward 6			
	No	% within	No	% within	No	% within	No	%
Very good	0	0	0	0	7	8.75	7	2.59
Good	6	7.14	0	0	46	57.5	52	19.25
Moderate	71	84.52	92	86.79	27	33.75	190	70.37
Bad	7	8.33	14	13.21	0	0	21	7.77
Very bad	0	0	0	0	0	0	0	0

participants can collect drinking water within a 30 min round trip, whereas 21% of respondents collect drinking water after spending >30 min round trip. People from Patharghata ward no. 9 collect water after spending >30 min time, and a few people bring water from >2 km away from their premises.

Peoples' perception about the availability of drinking water

Availability of drinking water is the 3rd criterion of SDG 6.1. It is depicted from Table 3.5 that 73.70% of respondents collect drinking water from a source that remains available throughout the year, while 26.30% households' drinking water sources remain partially available.

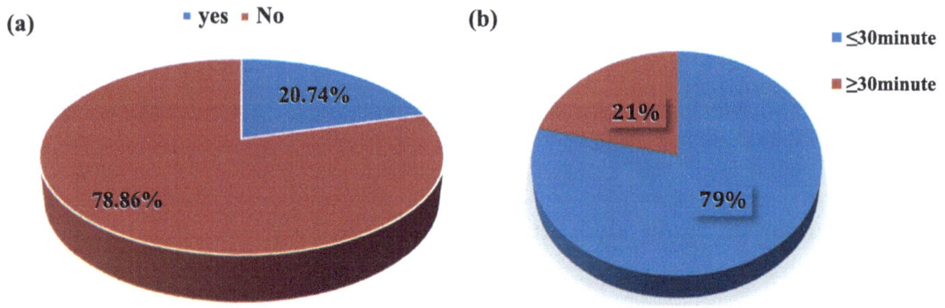

Fig. 3.9 **a** Accessibility of fresh drinking water on the premises, and **b** Accessibility based on collection time

Table 3.5 Drinking water availability during the year

Availability	Patharghata union				Charduani union		Total	
	Ward 9		Ward 2		Ward 6			
	No	% within	No	% within	No	% within	No	%
Whole year	79	82.14	106	100	24	30	199	73.70
Partially available	15	17.85	0	0	56	70	71	26.30

Drinking water quality based on laboratory analysis

Laboratory analysis is essential for ensuring the quality of water besides peoples' perception. Table 3.6 represents the observed values of the studied physicochemical properties of the water samples collected from source and households in the study area. The pH values ranged from 5.36 to 8.23 with an average of 7.1. According to the WHO (2017), the prescribed value of pH should be ranged from 6.5 to 9.2, while the Bangladesh standard of water quality ranged from 6.5 to 8.5. This study also gathered three RWHS samples and the pH of the two samples were 7.18 and 7.72, which are greater than average value of 7.0, and it indicates slightly alkaline water, while the pH of another sample was found 5.23 indicating acidic water. The DO is an important parameter for drinking water quality analysis. According to the Bangladesh standard, 6 mg/l is considered permissible, while this study measured the minimum as 0.8 mg/l and the maximum as 80.4 mg/l with an average of 12.1 mg/l. In most cases, higher DO was observed although comparatively lower DO values were measured for RWHS and VSST samples. In case of PSF, 3 out of 5 samples were found to have less DO than the standard. Besides, for the pond water, the DO values were less than standard in 11 samples out of 19.

The absence of fecal coliforms is considered as the primary indicator of suitability for water consumption. The investigation of fecal coliforms was done at both source and household level and a total of 60 samples (30 from the source and 30 from the households) were tested for the presence/absence of fecal coliforms in the drinking water. Among the 30 source samples, 9 samples were found to be contaminated with uncountable fecal coliforms, while no fecal coliforms were found in only 5 samples. The remaining 16 source samples were also contaminated with fecal coliform bacteria. Besides, at the household level, 5 samples were found to be contaminated with uncountable fecal coliforms, 8 samples with no bacteria, and the rest 17 specimens were found to be contaminated with fecal coliform bacteria. The results, therefore, indicated that a total of 25 and 22 samples among the 30 were affected by fecal bacteria at the source and household level, respectively (Table 3.6), indicating that the source water is more contaminated by fecal bacteria than the household drinking water. Among the four categories of water source, the PSF water was mostly affected by fecal coliform bacteria. In contrast, rainwater was less contaminated than the other water sources, while ponds and VSST were found to be partially contaminated.

The minimum, maximum and average concentrations of Fe was 0.104, 1.153 and 0.313 mg/l, respectively (Table 3.6). In 13 out of 19 pond samples, the concentrations of Fe resulted in less than 0.3, while the concentrations in the rest 6 samples were found within the allowable range. Besides, for the PSF, the Fe concentrations in 4 out of 5 samples were less than the lower limit, and 1 out of 5 were within the standard value. In addition, in the case of VSST and RWHS, all 6 samples showed a value of less than the lower allowable limit. The Cl^- is also considered as an important parameter for water supply, especially in the coastal areas, and its standard value for Bangladesh is 150 to 600 mg/l of water. Among all of the water quality samples, the minimum, maximum and average Cl^- concentrations were 10, 2134

Table 3.6 Physicochemical and biological properties of drinking water samples collected from different sources in the study area

No	Water source	pH	Faecal coliforms		DO	Fe	Cl^-	TDS	As	Turbidity
			Source	Household						
1	Pond	6.24	Uncountable	Uncountable	7.86	0.425	15	65	0.0023	6.1
2	Pond	6.29	60	44	4.21	0.396	20	67	0.0032	8.6
3	**Pond**	**6.56**	**90**	**56**	**4.88**	**0.115**	**20**	**154**	**0.0087**	**19.3**
4	**Pond**	**6.84**	**32**	**20**	**5.0**	**0.259**	**20**	**166**	**0.009**	**28.9**
5	Pond	7.2	35	0	4.7	0.128	30	215	0.004	2.2
6	Pond	7.21	15	0	4.77	<LOQ	30	220	0.0029	17.5
7	**Pond**	**7.49**	**Uncountable**	**30**	**5.15**	**<LOQ**	**10**	**91**	**0.0097**	**45.8**
8	Pond	7.06	Uncountable	80	5.1	0.630	50	253	<LOQ	43.2
9	Pond	7.15	Uncountable	Uncountable	3.8	0.677	60	174	0.0015	33.5
10	**Pond**	**7.15**	**70**	**20**	**4.75**	**1.153**	**89**	**257**	**0.0081**	**30.6**
11	Pond	7.58	Uncountable	56	5.29	0.416	35	104	0.005	17.8
12	Pond	7.71	26	0	9.1	<LOQ	993	3360	0.0020	9.1
13	Pond	7.33	0	35	14.7	<LOQ	45	515	<0.001	14.7
14	Pond	7.43	0	50	80.4	<LOQ	25	152	0.0028	80.4
15	**Pond**	**7.2**	**0**	**60**	**30.7**	**0.176**	**35**	**219**	**0.007**	**33.8**
16	**Pond**	**7.12**	**80**	**20**	**14.1**	**0.132**	**268**	**860**	**0.0071**	**14.1**
17	Pond	7.24	80	0	11.0	<LOQ	20	76	<0.001	54.4
18	Pond	7.19	105	105	54.4	0.203	25	149	0.0027	10.1
19	Pond	6.95	Uncountable	Uncountable	25.6	0.189	10	76	0.0043	25.6
20	PSF	6.11	44	0	6.43	<LOQ	25	178	0.0015	0.9
21	PSF	6.79	44	28	5.28	0.145	15	80	0.0029	1.8

(continued)

Table 3.6 (continued)

No	Water source	pH	Faecal coliforms		DO	Fe	Cl^-	TDS	As	Turbidity
			Source	Household						
22	PSF	7.2	100	56	5.12	<LOQ	20	171	0.0021	30.7
23	PSF	7.77	48	30	0.8	0.201	35	233	0.0020	0.8
24	PSF	7.39	Uncountable	Uncountable	27.9	0.349	30	267	0.0012	27.9
25	VSST	7.26	Uncountable	Uncountable	4.67	<LOQ	1077	2850	<0.001	–
26	VSST	8.23	0	0	4.56	0.170	2134	3820	<0.001	–
27	VSST	7.3	100	60	4.82	0.253	1340	3320	<0.001	–
28	RWHS	7.18	Uncountable	80	5.26	0.154	10	38	<0.001	–
29	RWHS	5.36	0	0	2.9	0.104	10	21	<0.001	20.9
30	RWHS	7.72	80	0	2.4	<LOQ	10	42	<0.001	–
Minimum		5.36	0	0	0.8	0.104	10	21	<0.001	0.8
Maximum		8.23	Uncountable	Uncountable	80.4	1.153	2134	3820	0.0097	80.4
Average		7.10	–	–	12.1	0.313	216.86	606.4	0.0043	24.05
Std. Dev		0.56	–	–	17.0	0.257	497.70	1108	0.0028	19.53
Unit		–	CFU		mg/l	mg/l	mg/l	mg/l	mg/l	NTU
BD Standard		6.5–8.5	No/100 ml		6.0	0.3–1	150–600	1000	0.005	10

and 216, respectively. Interestingly, the range of Cl^- in 3 VSST-samples was more than 1000 mg/l, while the values remained 10 mg/l for RWHS samples. It was less than the lower value in 22 out of 24 of both pond and PSF sample, whereas the remaining one was more than the upper limit and the other one was between the standard range. The minimum, maximum and average concentrations of As was <0.001, 0.0097 and 0.0043 mg/l, respectively. Among the tested water samples, 6 out of 19 pond samples contained As concentrations higher than the standard (Table 3.6), while lower level of As was detected in PSF and VSST samples. On the other hand, the minimum, maximum and average TDS values were observed as 21, 3820 and 606 mg/l, respectively. For 3 VSST samples, it was found nearly 3000 mg/l, while lower TDS than the Bangladesh standard was found in 26 out of 27 specimens (Table 3.6). Turbidity was also found to be higher in some of the pond, PSF, and RWHS water samples.

3.5 Challenges Towards Achieving SDG 6.1

Ensuring equal access to safely managed drinking water has become challenging in Bangladesh and in most other developing countries. Although plenty of safe water exists on the planet, millions of people die, most of them children, from water-borne diseases due to inadequate water supply, sanitation, and hygiene. Lack of safe water impacts on the quality of the life especially health and the people's productive capacity. Dangerous microbes, heavy metals, or salts in many areas pose a challenge to ensuring safe drinking water across in Bangladesh. Although the country achieved 98% of the MDG-drinking water target, while considering the SDG drinking water indicators, the coverage becomes 39% (World Bank 2018); 46% (MICS 2019); or 55% (JMP 2017). It shows that nearly 50% of the population lack access to safely managed drinking water in the country. Although Bangladesh has an abundance of water, lack of proper urban water management and ineffective single source drinking water has posed severe challenges.

The majority of the people of Bangladesh for drinking water depend on groundwater sources which are increasingly affected by both natural and anthropogenic factors likewise urbanization, rapid industrialization, and excessive irrigation activities (Kormoker et al. 2020, Idriss et al. 2020). Lack of safe water acquirer, higher salinity and unaffordable cost of water treatment, pose a challenge to ensure safe water supply in the study area (Islam et al. 2014). Salinity intrusion happens in surface and ground water aquifers due to climate change induced factor (Hoque et al. 2016). Azam and Sarkar (2012) show that 74% of the population in the coastal areas of Bangladesh has little or no access to safe drinking water. That is why, coastal households mainly depend on rainwater and surface water sources like ponds and rivers. Generally, pond water gradually reduces in the dry season and gets affected by pathogenic microbes. As a result of drinking water contamination, people suffer from water-borne diseases. In addition, rainwater can be useful in the rainy season which does not ensure availability as SDG indicator.

The government of Bangladesh has approved the SDG framework following the National Strategy for Water Supply and Sanitation which clearly describes the activity of different stakeholders for achieving the targets. Institutional service delivery gaps have not been identified following the national strategy. Local government institutions like the union parishad, municipalities, and city corporations lack the technical and financial capacity to provide and maintain high quality water services for the residents. As water become a multifunctional natural resource which creates obstacle for ensuring coverage throughout the country, the Department of Public Health Engineering (DPHE), responsible organization for water supply in Bangladesh, constructs desalination units in the acute water shortage areas but it is not the universally accepted solution due to dependency fossil fuel-derived energy and higher technical expertise.

3.6 Conclusion

As nearly 61% of the people depend on single water source, the rest 39% of inhabitants collect the drinking water from mixed sources throughout the year in the study area while more than 90% depend on pond water among the single source users. A few people collect water from VSST, PSF, and RWHS. Among the SDG indicators, 77% people are satisfied about the drinking water quality, nearly 79% of the people have no water sources on their premises, and 74% of the people have sources that are available throughout the year. Considering the accessibility, availability and quality indicators, only 21% people have water that is safely managed. The bacteriological analysis depicts that 83.33% of source water and 73.33% of household water were affected by fecal coliform, indicating huge challenges for acquiring SDG drinking water target within the 2030 timeframe. Besides, among the different drinking water sources, Cl^- was found in VSST at nearly 3000 mg/l whose acceptable limit for Bangladesh standard fixed as 600 mg/l. The excessive concentrations of As and Cl^- poses significant health risks for the users, especially for the children. In addition, the average value of turbidity was 24.05 NTU. Moreover, to obtain sustainable water services, it requires new tactics to construct infrastructure and to improve the current monitoring practices. One of the limitations of this study was that it has randomly chosen 3 wards among the 18 wards of the two unions. In addition, it selects 2 unions among the 7 unions in the upazila (sub-district), while the survey questionnaires were prepared very briefly targeting mostly the SDGs' respective target and indicators.

References

Azam M, Sarker T (2012) Adaptation strategies for safe drinking water supply in the low lyingcoastal regions in Bangladesh. Asian J Environ Disaster Manage 4(3)

Baird RB, Eaton AD, Rice EW (2017) Standard methods for the examination of water and wastewater. 23rd Edn., American Public Health Association, American Water Works Association, Water Environment Federation, Washington D.C., USA
Bangladesh Bureau of Statistic (2013) "District Statistic 2011 Bagerhat" statistics and Information Division (SID), Ministry of Planning, Dhaka, Bangladesh
Bangladesh Bureau of Statistic (2013) "District Statistic 2011 Barguna" statistics and Information Division (SID), Ministry of Planning, Dhaka, Bangladesh
Bangladesh Bureau of Statistic (2013) "District Statistic 2011 Khulna" statistics and Information Division (SID), Ministry of Planning, Dhaka, Bangladesh
Bangladesh Bureau of Statistic (2013) "District Statistic 2011 Pirojpur" statistics and Information Division (SID), Ministry of Planning, Dhaka, Bangladesh
Bangladesh Bureau of Statistic (2013) "District Statistic 2011 Satkhira" statistics and Information Division (SID), Ministry of Planning, Dhaka, Bangladesh
Bangladesh Bureau of Statistics (2015) Population and housing census 2011: socio-economic and demographic report. National Report, Volume-4. Ministry of Planning, Bangladesh Bureau of Statistics, Statistics and Information Division, Dhaka, Bangladesh
Banglapedia (2012) National encyclopedia of Bangladesh. Asiatic Society of Bangladesh, Dhaka, Bangladesh
Basar A (2012) Water security in coastal region of Bangladesh: would desalination be a solution to the vulnerable communities of the sundarbans? Bangladesh e-J Sociol 9(2):31
Galway LP, Bell N, Al Shatari SA, Hagopian A, Burnham G, Flaxman A, Takaro TK (2012) A two-stage cluster sampling method using gridded population data a GIS, and google earth TM imagery in a population-based mortality survey in Iraq. Int J Health Geogr 11(1):12
Hoque MA, Scheelbeek PFD, Vineis P, Khan AE, Ahmed KM, Butler AP (2016) Drinking water vulnerability to climate change and alternatives for adaptation in coastal South and South East Asia. Clim Change 136(2):247–263
Hoque SF, Hope R, Arif ST, Akhter T, Naz M, Salehin M (2019) A social-ecological analysis of drinking water risks in coastal Bangladesh. Sci Total Environ 679:23–34
Huq S, Rabbani MG (2014) Climate adaptation technologies in agriculture and water supply and sanitation practice in the coastal region of Bangladesh. CRC Press, Taylor & Francis Group, Boca Raton, FL, pp 185–202
Idriss IE, Abdel-Azim M, Karar KI, Osman S, Idris AM (2020) Isotopic and chemical facies for assessing the shallow water table aquifer quality in Goly Region, White Nile State, Sudan: focusing on nitrate source apportionment and human health risk. Toxin Rev 1–13
JMP (2017) Progress on drinking water, sanitation and hygiene: 2017 update and SDG baselines. The United Nations Children's Fund (UNICEF) and World Health Organization (WHO), Geneva, Switzerland
Islam MA, Karim MR, Higuchi T, Sakakibara H, Sekine M (2014) Comparison of the trace metal concentration of drinking water supply options in southwest coastal areas of Bangladesh. Appl Water Sci 4(2):183–191
Kormoker T, Idris AM, Khan MM, Tusher TR, Proshad R, Islam MS, Khadka S, Rahman S, Kabir MH, Kundu S (2020) Spatial distribution, multivariate statistical analysis, and health risk assessment of some parameters controlling drinking water quality at selected primary schools located in the southwestern coastal region of Bangladesh. Toxin Rev 1–14
MICS (2019) Multiple Indicator Cluster Survey 2019. Survey Findings Report, Progotir Pathey Bangladesh, Bangladesh Bureau of Statistics (BBS), Dhaka, Bangladesh
Rabbani G, Rahman A, Mainuddin K (2013) Salinity-induced loss and damage to farming households in coastal Bangladesh. Int J Global Warming 5(4):400–415
Sarker MH, Ahmed F (2015) Climate change vulnerability of drinking water supply infrastructure in coastal areas of Bangladesh. International Union for Conservation of Nature (IUCN), Bangladesh Country Office, Dhaka, Bangladesh
Scanlon J, Cassar A, Nemes N (2004) Water as a human right? IUCN Environmental policy and law paper no.51, IUCN, Gland, Switzerland and Cambridge, U.K, 1-53

Shamsuzzoha M, Rasheduzzaman M, Ghosh RC (2018) Building resilience for drinking water shortages through reverse osmosis technology in coastal areas of Bangladesh. Procedia Eng 212:559–566

Sultana S, Ahmed KM, Mahtab-Ul-Alam SM, Hasan M, Tuinhof A, Ghosh SK, Rahman MS, Ravenscroft P, Zheng Y (2015) Low-cost aquifer storage and recovery: Implications for improving drinking water access for rural communities in coastal Bangladesh. J Hydrol Eng 20(3), 10.1061/(ASCE)HE.1943-5584.0001100

United Nations (2018) Transforming our world: the 2030 agenda for sustainable development. Sustainable development goals knowledge platform, United Nations, New York, USA

World Bank (2018) Promising progress a diagnostic of water supply, sanitation, hygiene, and poverty in Bangladesh. Washington, DC

WHO (2017) Guidelines for drinking-water quality, 4th edn., incorporating the 1st addendum, World Health Organization (WHO), Geneva, Switzerland

Chapter 4
Modelling of Antimicrobial Agent Triclosan (TCS) Transport in Nanofiltration System for Sustainable Design of Drinking Water Treatment Systems for South Asian Countries

Dina Dzafira Ramlan, Antonios Parasyris, Marco Discacciati, and Diganta Bhusan Das

Abstract The continued inclusion of antimicrobial agent triclosan (TCS) in heavily used pharmaceutical and personal care products (PPCPs), especially during disease outbreaks such as the COVID-19 pandemic in several south Asian countries, have raised concerns among healthcare professionals, researchers and the general public. The side effects of TCS' presence in water sources, particularly drinking water, have now encouraged further studies in water treatment. The research in this work encompasses a mathematical model-based prediction of TCS-water crossflow nanofiltration (NF) behaviour through coupled MATLAB image processing and COMSOL multiphysics simulations. The NF system geometry focused on was the spiral channel at the base of the filtration cell where the free and porous flow domains were modelled through a coupling of Navier–Stokes equations and Darcy's law using the Beavers-Joseph interfacial condition. The simulation demonstrated the correlation between the TCS concentration distribution and the dissipation of crossflow velocity on the membrane surface. Altering process parameters such as inlet pressure, inlet concentration and porosity were also found to distinctively influence the spread of TCS on the membrane surface, while permeate flux was mainly affected by membrane permeability. The results of this study suggest that optimisation of these process parameters following the patterns observed through the model would be highly beneficial for the improvement and sustainable design of NF systems for TCS-water separation.

D. D. Ramlan · D. B. Das (✉)
Department of Chemical Engineering, Loughborough University, Loughborough LE113TU, UK
e-mail: D.B.Das@lboro.ac.uk

A. Parasyris · M. Discacciati
Department of Mathematical Sciences, Loughborough University, Loughborough LE113TU, UK

A. Parasyris
Institute of Applied and Computational Mathematics, Foundation for Research and Technology (FORTH)-Hellas, 71601 Heraklion, Greece

S. Anand et al. (eds.), *Sustainability in South Asian Cities*, Advances in Geographical and Environmental Sciences, https://doi.org/10.1007/978-981-97-7455-5_4

Keywords Triclosan · Crossflow filtration · Nanofiltration · COMSOL modelling · Navier–Stokes equations · Darcy's law

4.1 Introduction

The antimicrobial agent triclosan (TCS; 5-chloro-2-(2,4-dichlorophenoxy)phenol) has been a commonly encountered component in pharmaceutical and personal care products (PPCPs), herbicides, household essentials, clothing and children's toys since the 1960s (Aiello and Larson 2003; Yueh and Tukey 2016). It functions as an antibacterial and is industrially recognised as Irgasan DP 300 (Adolffson-Erici et al. 2002). Bacteria are eliminated by this chemical through their inhibition of a specific bacterial site; enoyl-acyl carrier protein reductase (Fab 1 enzyme) which consequently stifles the growth of bacteria but also may result in cross-resistance of the bacteria to antibiotics that share the same target (Aiello and Larson 2003). Thus, TCS has been identified as a potential risk factor for antibiotic resistance in addition to its association with endocrine disruption, developmental disorders and liver carcinogenesis (Yueh and Tukey 2016; Quan et al. 2019).

However, most of these harmful effects of TCS have only been studied in smaller organisms (Adolfsson-Erici et al. 2002; Lin et al. 2010; Belosludtsev et al. 2018) and it is often deemed safe for humans at the low concentrations it has been detected in. Although certain organisations halted the involvement of TCS in their PPCPs altogether (Unilever 2020), others have justified its continued use due to its effectiveness and the lack of evidence of its consequences on humans (Riley and Lamont 2013). Without universal regulations set in place especially during disease outbreaks such as the ongoing COVID-19, an aggravation of TCS concentrations in sewage systems, soil, environmental waters (surface water, groundwater, etc.) and eventually drinking water is expected to ensue including the South Asia (Kumar et al. 2021; Lu et al. 2021; Sharma et al. 2021; Tan et al. 2021; Adhikari et al. 2022; Dai et al. 2022; Wang et al. 2022a, b). Further bioaccumulation of TCS in humans would occur due to its lipophilic nature, which has been confirmed by its contamination of human milk in Sweden in an earlier study (Adolfsson-Erici et al. 2002).

Concerns over the escalation of this situation have led to the development of separation techniques which could contribute to higher TCS removal efficiencies when added to current conventional water treatments. Membrane filtration is a leading option as it is accessible, easy to maintain and does not result in harmful by-products (Noguera-Oviedo and Aga 2016; Ozturk 2018). Nanofiltration (NF) in particular, a pressure-driven process with smaller membrane pore sizes, is often explored in the removal of antimicrobial agents. This is owed to its low operational cost, energy consumption and environmental impacts while maintaining high removal efficiencies and membrane fluxes (Taheran et al. 2016; Yang et al. 2017; Zepon Tarpani and Azapagic 2018).

Traditionally, two modes of operation are applied in NF: dead-end or crossflow. In dead-end filtration, the feed solution flows normal to the membrane surface while in crossflow filtration, the feed flows tangentially to the membrane surface (Maloney 2008). In the latter, the shear force exerted on the membrane surface is able to flush accumulated components from the membrane. This gives a more stable flux decline and controls fouling and concentration polarisation effects in crossflow configurations; however, dead-end configurations are known for being more economical (Van der Bruggen 2018). Current crossflow filtration technologies have managed to combine the elements of both configurations (Parasyris et al. 2020; Seshasayee et al. 2020), thus, benefiting from the advantages of both and this is the technology that will be considered in this work.

Although experiments involving the crossflow NF of TCS from water have been conducted, there is a gap in research in terms of computational modelling of the process using this chemical. The in-depth understanding of a system's behaviour and the simplicity of generating varying conditions achieved via modelling saves time and cost in developing optimised physical NF processes and selecting appropriate operating conditions. Thus, the overall aim of this research is to develop an accurate simulation model of the separation of TCS through an existing NF system (Seshasayee et al. 2020) over a range of pressure and concentration values.

The specific objectives of the paper are to: (i) substitute the sub-microscopic examination of the NF membranes via an image processing tool, (ii) estimate practical values of parameters for the numerical simulation through empirical equations, (iii) develop a mathematical model of the NF process via COMSOL Multiphysics and (iv) simulate the distribution of TCS and permeate water flow in the system at varying pressures and concentrations in the NF system (Seshasayee et al. 2020). In the following sections, the methodologies adopted for achieving the above goals, obtained results and main conclusions are discussed. We envisage that the developed methodology will help design more field-ready NF filtration system for the treatment of drinking water in TCS-affected South Asian countries, which in turn will help meet the Sustainable Goals of the United Nation (UN), e.g., goal 3 (good health and well-being) and 6 (clean water and sanitation for all).

4.2 Methodology

4.2.1 Image Processing

At the time of developing the mathematical model, no laboratory work could be done due to the COVID-19. Therefore, as an alternative to typical microscopic and sub-microscopic experimental examinations of membrane nanopores (Gnanasekaran et al. 2018; Seshasayee et al. 2020), an image processing tool was applied to examine the pore space, find the average porosity of the membranes and relate these to numerical simulations. The membranes considered in this work (Gnanasekaran et al. 2018; Seshasayee et al. 2020) were prepared for water contamination using locally available materials in South Asian countries (e.g., membrane fillers and polymers). Scanning electronic microscope (SEM) images of selected membranes from Gnanasekaran et al. (2018) were utilised as the input to an open-source MATLAB code that was designed for the processing of membrane pore-space, and in particular, to detect average porosity through a series of steps (Rabbani and Salehi 2017; Ezeakacha et al. 2018). The membranes selected were the metal organic framework (MOF) membranes for polyethersulfone (PES), cellulose acetate (CA) and polyvinylidene fluoride (PVDF) and the neat PES, CA and PVDF membranes. As elaborated in Rabbani and Salehi (2017), these key steps consist of (i) multi-level thresholding of the raw SEM images (Otsu 1979), (ii) binary image reduction and median filtering, (iii) segmentation of the porous space using the Watershed algorithm (Rabbani and Ayotallahi 2015) and (iv) labelling of pores and calculation of average porosity.

The progression of typical raw SEM image for the PVDF/MOF membrane through each step is illustrated in Fig. 4.1. For the set of SEM images of the membranes used, it was found that tuning the number of intensity levels on the MATLAB code to 1 or 2 produced the highest accuracy to the porosities of NF membranes reported by Gnanasekaran et al. (2018).

A graph of the porosity values obtained via image processing plotted against the experimental porosity values in Gnanasekaran et al. (2018) is displayed in Fig. 4.2 and the relative errors of the estimated values compared to the experimental (true) values is presented in Table 4.1. From Fig. 4.2, it is observed that the developed image processing code has a propensity to underpredict the porosities of the NF membranes. The relative errors for 4 out of 6 of the membrane porosities in Table 4.1 remain below

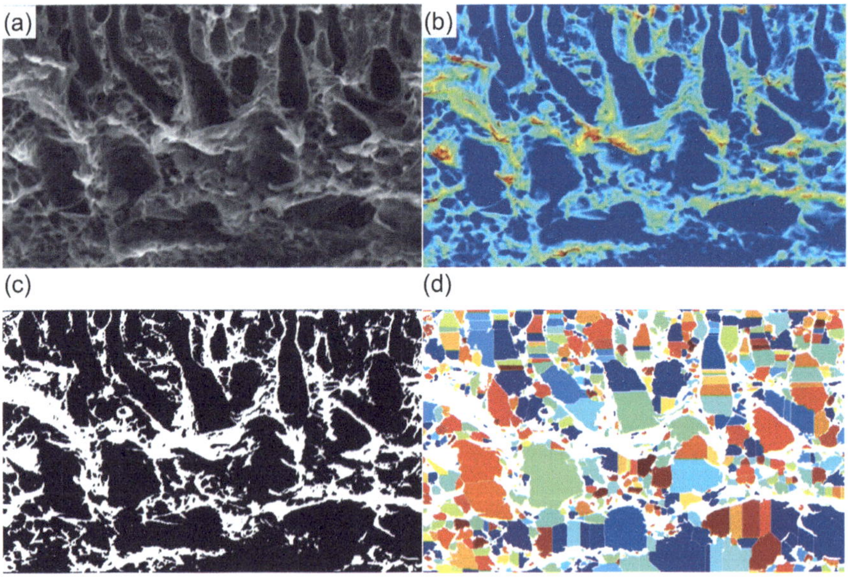

Fig. 4.1 **a** PVDF/MOF membrane SEM image, **b** image after multi-level thresholding, **c** image after binary reduction and median filtering and **d** final image after pore segmentation

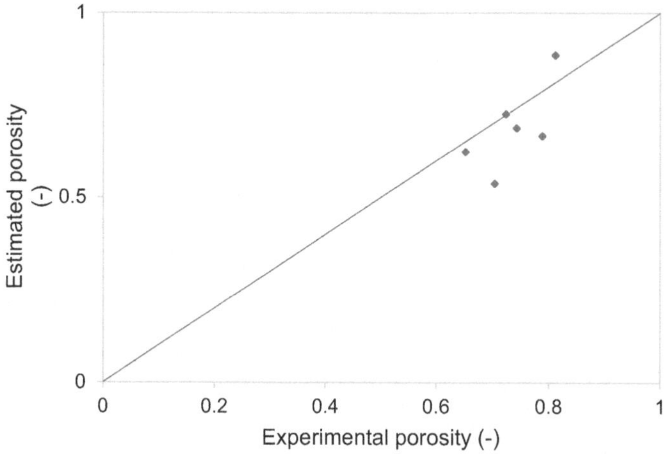

Fig. 4.2 Scatter plot of estimated porosity via image processing against experimental porosity (Gnanasekaran et al. 2018)

Table 4.1 Relative error in the calculation of membrane porosities using image-processing tool

Membrane	Relative error (%)
PES	23.6
CA	0.3
PVDF	4.3
PES/MOF	15.4
CA/MOF	9.1
PVDF/MOF	7.3

10%, providing confidence in the suitability of the code developed in Rabbani and Salehi (2017) for these membranes. It is proposed that the accuracy of estimated values in future studies may be improved through analysis from multiple angles and the usage of higher quality SEM images.

4.2.2 Estimation of Parameters

In order for the numerical simulation of the NF process to resemble physical experiments as closely as possible, reasonable values of the parameters were selected through a review of previous TCS NF studies and parameter correlations. The pressures and concentrations applied in the papers utilising NF in the removal of TCS from water are summarised in Table 4.2.

Table 4.2 Inlet pressures and TCS concentrations used in previous TCS crossflow NF papers

Pressure (bar)	TCS concentration (molm^{-3})	References
5.0	1.04×10^{-6}	Garcia-Ivars et al. (2017)
6.9	2.76×10^{-3}	Lin et al. (2018)
	1.73×10^{-3}	Chang et al. (2012)
	1.73×10^{-2}	Ogutverici et al. (2016)
18.0	2.59×10^{-3}	Simon et al. (2013), Nghiem and Coleman (2008)

Using the cited papers as reference, it was decided that it would be practical for the COMSOL process simulations to be run with inlet pressures ranging from 4 to12 bar and inlet concentrations ranging from 0.003 to 0.015 molm^{-3}.

Diffusional tortuosity through the membrane was found through an expression derived by Weissberg (1963) for random overlapping spheres of uniform and non-uniform sizes (4.1).

$$\tau = 1 - 0.5ln(\phi) \tag{4.1}$$

Due to the solvent association factor for water being a known value, the diffusivity of TCS in this solvent could be determined from the general empirical correlation proposed by Wilke and Chang (1955) for dilute binary solutions (4.2).

$$D_{AB} = 7.4 \times 10^{-8}\left[\frac{(\varphi_B M_B)^{\frac{1}{2}}T}{\mu V_A^{0.6}}\right] \tag{4.2}$$

The subscripts A and B denote the solute (TCS) and the solvent (water), respectively. The molar volume of TCS at boiling point under normal conditions was quantified using a generalised Eq. (4.3) validated through comparisons with empirical data that gave an absolute error of 2.1% (Maloka 2005).

$$V_A = 7.047345 + 0.4V_c + (0.01724 + \frac{15.3765}{T_c} + 0.004387\omega)T_b \tag{4.3}$$

Critical data for TCS (temperature (4.4), pressure (4.5) and volume (4.6)) were obtained via the Joback method (Maloney 2008) which are expressed as follows:

$$T_c = T_b\left(0.584 + 0.965\sum \Delta_T - \left(\sum \Delta_T\right)^2\right)^{-1} \tag{4.4}$$

$$P_c = \left(0.113 + 0.0032n_A - \sum \Delta_P\right)^{-2} \tag{4.5}$$

$$V_c = 17.5 + \sum \Delta_V \tag{4.6}$$

Table 4.3 Estimated TCS diffusivities in water and effective diffusivities in the membranes

Membrane	TCS diffusivity in water (m^2/s)	TCS effective diffusivity in membrane (m^2/s)
PES	1.49×10^{-9}	6.14×10^{-10}
CA		9.36×10^{-10}
PVDF		7.54×10^{-10}
PES/MOF		8.29×10^{-10}
CA/MOF		1.25×10^{-9}
PVDF/MOF		8.67×10^{-10}

The Pitzer acentric factor introduced in Pitzer (1955) and emphasised in Velasco et al. (2015) is defined as (4.7):

$$\omega = -log\frac{P_{\sigma}}{P_c} - 1.000 \quad (4.7)$$

With (4.3)–(4.7) solved and the diffusivity of TCS in water deduced, the effective diffusivity of TCS as it flows through the membrane was calculated through expression (4.8).

$$D_E = \frac{\phi}{\tau} D_{AB} \quad (4.8)$$

The final predicted values for diffusivity and effective diffusivity of TCS using (4.2) and (4.8) respectively are shown in Table 4.3. Due to the scarcity of available data on the diffusivity of TCS through NF membranes, the estimated values were compared with data from NF literature on other solutes with varying molecular weights (Keucken et al. 2016; Abdellah et al. 2018; Werner et al. 2019). From the comparison, it was deduced that the estimated diffusivities are sensible as they are within the same order of magnitude (10^{-9}–10^{-10}) as the calculated diffusivities in the aforementioned studies.

4.2.3 COMSOL Process Simulation

4.2.3.1 Filter Geometry

The foundation of the model's geometry is the crossflow filtration rig shown in Fig. 4.3a with a schematic diagram of an identical rig used in Parasyris et al. (2020) and Seshasayee et al. (2020) illustrated in Fig. 4.3b which combines dead-end and crossflow configurations.

The membrane is pressed together between two plastic plates at the base of the filtration cell that are shaped to create a spiral path. The TCS solution is introduced

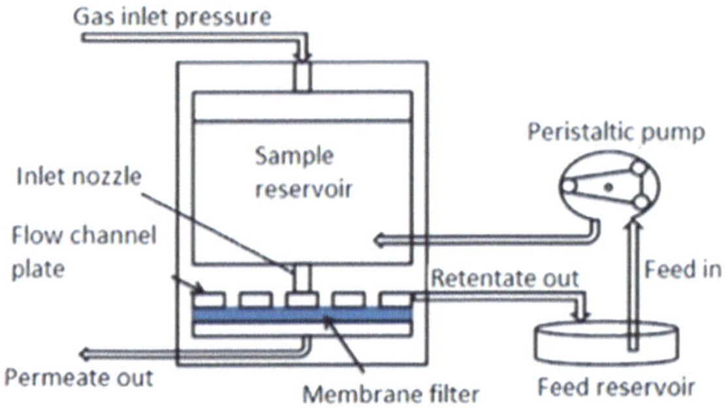

Fig. 4.3 **a** Schematic diagram of the crossflow filtration system (Parasyris et al. 2020; Seshasayee et al. 2020)

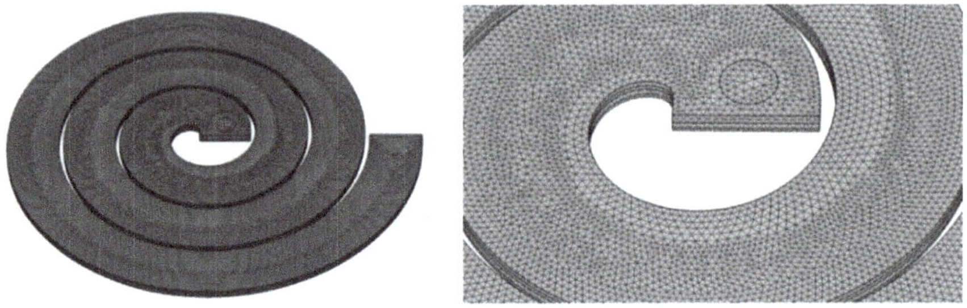

Fig. 4.4 **a** Element mesh of 3D model and **b** magnified view of the mesh

through an inlet at the middle of the top plate and is filtered as permeate as it flows through the spiral channel. Retentate leaves at the end of the path and is recycled back into the feed reservoir.

The crossflow filtration cell was reproduced in COMSOL Multiphysics 5.5 as a spiral channel with an inlet and a compacted membrane. The structure was simplified by eliminating the cylindrical tank to reduce the complexity of the simulation. However, the additional pressure from the tank height was accounted for at the inlet of the spiral channel through suitable inflow boundary conditions. The default physics-controlled finite element mesh with uniform element size was used in solving the problem within lower computational time than the non-uniform adaptive mesh considered in Parasyris et al. (2020). The resulting mesh is displayed in Fig. 4.4a with a magnification of the mesh structure in Fig. 4.4b.

4.2.3.2 Governing Equations

The modelling of flow conditions that are present in membrane filtration systems have incorporated Navier–Stokes equations to represent the laminar, incompressible, free flow that exists within the channel, coupled with Darcy's law for the slow, non-inertial, incompressible flow through the porous membrane (Damak et al. 2004; Parasyris et al. 2020). The integration of these coupled equations in fluid domains are achieved through application of an interfacial boundary condition discussed in Sect. 2.3.3.

The free flow regime is expressed in terms of fluid mass continuity and momentum Navier–Stokes equations expressed as (4.9) and (4.10), respectively:

$$\nabla \cdot \boldsymbol{u}_f = 0 \tag{4.9}$$

$$\rho\left(\boldsymbol{u}_f \cdot \nabla\right)\boldsymbol{u}_f = \nabla \cdot \left(-P_f \boldsymbol{I} + \mu\left(\nabla \boldsymbol{u}_f + \left(\nabla \boldsymbol{u}_f\right)^T\right)\right) + \rho g \tag{4.10}$$

The porous flow regime is governed by mass continuity and Darcy's law as shown in (4.11) and (4.12), respectively:

$$\nabla \cdot \boldsymbol{u}_p = 0 \tag{4.11}$$

$$\boldsymbol{u}_p = -\frac{\boldsymbol{k}}{\mu}(\nabla P_p + \rho \mathbf{g}) \tag{4.12}$$

In Eqs. (4.9)–(4.12), $\boldsymbol{u}_f$, $\boldsymbol{u}_p$, P_f, P_p denote the fluid velocity and pressure, ρ the density of the fluid, $\boldsymbol{k}$ the permeability tensor, μ the fluid viscosity, and $\boldsymbol{g}$ the gravity acceleration. The transport of TCS through the system is modelled through the following physics interfaces available in COMSOL which assume transport through diffusion and convection: Transport of Diluted Species (tds) in the free flow domain and Transport of Diluted Species in Porous Media (tdspm) in the porous domain. These interfaces correspond to the convection–diffusion Eq. (4.13) (Atangana 2018) and are applicable in this scenario as all the TCS concentrations selected are low values. More precisely, the convection–diffusion equation reads

$$\frac{\partial c}{\partial t} = \nabla \cdot (D_{AB}\nabla c) - \nabla \cdot (\boldsymbol{u}c) + R \tag{4.13}$$

where c is the TCS concentration in the feed channels and through the membrane, $\boldsymbol{u}$ is the fluid velocity which is equal to $\boldsymbol{u}_f$ in the feed channels and to $\boldsymbol{u}_p$ inside the membrane, D_{AB} is the diffusion coefficient, and R represents a source term.

4.2.3.3 Free/porous Domain Interfacial Conditions

In setting the stress conditions at the interface between the free flow region and the porous region of the model, the Beavers-Joseph condition was applied (Beavers and Joseph 1967), as defined in (4.14).

$$\left(\frac{\partial u_f}{\partial y}\right) = \frac{\gamma}{\sqrt{k}}(u_f - u_p) \tag{4.14}$$

In this formula, u_f, u_p, and k represent the horizontal components of the velocity fields $\boldsymbol{u}_f$, $\boldsymbol{u}_p$, and of the permeability tensor $\boldsymbol{k}$. More details on the solution algorithms can be found in Parasyris et al. (2020).

It is suggested in this condition that the slip velocity from free flow at the membrane surface and the mean velocity within the membrane from the porous flow are allowed to vary significantly through the presence of a nominal boundary layer. From the analysis in Beavers and Joseph (1967), the boundary layer is deduced to be formed from the rectilinear flow of a viscous fluid over the surface of the permeable material. A continuity of mass is defined at the free/porous domain interface which is used to couple the mass transport equations in the free and porous flow side. Finally, the TCS concentration c is assumed to change continuously in the interfacial region between feed channels and porous membrane.

4.3 Results and Discussion

4.3.1 Pressure and Velocity Distribution

The base simulation, which will be discussed in Sects. 3.1, 3.2 and 3.3, was performed using estimated data from the PES/MOF membrane with an inlet pressure of 6 bar and inlet TCS concentration of 0.003 molm^{-3}. The pressure and free flow velocity distributions throughout the surface of the channel are presented in Fig. 4.5a and b, respectively while the direction of flow is displayed in Fig. 4.5c.

The pressure and velocity distributions throughout the structure follow a smooth gradient that gradually declines from the channel inlet (centre of spiral) to the retentate outlet (end of spiral). It is also observed in Fig. 4.5b that higher magnitudes velocity occur at the outer wall of the spiral channel, which is likely caused by the Dean vortices induced by the presence of centripetal forces (Parasyris et al. 2020) as a result of the channel geometry. The fluid flowing tangentially to the membrane as depicted in Fig. 4.5c is the phenomenon that aids in the mitigation of fouling in crossflow configurations.

4.3.2 Proof of Coupling

In order to examine the success of pairing Darcy's law with COMSOL's tdspm physics interface, both the surface plots in the porous domain for Darcy velocity and the average linear velocity from the interface were superimposed as in Fig. 4.6.

The distribution of the distinct velocities appears to blend seamlessly together and possess identical ranges. There were also no clear discrepancies detected in the superimposed plot, showing that Darcy's law and the interface complement each other and the success of their combined application in the model.

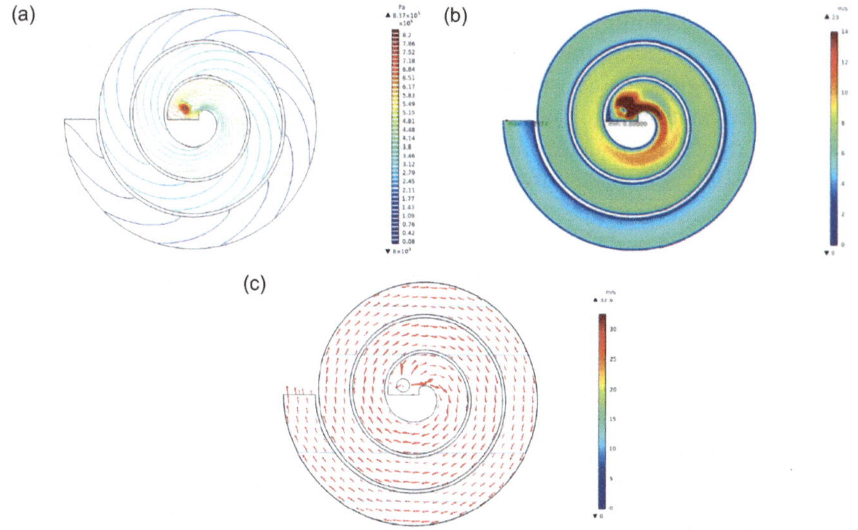

Fig. 4.5 **a** 2D pressure contour plot, **b** surface plot of free flow velocity and **c** direction of flow using the PES/MOF membrane at 0.003 $molm^{-3}$ and 6 bar

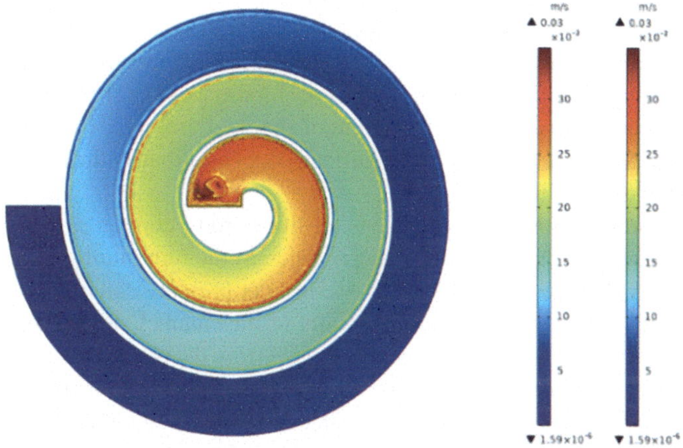

Fig. 4.6 Superimposed surface plot of Darcy velocity (left legend) magnitude and average linear velocity magnitude (right legend)

4.3.3 Concentration Distribution

The concentration distributions for TCS in the free flow domain and the porous domain are demonstrated in Fig. 4.7a and b respectively. The effect of the pressure and velocity distributions on these plots are apparent through the concentration gradient.

The concentration profile appears to be the inverse of the pressure and velocity profiles, with lower concentrations noticed in areas of high pressures and velocities.

In areas of high velocities, there is an enhanced driving force for the filtration of TCS which contributes to an improvement in performance, thus lower concentrations. As the flow circulates following the path of the spiral channel as shown in Fig. 4.5c, TCS is dispersed from the inlet and settles towards the outlet, where higher concentrations are present. TCS concentrations that are higher than the inlet concentration (0.003 $molm^{-3}$) are detected at the walls along the spiral channel in the porous domain where the shear forces are high, as seen in Fig. 4.7b. This result illustrates the occurrence of concentration polarisation in which there is a deposition or accumulation of TCS on the membrane surface.

4.3.4 Effects of Varying Process Parameters

4.3.4.1 Inlet Pressure

In assessing the impact of inlet pressure on the distribution of surface velocity and concentration, the simulation was run separately at 4, 6, 8, 10 and 12 bar (within the range specified in Sect. 2.2) while maintaining the concentration, porosity and both TCS diffusivities at constant values. The maximum and minimum values of surface concentration and surface velocity were plotted against the inlet pressure in Figs. 4.8a and b, respectively. The surface concentrations in the free flow domain of the spiral channel appear to be independent of the inlet pressure. Conversely, there are subtle changes to the range of Darcy surface concentrations in the porous domain as the inlet pressure is increased.

The maximum surface concentrations in the porous domain are observed to steadily increase over the range of pressures whereas the minimum surface concentrations steadily decline after 8 bar. The rise in inlet pressure causes an increase in surface velocity at both the inlet and the outlet of the channel, as seen in Fig. 4.8b. Consequently, this enhances the dispersion of TCS from the centre of the channel and accumulation of TCS at the walls of the channel. The escalation of surface velocity due to inlet pressure is more drastic at the inlet than it is at the outlet, indicating that the velocity dissipates rapidly throughout the channel.

4.3.4.2 *Inlet Concentration*

The simulation was then run separately at inlet concentrations of 0.003, 0.006, 0.009, 0.012 and 0.015 $molm^{-3}$ without changing the other parameters. There were no visible changes in surface velocity hence only the effects on surface concentration were analysed and plotted in Fig. 4.9. The plot reveals that the maximum and minimum surface concentrations in the free flow domain nearly overlap and hence rise at the same magnitude.

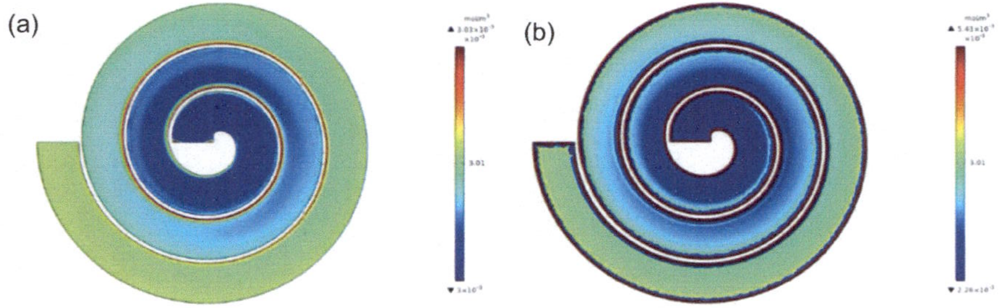

Fig. 4.7 **a** Surface plot of concentration in the free flow domain and **b** surface plot of concentration in the porous domain using the PES/MOF membrane at 0.003 molm^{-3} and 6 bar

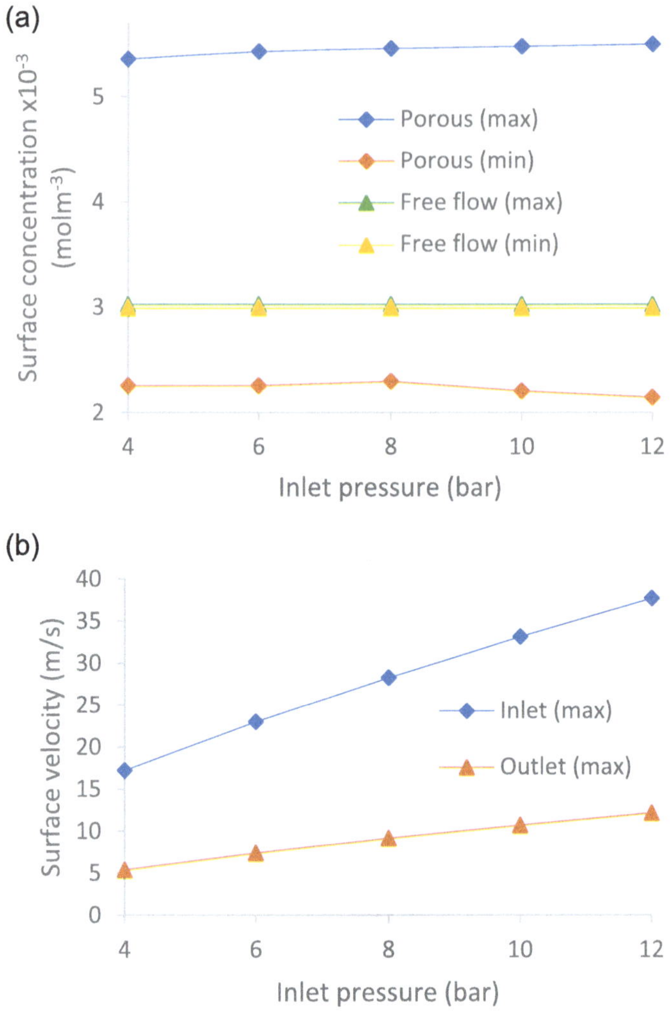

Fig. 4.8 **a** Variation of surface concentrations with changes in inlet pressure and **b** variation of surface velocities with changes in inlet pressure

However, in the porous domain, the maximum Darcy surface concentration has a steeper rise than the minimum value. Therefore, the range of values in Darcy surface concentrations gradually expands as the inlet concentration is increased. This result can be associated with an intensification in concentration polarisation on the membrane surface which may lead to fouling and reduce the performance of the membrane. In order to balance this, the inlet pressure or crossflow velocity should be adjusted accordingly.

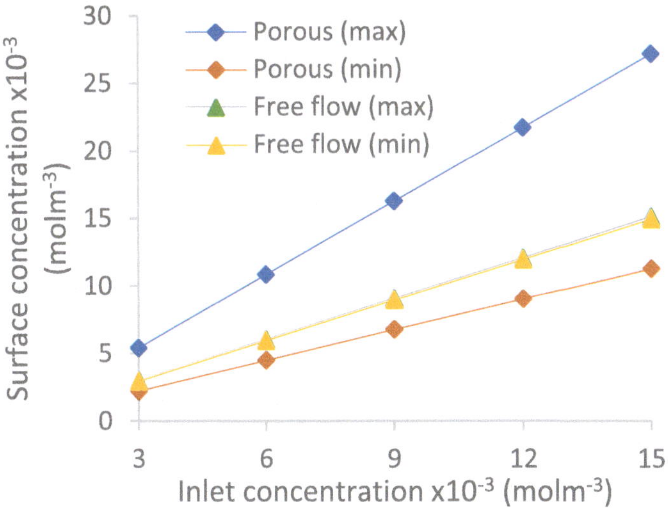

Fig. 4.9 Variation of surface concentrations with changes inlet concentration

4.3.4.3 *Membrane Porosity and Permeability*

Only 4 out of 6 of the membrane porosities were studied as the remaining 2 are very close in value to the ones selected. The selected porosities are the ones of the PES, CA, PES/MOF and PVDF/MOF membranes. The COMSOL model was simulated with the porosities of these membranes and their respective effective TCS diffusivities (outlined in Sects. 2.1 and 2.2). The permeability of the membrane was assumed to be independent of porosity throughout these set of simulations. The surface concentrations plotted against the porosities are displayed in Fig. 4.10a.

Following the trend in Sect. 3.4.1, the TCS surface concentration in the free flow domain remains uninfluenced by membrane porosity but slight shifts are seen in Darcy surface concentration in the porous domain. This finding was predicted as the effective diffusivity of TCS in the membrane does not play a factor in the free flow domain. In the porous domain, the maximum concentration at the channel walls show a slight decline whereas the minimum concentration surrounding the channel inlet experiences a gradual rise. The increase in effective diffusivity linked to the increase in porosity is likely the cause of this trend, as more TCS is allowed to pass through the membrane surface. However, similar to when the inlet pressure was increased, the consequences of increasing porosity in this range are nearly negligible.

In another set of simulations, the effect of permeability on permeate flux was studied and plotted as in Fig. 4.10b. In these simulations, porosity was assumed to be constant as the porous domain is idealised as a continuum (Das et al. 2002). The results suggest that increments in membrane permeability enhances the permeate flux obtained but that it gradually reaches a plateau at higher values. Hence, optimisation

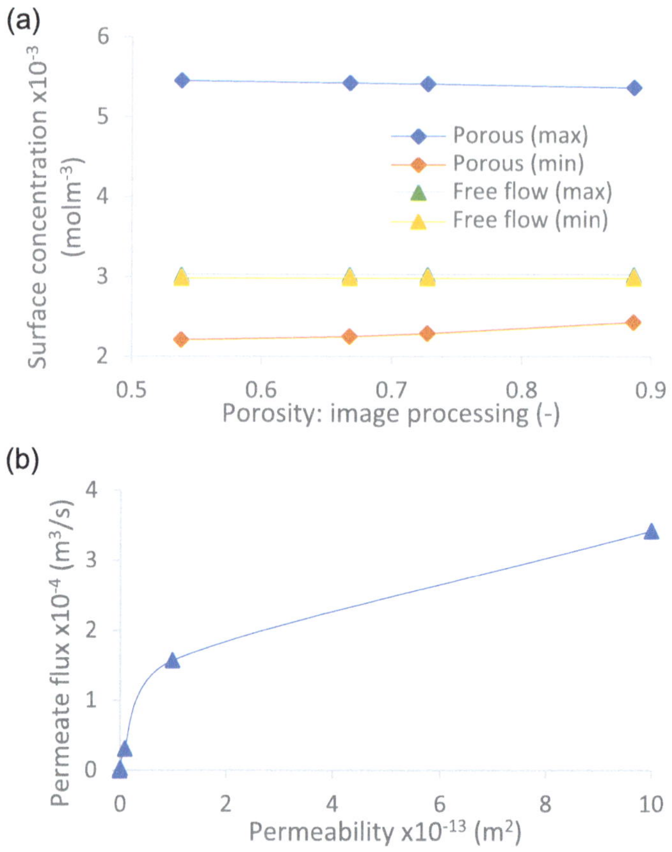

Fig. 4.10 **a** Variation of surface concentrations with changes in porosity and **b** variation of permeate flux with changes in permeability

of permeate flux through membrane permeability can only be achieved up until a specific limit.

4.3.5 Conclusion

The coupling of the Navier–Stokes and Darcy's law along with solute transport interfaces enabled the simulation of TCS-water crossflow in a NF system in this work. The modelling work presented in this paper correspond to an existing laboratory-scale NF water treatment system. The pairing of Darcy's law with COMSOL's physics interface was proven to be suitable through the superimposition of their respective velocity profiles. Distributions of pressure, velocity and concentration on the membrane, which would not be visible experimentally, were found to be correlated.

The change in parameter which produced the largest impact on the modelling results was inlet concentration. Varying inlet pressure and porosity led to minor shifts in the concentration in the porous domain but had no effect in the free flow domain. Overall, the TCS NF simulation showed rational patterns that could assist in the selection of operating conditions and membranes that would be optimal for future NF systems aiming to minimise the presence of TCS in the environment particularly in South Asian Counties where there is an increase in TCS concentration since the Cov-d-19. The simulations were performed without the consideration of porosity-permeability correlations which will be explored in a future work along with the outcomes of changing the geometry of the filtration cell. We envisage that the work presented in this paper help design field-ready NF filtration system for the treatment of drinking water in TCS-affected South Asian countries, which in turn will help meet the Sustainable Goals of the United Nation (UN), e.g., goal 3 (good health and well-being) and 6 (clean water and sanitation for all).

Nomenclatures

ϕ	Porosity (-)
ρ	Water density (kg/m^3)
n_A	Number of atoms in molecule (-)
P	Operating pressure (Pa)
P_σ	Vapour pressure where T/T_c is 0.7 (Pa)
Δ_P	Joback group contributions to P_c (Pa)
P_f	Pressure in free flow domain (Pa)
P_p	Pressure in porous domain (Pa)
R	Species reaction rate (mol/ m^{3}s)
τ	Tortuosity (-)
μ	Water viscosity (Pa s)
M_B	Water molecular mass (g/mol)
T	Operating temperature (K)
T_b	Normal boiling temperature (K)
Δ_T	Joback group contributions to T_c (K)
D_{AB}	TCS diffusivity in water (m^2/s)
D_E	TCS effective diffusivity in membrane (m^2/s)
c	Species concentration (mol/m^3)
k	Intrinsic permeability (m^2)
Ω	Pitzer acentric factor (-)
φ_B	Solvent association parameter (-)
t	Time (s)
V_A	Molar volume at normal boiling point (m^3/mol)
Δ_V	Joback group contributions to V_c (m^3/mol)
u_f	Velocity in free flow domain (m/s)
u_p	Velocity in porous domain (m/s)
γ	Slip coefficient at boundary layer (-)

References

Abdellah M, Scholes C, Freeman B, Liu L, Kentish S (2018) Transport of terpenes through composite PDMS/PAN solvent resistant nanofiltration membranes. Sep Purif Technol 207:470–476

Adhikari S, Kumar R, Driver EM, Perleberg TD, Yanez A, Johnston B, Halden RU (2022) Mass trends of parabens, triclocarban and triclosan in Arizona wastewater collected after the 2017 FDA ban on antimicrobials and during the COVID-19 pandemic. Water Res 222, art. no. 118894. https://doi.org/10.1016/j.watres.2022.118894

Adolfsson-Erici M, Pettersson M, Parkkonen J, Sturve J (2002) Triclosan, a commonly used bactericide found in human milk and in the aquatic environment in Sweden. Chemosphere 46(9–10):1485–1489

Aiello A, Larson E (2003) Antibacterial cleaning and hygiene products as an emerging risk factor for antibiotic resistance in the community. Lancet Infect Dis 3(8):501–506

Atangana A (2018) Fractional operators with constant and variable order with application to geo-hydrology, 1st edn. Academic Press

Beavers G, Joseph D (1967) Boundary conditions at a naturally permeable wall. J Fluid Mech 30(1):197–207

Belosludtsev K, Belosludtseva N, Tenkov K, Penkov N, Agafonov A, Pavlik L, Yashin V, Samartsev V, Dubinin M (2018) Study of the mechanism of permeabilization of lecithin liposomes and rat liver mitochondria by the antimicrobial drug triclosan. Biochim Biophys Acta (BBA)—Biomembr 1860(2):264–271

Chang E, Chang Y, Liang C, Huang C, Chiang P (2012) Identifying the rejection mechanism for nanofiltration membranes fouled by humic acid and calcium ions exemplified by acetaminophen, sulfamethoxazole, and triclosan. J Hazard Mater 221–222:19–27

Dai H, Gao J, Li D, Wang Z, Cui Y, Zhao Y (2022) Family Sphingomonadaceae as the key executor of triclosan degradation in both nitrification and denitrification systems. Chem Eng J 442, art. no. 136202. https://doi.org/10.1016/j.cej.2022.136202

Damak K, Ayadi A, Zeghmati B, Schmitz P (2004) A new Navier-Stokes and Darcy's law combined model for fluid flow in crossflow filtration tubular membranes. Desalination 161(1):67–77

Das D, Nassehi V, Wakeman R (2002) A finite volume model for the hydrodynamics of combined free and porous flow in sub-surface regions. Adv Environ Res 7(1):35–58

Ezeakacha CP, Rabbani A, Salehi S, Ghalambor A (2018) Integrated image processing and computational techniques to characterize formation damage. In: SPE-189509, presented at the SPE international conference on formation damage control, pp 1–15

Garcia-Ivars J, Martella L, Massella M, Carbonell-Alcaina C, Alcaina-Miranda M, Iborra-Clar M (2017) Nanofiltration as tertiary treatment method for removing trace pharmaceutically active compounds in wastewater from wastewater treatment plants. Water Res 125:360–373

Gnanasekaran G, Balaguru S, Arthanareeswaran G, Das D (2018) Removal of hazardous material from wastewater by using metal organic framework (MOF) embedded polymeric membranes. Sep Sci Technol 54(3):434–446

Kumar S, Paul T, Shukla SP, Kumar K, Karmakar S, Bera KK, Bhushan kumar C (2021) Biomarkers-based assessment of triclosan toxicity in aquatic environment: a mechanistic review. Environ Pollut 286, art. no. 117569. https://doi.org/10.1016/j.envpol.2021.117569

Keucken A, Wang Y, Tng K, Leslie G, Spanjer T, Köhler S (2016) Optimizing hollow fibre nanofiltration for organic matter rich lake water. Water 8(10):430

Lin D, Zhou Q, Xie X, Liu Y (2010) Potential biochemical and genetic toxicity of triclosan as an emerging pollutant on earthworms (Eisenia fetida). Chemosphere 81(10):1328–1333

Lin Y, Tsai J, Hung C (2018) Using in situ modification to enhance organic fouling resistance and rejection of pharmaceutical and personal care products in a thin-film composite nanofiltration membrane. Environ Sci Pollut Res 26(33):34073–34084

Lu J, Guo Z, Wang S, Li M, Wang N, Zhou L, Wu H, Zhang J (2021) Remove of triclosan from aqueous solutions by nanoflower MnO_2: insight into the mechanism of oxidation and adsorption. Chem Eng J 426, art. no. 131319. https://doi.org/10.1016/j.cej.2021.131319

Maloka I (2005) Generalized liquid molar volume at the normal boiling point correlation. Pet Sci Technol 23(2):133–136

Maloney J (2008) Perry's chemical engineers' handbook, 8th edn. McGraw-Hill, New York

Nghiem L, Coleman P (2008) NF/RO filtration of the hydrophobic ionogenic compound triclosan: transport mechanisms and the influence of membrane fouling. Sep Purif Technol 62(3):709–716

Noguera-Oviedo K, Aga D (2016) Lessons learned from more than two decades of research on emerging contaminants in the environment. J Hazard Mater 316:242–251

Ogutverici A, Yilmaz L, Yetis U, Dilek F (2016) Triclosan removal by NF from a real drinking water source—effect of natural organic matter. Chem Eng J 283:330–337

Otsu N (1979) A threshold selection method from gray-level histograms. IEEE Trans Syst Man Cybern 9(1):62–66

Ozturk E (2018) Applying analytical decision methods for determination of the best treatment alternative to remove emerging micropollutants from drinking water and wastewater: triclosan example. Environ Sci Pollut Res 25(30):30517–30546

Parasyris A, Discacciati M, Das D (2020) Mathematical and numerical modelling of a circular cross-flow filtration module. Appl Math Model 80:84–98

Pitzer K (1955) The volumetric and thermodynamic properties of fluids. I. Theoretical basis and virial coefficients1. J Am Chem Soc 77(13):3427–3433

Quan B, Li X, Zhang H, Zhang C, Ming Y, Huang Y, Xi Y, Weihua X, Yunguo L, Tang Y (2019) Technology and principle of removing triclosan from aqueous media: a review. Chem Eng J 378:122185

Rabbani A, Ayatollahi S (2015) Comparing three image processing algorithms to estimate the grain-size distribution of porous rocks from binary 2D images and sensitivity analysis of the grain overlapping degree. Spec Top Rev Porous Media: An Int J 6(1):71–89

Rabbani A, Salehi S (2017) Dynamic modeling of the formation damage and mud cake deposition using filtration theories coupled with SEM image processing. J Nat Gas Sci Eng 42:157–168

Riley P, Lamont T (2013) Triclosan/copolymer containing toothpastes for oral health. Cochrane Database Syst Rev 12:1465–1858

Seshasayee M, Yu Z, Arthanareeswaran G, Das D (2020) Preparation of nanoclay embedded polymeric membranes for the filtration of natural organic matter (NOM) in a circular crossflow filtration system. J Water Process Eng. https://doi.org/10.1016/j.jwpe.2020.101408

Sharma S, Dar OI, Andotra M, Sharma S, Kaur A, Faggio C (2021) Environmentally relevant concentrations of triclosan induce cyto-genotoxicity and biochemical alterations in the hatchlings of labeo rohita. Appl Sci (Switzerland) 11(21), art. no. 10478. https://doi.org/10.3390/app112110478

Simon A, McDonald J, Khan S, Price W, Nghiem L (2013) Effects of caustic cleaning on pore size of nanofiltration membranes and their rejection of trace organic chemicals. J Membr Sci 447:153–162

Taheran M, Brar S, Verma M, Surampalli R, Zhang T, Valero J (2016) Membrane processes for removal of pharmaceutically active compounds (PhACs) from water and wastewaters. Sci Total Environ 547:60–77

Tan Q, Chen J, Chu Y, Liu W, Yang L, Ma L, Zhang Y, Qiu D, Wu Z, He F (2021) Triclosan weakens the nitrification process of activated sludge and increases the risk of the spread of antibiotic resistance genes. J Hazard Mater 416, art. no. 126085. https://doi.org/10.1016/j.jhazmat.2021.126085

Unilever (2020) Triclosan And Triclocarban. Unilever global website. https://www.unilever.com/brands/Our-products-and-ingredients/Your-ingredient-questions-answered/Triclosan-and-triclocarban.html. Accessed 15 May 2020

Van der Bruggen B (2018) Chapter 2—Microfiltration, ultrafiltration, nanofiltration, reverse osmosis, and forward osmosis. In: Fundamental modelling of membrane systems, 1st edn. Elsevier, pp 25–70

Velasco S, Santos M, White J (2015) Extended corresponding states expressions for the changes in enthalpy, compressibility factor and constant-volume heat capacity at vaporization. J Chem Thermodyn 85:68–76

Wang M, Hu B, Zhou W, Huang K, Fu J, Zhang A, Jiang G (2022a) Enhanced hand-to-mouth exposure from hand sanitizers during the COVID-19 pandemic: a case study of triclosan. Sci Bull 67(10):995–998. https://doi.org/10.1016/j.scib.2022.03.016

Wang Y, Cun D, Zhang Z, Pu D, Li X, Liang W, Fang T (2022b) Occurrence and risk assessment of triclosan in freshwater lakes in the middle Yangtze River basin (Wuhan, Central China). Water Biol Secur art. no. 100063. https://doi.org/10.1016/j.watbs.2022.100063

Weissberg H (1963) Effective diffusion coefficient in porous media. J Appl Phys 34(9):2636–2639

Werner A, Rieger A, Helbig K, Brix B, Zocher J, Haseneder R, Repke J (2019) Nanofiltration for the recovery of indium and germanium from bioleaching solutions. Sep Purif Technol 224:543–552

Wilke C, Chang P (1955) Correlation of diffusion coefficients in dilute solutions. AIChE J 1(2):264–270

Yang Y, Ok Y, Kim K, Kwon E, Tsang Y (2017) Occurrences and removal of pharmaceuticals and personal care products (PPCPs) in drinking water and water/sewage treatment plants: a review. Sci Tot Environ 596–597:303–320

Yueh M, Tukey R (2016) Triclosan: a widespread environmental toxicant with many biological effects. Annu Rev Pharmacol Toxicol 56(1):251–272

Zepon Tarpani R, Azapagic A (2018) Life cycle environmental impacts of advanced wastewater treatment techniques for removal of pharmaceuticals and personal care products (PPCPs). J Environ Manage 215:258–272

Chapter 5
Culinary Landscapes—Understanding the Intersections of Food Cultures in South Asia

Jyotishma Bordoloi and Madhushree Das

Abstract Food is intertwined with personal preferences, choices, socio-economic patterns and traditional beliefs. This paper attempts to explore the food culture, history, associated patterns, changing perspectives and the role of food tourism in South Asian countries. With the recurrent debates around globalization, food has undergone changes in its culinary landscape. The South Asian food culture is now being explored all across the world and hence the study is significant to develop a better understanding of the patterns of food culture and the dimensions that each country have succeeded in maintaining their own identity.The paper underscores a content analysis of various journals, articles, social media platforms and blogs to look into the food culture of the South Asian region. The paper provides an understanding of how people eat, what they eat, with whom, when, and why they eat for understanding the social landscape, as well as the connections, emotions, statuses, and transactions of people within it. The paper emphasizes the significant role of food in discourses around identities, culture and everyday life.

Keywords Culinary landscapes · Consumption patterns · Globalization · Food tourism · Curried culture

5.1 Introduction

Food is a crucial element in maintaining world order, society and the identity of people. Exploration of food culture provides clues to how "global flows" and "new notions of identity" are implicated in everyday food consumption practices (Watson and Caldwell 2005). When deciding a destination, food rises above factors like climate, accomodation, and landscape (Hu and Ritchie 1993). Food is also considered a "window on the political" (Zaman et al. 2013). The food ideology of "cultural globalization" results in a surge of preference for fast food. The process of

J. Bordoloi · M. Das (✉)
Department of Geography, Gauhati University, Guwahati, Assam, India
e-mail: madhushreedas@gauhati.ac.in

S. Anand et al. (eds.), *Sustainability in South Asian Cities*, Advances in Geographical and Environmental Sciences, https://doi.org/10.1007/978-981-97-7455-5_5

"McDonaldization," is described by Ritzer (2010) as a sensitive marker of globalization. Food is an integral part of the overall tourism experience and major reason for travelling (Hall et al. 2004). The numerous food, wine, and travel magazines, as well as the constant food-focused television channels, are evidence of the rising popularity of culinary tourism (Hashimoto and Telfer 2003). Food has a crucial part in tourism, making it an effective tool for promoting travel (Huang 2009).

The South Asian region consists of the countries of Afghanistan, Bangladesh, Bhutan, Maldives, Nepal, Pakistan, Sri Lanka and India. South Asian native and traditional dishes are a great indicator of how cultural diversity and human wisdom have coevolved through millennia to create a shared cultural legacy and identity. Human food consumption patterns have changed recently as a result of economic growth, urbanisation, changes in occupational structure, the influx of women into the workforce, the advent of processed food companies/manufacturers, and technological advancements (Bandara et al. 2021). South Asian countries are currently undergoing a "nutrition transition" characterized by the over-consumption of food that is high in sugars and fat concurrent with a decline in cereal, fruit and vegetable consumption (Popkin 1998), especially among the affluent class. Food provides a dimension of authenticity and enriches the experience (Scarpato 2002) hence this can help to develop a regional food brand in the region of South Asia.

Food is one of the most traded commodities, armed conflicts, climate shocks and pandemic have caused major disruptions in the world's food systems (World Food Programme 2022). Food supply has been adversely impacted by the Russian-Ukrainian conflict, which has had considerable repercussions on production, sourcing, manufacturing, processing, and demand shifts between countries that depend on imports from Ukraine (Jagtap 2022). The reduction in global food production, supply networks, cost, and availability has a long standing impact on the food crisis and insecurity in South Asia. In addition to rapid population expansion, unplanned urbanisation due to rural-to-urban migration, shrinking arable land, decreased average farm size, stagnant structural reforms, and weak institutions are further factors contributing to food insecurity in South Asia (Ahmad et al. 2015). Despite the challenges and constraints, South Asia's culinary and gastronomy have revolutionized the world's food landscape and it necessitates examining South Asia through food, covering the cuisine, delving into its resources, and understanding its symbolic connotations.. The kind of food that is chosen continues to be a significant indicator in daily life because it is influenced by culture, geography, and history. This chapter attempts to explore the food culture, associated patterns, changing consumer behavior, and the role of food tourism in South Asian countries.

5.2 Background of the Study

Insofar, research on food has explored several dimensions of globalization of food, food security, food consumption and production, nutritional aspects of food, food tourism and the role of food in the making of identity. This paper acknowledges the resourceful work in this field and tries to extend upon the subject matter by discussing nuances of food culture in South Asia. The countries in this region are not homogenous and bring about different flavors, cuisine and identity. The region has been able to make a niche for itself in terms of food identity despite the wave of globalization. This becomes a strong case in point to study the region as a whole in order to understand how food culture and identity transformations have taken place and the elements of resistance within the culture that have made way for the perseverance of the culinary community in the region. The study encompasses the authentic food itineraries of the countries in the region and explores the associated socio-political, economic and geographical reasons that have shaped them. In doing so, the paper makes an argument that the food culture is deeply rooted in traditional norms, it has experienced changes in the contemporary world and undergone evolution despite being able to persist in the global world.

5.3 Research Methodology

This research is built out of secondary sources with content analysis of journals, research articles, social media platforms and documentaries. Looking at these resources gives us a birds-eye view of the food culture and helps to make arguments around the dynamics and perspective on the food culture of the region. In the next sections that follow, the study brings about a country-specific understanding of food culture.

5.4 Exploring Food in South Asia: Country Specific Perspectives

Maldives

Maldivian food culture, commonly known as Dhivehi cuisine, is the cuisine of the Maldives and of Minicoy, India. Because the Maldives are almost entirely covered by water, fish (mostly tuna) is the most important food source (Ellis 2008). The island nation's cuisine is influenced by Arabic, Indian, Sri Lankan, and East Asian influences (Fathimath and Milne 2011). Maldivian cuisine is built on three main ingredients and their derivatives: coconuts, fish, and starches (Hussain et al. 2012). The Maldives' main food are fish and rice, with meat and chicken reserved for special occasions. Fried fish, fish curry, and fish soup are all popular national foods. Local foods include garudhiya, mas huni, fihunu mas, gulha, keemia, kulhi boakiba, and masroshi. After dinner, arecanut is used like a mint. Banbukeyo Bondibai and Dhonkeyo Kajuru are popular desserts in the Maldives (Food n.d.). However, in comparison to the rest of South Asia, the use of spices is considered to be extremely modest (Husaain et al. 2012).

Gastronomy in the Maldives has a long and illustrious history. During his travels in the area in the fourteenth century, Ibn Battuta, the renowned traveler, wrote about the use and importance of the coconut tree to the people and traditions of the Maldives in his work The Rehla of Ibn Battuta—India, Maldive Islands, and Ceylon (Food n.d.).

The evolution of the cuisine culture in the Maldives has been largely influenced by tourism. The Maldives' food tourism industry has experienced exceptional expansion due to the development of premium resorts. The tourist resorts strewn across the islands serve a wide range of international food. In the Maldives, seafood plays a starring role. The Maldives offers unique dining choices. Several resorts include underwater restaurants where diners may completely 'immerse' themselves in the Maldives. Some resorts bring in celebrity chefs or provide cooking lessons (Woźniczko and Orłowski 2017). Maldivian tourism does not prioritize local food as a key component of the tourist experience, instead focuses on international cuisine and presentation styles. The foods that are given the highest priority reflect the two most important tourist markets, Europe and Asia (Fathimath and Milne 2011). As a result, local food has the potential to contribute to the tourism experiences, with a diverse choice of local produce and distinct Maldivian food culture being promoted locally and worldwide through tourism.

Afghanistan

Afghanistan is a landlocked country, seafood is often not consumed. Meat tends to be more expensive and scarce than vegetables. Afghanistan's cuisine mostly takes its inspiration from Persia (Iran), India, and Mongolia (Special Broadcasting Service Corporation 2013). Some Afghans believe that food is elemental in nature and can be categorized as 'hot' (high caloric density) or 'cold' (low caloric density) (low caloric density) (Afghan Food and Cultural Profile: Dietic Consultation Guide 2016). Afghans often follow the rigorous restrictions of the Islamic faith in all aspects of their lives, including food and dietary patterns (Banting 2021). If meat items containing gelatins and food additives are not halal, Afghans avoid them. They fast during Ramadan, as do other Muslims around the world. When Ramadan concludes for the year (Eid), Afghans commonly consume homemade sweets, cakes, nuts, and fruits (Every Culture. Culture of Afghanistan—history, people, clothing, women, beliefs, food, customs, family, social 2015). A big spread of food for ceremonial purposes is referred to as Dastarkhan. Afghans always greet visitors with respect and deliver the best food (Special Broadcasting Service Corporation 2013).

Afghans consume a broad range of fruits and vegetables. In this country of harsh landscapes and severe winters, fatty foods are a vital source of fuel. Mutton fat is customarily added to numerous meals, and oil is used liberally (Special Broadcasting Service Corporation 2013). Eggplant, spinach, potato, carrot, peas, onion, beans, tomato, cucumber, and lettuce are all common vegetables. Fresh mint and coriander are also commonly used herbs. In the Afghan community, sugar-sweetened tea is a popular beverage. Even though it is against Islam, some young Afghans may consume alcohol (Afghan Food and Cultural Profile: Dietic Consultation Guide 2016).

.Afghan dishes include Mastawa (a traditional winter dish made with rice, chickpeas, and sun-dried salted mutton), Aushak (leek and scallion dumplings sometimes served with tomato sauce and dried mint), Kabuli Pulaw (a national dish made with meat and stock topped with fried raisins, pistachios, and slivered carrots), and Afghan Kebab (a traditional summer dish made from lamb meat, served with naan instead of rice) (Abdullah 2021).

Afghanistan has an unstable condition of food insecurity due to decades of conflict, political instability, and repeated droughts. Various policymakers and organizations are attempting to resolve the country's food scarcity and availability issues (D' Sourza and Jolliffe 2016).

Nepal

Nepal food culture is influenced by Indian and Tibetian food. Nepali cuisine is divided into three categories: (a) Madhesi cuisine, which includes Maithili, Bhojpuri, Awadhi, and Lohorung ethnic groups from Eastern Nepal; (b) Janajati cuisine from the middle hills; (c) Urbanized Newari cuisine from Kathmandu Valley; (d) Thakali cuisine from the Himalayan lowlands; and (e) Himalayan cuisine, culturally close to trans Himalayan and Tibetan ethnic groups. Newari and Thakali foods, which are accessible in a wide number of eateries around the valley, have an authentic Nepali essence (Nepal Tourism Board 2022).

Dal Bhat is a common South Asian dish that is both simple and affordable which is widespread especially in Nepal. Chukauni, Timmure Alu, Phulaura, Sekuwa, Quanti, Yomari, Wo or Bara are some authentic dishes of Nepal. Momo is a popular food and dhido is a traditional Nepalese dish popular in Nepal's highlands and mountains, where rice and wheat are scarce. The local beverages include Aaila, Tongba, Chhyang which holds traditional significance (Nepal Tourism Board 2022).

Nepal, and particularly Kathmandu, is known as Asia's budget eating capital. Unfortunately, it is known for pseudo-Western food such as pizza, chips (fries), "sizzling" steaks, and apple pie, rather than Nepali food culture (Guides 2015).

Nepali cuisine has also been influenced by cross-border trade and migration. Momo (meat dumplings) were brought to Kathmandu by Newari traders who had a flourishing trading and cultural relationship with Tibet. Momo (meat dumplings) were adapted to local ingredients and flavor. They also acquired the Tibetan butter tea drinking tradition. The migration of Tibetans, particularly during the 1960s conflict, carried their gastronomic culture with them, significantly expanding the Nepali palate. The dietary culture and legacy of Nepal's southern plains are similar to those of North India. Some Nepalis who had spent generations in Burma returned to Nepal with Burmese cuisine. Muslims who have lived in Nepal for decades have developed their own eating culture and contributed to the enrichment and diversity of Nepali cuisine. As a result, Nepali food can be regarded as heaven for culinary tourism among the many possibilities in Nepal's tourism products (Khanal 2018).

Pakistan

The food culture of Pakistan is the lesser-known cuisine of the subcontinent, although it is rich in tradition and offers a wide range of exquisite delicacies (Juni 2014). Despite the fact that Pakistan is a young nation, its cuisine has absorbed flavors from its neighbors, India, Afghanistan, and Iran, over the long period of time. A unique blend of complex flavors is created by combining Indian, Far Eastern, and Middle Eastern culinary traditions (Ali et al. 2019).

Pakistan is generally known for its bread culture, with meals being eaten with the right hand and naan bread or roti used to scoop up curries and accompaniments as is the practise in Muslim culture. Some important dishes are slow cooked such as the famous haleem, which is a blend of legumes, meat, and spices. It is referred as 'Haleem, the King of Curry' in Pakistan. (About Pakistani food 2008). Lamb is the most consumed meat, followed by beef, chicken, and goat. Sweets abound, with substantial amounts of ghee, sugar, and nuts like pistachios and almonds being used. Halva (sweet) is a popular dessert that can be made using wheat or semolina, but can also be made with carrots or pumpkin. Many sweets contain fragrant essences, such as rosewater (Ali et al. 2019).

In urban Pakistan, changes in the food system and diet have been seen, which have recently been connected to health issues (Jafar et al. 2004). The preference for Pakistani food over others demonstrates that local culinary cultures provide a stabilizing response against globalization-related issues. Nestle sells juice made from Pakistani chaunsa mangoes during the summer months, and it is more expensive than other Nestle juices. Maintaining a sense of 'Pakistani-ness' in food is also linked to conviviality and spaces (Hasnain 2018).

Bhutan

Bhutan's location, culture, and history have all played a part in establishing the country's incredible culinary culture. The spiciness of Bhutanese cuisine is its most distinguishing feature. Bhutanese cuisine is based on rice. Rice is grown in both white and red varieties. Tomatoes, turnips, onions, greens like spinach, green beans, chili peppers, and radishes are among the vegetables (Dendup 2018). Meats such as beef, chicken, hog, yak, and goat are prevalent in Bhutanese cuisine, but vegetarian meals are also popular among Buddhists.

The food 'Ema datshi' is considered a national dish. 'Jasha Maru' is a rice dish with spicy chicken curry and tomatoes. Butter tea, cereal grain brews, and rice wine are among the beverages available. Snack foods such as dumplings and noodles are popular. Bhutanese cuisine has been heavily influenced by Indian and Chinese cuisines (Brokalaki and Murray 2019). Consumption of dried vegetables and dried meat is another traditional Bhutanese food culture. Bhutan's history is intertwined with its culture. Bhutan is a landlocked country that did not open to the outside world until the 1970s, resulting in a lack of vegetables, particularly in higher altitude locations. Bhutanese dried vegetables were used to keep warm during the chilly winter days, and it is now common practise for Bhutanese to eat dry vegetables (Lhendup 2008).

Until the mid-twentieth century, Bhutan remained hidden in the Himalayas. Years of seclusion enhanced the country's traditional culture and beliefs, thus traditional culinary practices still exist (Lhendup 2008).

Sri Lanka

Sri Lanka is often referred to as the Teardrop of India and the Pearl of the Indian Ocean (Bandara and Naranpanawa 2007). Each region of Sri Lanka has its own distinct cuisine and culinary style. Rice, coconut, and spices are hallmarks of contemporary Sri Lankan cuisine (Jayatissa et al. 2014). There are many different edible food species in Sri Lanka. Food culture and customs in Sri Lanka have grown in tandem with cultivated crops, daily activities, beliefs, and the seasonality of food sources because the region is primarily agrarian (Weerasekara 2020). Milk rice and curry, Mun Keum, Kokis, Aasmi, Aluwa, Undu Wal, Saw dodol, Helapa, Pongal (Pongal Rice), Badusha, Watalappan, Daging Chukka (Beef Chukka), and Sookun Goring (Fried Breadfruit) are among the most popular cultural foods (Jayatissa et al. 2014). A specialty of Sri Lankan cuisine is milk rice. Lingus, a special form of sausage, is a specialty in Sri Lanka and a delicacy inherited from Portuguese (Mihiranie et al. 2020). Portuguese, Dutch, Malays, Arabs, and South Indians have all left their culinary imprints and influenced Sri Lankan cuisine (Ekanayake 2016). The Sinhalese, Buddhists, Christians and Catholics, Tamil Hindus and Tamil Christians, Muslims and Burghers who live on the island consume these foods (Ekanayake 2016).

Sri Lankan traditional foods are a mix of indigenous dishes and modifications from diverse non-native civilizations. Changes in food production, dietary behaviors, and food systems have resulted from urbanization, food preferences and sedentary lifestyles (Weerasekara 2020).

Sri Lanka's history and economics are intertwined with tea. Given the decline of tea as a core sector, tea-related tourism could be a viable alternative for economic development. In Sri Lanka, there are chances for tea-related tourism. Tea landscapes, tea-related architecture, and the legacy traditions of tea picking (plucking), processing, selling, and consumption are among the heritage resources for tea tourism in Sri Lanka (Jolliffe and Aslam 2009).

Bangladesh

The culture of Bangladesh is rich in a vast and magnificent array of cuisines. Bangladesh is a festival-loving country and food is an integral part of it. "Baro mase tero parban" is a well-known adage, which means "thirteen festivals in twelve months" (Islam 2012). Bangladeshi food is a remarkable blend of spices, rice, fish, and desserts. Depending on the texture, size, amount of fat, and bones, fish can be cooked in a variety of ways. Bangladesh, despite its youth, is a country steeped in millennia of culture and tradition (Hashanat 2019). Bangladesh is largely a Muslim country, and this influences every element of its culture. For example, pork will never be served in Bengali cuisine, and wine will never be used as an ingredient (Islam 2012).

Spice is a common component in Bengali cooking. Common ingredients include dried red chilli and bay leaves, cumin powder, cumin, turmeric powder, 'panch phoran,' and coriander powder (Hashanat 2019). All around the world, Bengalis are known as "Mase (Fish) bhate (rice) Bangali" (Islam 2012). However, the proverb no longer holds true because the availability of fish has severely decreased, owing primarily to chemical intensive agriculture cultivation and water scarcity.

Sarbat is a popular drink in Bangladesh, made with water, sugar, and lemon. In the countryside, green coconut water is also available as a fresh ready-to-drink option. Corporate drinks such as Pepsi, Coca-Cola, Fanta, 7-up, Mirinda, Sun Crest, and Sprite are now widely available and have replaced traditional drinks as a symbol of royalty. The country restricts the drinking of alcoholic beverages in public (Islam 2012).

The traditional food culture system is changing. Traditional rice, chicken, and fish have been replaced by exotic varieties. Since the 1960s and 1970s, Bangladesh has experienced tremendous urbanization (Hossain 2006). Transnational fast-food companies have been expanding across the country since the late 1990s. Fast-food culture is a relatively recent phenomenon (Ashraf et al. 2014). Growth of fast food preference can be attributed to factors such as gathering spots for teenagers, changes in family structure and values, media influence, and peer pressure. A commercial approach to food culture has taken place in this region (Islam and Ullah 2010).

Bangladesh's traditional cuisine culture contains several discriminating characteristics. Women are the most susceptible group to discrimination. While the majority of Bengali castes ate meat and fish, widows were prohibited from doing so. The women eat the leftover food after taken by the male members of the family (Islam 2012).

Bangladesh's epicurean heritage remains completely unexplored. There are opportunities to promote Bangla food heritage to attract epicurean tourists. The local communities can play a role in this creating partnership with diverse groups of businesses, providing training on heritage cooking and introducing village tourism/ agro-tourism along with culinary tourism (Hashanat 2019).

India

In India, food is used to identify caste, class, family, kinship, tribe connection, lineage, religiosity, ethnicity, and, increasingly, secular group identification. Rice represents hunger and want, as well as fulfillment and fertility. Only the elite ate rice until the late nineteenth century, and the majority of Indians ate millet and sorghum (Srinivas 2011).

India's cuisine is known for its spiciness. Kashmiri cuisine, which has strong Central Asian influences, is popular in north India. In Kashmir, almost every cuisine revolves around the main course of rice, which is plentiful in the beautiful valley. States such as Punjab, Haryana, and Uttar Pradesh, on the other hand, have a large consumption of chapatis as a staple food. Punjab is referred to as India's bread-basket. However, in the north, the influence of Mughlai cuisine is evident. Curry is an important segment of Indian food. The desert cuisine of western India is known for its distinct flavor and variety of foods. The shoreline of Mumbai is home to a wide variety of fish. Dishes like Bombay Prawn and Pomfret are just a few of the delectable options. Portuguese influence can be evident in both the cooking style and the foods in Goa. The Bengali and Assamese cooking traditions are evident throughout eastern and north eastern India. The combination of rice and fish is Bengalis' favorite meal.

The 'Bamboo Shoot' is another unique element widely utilized in North Eastern cooking. The states of southern India make extensive use of spices, fish, and coconuts. In Tamil Nadu, tamarind is used to add sourness to meals (Banerji 2010).

In contemporary India, continental cuisine consists of a mix of English breakfast meals, Western dishes, French cuisine, Italian, and Mexican cuisine spiced to suit the Indian palette. Street vendors serve vernacular versions of contemporary food such as 'gobi manchurian'—a unique Indian-Chinese dish of fried spiced cauliflower (Srinivas 2007). In today's India, regional food and culinary practices are celebrated. Train travel is a culinary adventure, with stations stocked with local delicacies, making it necessary for travelers to stock up on legendary specialties (Srinivas 2011). Domestic food tourism in India, both within and outside India, fosters a dynamic culinary imagination and a gastronomic environment. Food festivals are an important component of promoting tourism in India. Delhi Tourism's Dilli Ke Pakwaan Festival, International Mango Festival, Kashmir Food festival, Sea Food Festival of West Bengal are some examples of food festivals which are vibrant representations of the myriad tastes of the country. In International Cookery Shows, Indian cuisine is gaining prominence (Banerjee 2015). Indian food and ingredients have made an impact on the greater gastronomical map, from street vendors to subway cafeterias to Michelin-starred restaurants (Banerjee 2015).

5.5 Analysis and Discussion

It can be realized that food choices in South Asian countries are influenced by three main factors i.e. geography, socio-political conditions and globalization. Looking into the geographical perspective, since Maldives and Sri Lanka are island nations, the food culture revolves around sea food. While countries like Afghanistan and Bhutan are landlocked, seclusion, traditional norms and religion emerge as dominant factors influencing food. Insights from the region include Islamic faith in Afghanistan and Pakistan, influence of Buddhism in Bhutan, presence of Sinhalese, Catholics, Christians and Tamils in Sri Lanka.

Socio-economic and political factors also have important roles to play. The Pakistani food culture is lesser known compared to other countries of the region. This can be attributed to violence and terrorism in the country. Afghanistan is also a region of political instability and turmoil hence the region's culinary landscape is not much explored and the impact of globalization is not significant.

From the globalization perspective, the local food culture of countries in the South Asian region has received a global outlook. This has enhanced the scale of food culture from regional to a global level. The food of the region can now be found across the globe. This has happened because of migration, cultural appropriation and confluence of cultures around the world. This has extended the culinary landscape of food of the South Asian countries. At the same time, it has instilled confidence within the local people around their own food culture. This feeling of universal acceptance and economic calibrations has enabled a counter approach against homogenization of youth food culture in the region.

The changing dietary behavior of youths and the recent growth of the fast food culture are two further implications of food globalisation. It is crucial to realize that restaurants are now a consequence to the changing lifestyle of the urban population in South Asian developing countries. Restaurants and cafes are used for more than just eating and having fun; they are also used for social gatherings and business meetings (Astuti and Hanan 2010). These spaces provide as platforms for people to showcase their talents, engage in innovative conversations, and unwind while also allowing individuals to experience a different environment. The largest cities in the South Asian region have seen a significant culinary revolution as a result of the digital transformation. Consumers' changing lifestyles and the growth of digital engagement in nation like India have unquestionably changed the patterns in the digital food ordering environment (Badhusa 2019). With the introduction of food delivery services, the culture of takeout and home deliveries has increased in South Asian nations. Because social media platforms like Facebook, Instagram, and others are becoming more and more popular, media is becoming increasingly crucial in defining culinary places. User-generated content, incentives and giveaways, and partnerships with delivery partners all have an impact on how young people's eating habits are changing impacting social structures and personal experiences of the people of these developing nations.

Despite being a part of urban cultures around the world, 'curried culture' continues to be associated with South Asian cuisine (Ray and Srinivas 2012). There is a pattern of similarity observed in the food culture. Most countries' cuisine in the region is influenced by the food history of other Asian countries. There is usage of similar kinds of ingredients, fruits and vegetables and uses of spices in the cuisines though the proportions vary with countries. As observed, each country has its specialty curry prepared in its own unique style. This curry culture cements social relationships and builds cultural efflorescence between the countries and also among different places, the local, and the supra-local.

Furthermore, food tourism can emerge to be the most important part of the culture as these regions present broad diversity and richness in its culinary landscape. In a few South Asian nations, such as Nepal and India, food tourism has grown as a means of diversifying food options and generating viable, long-term solutions such as India's food festivals, Sri Lanka's tea tourism, Nepal's and Bhutan's authentic unique dishes. However, a challenge remains as to how places like the Maldives, Pakistan and Afghanistan which have not been explored for their food tourism, can improve the role of food in tourism. Overall, looking at the three perspectives of geography, socio-political influences and globalization, the study points out that despite so many variables impacting the region, food and its dimensions of each country have succeeded in maintaining their own identity in its own unique way.

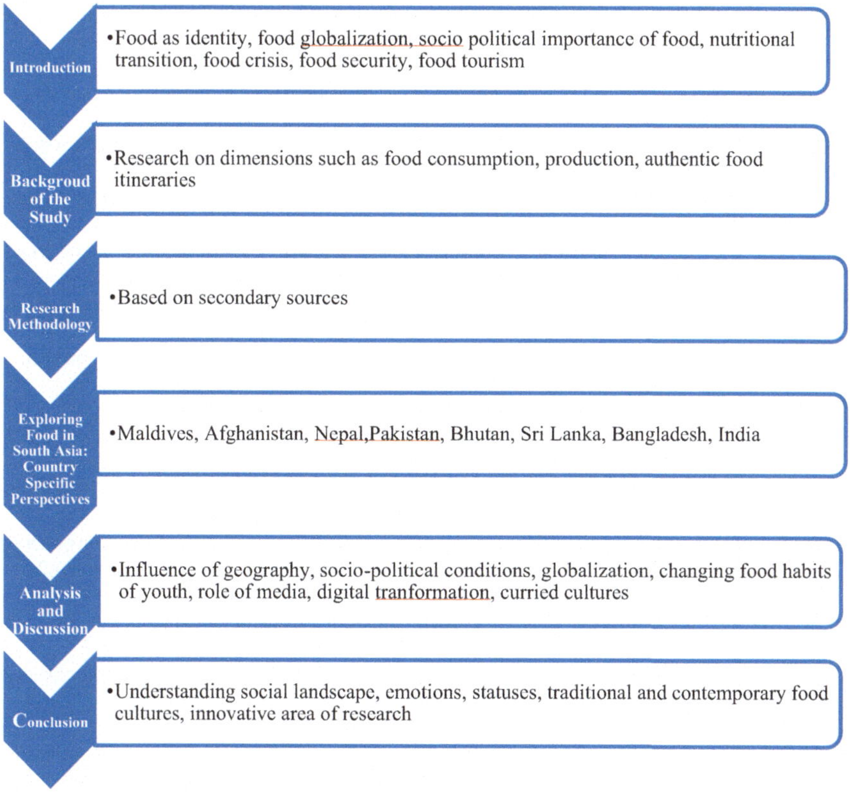

Fig. 5.1 A roadmap of the research

5.6 Conclusion

The research paper made an attempt to understand the existing food culture, consumption patterns, culinary choices and perceptions of the people in the South Asian region. Understanding the social landscape, as well as the connections, emotions, statuses, and transactions of people within it, requires an understanding of how people eat, what they eat, with whom, when, and why they eat. Food would seem to have a significant role in discussions about nationalism and identities in the recurring identity crises that globalization tends to induce. Traditional and contemporary food cultures, encompassing politics, food processing, production, and consumption, are all evolving and exhilarating at the same time. Food will almost probably see much more innovation and attention in the future, and it promises to be an interesting area of innovation and critical research.

References

A global food crisis. World Food Programme (2022) https://www.wfp.org/global-hunger-crisis. Accessed 13 Oct 2022

Abdullah R (2021) Afghanistan culture, food, and holidays. gleaners nutrition hub. http://www.gleanersnutritionhub.org/blog/afghanistan-culture-food-and-holidays

About Pakistani food (2008) SBS food. https://www.sbs.com.au/food/article/2008/07/01/about-pakistani-food

Afghan Food and Cultural Profile: Dietic Consultation Guide (2016) Metrosouth Heath. https://metrosouth.health.qld.gov.au/sites/default/files/dietetic-profile-afghan.pdf

Ali M, Puah CH, Ayob N, Raza SA (2019) Factors influencing tourist's satisfaction, loyalty and word of mouth in selection of local foods in Pakistan. Brit Food J

Ashraf MA, Akhter S, Noor SI (2014) Consumer behavior in fast food marketing in Bangladesh: a case study. Dev Country Stud 4(9):34–44

Astuti HH (2010) The behaviour of consumer society in consuming food

Badhusha DM (2019) Consumers' satisfaction towards digital food ordering in Tiruchirappalli city. J Compos Theory 12(9):1460–1471

Bandara JS, Naranpanawa A (2007) The economic effects of the Asian Tsunami on the 'Tear Drop in the Indian Ocean' a general equilibrium analysis. South Asia Econ J 8(1):65–85

Bandara S, Kumara T, Dharmadasa S, Samaraweera R (2021) Changes in food consumption patterns in Sri Lanka: food security and sustainability: a review of literature. Open J Soc Sci 9(10):213–237

Banerjee M (2015) Food tourism: an effective marketing tool for Indian tourism industry. Int J Sci Technol Manage 4(2):2394–1529

Banerji C (2010) Eating India: exploring the food and culture of the land of spices. Bloomsbury Publishing

Banting E (2021) Afghanistan: the people. Crabtee Publishing Company

Brokalaki Z, Murray JB (2019) The method of creative ethnography: exploring food consumption and pleasurable experiences. In: Food and experiential marketing. Routledge, pp 95–118

D'Souza A, Jolliffe D (2016) Coping with food price shocks in Afghanistan. In: Food price volatility and its implications for food security and policy. Springer, Cham, pp 543–581

Dendup T (2018) Agricultural transformation in Bhutan: from peasants to entrepreneurial farmers. Asian J Agric Extension, Econ Sociol 23(3):1–8

Ekanayake S (2016) Ethnic fermented foods and beverages of Sri Lanka. In: Ethnic fermented foods and alcoholic beverages of Asia. Springer, New Delhi, pp 139–150

Ellis R (2008) Maldives. Bradt travel guides

Every Culture. Culture of Afghanistan—history, people, clothing, women, beliefs, food, customs, family, social [Internet]. 2015. http://www.everyculture.com/A-Bo/Afghanistan.html

Fathimat A, Milne S (2011) Marketing Maldives beyond 'The Sunny Side of Life': a role for food?. E-review of Tour Res 9(5)

Food and Culinary (2022) Nepal Tourism Board. https://ntb.gov.np/things-to-do/food-&-culinary

Food (n.d.) SAARC Cultural Centre; saarcculture.org. http://saarcculture.org/. Accessed 23 April 2022

Guides R (2015) The rough guide to Nepal. Rough Guides UK

Gupta DR, Paper 11: Special interest tourism module 33: popular travel itineraries of Special Interest tourism in India and in the world. Doctoral dissertation, Central University of Jammu

Hall CM, Mitchell R, Sharples L (2004) Consuming places the role of food, wine and tourism in regional development: the role of food, wine and tourism in regional development. In: Food tourism around the world. Routledge, pp 25–59

Hashanat IM (2019) Developing epicurean heritage tourism: the role of local food as a component of tourism in Bangladesh. Econ Soc Dev: Book of Proc 123–141

Hashimoto A, Telfer DJ (2003) Positioning an emerging wine route in the Niagara region: understanding the wine tourism market and its implications for marketing. J Travel Tour Mark 14(3/4):61–76

Hasnain S (2018) "Everyone just ate good food":'Good food'in Islamabad, Pakistan. Appetite 127:1–9

Hu Y, Ritchie JRB (1993) Measuring destination attractiveness: a contextual approach. J Travel Res 32(2):25–32

Huang R (2009) The role of food in promoting Chinese regions on the web. Int J Tour Policy 2(4):289–305

Hussain Z, Lema J, Agrusa J (2012) Enhancing the cultural tourism experience through gastronomy in the Maldives. J Tour Challenges Trends 5(2):71–84

Islam MS (2012) Traditional food culture & food security in Bangladesh

Islam N, Ullah GM (2010) Factors affecting consumers' preferences on fast food items in Bangladesh. The J Appl Bus Res 26(4):131

Jafar TH, Levey AS, White FM, Gul A, Jessani S, Khan AQ, Jafary FH, Schmid CH, Chaturvedi N (2004) Ethnic differences and determinants of diabetes and central obesity among South Asians of Pakistan. Diab Med 21(7):716–723

Jagtap S, Trollman H, Trollman F, Garcia-Garcia G, Parra-López C, Duong L, Afy-Shararah M et al (2022) The Russia-Ukraine conflict: its implications for the global food supply chains. Foods 11(14):2098

Jayatissa RLN, Wickramasinghe WD, Piyasena C (2014) Food consumption patterns in Sri Lanka. Hector Kobbekaduwa Agrarian Research and Training Institute, 114, Wijerama Mawatha Colombo 7, Sri Lanka

Jolliffe L, Aslam MS (2009) Tea heritage tourism: evidence from Sri Lanka. J Herit Tour 4(4):331–344

Juni MS (2014) Impact of global media on the culture of Pakistan: a case study of youth of Layyah city. Int J Innov Appl Stud 8(3):1008

Khanal P (2018) Defining our food culture. The Kathmandu Post. https://kathmandupost.com/opinion/2018/08/12/defining-our-food-culture

Lhendup K (2008) Food crops of Bhutan. Bhutan; Ways of Knowing

Mihiranie S, Jayasinghe JK, Jayasinghe CV, Wanasundara JP (2020) Indigenous and traditional foods of Sri Lanka. J Ethnic Foods 7(1):1–19

Popkin BM (1998) The nutrition transition and its health implications in lower income countries. Public Health Nutr 1(1):5–21

Ray K, Srinivas T (eds) (2012) Curried cultures: globalization, food, and South Asia, vol 34. Univ of California Press

Ritzer G (2010) McDonaldization: the reader. Pine Forge Press, Thousand Oaks, CA

Scarpato R (2002) Gastronomy as a tourist product: the perspectives of gastronomy studies. In: Hjalager A-M, Richards G (eds) Tourism and gastronomy. Routledge, London, pp 51–70

Special Broadcasting Service Corporation. About Afghan food [Internet] (2013) http://www.sbs.com.au/food/article/2013/02/14/about-afghan-food

Srinivas T (2007) Everyday exotic: transnational space, identity and contemporary foodways in Bangalore city. Food, Culture & Soc 10(1):85–107

Srinivas T (2011) Exploring Indian culture through food. Educ About Asia 16(3):38–41

Watson JL, Caldwell ML (2005) The cultural politics of food and eating: a reader. Blackwell Publishing

Weerasekara PC, Withanachchi CR, Ginigaddara GAS, Ploeger A (2020) Understanding dietary diversity, dietary practices and changes in food patterns in marginalised societies in Sri Lanka. Foods 9(11):1659

Woźniczko M, Orłowski D (2017) Unusual gastronomic establishments in the world as destinations of culinary tourism. Geogr Tour Semi-Annual J 5(1):113–127

Zaman S, Selim N, Joarder T (2013) McDonaldization without a McDonald's: globalization and food culture as social determinants of health in urban Bangladesh. Food, Culture Soc 16(4):551–568

Chapter 6
Towards Solving Environmental, Societal, and Economic Teleconnections Between Delhi, Punjab, and Assam

Haruhisa Asada

Abstract In March 2020, in response to the spread of Covid-19, the Prime Minister of India declared a three-week nationwide lockdown. Immediately afterward, economic activities were suspended, and people were forced to stay at home. However, in the cities of North India, atmospheric visibility drastically improved, and residents were able to see the Himalayan ridges for the first time in a long while. Through this impressive event, the local population became aware of the damaging impacts of economic activities on the environment. It was revealed that recent air pollution in North India was due not only to emissions from vehicles and industries but also to the expansion of intensive agriculture in Northwest India. The Green Revolution had depleted groundwater and soil nutrition, leading to agricultural residue burning and air pollution in wider regions. Rice and wheat produced in Punjab are distributed to other states through the central government's scheme of Public Distribution System. Rural residents of Assam also indirectly contribute to environmental pollution by consuming rations from Punjab. While the rations have saved many families below the poverty line from hunger, their expanded coverage has made more people dependent on them. Consequently, the number of rice farmers is declining, and the sustainable production system is being lost in rural Assam. Northeast India is the region most heavily impacted by the central government's policies. Environmental problems in the twenty-first century are not always caused by economic activities in the region but by the connection with remote regions through the modern capitalist economy. It is, therefore, necessary to investigate the teleconnection of environmental issues across regions to achieve the SDGs.

Keywords Rural land use · Government policy · Air pollution · Green revolution · Sustainable development goals

H. Asada (✉)
Department of History, Sociology and Geography, Nara Women's University, Nara, Japan
e-mail: asada@cc.nara-wu.ac.jp

S. Anand et al. (eds.), *Sustainability in South Asian Cities*, Advances in Geographical and Environmental Sciences, https://doi.org/10.1007/978-981-97-7455-5_6

6.1 Introduction

In March 2020, the Prime Minister of India, Narendra Modi, declared a three-week nationwide lockdown in response to the spread of the novel coronavirus (Covid-19). Except for essential services, most economic activities were suspended, and the national population was instructed to stay at home. However, atmospheric visibility drastically improved in the cities of North India due to the significant reduction in atmospheric pollution. Compared to pre-lockdown levels, the concentration of atmospheric fine particulate matter ($PM_{2.5}$), an indicator of air pollution, was reduced by 40–70% during the first week of lockdown (March 25–31, 2020) (Dhaka et al. 2020). For the first time in 30 years, residents of Jalandhar, Punjab, were able to see the Himalayan range at a distance of 200 km. As a result, the local population understood the damaging impacts of economic activities on the global environment as well as the importance of clean air.

Air pollution is one of the most significant environmental problems in contemporary India. In the Delhi National Capital Region (NCR), in particular, air pollution results from decades of unsustainable economic development. Approximately 1.1 million people died from air pollution in India in 2015, which was the highest death rate in the world (Health Effects Institute 2017). India must, therefore, focus on achieving its Sustainable Development Goals (SDGs) by identifying and solving the dominant causes of air pollution. The problem of air pollution can be linked to Goal 3 of the SDGs (ensure healthy lives and promote well-being for all at all ages) and Target 3–9 (by 2030, substantially reduce the number of deaths and illnesses from hazardous chemicals and air, water, and soil pollution and contamination).

In late 2020, another problem arose in Delhi after the Covid-19 pandemic: farmers' large-scale protests. Following the passage of the new agricultural laws in September 2020, farmers in Punjab, Haryana, Uttar Pradesh, and Rajasthan, who opposed agrarian reforms by the central government, have started to block national highways around Delhi NCR for about a year. They worried that the laws would allow private firms to enter the agricultural sector, making local farmers in Punjab more financially distressed. It implies that agriculture in Punjab, despite being its highest productivity in India, is gradually becoming less sustainable. Problems in agriculture can be linked to Goal 2 of the SDGs (End hunger, achieve food security and improved nutrition and promote sustainable agriculture).

Air pollution in the Delhi NCR and agricultural reforms in Punjab are not localized environmental problems. Northeast India, which is more than 1,000 km from Delhi and where the productivity in agriculture is the lowest level among Indian states, is also linked to those problems. The region is geographically separated from mainland India and suffers significantly less air pollution than other parts of the country. However, problems of environmental degradation and societal transformation in this region are indirectly linked to those in Delhi NCR and Punjab because of the economic policies of the central government. Although the Net State Domestic Product (NSDP) has steadily grown (Rs 173,050.49 crore during 2016–17 at a growth rate of 6.43%), and the population below the poverty line has decreased over the last several decades (34.40% during 2004–04 and 31.98% during 2011–12) (Government of Assam 2017), sustainable development has not been fully achieved in Assam. The performance of SDGs is higher in some districts (i.e., Nalbari, Jorhat, and Tinsukia), but lower in others (i.e., Karimganj, Dhubri, and Hailakandi) (Government of Assam 2017). It is especially lower in rural villages, where most people depend on their livelihoods from subsistence agriculture.

Some problems in rural Assam arise from transforming the relationship between the environment and society at the local level, but others are linked to problems in other states on the mainland of India. These include air pollution in Delhi and agricultural transformation in Punjab, which cause environmental degradation and societal problems in rural Assam. This paper explores the hidden links between the environmental problems of Delhi, Punjab, and Assam. In Sect. 6.2, the causes and historical background of air pollution in Delhi are explained. In Sect. 6.3, environmental problems caused by the Green Revolution in Punjab are discussed. In Sect. 6.4, the transformation of agriculture in Assam and its linkage with other parts of India are revealed. In the final section, the importance of linking the problems of Assam with those of Delhi and Punjab is discussed.

6.2 Air Pollution in Delhi

According to the 2016 WHO database (Global Urban Ambient Air Pollution Database), Indian cities account for half of the 20 most polluted cities in the world. Delhi NCR is one of the most polluted cities in India (Fig. 6.1), with an annual average $PM_{2.5}$ concentration of 153 $\mu g/m^3$, which is significantly higher than the WHO standard value of 10 $\mu g/m^3$ and the Indian government standard of 40 $\mu g/m^3$. Air pollution typically shows high seasonal fluctuations, with $PM_{2.5}$ values often exceeding 1,000 $\mu g/m^3$ during the dry season. This is because intermittent rainfall during the rainy season (June to September) washes out fine particles in the atmosphere. In contrast, the absence of rainfall in the dry season results in severe air pollution. Delhi's main sources of air pollution are diesel, gasoline, and coal combustion in vehicles, factories, and power stations (Government of India 2015). India is an emerging economic power, having experienced rapid economic growth since economic liberalisation in 1991. In particular, industries in Delhi and its satellite cities have experienced rapid growth during the last three decades. Moreover, Delhi is situated away from the coast and is unaffected by sea breezes. Therefore, convection predominantly influences the weather conditions in Delhi, which reduces in the dry season due to lower surface temperatures. This process forms a stable layer conducive to accumulating atmospheric pollutants.

Significant initiatives were taken by the Delhi Government in the past decades, such as the conversion of three-wheelers and the bus fleet from petrol/diesel to compressed natural gas (CNG), construction of a 180 km metro rail system, and flyovers in Delhi NCR. However, the benefits rapidly receded with the overall increase in the number of vehicles and other urban emission sources (Chandra et al. 2018). In

Fig. 6.1 Hazy sky in Delhi. *Source* Taken by Author, November, 2018

2016, the Delhi Government introduced the odd–even traffic scheme, which allows only odd (even) license-plate private four-wheelers/cars to travel across the city on odd (even) days. As a consequence, the air quality over Delhi was only marginally improved during the scheme (Dhaka et al. 2020).

Other hidden but important factors have also worsened air pollution in Delhi in the 2010s. For example, biomass is the crop residue (mainly paddy straw and stubble) in fields after harvesting (Fig. 6.2). Biomass burning is the leading cause of air pollution in the post-monsoon season in October and November. Black carbon released from burning undergoes a chemical change in the atmosphere to produce $PM_{2.5}$, which enters the human body and damages respiratory organs.

Due to changes in India's agricultural sector, air pollution caused by stubble burning has only received attention since the beginning of the twenty-first century. The traditional rice cultivation areas are mainly located in the eastern and southern parts of the country, which receive significant amount of rainfall during the rainy season. In Northeast India, rice is even cultivated in rain-fed conditions without irrigation. In contrast, the amount of rainfall in Delhi and Northwest India during the rainy season is less than 1,000 mm, which is insufficient for rice cultivation. However, Haryana and Punjab in Northwest India near Delhi are the largest rice-growing regions in the country now. The Green Revolution, which is a global agricultural innovation, enabled intensive rice production in these semiarid regions of India.

Fig. 6.2 Agricultural laborer burning rice straw in fields. *Source* Taken by Author, November, 2018

During the Green Revolution in the late 1960s, high-yielding varieties (HYV) of main grains, such as wheat, rice, and maize, were developed at the International Rice Research Institute in Philippines. In addition to HYVs and agricultural machineries, chemical fertilizers and irrigation technology by tubewells were introduced and spread to developing countries' rural areas, rapidly increasing their food grain production. This immediately solved the food shortage problem and allowed food self-sufficiency to be achieved in India in 1978. The Green Revolution in India first began and was most widespread in Punjab and Haryana.

Punjab covers only 1.5% of India's land area, but wheat and rice production in the state accounts for 19% and 11% of the country's total production, respectively (Government of India 2013). The widespread use of groundwater irrigation has enabled agricultural production throughout the year, with the double cropping of rice in the rainy season and wheat in the dry season. Although groundwater extraction has solved the initial water problem, solving the temperature barrier for wheat crops is proving more difficult.

In North India, the hot season begins in March and peaks in May, immediately before the rainy season. Wheat crops require cold temperatures during the growing period; therefore, any delays in harvest time expose the crops to higher temperatures in May, which increases the risk of low crop yield. The wheat seeds must be sown by November 15th to harvest the wheat in time. It is, therefore, necessary to quickly complete the rice harvesting work to clear the field for the sowing of wheat seeds. Unfortunately, burning rice straw and stubble residues in the field is the cheapest and least labor-intensive method for clearing fields after the rice harvest. Agricultural residues can be burned in less than an hour as the soil is drier and less humid.

Moreover, double cropping is necessary for farmers' subsistence in Punjab as the entire amount of the production are assured to be purchased by the government agencies at Minimum Support Price. Biomass burning is therefore considered a new environmental problem caused by the unprecedented relationship between nature and society. Subsistence activities conducted by farmers in Punjab, thus, directly threaten the health of residents in Delhi, approximately 300 km away.

6.3 End of the Green Revolution in Punjab

Air pollution caused by agricultural residue burning became apparent in the 2010s, half a century after the start of the Green Revolution in the 1960s. This suggests that the effects of the Green Revolution in increasing food production and reducing hunger are declining. The advancement of new agricultural technology in the last 50 years has exceeded the environmental capacity and caused various environmental problems. The Green Revolution modified the natural environment by introducing new technologies (such as chemical fertilizers, pesticides, and irrigation pumps) from developed countries into arid areas in India that were originally unsuitable for food grain production. As a result, in Punjab, where only pulses and other grains could be cultivated, the use of irrigated groundwater for agricultural purposes throughout the year resulted in the highly intensive double-cropping system of rice and wheat. Despite the success of the Green Revolution, its environmental side-effects resulted in various forms in the local ecosystem (e.g. Shiva 1991).

First, the intensive use of chemical fertilizers has caused severe soil deterioration. Unlike conventional crop varieties that do not respond to chemical fertilizers, the yield per unit area of HYVs (introduced in the Green Revolution) increases proportionally to the input amount of chemical fertilizers. As a result, farmers must apply significant amounts of chemical fertilizers each year to increase the yield of HYVs. Chemical fertilizers replenish the soil with significant nutrients, such as nitrogen, potassium, and phosphorus, but do not contain other trace elements, such as zinc, iron, manganese, and copper, which are necessary for crop growth. In traditional agriculture, cow dung is typically applied to replenish essential nutrients in the soil of cultivated land. However, using chemical fertilizers in modern agriculture has disrupted the biogeochemical cycles of the local ecosystem.

The excessive extraction of the groundwater for irrigation has also led to a local environmental problem. HYVs require more irrigation water than local varieties. Irrigation must also be applied at the right time, which is only made possible by groundwater extraction via tubewells. Unlike canal irrigation, which requires large-scale civil engineering, individual farmers can install tubewells themselves on cultivated land. However, these installations began to affect groundwater, which has been declining significantly across Punjab due to long-term pumping. Studies have predicted that even fossil water deep underground will be depleted in the near future at the current extraction rates. As a result, the state government set the Punjab Preservation of Subsoil Water Act, 2009, which banned any rice transplantation before June

10 (Vatta 2016) to minimize the amount of groundwater extraction. However, this policy delayed the rice harvest period, which shortened the interval leading up to the wheat sowing period. Consequently, significantly more farmers conducted stubble burning, causing severe atmospheric pollution.

The adverse effects of the Green Revolution on soil and water have been frequently reported over the past decades. However, the relatively new air pollution problem suggests that the Green Revolution also impacts the atmosphere. This indicates that human activity in the twenty-first century significantly influences all aspects of the environment. The various impacts of the Green Revolution are typical of the Anthropocene, the current era in which human activities influence the global environment (Crutzen and Stoermer 2000).

The Green Revolution has had negative impacts on local farmers over the decades. The yields of rice and wheat have gradually decreased since the 1990s in response to damage to soil nutrients and groundwater. Therefore, it is no longer possible to maintain high yields even with significant chemical fertilizer input, which has significantly reduced the income of local farmers. Unlike traditional farming methods that use rainfall and cow dung to maintain crop yields, the input costs of modern agriculture are high, including chemical fertilizers, pesticides, irrigation pump fuel, and agricultural machinery expenditures. Local farmers are unable to meet these significant costs and must therefore borrow money from formal institutions or informal money lenders. However, these debts cannot be repaid if the income is lower than expected, which ultimately leads to farm management failure. Indebted farmers are forced to give up their land, and some even commit suicide. It has been reported that 14% of farmers in Punjab have abandoned farming since 1991 (Singh and Bhogal 2014). Since 2006, the number of farmer suicides has rapidly increased and is currently at a rate of three suicides every two days in Punjab (Kaur and Kaur 2016). Despite the enormous impact on local farmers' environment and economic status, the Green Revolution continues in Punjab because the state's agricultural system is also closely linked to other states in India.

Agriculture has become more intensive in Punjab since the Green Revolution in the 1960s, which has increased the labor force in several agricultural processes, such as rice transplantation, timely irrigation, and post-harvest processes. In particular, increased labor is required for harvesting rice and sowing wheat during the one-month period from mid-October to mid-November. Moreover, the demand for labor increased beyond the local labor force after establishing the double-cropping system for rice and wheat. Therefore, the number of migrant workers from other states, such as Uttar Pradesh and Bihar, increased to 1.26 million in 2006, including both on- and off-season (Sidhu et al. 2009). Currently, most of the agricultural work in Punjab is conducted by migrant workers. This work adequately supports their livelihoods as well as their families in their home states.

6.4 Agricultural and Societal Transformation in Rural Assam

The natural environment and societal characteristics of Assam, located in Northeast India, significantly differ from those of Punjab. Assam is one of the wettest states in India, with an annual rainfall of approximately 3,000 mm, which is more than three times the amount in Punjab. In addition, the annual flooding of the Brahmaputra, a mighty river with abundant water flow, supplies fertile soil to the surrounding land. Assam, therefore, has the best agricultural conditions in the country, especially for rice cultivation. However, as groundwater irrigation and chemical fertilizers are not required for rice cultivation, the spread of the Green Revolution in this region is much slower, and many traditional farming technologies are still used by local farmers (Asada 2012).

Although rice yields are relatively low, rice production in Assam is mainly for self-consumption, and any surplus is sold to earn income. Rice cultivation has always been the center of agriculture in Assam. However, the number of rice farmers in its rural villages has been gradually decreasing since 2000. As a result, some farmers have instead planted trees such as teak (e.g., *Tectona grandis* and *Gmelina arborea*) with high market values in paddy fields (Fig. 6.3). In contrast, others have converted their fields into ponds for fish cultivation (e.g., *Labeo rohita* and *Cirrhinus mrigala*) (Fig. 6.4). In the past, they produced their staple food from paddy fields, but now they try to earn more income by converting the fields into other land uses for their economic necessity.

There are several reasons for the decreasing number of rice farmers in recent years, including low productivity, unstable weather conditions, increased wages for agricultural laborers, and increased opportunities for non-farm jobs in urban regions. However, the government's food distribution policy is the most significant reason. The state in Northeast India are linked to modern agriculture in Punjab via food supply rather than the labor force, as is the case for Uttar Pradesh and Bihar.

In India, the central government introduced a public distribution system (PDS) in 1965 as a supporting measure for households below the poverty line (BPL). As part of this system, subsidies are provided for housing and toilet construction, and daily commodities such as rice, dal, atta, and kerosene are provided at low prices every month. Most of the rice that is distributed to BPL households is produced in Punjab. Rice is not a staple food in Punjab and is mainly produced as a commercial crop, which government organizations then purchase (e.g., Food Corporation of India), store in warehouses, and ship to other states. In Assam, the market price of rice is approximately Rs 25 per kg. The PDS rice ration can be purchased at Rs 5 per kg for BPL househlds and at Rs 3 per kg for ultra-poor households (Antyodaya Anna Yojana: AAY) under the National Food Security Act (NFSA), 2013. Then the distribution of 5 kg of free rice/wheat per person per month in addition to their normal PDS entitlement started in 2020 following the outbreak of the Covid-19 pandemic. Due to their low income, many farmers belong to BPL or AAY households and therefore

Fig. 6.3 Trees planted in agricultural fields. *Source* Taken by Author, August 2018

Fig. 6.4 Fish ponds created in agricultural fields. *Source* Taken by Author, August 2018

frequently consume rice produced in Punjab. These rice varieties differ entirely from locally produced rice, but the quality and taste are adequate.

Compared to Punjab, the rice market in Assam is relatively underdeveloped, resulting in farmers being unable to sell their surplus rice in markets even if production rates are high. If rice continues to be distributed at meager costs, farmers are likely to abandon rice cultivation due to low profits. Moreover, the government provides minimum food to sustain the livelihoods of rice farmers, who do not have to work on transplanting and harvesting rice fields. Consequently, it is more economical for farmers to convert some of their paddy fields into forests or fish ponds to raise their income without hard labor. As a result, ecosystem functions and rural landscapes have been significantly transformed. Specifically, Punjab's agriculture is transforming Assam's rural ecosystems through the invisible PDS. Traditional agro-ecosystems in rural villages in Assam have significant biophysical diversity, comprising numerous varieties of domesticated and wild plants and animals (Asada et al. 2022). These systems maintain low-input rice cultivation without introducing new technologies, such as chemical fertilizers or agricultural machineries. The productivity of these systems is relatively low, but the sustainability of agriculture may be lost through the conversion of paddy fields to other land use types.

The conversion of paddy fields to fish ponds in rural villages in Assam transforms the local ecosystem and affects rural societies. For example, Hindu landowners invite fish merchants from Muslim or tribal communities in surrounding villages to catch fish in their ponds. Hindus engaged in any fish business are considered to belong to a low caste and typically choose not to catch and sell fish themselves. Instead, Muslim fish merchants usually cycle around villages in groups of five or six and arrive at the ponds when called by the landowners. The merchants immediately pay the landowner according to the weight of the caught fish, which are then sold in nearby markets. As a result, landowners can earn cash income from fish ponds at a far lower cost and effort than rice cultivation. In the past, neighboring farm households jointly cooperated to transplant and harvest rice, but this mutual cooperation has gradually diminished. Instead, individual farmers are now directly connected to merchants outside their villages, which may hinder progress towards sustainability in rural societies in Assam.

Farmers are fully aware of the indirect impacts of cheap rice on the surrounding ecosystem and rural societies. For example, some farmers have noticed significant changes in the species of wild vegetables and fish in their villages (Asada et al. 2022). Other farmers have observed changes in the working culture of younger generations who appear to be uninterested in agricultural labor. These changes are direct results of the economic policies implemented by the central government, and individual farmers are unable to prevent these changes themselves. Northeast India, including Assam, is the region that is most heavily impacted by the central government's policies.

6.5 Towards a Sustainable Relationship Between Nature and Society

People in India must work towards harmonizing the environment, economy, and society to achieve the SDGs. However, as the links between farmers and the macro-economy continue to strengthen in rural Assam, the relationship between nature and society might collapse in the near future. The ecological environment of cultivated land is changing in response to the ongoing conversion of paddy fields to forests and fish ponds. Farmers' income depends on cultivated land, but their motivation for agricultural work is declining with the increasing need for cash income. As a result, more farmers are seeking non-farm income in urban areas. In the past, farm households in these villages cooperated jointly in rice cultivation work, but this mutual cooperation has weakened in recent years.

In addition, cooperative efforts at regional, national, and global scales are necessary to improve the harmonization between the environment, economy, and society. Modern environmental problems are caused both by economic activities within the region and by the economic influences of external regions. Distorted resource development and distribution systems are causing regional vulnerability in Assam and throughout India. Therefore, analyzing the problems in individual states cannot effectively elucidate the overall impacts of agricultural development in India. To understand the complex problems in one state, it is necessary to integrate the results across other states.

In the current era of the Anthropocene, all regions on Earth are linked by the global chain of the modern capitalist economy. Human activities in a particular region can indirectly and directly impact the environment of other regions. The global system of the capitalist economy, in which the hidden chains of nature, society, and economy interconnect across regions, causes air pollution in Delhi, groundwater reduction in Punjab, and environmental and social problems in Assam at the same time. As a result, individual efforts within a particular region cannot solve all of India's environmental issues. It is, therefore, necessary to further investigate the teleconnection of environmental issues across states to achieve India's SDGs as well as the SDGs of Northeast India. Now that the importance of environmental conservation has been realized among many people across the world due to the spread of Covid-19, it is the good opportunity to share information and knowledge on environmental problems in various regions to consider about sustainable solutions.

Acknowledgements The paper is the results of the Research Project, An Interdisciplinary Study toward Clean Air, Public Health and Sustainable Agriculture: The Case of Crop Residue Burning in North India (No. 14200133), in Research Institute for Humanity and Nature (RIHN), Japan. The author appreciates to the project leader, Dr. Sachiko Hayashida and Dr. Prabir Patra for their academic supports.

References

Asada H (2012) Climate and rice cropping systems in the Brahmaputra Basin: an approach to area studies on Bangladesh and Assam. Rubi Enterprise

Asada H, Yamane Y, Deka N (2022) Environment changes and rural livelihoods in the Brahmaputra valley, Assam, India: A micro observation. In: Saikia A, Thapa P (eds) Environmental change in South Asia: essays in honor of Mohammed Taher. Springer, pp 51–69

Chandra BP, Sinha V, Hakkim H, Kumar A, Pawar H, Mishra AK, Sharma G, Garg PS, Ghude SD, Chate DM, Pithani P, Kulkarni R, Jenamani RK, Rajeevan M (2018) Odd-even traffic rule implementation during winter 2016 in Delhi did not reduce traffic emissions of VOCs, carbon dioxide, methane and carbon monoxide. Curr Sci 114(6):1318–1325

Crutzen PJ, Stoermer EF (2000) The "Anthropocene." IGBP Newsl 41:17–18

Dhaka SK, Chetna, Kumar V, Panwar V, Dimri AP, Singh N, Patra PK, Matsumi Y, Takigawa M, Nakayama T, Yamaji K, Kajino M, Misra P, Hayashida S (2020) $PM_{2.5}$ diminution and haze events over Delhi during the covid-19 lockdown period: an interplay between the baseline pollution and meteorology. Sci Rep 10:13442

Government of Assam (2017) Economic survey Assam 2017–18. Directorate of Economics and Statistics

Government of India (2013) Cost of cultivation of principal crops, unit level data. Ministry of Statistics and Programme Implementation

Government of India (2015) Report of the steering committee on air pollution and health related issues. Ministry of Health and Family Welfare

Health Effects Institute (2017) State of global air 2017. Special Report. Health Effects Institute

Kaur P, Kaur M (2016) Indebtedness and farmer suicides in Punjab. Agric Situat India

Shiva V (1991) The violence of the green revolution: third world agriculture, ecology, and politics. Zed Books

Sidhu MS, Joshi AS, Kaur IP (2009) Migrant agricultural lbour in Punjab. Man Dev 31(4):67–83

Singh S, Bhogal S (2014) Depeasantization in Punjab: status of farmers who left farming. Curr Sci 106(10):1364–1368

Vatta K (2016) Sustainability of groundwater use in Punjab agriculture: issues and options. J Jpn Assoc South Asian Stud 28:211–212

Chapter 7
Making Cities Sustainable: Tracking Economic Sustainability in Urban India

Krittika Roy

Abstract The United Nation's 11th Sustainable Development Goal (SDG) to "make cities inclusive, safe, resilient and sustainable" is one which encompasses the idea of sustainable urban planning and building safe, inclusive and resource-efficient cities. For cities to become sustainable, it is imperative that they relieve themselves of problems like poverty, hunger, inequality, illiteracy, drinking water and sanitation problems, energy poverty, and lack of innovation and infrastructure, which are intrinsic to the less-developed and low-income countries. The paper aims to study and analyse changes in urban housing and living conditions and the lives of urban dwellers and migrants in urban areas between 2012 and 2018. Using secondary data on drinking water, sanitation, hygiene and housing condition from the National Sample Survey Reports for the 69th Round (July to December 2012) and the 76th Round (July to December 2018), this paper tries to observe and analyse differences and also study the settlement and migration patterns in slums and squatters in urban areas. This paper also tries to assert that government intervention in the form of well-planned national schemes and programmes are effective in pushing the country on the path of sustainable urban development to achieve international objectives like building sustainable cities. Lastly, this paper makes a note that to achieve the goal of sustainable cities, there is need for international recognition of the problem as well as inclusive policies by the government for better access of amenities in urban slums and squatters.

Keywords Access · Urban · Amenities · Slums · Squatters · Government

7.1 Introduction

'A sustainable city should promote economic growth and meet the basic needs of its inhabitants, while creating sustainable living conditions for all'—UN Environment Programme (UNEP, 2018).

K. Roy (✉)
Department of Economics, University of Calcutta, Kolkata 700073, West Bengal, India
e-mail: roykrittika26@gmail.com

S. Anand et al. (eds.), *Sustainability in South Asian Cities*, Advances in Geographical and Environmental Sciences, https://doi.org/10.1007/978-981-97-7455-5_7

The United Nation's 11th Sustainable Development Goal (SDG) to "make cities inclusive, safe, resilient and sustainable" is one which encompasses the idea of sustainable urban planning and building safe, inclusive and resource-efficient cities. According to the UNDP, '*Making cities sustainable means creating career and business opportunities, safe and affordable housing, and building resilient societies and economies. Investment in public transport, creating green public spaces, and improving urban planning and management are essential pre-requisites for building sustainable cities.*' Sustainable cities are cities that aim to achieve social, economic and environmental sustainability by building a resilient habitat for the urban population. However, sustainable development cannot be achieved without significantly transforming the way urban spaces are built and ensuring access to decent living and housing conditions (Trindade et al. 2017). With growing migration from rural areas to cities and rising population, there has been a rapid growth of cities as well as a boom in mega-cities, especially in the developing world. But this also means that urban slums have become more and more significant to accommodate this rising urban population. In order to uphold the principle of sustainability in urban planning, there needs to be more focus on whether the growing urban population can make a transition to a better quality of life over time.

It was in 2007, according to UN estimates, the urban population surpassed the rural population in the world for the first time. According to the World Bank, the percentage of the total world population living in urban areas has grown from 33.6 per cent in 1960 to 55.7 per cent in 2019, and the United Nations estimates that by the year 2050, that number will rise to 70 percent. Most high-income countries in Western Europe, the Americas, Australia, Japan and the Middle East have more than 80 per cent of the population living in urban areas and most upper-middle-income countries in Eastern Europe, East Asia, North and Southern Africa, and South America between 50 and 80 per cent live in urban areas. The majority of the population in many low to lower-middle-income countries still live in rural areas, but this proportion of the urban population has been rising. For example, although only 35 per cent of the people live in urban areas in India, the proportion residing in urban areas has doubled in 2020 from 17 per cent in 1950. It is expected to reach the 50 per cent mark by 2050 (UN World Urbanisation Prospects 2018 revision).

When dealing with sustainable development and the enlisted goals, it becomes clear that they are somewhat interrelated. For cities to become sustainable, it is imperative that they relieve themselves of problems like poverty, hunger, inequality, illiteracy, drinking issues of water and sanitation, energy poverty, and lack of innovation and infrastructure, which are intrinsic to the less-developed and low-income countries (Smets and Lindert 2016; Ahvenniemi et al. 2017). The UN-Habitat's complementary New Urban Agenda, adopted as the outcome document from the Habitat III Conference in 2016, offers national and local guidelines on the growth and development of cities through 2036. To meet the goal of sustainable cities, improving living and housing conditions and access to basic amenities can go a long way.

Even before the Sustainable Development Goals were implemented, the Millennium Development Goal 7, Target 11, made a call for efforts to achieve "a significant improvement in the lives of at least 100 million slum dwellers" by 2020. Promoting sustainable human settlements development has been a topic of discussion in the second and third sessions of the Commission on Sustainable Development, calling for adequate shelter for all; sustainable land-use planning and management; integrated provision of environmental infrastructure: water, sanitation, drainage and solid waste management; sustainable energy and transport systems in human settlements; human settlements planning and management in disaster-prone areas; sustainable construction industry activities etc. In cities in low- and middle-income nations, it is common for most urban populations to live in informal settlements lacking adequate water, sanitation, health care, and other essential basic amenities. However, apart from international concern regarding slums and unsustainable living conditions in cities and towns, the eleventh five-year plan in India recognises that the growth of rural settlements which are acquiring urban characteristics, is very slow. This reflects a lack of growth dynamics in the rural economy and smaller towns. The urbanisation process has, thus, become concentrated in developed regions and larger cities in recent years, with backward areas and smaller towns tending to stagnate (Kundu 1989, 2009; Kundu et al. 2002). According to the 2011 Census, 17.4 per cent of urban households lived in slums. Although more than 70 per cent have permanent, owned households with water availability from tap and access to electricity and LPG, about one-third were located in 46 million plus cities. Among the metro cities, Mumbai had the highest number of slum-dwelling households (41.3 per cent of its population) followed by Kolkata at 29.6 per cent, Chennai at 28.5 per cent and Delhi at 14.6 per cent (Census, 2011).

With growing migration in large cities and metropolitan cities, there has been a growth in slum, and squatter settlements as the majority that migrate to rural and backward regions do not have enough resources to afford decent living and housing facilities. Hence, this paper tries to compare the access to basic amenities, housing conditions, and slum and squatter settlement patterns in the period before and after the Sustainable Development Goals were adopted (Fig. 7.1).

7.2 Objective

The fundamental objective of the paper is to study and analyse changes in the urban housing and living conditions and the lives of urban dwellers and migrants in urban areas between 2012 and 2018. Using secondary data on drinking water, sanitation, hygiene and housing condition from the National Sample Survey Reports for the 69th Round (July to December 2012) and the 76th Round (July to December 2018), this paper tries to observe and analyse differences and also study the slum and squatter settlement patterns in urban areas. This paper also tries to assert that government intervention in the form of national schemes and programmes is more effective in pushing the country toward sustainable urban development rather than international objectives like the SDGs.

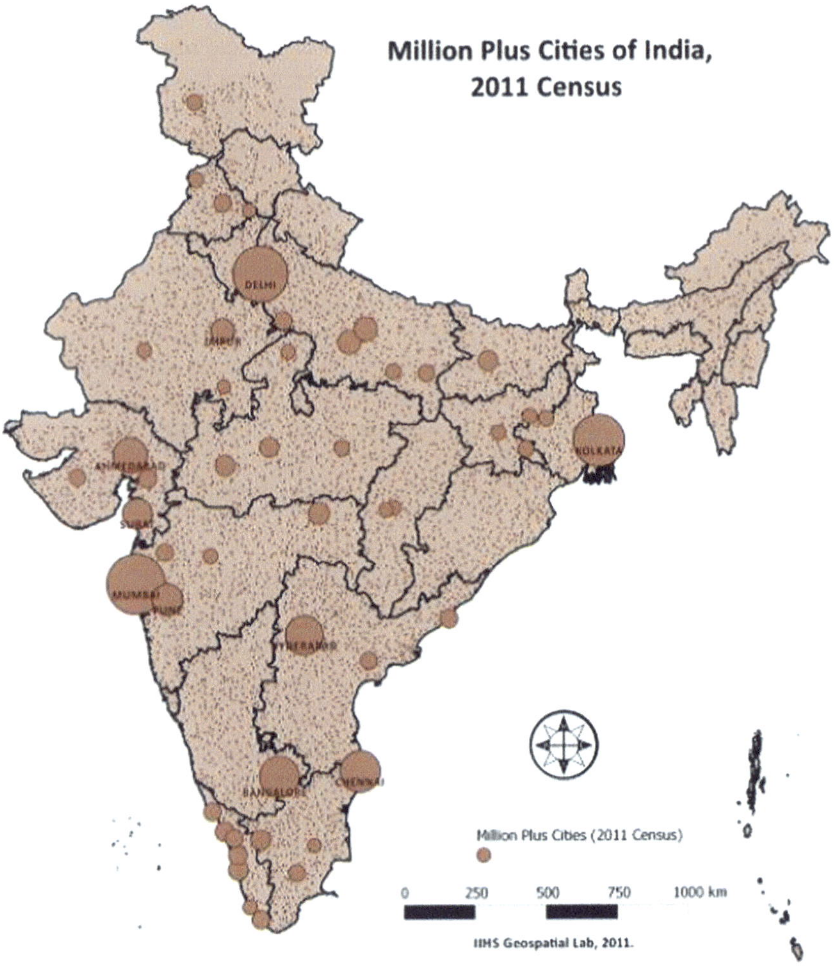

Fig. 7.1 Map of India showing million-plus cities in 2011. Source: IIHS Analysis 2011 (built-up area), Census 2011 (population), Planning Commission 2011 (DPP Estimates 2005–06) from Urban India 2011: Evidence (IIHS, 2011)

7.3 Brief Review of Literature

Just like the concept of sustainable development that implies meeting the needs of the present generation without compromising the ability of future generations to meet their own needs, the term sustainable city or urban sustainability denotes a city planned based on the three aspects of sustainability: social, economic and environmental, in a resilient manner, not compromising with the ability of the future generations to experience the same. In the 1990s, the term sustainable city gained popularity (Roy 2009) but in 2015, it was made an integral part of the 2030 Agenda by including it as one of the Sustainable Development Goals.

According to Hiremath et al. (2013), sustainable urban development is defined as the development of urban areas while safeguarding the environment. They emphasize the equity in income, employment, basic services like shelter and transportation, as well as social infrastructure in urban areas. Dhingra and Chattopadhyay (2016) define 'sustainable city' as one which prioritizes the quality of life and well-being of its citizens. This includes access to social and community services and amenities, basic services and infrastructure such as public transportation, water supply and drainage, telecommunication and other utilities, establishing an environmentally responsible and sustainable approach to development. Additionally, it requires strong local governance with fair policies that can adapt to changing needs and ensure long-term success.

Ebong (1983) identified several factors that contribute to both sustainable and decent quality housing, like aesthetics, proper sanitation and drainage, the building's age, access to basic amenities, enough space, manageable noise levels, effective waste disposal, and good air quality. He also highlighted the importance of easy movement within the neighborhood. Hanmer et al. (2000) took a different approach, focusing on infrastructure services. They argued that high-quality housing is linked to infrastructure that promotes sustainable growth and development. Even according to the World Bank, the efficient and equitable provision of improved sanitation, safe water and solid waste collection contribute to sustainable settlements (World Bank 2013, p. 33).

Neilson (2004) stipulated that access to basic housing and community facilities, the quality of infrastructural amenities, spatial adequacy and quality of design, fixtures and fittings, building layout and landscaping, noise and pollution control, and security are all indicators of the quality and sustainability of housing. In fact, Ferguson et al. (2014) assert that housing in low-income settings cannot be made sustainable unless proper urban planning is done to avoid problems such as exhaustion of natural resources, ecosystem destruction, pollution, overpopulation, growing inequality, and degradation of human living conditions.

The less economically developed countries often lack the ability to provide the infrastructure (e.g., roads and affordable housing) and basic services (e.g., water and sanitation) to provide adequately for the increasing influx of people into cities, but they are also home to the vast majority of slums in the world. Slums have existed as long as cities, and their presence has been documented. However, the word 'slum' is primarily used to describe '*people living under substandard conditions and squalor*' (Mahabir et al. 2016). The growth of slums is expected to have major consequences on both humans and the environment, with slum dwellers living under physical threat from natural and manmade disasters and improper housing have direct impacts on their residents (Napier 2007) and impacting their environment due to lack of basic services, resulting in contaminated soil and polluted air and waterways. This creates a self-reinforcing cycle of degradation for slum residents and the surrounding environment; the consequences of which can spread beyond the slums, potentially affecting other communities (Ali and Sulaiman 2006).

Patel (2012) argues that uncontrolled slum growth undermines sustainable urban development on local, national, and regional levels. Residents in slums often face deplorable conditions with limited access to basic necessities like healthcare, clean water, and sanitation. This vulnerability makes them more susceptible to disease outbreaks and hinders their overall well-being. Their low socioeconomic status, limited education, and high birth and death rates highlight the need for targeted interventions in public health, family planning, and reproductive health services.

7.4 Data Analysis

(a) *Data and Methodology*

The data sources used for analysing urban settlements and amenities in urban areas is secondary data observed in the National Sample Survey reports on Drinking Water, Sanitation, Hygiene and Housing Conditions for July to December 2012 and July to December 2018. The National Sample Survey Office (NSSO) conducted a nationwide survey on 'Drinking water, Sanitation, Hygiene and Housing Conditions' in its 69th Round (July 2012 to December 2012) and 76th round (July 2018 to December 2018) of operations. The objective of both these surveys was to examine and study different aspects of living and housing conditions necessary for a decent and healthy living of the household members. The sampling was multistage sampling, with households being the last stage of sampling. The recall period for the survey was 365 days before the survey.

The Sustainable Development Goals were adopted in 2015. The data for 2012 and 2018 give us a good picture of the access to basic amenities, housing conditions, settlement patterns and drinking water and sanitation conditions before and after the Sustainable Development Goals were implemented. There are tables constructed to compare how access to amenities, access to government support and settlement patterns have changed in the six-year period.

The access to basic amenities covered variables like access to improved drinking water sources, access to the bathroom, access to latrine facility, access to electricity, access to LPG and access to waste disposal services for urban households in 2012 and 2018. The housing conditions observe the household structure, access from roads, ventilation, drainage facility, separate kitchen, and ownership of dwelling units. Furthermore, the migration in the last year to urban areas is observed, and the reasons for migration are studied. Lastly, the benefits received from the government between 2015 and 2018 are enumerated to observe how government interventions have impacted drinking water, sanitation and housing conditions.

In 2014, the country's political landscape changed with the National Democratic Alliance (NDA) government coming to power. Since 2015, the government has enacted and implemented several missions and programmes to improve urban households' drinking water, sanitation, energy access and housing conditions. Along with the Sustainable Development Goals with the 2030 Agenda, government programmes have played a part in influencing access to basic amenities and housing conditions. The comparative analysis of slum and squatter settlement patterns in 2012 and 2018 can illuminate whether living conditions in urban areas have become better off or worse off in the country over time and what reasons migration has been guided by.

(b) *Access to basic amenities in 2012 and 2018*

Regarding the drinking water supply to households in urban India, 95.3 percent households had improved source of drinking water during 2012 which increased to 97.4 per cent in 2018; where, the 'improved source' of drinking water includes: 'bottled water', 'piped water into dwelling', 'piped water to yard/plot', 'public tap/standpipe', 'tube well/borehole', 'protected well', 'protected spring', and 'rainwater collection'. About 89.6 percent of households in urban India had sufficient drinking water, and 76.8 percent of households in urban India got drinking water within the premises in 2012. In 2018, 90.9 percent of the households in the urban areas had sufficient drinking water throughout the year from the principal source, and 80.7 percent had drinking water facilities within the household premises. The percentage of urban households with exclusive access to the principal source of drinking water increased by a little more than 10 per cent in the period between 2012 and 2018, from 46.8 per cent to 57.5 per cent.

In the period from 2012 to 2018, sanitation and waste disposal in urban households has shown some remarkable improvement. The access to bathroom facilities increased from 83.3 percent of households in 2012 to 91.2 per cent of households in 2018, and the access to latrine facility increased from 91.2 per cent in 2012 to 96.2 per cent in 2018. The percentage of households with attached bathrooms increased by almost 20 per cent between 2012 and 2018, from 55.4 per cent to 74.8 per cent. The urban households with exclusive access to latrines rose from 63.9 per cent to 77.6 per cent, and access to improved latrines was possible for 96.2 per cent of households in urban India. 75.8 percent of households, respectively, had some garbage disposal arrangement during 2012, which rose to 82.4 per cent in 2018.

Apart from drinking water, sanitation and waste disposal, electricity access for domestic use increased from 97.9 per cent in 2012 to 99.1 per cent in 2018, and access to LPG as a primary cooking fuel in urban homes increased from 68 per cent in 2012 to 86.6 per cent in 2018. This shows that household access to cleaner energy has improved in this period aiding the process of energy transition from traditional to modern energy in urban India (Fig. 7.2).

The point that is highlighted in the table above (Table 7.1) is that access to basic amenities like drinking water, sanitation, waste disposal, and modern energy for cooking and lighting has improved over the time period between 2012 and 2018, making lives cleaner and easier for people living in urban areas. A considerable part of this progress in living and housing conditions has resulted from the government's proactive policies that came to power in 2014. The government policies that could have contributed to ameliorating amenities and living conditions in urban areas have been discussed later in this paper.

(c) *Access to proper housing conditions in 2012 and 2018*

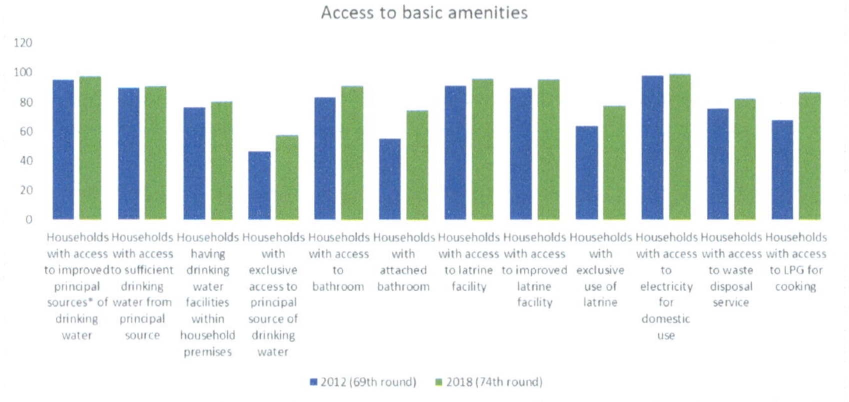

Fig. 7.2 Access to basic amenities in 2012 and 2018 for urban households in India, in per cent. Source: Compiled by author from NSS 69th Round (July 2012–December 2012) Drinking Water, Sanitation, Hygiene and Housing Condition in India Report no. 556 and NSS 76th Round (July 2018–December 2018) Drinking Water, Sanitation, Hygiene and Housing Condition in India Report no. 584, MoSPI, GOI (MoSPI 2012; 2018)

Table 7.1 Access to basic amenities for urban households in India in 2012 and 2018, in per cent

Access to basic amenities		
Urban households with access (in per cent)	2012 (69th round)	2018 (74th round)
Households with access to improved principal sources* of drinking water	95.3	97.4
Households with access to sufficient drinking water from principal source	89.6	90.9
Households having drinking water facilities within household premises	76.8	80.7
Households with exclusive access to principal source of drinking water	46.8	57.5
Households with access to bathroom	83.3	91.2
Households with attached bathroom	55.4	74.8
Households with access to latrine facility	91.2	96.2
Households with access to improved latrine facility	89.6	95.4
Households with exclusive use of latrine	63.9	77.6
Households with access to electricity for domestic use	97.9	99.1
Households with access to waste disposal service	75.8	82.4
Households with access to LPG for cooking	68	86.6

**improved principal sources of drinking water include piped water, bottled water, public tap, tubewell/borehole, protected well, protected spring, rainwater collection*

Source: Compiled by author from NSS Report 556 and NSS Report 584, MoSPI, GOI (MoSPI, 2012; 2018)

The percentage of households with owned dwelling units and hired dwelling units has not changed much in the period between 2012 and 2018, with the majority of households being owned dwelling units. In terms of the household structure, housing conditions in urban India were never deplorable for the majority. Even in 2012, 93.6 percent of households in urban India lived in houses with pucca structures, 5.0 percent in urban India lived in houses with semi-pucca structures, and only 1.4 percent of households in urban areas lived in katcha houses (Table 7.2). In 2018, that percentage increased to 96 per cent living in pucca households, with 3.2 per cent living in semi-pucca households and only 0.8 per cent were katcha households. But when it comes to having a separate kitchen and good ventilation, 66.0 percent of households in urban India had a separate kitchen in their dwellings in 2012, which increased to 75.3 per cent in 2018, and 47.1 percent of households had 'good ventilation' in 2012 which increased to 53.8 per cent households in 2018 (Table 7.2). Therefore, even in pucca households, many houses had no separate kitchen or good ventilation.

But, 87.5 percent of households in urban areas in 2012 and 92.0 percent of the urban households in the urban areas had drainage systems in the house for disposal of wastewater/liquid waste. Along with that around 95 per cent of the households had a direct opening to road/lane/constructed path in 2012 as well as in 2018. The absence of a direct opening to the road from the house indicates housing congestion. A household living in a house without any direct opening to the road is deprived of easy access to surrounding areas and will face problems accessing those many services dependent on road transport. Although most urban households have a drainage system and direct access to roads, there is the possibility of congestion within the households with no separate kitchen and poor ventilation.

In the period between 2012 and 2018, there have been more pucca houses, more houses with access to a drainage system and a direct approach to the road, which implies that the accessibility of urban households and the condition of the dwelling units have both improved. This is an aspect of better living conditions because pucca houses guard against rainfalls, storms and minor natural calamities, and direct opening to a road improves chances of availing services that are dependent on transport. But, the main concern is, the pucca houses do not necessarily have good ventilation inside. Despite road accessibility, they may be located in congested localities, which makes conditions less optimal for better living (see Fig. 7.3).

(d) *Slum and squatter settlement patterns in 2012 and 2018*

Slums are any densely packed settlement with at least 20 households. These areas typically consisted of poorly constructed, temporary housing units crowded together with inadequate sanitation, hygiene facilities, and access to clean drinking water. Such slum settlements were considered *notified slums* if they were notified by the concerned State governments, municipalities, corporations, local bodies or development authorities; otherwise, they were considered *non-notified slums*. In this survey, 'slums' encompassed both officially recognized settlements (notified slums) and those that have not been formally designated (non-notified slums). Studying how and where urban households move, we can understand migration patterns in

Table 7.2 Housing conditions in urban households in India for 2012 and 2018, in per cent

Housing conditions		
Urban households with access (in per cent)	2012 (69th round)	2018 (74th round)
Households with owned dwelling units	61.1	63.8
Households with hired dwelling units	35.4	32.8
Households living in pucca unit	93.6	96
Households living in katcha unit	1.4	0.8
Households with good ventilation of dwelling unit	47.1	53.8
Households having separate room as kitchen	66	75.3
Households with drainage facility	87.5	92
Households with direct opening of approach road/lane/ constructed path to the house	95	95.7

Source: Compiled by author from NSS Report 556 and NSS Report 584, MoSPI, GOI (MoSPI 2012; 2018)

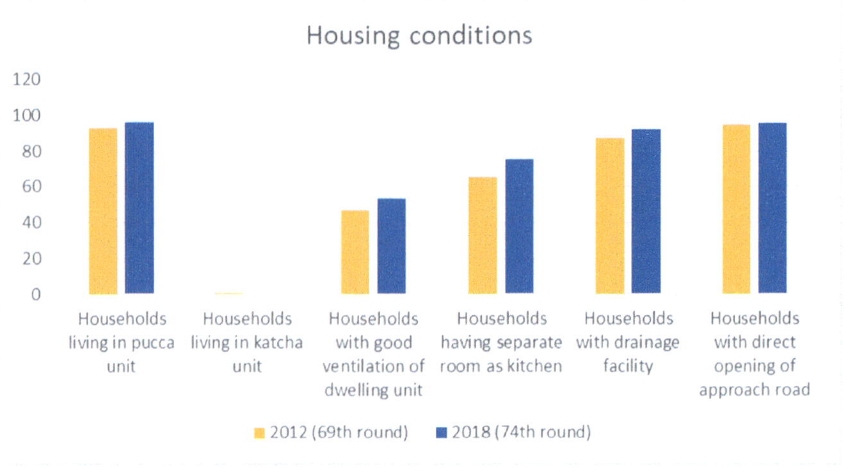

Fig. 7.3 Housing conditions in 2012 and 2018 for urban households in India, in per cent. Source: Compiled by author from NSS 69th Round (July 2012–December 2012) Drinking Water, Sanitation, Hygiene, Housing Condition in India Report no. 556 and NSS 76th Round (July 2018–December 2018) Drinking Water, Sanitation, Hygiene, Housing Condition in India Report no. 584, MoSPI, GOI (MoSPI, 2012; 2018)

different areas, including notified slums (officially recognized) and non-notified slums (unofficially recognized).

The NSS 69th round survey reveals that during last 365 days from the date of survey, 4.9 per cent of urban households moved in to the present area, out of which 3.1 per cent moved into notified slum areas, 4.3 per cent moved into non-notified slum areas, 21.3 per cent moved into squatter settlement areas and 5 per cent moved to other areas (shown in Table 7.3). 37.3 percent of urban households moved into the present area from 'other areas of the same town' followed by 32.4 percent of households from 'other areas of the other town'. The survey also estimated that, during the studied period, 23.7 percent of urban households moved into urban areas from villages. Among those who moved into notified slums, 29.5 per cent was from other areas of the same town, 22.8 per cent came from villages and 11.5 per cent from other areas of other towns. Among those who moved to non-notified slum areas, 49.2 per cent came from villages,12.1 per cent came from other areas of the same town and 7.3 per cent came from other towns. Among those who settled in squatters, 60.5 per cent came from villages, 27.8 per cent came from other areas of other towns. This means that about 41 per cent of notified slum dwellers, 19.4 per cent of the non-notified slum dwellers and 29.7 per cent of the squatter settlers between July 2011 and June 2012 migrated from areas other than slums and squatters.

The NSS 76th round survey reveals that during the last 365 days from the date of the survey, 5.4 per cent of urban households moved into the present area, out of which 4.9 per cent moved to notified slums, 10.9 moved into non-notified slums, 2.8 per cent moved into squatters and 5.3 into other areas (shown in Table 7.3). This survey showed 20.5 per cent of households moved from villages. This shows that compared to 2012, more households have moved into slums rather than squatter settlements. In the last 365 days before the survey, 20.1 per cent of those households which moved into slums, 20.3 per cent of those households that moved into non-notified slums and 30.6 per cent of those which moved into squatters migrated from other areas of either the same town or other towns. 30.5 per cent of notified slum dwellers, 58.6 per cent of the non-notified slum dwellers and 56.5 per cent of squatter settlers were migrant households from villages. In the year preceding the survey in 2018, more urban households settled in non-notified slums or squatter settlements, but among those households migrating from villages, most of them settled in non-notified slums followed by squatters and notified slums which is only slightly different from the year preceding the 69th Round when village households mostly settled in squatters followed by non-notified slums and notified slums.

What is worth noting in the migration to slum and squatter settlements in 2012 was that the majority of the urban households that migrated to these areas came from areas other than slums and squatters. Although there is insufficient evidence in the data to indicate whether the living conditions in the other areas, where these households lived prior to their present area of settlement, were conducive to a decent and healthy life, moving into slums and squatters can be considered a step down in the living conditions of the migrating households. In the year before 2018, however, the pattern changed slightly when most of the households migrating from other areas began settling in other areas rather than slums or squatters. The percentage of households moving from other areas to non-notified slums and squatters did not change significantly, but those moving into notified slums decreased. When considering the migration from villages, in the year before 2012, most of the migrant households were settling in squatters and non-notified slums and this pattern remained unchanged in 2018. However, in 2012, there were more migrant households from villages that had settled in other

Table 7.3 Slum and Squatter settlement in 2012 and 2018 in urban areas in India, in per cent.

Slum settlement					
Present area of settlement	Notified slum	Non-notified slum	Squatter settlement	Other areas	All
69th round July to December 2012					
Moved into the area in the last one year	3.1	4.3	21.3	5	4.9
(i) Moved in from slum/squatter of same town	34.1	28.7	9.8	2.4	4.7
(ii) Moved in from other areas of the same town	29.5	12.1	1.9	38.9	37.3
(iii) Moved in from slum/squatter of other town	2.1	2.8	0	0.9	1
(iv) Moved in from other areas of other town	11.5	7.3	27.8	34.4	32.4
(v) Village	22.8	49.2	60.5	22.4	23.7
76th round July to December 2018					
moved into the area in the last one year	4.9	10.9	2.8	5.3	5.4
(i) Moved in from slum/squatter of same town	28.7	19.9	0	5.4	6.9
(ii) Moved in from other areas of the same town	2.2	16.4	0	43.5	41.2
(iii) Moved in from slum/squatter of other town	0.9	1.1	12.8	1	1
(iv) Moved in from other areas of other town	17.9	3.9	30.6	32.2	30.3
(v) Village	30.5	58.6	56.6	18	20.5

Source: Compiled by author from NSS Report 556 and NSS Report 584, MoSPI, GOI (MoSPI 2012; 2018) areas (22.4 per cent) compared to the migrant households from villages occupying in other areas in 2018 (18 per cent).

The movement of the households into their present place of residence (see Table 7.4 below) shows that at the all-India level during 2012, the primary reason for such movement of the urban households was 'others' (33.2 per cent) followed by 'other employment-related reason' (21.6 per cent) and proximity to the workplace (18.5 per cent). Among households which moved into 'notified slum' areas, the same pattern was observed; 'others' (39.1 per cent) followed by 'other employment-related reason' (19.3 per cent). On the other hand, for households presently residing in 'non-notified slum' areas, the reason 'others' (44.8 percent) was cited as the main reason, followed by the reason 'proximity to workplace' (18.3 percent). Among those moving into squatters, the majority stated 'other employment-related reasons' (86.7 per cent) as their main reason to move. The pattern was almost similar in 2018, when 'others' was the most common response (36.3 per cent) followed by 'proximity to workplace' (19.6 per cent) and 'other employment-related reasons' (12.3 per cent). Among those who moved into 'notified' slums and 'non-notified' slums, 55.9 per cent stated 'others' and 71 per cent stated 'others' respectively. Among those households that moved into squatters, 45.4 per cent stated 'others', and 30.6 per cent stated 'better locality' as their reason for movement.

(e) *Government schemes between 2014 and 2018*

In 2014, when the NDA government came to power, they introduced a variety of schemes for transforming urban India. The Swachh Bharat Mission was launched on 2 October 2014, to make India a clean country by October 2019. Under this scheme, there were efforts to achieve universal sanitation coverage and waste management. The aim was to make cities and villages open and defecation free by constructing toilets and solid and liquid waste management facilities. The Smart City scheme and the AMRUT Mission came in June 2015 with 100 selected cities to be made into smart cities between 2015 and 2020 under the former scheme and development of basic amenities in more than 500 cities with a population of more than one lakh to be done as a part of the AMRUT Plan. Apart from this, there have been initiatives to increase domestic electrification and access to clean cooking fuel like LPG under the PM Ujjwala Scheme. These schemes highlight how access to basic amenities had been an important objective for the government. From 2015 to 2018, there has been progress in drinking water, sanitation, hygiene, and waste management in urban areas due to the implementation of these schemes. Moreover, the PM Awas Yojana, launched in June 2015, has helped households convert their kutcha dwelling units to pucca houses, providing urban households with more durable housing facilities (see Table 7.5 below).

Table 7.4 Reason to move into slums, squatters and other areas in 2012 and 2018, urban India, in per cent

Reason to move to the present area					
Present area of settlement	Notified slum	Non-notified slum	Squatter settlement	Other areas	All
69th round July to December 2012					
Free/low rent	2	4	0	4.5	4.4
Independent accommodation	11.2	6.9	5.4	9.4	9.4
Better locality	16.6	12.2	0	12	12.1
Proximity to workplace	11.7	18.3	5.9	18.9	18.5
Other employment related reasons	19.3	13.9	86.7	21.3	21.6
Others	39.1	44.8	1.9	32.8	33.2
76th round July to December 2018					
Free/low rent	11	2.5	0	8.4	8.2
Independent accommodation	6.4	6.3	0	9.4	9.6
Better locality	9.3	3.6	30.6	14.6	14
Proximity to workplace	12.5	13.3	11.2	20.3	19.6
Other employment related reasons	4.8	3.3	12.8	13.1	12.3
Others	55.9	71	45.4	33.7	36.3

Source: Compiled by author from NSS Report 556 and NSS Report 584, MoSPI, GOI (MoSPI, 2012; 2018)

7.5 Discussion

The availability of basic living facilities like drinking water, bathroom, latrine, housing etc., to each individual of a nation is an essential requirement for its growth and development. With a rapidly growing population, it is one of the most challenging tasks of the respective governments, especially in developing countries, to create infrastructure to ensure the decent living of its inhabitants. Along with the availability of a basic amenity, the amenities must be available sustainably.

Between 2012 and 2018, access to basic amenities, living conditions and housing conditions have improved, although not considered in the indicators. By 2018, more than 97 per cent of urban households had access to improved principal sources of drinking water, more than 91 per cent had access to a bathroom, more than 95 per cent had access to improved latrine facilities, and more than 82 per cent had access to waste disposal services. The year 2015 saw the launching of government schemes like the AMRUT Mission and the Swachh Bharat Mission, which aimed to improve urban India's drinking water, sanitation and waste management. Although the timing

Table 7.5 Urban households receiving government benefits in India between 2015 and 2018, in per cent

Benefits received from the government in the last 3 years	
Scheme	Percentage of urban households receiving benefits
(i) For drinking water	
Atal Mission for Rejuvenating and Urban Transformation (AMRUT)	17.8
Smart City Mission	0.3
Others	62.1
(ii) For sanitation	
Swachh Bharat Mission	81.8
AMRUT Mission	1.4
Smart city mission	0.5
Others	11.6
(iii) For housing	
PM Awas Yojana	53.3
Others	41.6
Electrification schemes	85.1
PM Ujjwala Yojana	55.9
Other LPG schemes	28.4

Source: Compiled by author from NSS Report 584, MoSPI, GOI (MoSPI, 2018)

and objectives of these schemes coincide with the time when the Sustainable Development Goals were adopted, it makes a stronger case to assert that the improvement in basic amenities was primarily due to national initiatives rather than international objectives. However, stating that Sustainable Development Goals had no role in this improvement may also be wrong. These goals can influence the programmes and schemes launched by the NDA government, and achieving these goals through country-wide programmes and initiatives can show the country and the government in a better light.

Regarding clean energy access, the government has been highly proactive in launching flagship programmes like the PM Ujjwala Yojana, and UJALA scheme that has favoured LPG access and electricity access in urban households belonging to even the lowest income categories. Ninety-six per cent of households have pucca dwelling units, 92 per cent have drainage facilities, and more than 95 per cent have a direct opening to a road or constructed path as of 2018. The Smart City Mission, AMRUT Plan, and PM Awas Yojana have all significantly achieved these numbers since 2015. However, there were a few aspects where even government programmes did not contribute favourably. About 25 per cent houses did not have a separate kitchen and about 46 per cent houses did not have good ventilation, which implies that there was possibility of congestion and indoor air pollution from cooking or smoking possible within the dwelling units, even if they had direct access to roads or had a pucca structure.

The migration and settlement pattern in slums and squatters revealed that in the year before 2012 and the year before 2018, urban households did move from 'other areas' to slums and squatters. However, even migration from villages led households to settle mostly in non-notified slums and squatters, which meant they were further away from being recognised by the government as those that needed support. This pattern throws light on the fact that migration of urban households happens for reasons related to employment as well as for reasons other than employment, rent or locality. It cannot be understood whether migrating to slums and squatters makes urban dwellers from other areas better off or worse off because the report does not define the conditions in other areas, and all slums and squatters may not have worse conditions than other areas. However, the existence of slums and squatter settlements shows an informal way of living and housing without access to improved basic amenities, housing and living conditions, often outside the purview of the government schemes that aim to provide better amenities and living conditions.

The plight of slum dwellers gained importance in the Millenium Development Goals Agenda in 2000. In the Sustainable Development Goal of Building Sustainable Cities, the amelioration of living conditions in slums and the lives of slum dwellers is an implicit and essential pre-requisite. Until improved basic amenities reach slums and squatter settlements and improve the lives of the millions of urban slum dwellers, sustainable cities cannot be achieved. Over the past decade, all the schemes planned and implemented by the government have treated urban India as a whole, rather than looking at slum and non-slum areas differently. As a result, there is a lack of recognition of the differences between other areas and slums and squatters. National policies keeping with the international objective of sustainable development have helped increase access to basic amenities and improved housing in urban areas. International recognition of the condition of slums and slum-dwellers can translate into more policies explicitly aimed at their betterment.

7.6 Conclusion and Policy Implications

The access to basic amenities, living and housing conditions have improved in urban India between 2012 and 2018. Part of the credit goes to proactive government initiatives and schemes implemented since 2015. In all these government initiatives, the slums, squatters and lives of urban dwellers in slums and squatters have been neglected to some extent by trying to view urban India as a single homogenous unit. Adopting the 2030 Agenda for meeting the goals of sustainable development, including the goal of building sustainable cities, is impossible with the existence of slum and squatter settlements and the less-than-optimal living and housing conditions in these settlements. This paper has shown that with the help of national schemes, the access to basic amenities and housing conditions has improved between 2012 and 2018. The importance of international objectives comes into the picture when a country adopts the agenda and national policies to achieve the goals.

This paper has also shown that the migration to slums and squatters has continued in urban India and primarily due to reasons other than rent, employment and locality. There is a need to find the other reasons, whether social, economic or psychological, that lead people to migrate. Moreover, policies for urban India may or may not apply to slum and squatter settlements. As a part of a more inclusive policy-making approach, there have to be more specific and well-implemented policies and schemes for slum and squatter dwellers to contribute positively towards sustainable urban development. India needs the international agenda of building sustainable cities to provide the backdrop for more inclusive and well-planned policies and programmes aiding sustainable urban planning and development.

Acknowledgements I would like to extend my gratitude to my supervisor, Dr. Madhura Swaminathan, for her constant guidance and encouragement to work on sustainable development issues. I would also like to thank Indian Statistical Institute, Bangalore, for providing me with the resources and support I needed for working on this paper.

References

Ahvenniemi H, Huovila A, Pinto-Seppä I, Airaksinen M (2017) What are the differences between sustainable and smart cities? Cities 60:234–245

Ali MH, Sulaiman MS (2006) The causes and consequences of the informal settlements in Zanzibar. In: XXIII congress of the international federation of surveyors. Munich, Germany. https://www.fig.net/resources/proceedings/fig_proceedings/fig2006/papers/ts35/ts35_01_ali_sulaiman_0320.pdf

Dhingra M, Chattopadhyay S (2016) Advancing smartness of traditional settlements-case of Indian and Arab old cities. Int J Sustain Built Environ 5(2):549–563. https://doi.org/10.1016/j.ijsbe.2016.08.004

Ebong MO (1983) The perception of residential quality: a case study of Calabar, Nigeria. Third World Plann Rev 5(3):273–284

Ferguson B, Smets P, Mason D (2014) The new political economy of affordable housing finance and urban development. In: Bredenoord J, van Lindert P, Smets P (eds) Affordable housing in the urban global south: seeking sustainable solutions. Routledge/Earthscan, London, pp 102–116

Hanmer L, Booth D, Lovell E (2000). Poverty and transport, a report prepared for the world bank in collaboration with DFID. Overseas Development Institute

Hiremath RB, Balachandra P, Kumar B, Bansode SS, Murali J (2013) Indicator-based urban sustainability: a review. Energy Sustain Dev 17(6):555–563

Indian Institute for Human Settlements (IIHS) (2011) Urban India 2011: evidence. In: India urban conference 2011: evidence and experience. https://smartnet.niua.org/sites/default/files/resources/IUC%20Booklet%20on%20Indian%20cities.pdf

Kundu A (2009) Exclusionary urbanisation in Asia: a macro overview, Econ Polit Wkly XLIV(48):48–58

Kundu A (1989) The dynamic of city hinterland relationship: the Indian case. In: Vennetier P (ed) La Peri-Urbanisation Dans Les Pays Tropicaux, CEGET-CNRS, Talence

Kundu A, Pradhan BK, Subramanium A (2002) Dichotomy or continuum: analysis of impact of urban centres on their periphery. Econ Polit Wkly 37(14)

Mahabir R, Crooks A, Croitoru A, Agouris P (2016) The study of slums as social and physical constructs: challenges and emerging research opportunities. Reg Stud Reg Sci 3(1):399–419

MOSPI, GOI (2012) Drinking water, sanitation, hygiene and housing condition in India (69th Round), NSS Report 556, MOSPI and NSO

MOSPI, GOI (2018) Drinking water, sanitation, hygiene and housing condition in India (76th round), NSS Report 584, MOSPI and NSO

Napier M (2007) Informal settlement integration, the environment and sustainable livelihoods in Sub-Saharan Africa. University of Montreal, Montreal, Canada. http://www.grif.umontreal.ca/pages/i-rec%20papers/napier.pdf

Neilson M (2004) Scottish housing quality Standard (SHQS). Scottish Executive Development Department

Patel A (2012) Slumulation: an integrated simulation framework to explore spatio-temporal dynamics of slum formation in Ahmedabad, India PhD Dissertation. George Mason University, Fairfax, VA

Roy M (2009) Planning for sustainable urbanisation in fast growing cities: mitigation and adaptation issues addressed in Dhaka, Bangladesh. Habitat Int 33(3):276–286

Smets P, Van Lindert P (2016) Sustainable housing and the urban poor. Int J Urban Sustain Dev 8(1):1–9. https://doi.org/10.1080/19463138.2016.1168825

Trindade EP, Hinnig MPF, da Costa EM et al (2017) Sustainable development of smart cities: a systematic review of the literature. J Open Innov 3:11. https://doi.org/10.1186/s40852-017-0063-2

UNEP (2018) Environment UN (2018) Sustainable cities. UNEP—UN Environment Programme. Accessed 22 Sept 2020

World Bank (2013) Building sustainability in an urbanizing world. A partnership report. Urban development series 13. World Bank, Washington (DC)

Chapter 8
Living by the Edges: Existential Vulnerabilities among the Slum-Dwelling Women of Delhi, India

Shriya Thakkar and Rituparna Bhattacharyya

Abstract The sudden spurt of rural to urban migration has left India grappling with the intermittent issue of over urbanization leading to sprawling slums that are characterized by congestion, multi-dimensional poverty, overcrowding, insecure residential status and inadequate resources. Women, who contribute indispensably toward internal migration often live by the edges and margins of higher risks associated with lack of sanitation and ill health. Additionally, the challenges they confront in their everyday lives within cities are manifold compared to men. While sanitation and hygiene in India have remained the cornerstone of multiple discourses, little has been focused on the contention of important issues such as violence against women (VAW), migratory advancements, sanitation-related psychosocial stresses and their implications on women. Through narratives and qualitative interviews, this research draws on the lived experiences of women dwelling in the slums of Delhi. It attempts to capture the multiple levels of vulnerabilities enveloped within the beguiled course of migration that these women often come across in the slums of urban cities, and, how, such stories essentially influence in shaping their identities as well as deploying coping mechanisms to tackle the myriad customary challenges and stresses that lie ahead of them.

Keywords Vulnerability · Women · Slums · Health · Sanitation · Violence · Delhi

---

S. Thakkar (✉)
Louisiana State University, Baton Rouge, LA 70803, USA
e-mail: sthakk1@lsu.edu

R. Bhattacharyya
Senior Fellow Advance HE (Formerly Higher Education Academy), London, UK
e-mail: rituparna.bhattacharyya@accb.org.uk

Centre for Indian Knowledge Systems, Indian Institute of Technology, Guwahati, India

Research Consultant and Editor-in-Chief (Joint), Space and Culture India, North Shields, UK

Department of Earth Science, University of Science and Technology, Meghalaya, India

Gender and Violence, Routledge, Taylor and Francis, Milton Park, UK

S. Anand et al. (eds.), *Sustainability in South Asian Cities*, Advances in Geographical and Environmental Sciences, https://doi.org/10.1007/978-981-97-7455-5_8

8.1 Introduction

With a recent spurt of urbanization and increased spatial variability, urban India is observing a steady upsurge in terms of infrastructure, housing, water, and food security challenges. The causes of urbanization have been attributed to the much known and obvious factors relating to increased internal labor migration from rural to urban India (Bhagat and Mohanty 2009; Datta 2007; Chandrasekhar and Sharma 2015; Mukherji 2002; Jaysawal and Saha 2014; Kapoor and Thakur 2020). A considerably large segment of this migrant population remains confined to the informal sector of work and employment activities wherein earlier studies have found that the incidence of migrants are the highest within the bracket of below poverty line (BPL) households (Gupta and Mitra 2002; Kone et al. 2018; Deshingkar 2006). The labour migrant crisis that emerged few days after the start of the lockdown on 24 March 2020 further reinforce the BPL status of the households (Bhattacharyya et al. 2020).

Women essentially comprise a greater proportion of this migrant population wherein their out-migration could be related to causes pertaining to better income and employment opportunities, maximization of their labor power for higher productivity or to join their spouses post marriage in urban cities (Breman 1996; Mosse et al. 2002; Singh et al. 2016; Mazumdar et al. 2013). According to recent statistics, approximately 70 percent of women cited 'marriage' as the predominant reason for migration whereas 'work and employment' topped the list for migration among men (Bansal 2016).

This steady rise in urbanization from 27.8 percent in 2001 to 31.1 percent in 2011 (Census, 2011) is often accompanied by the restructuring of urban boundaries to accommodate the increasing migrant population, and, hence, urbanization and growth of slums is often observed as a gradual and simultaneous process (Chimankar 2016). With the rising migrant population essentially being confined to slums, a vast number of studies have been conducted on the slum dwellers stating the multiple challenges in terms of their poor living conditions, sanitation, hygiene, water insecurity and other plausible vulnerabilities associated with their living (Hossain 2008; Babu et al. 2010; Datta 2007; Murthy 2012; Nijman 2012; Agarwal et al. 2007). There has been no standard worldwide definition of 'slum' as the type and nature of slums differ considerably from cities, nations, and states and city (O'Hare et al. 1998; Ministry of Housing and Urban Poverty Alleviation 2008; UN 2014). According to some researches, it is believed that slums can be more heterogeneous in nature than it is often assumed (Goli et al. 2011; Chandrasekhar and Montgomery 2009; Agarwal and Taneja 2005; Nolan 2015).

However, the UN-HABITAT operationally defines a slum household as a 'group of individuals living under the same roof in an urban area who lack one or more of the following five amenities": (1) A durable housing (a permanent structure that protects against extreme climatic conditions; (2) Living space of sufficient nature (allows a maximum of three people to share the same room); (3) Easy access to pure water (should be available at sufficient amounts at affordable rates); (4) Sufficient access to clean sanitation (could be in the form of public or private toilets which can be shared by a suitable number of people); (5) A security of tenure (in order to prevent forced evictions) (UN-HABITAT 2006/7).

Within contemporary India, the notion and names of slums vary by and large given the regional and linguistic differences or diversities from one state to another. Slums and clusters are commonly referred to as JJ Colonies (*Jhuggi Jhopadi* colony) in Delhi whereas they are popular as '*Jhopadpatti or chawls*' in Mumbai.[1] Likewise, other states have associated slums with various names such as, '*cheris*', '*bustees*', '*ahatas*', and '*keris*' in Chennai, Kolkata, Kanpur, and Bengaluru, respectively. Nevertheless, dilapidated and tottering structures with no toilet facilities and lack of basic hygiene and sanitation, deficient and open drainage systems with inadequate waste disposal management mark the customary sight and meaning associated with the term 'slums' in India.

Against this backdrop, the present research attempts to capture the multiple patterns of vulnerabilities among the female domestic workers dwelling in the slums of Delhi (Thakkar and Smiley 2024). In doing so, it seeks to reflect upon the lives of female migrants emerging as breadwinners for their family as well as their perpetual challenges of enduring their struggle within the urban slums (Thakkar 2024). Having been raised in their respective villages which warrants their homes amidst green fields, lands and vast expanses; how do these migrant workers adapt to the new environment in cities which grants them a thoroughly contrasting lifestyle as compared to their lives in villages. Further, how do they shape their lives around uncertainties of leaving their homes behind, setting up a new home and more importantly, does the process empower such women along the way? The study begins with the review of literature on challenges and notion of empowerment among female slum dwellers followed by the methodology utilized. The findings of this research have been discussed within the subsequent sections.

8.2 Review of Literature

There has been a wealth of literature focusing on the challenges and vulnerabilities of women slum dwellers who migrated to cities for employment and a comfortable lifestyle (Pryer 2017; Wood and Salway 2000; Rashid 2006; Bhagat 2017; Chant and Mcllwaine 2015; Grover and Parthasarathy 2012). Urbanization within India has been attributed to multi-dimensional poverty[2] induced migration of migrants hoarding in large numbers from rural areas to work in cities which has only led to

---

[1] Dharavi, which is considered to be one of the Asia's largest slums is located in Mumbai, Maharashtra. With a population density of over 277,136 /km^2, Dharavi is one of the most densely populated areas in the world.

[2] According to Oxford Poverty & Human Development Initiative (OPHI), multidimensional poverty encompasses the "various deprivations experienced by the poor in their daily lives—such as poor health, lack of education, inadequate living standards, disempowerment, poor quality of work, the threat of violence, and living in areas that are environmentally hazardous, among others". To read more, please visit: https://ophi.org.uk/policy/multidimensional-poverty-index/. The 2020 Global

emerging crises and issues in urban India (Mukherjee 1993; Kundu et al. 2001). This has been also termed as 'lopsided urbanization and overurbanization' wherein rural poverty complements urban misery (Datta 2007; Jaysawal and Saha 2014; Xu and Hou 2019). Other terminologies include pseudo urbanization (Yew 2012) and dysfunctional urbanization (Jaysawal and Saha 2014; Lauermann 2018) which is caused due to the congregation of the population in a few urban cities with no concomitant rise in the economic base of the cities. This, in turn, has led to the impairment within the urban infrastructure, which is predominantly marked by the lack of appropriate planning due to the unvarying surge in internal migration.

While it has been found that women dominate the rural to urban migration as part of 'marriage migration' or 'associated migration' (to accompany their migrant husbands) (Bansal 2016), few studies indicated that young women have moved on to join the migration flow into cities on their own where they aim to find employment within the manufacturing and informal sector of work in the recent times (Khoo et al. 1984; Chant 1998; Haan 1997; Mitra and Murayama 2009). It is vital to note the fundamental transformation within the lives of migrant workers when they move cities which involve the challenges of setting up a new home away from home and securing employment; further, the challenges in the case of women are manifold. Slums within developing nations such as India are characterized by poor living standards, illegal land tenure, exiguous infrastructure and low quality of housing stock (Gulyani and Talukdar 2008). Additionally, women of the households are predominantly responsible for ensuring the availability of fresh drinking water, arranging the charcoal or firewood for cooking hearth etc. Scarcity of freshwater is a customary problem within the slums and women tend to bear the brunt of it as they end up spending time on collection of water, ensuring safe disposal of waste and garbage as well as arranging fuelwood for the undisrupted provision of energy (Floro and Swain 2013; Parikh et al. 2015; UNDP 2006a, b; see also, Singh 2014; 2015).

The lack of proper sanitation in slums and rural areas has been one of the longstanding problems that India faces till date. To tackle the issue of sanitation and hygiene, the Government of India launched the Swachh Bharat Abhiyaan in 2014 (Clean India Mission) which aimed to construct around 120 million toilets in rural as well as urban India and establish an open defecation free (ODF) India by 2022. This ambitious development intervention, however, accounted for failures in multiple regions due to the nature of its top-down approach wherein the schemes are undertaken by the government were centrally endorsed and implementation strategy was from the central to the local level as opposed to vice versa (Thakkar 2016). Despite several sanitation campaigns undertaken by the government during the past years such as Total Sanitation Campaign (TSC) in 1999, Nirmal Bharat Abhiyaan (NBA) in 2009 and Swachh Bharat Abhiyaan (SBA) in 2014 within India; sanitation, unfortunately remains one of the worst problems when it comes to urban (slums) and rural India.

Multidimensional Poverty Index (MPI) developed by OPHI and UNDP measures the complexities of poor people's lives, individually and collectively, each year. http://hdr.undp.org/en/2020-MPI.

The interlinking of water, sanitation and hygiene (WASH) with individuals' health, status, education and socioeconomic development has been well-researched and documented (Prakash et al. 2024). Khosla (2000) has explained the necessity of functional toilets for the urban poor (especially women) as a matter of immediate urgency given the need to preserve a woman's safety and dignity as a priority apart from other health concerns (Koner 2018). Thus, when it comes to issues as above mentioned, the challenges faced by women and girls become twofold as its adverse impacts on chronic stress, mental health, and psychological wellbeing cannot be discounted. Factors such as privacy and time of day often contribute to sanitation-related stress (Sultana 2011). Further, women may have to develop various coping mechanisms to tackle violent or sexual assault and rape during private activities relating to sanitation (Abrahams et al. 2006; Massey 2011a, b; Koner 2018). A study in Bangladesh found that women have to undergo constant struggles with insufficient access to resources, social status and social conflicts, and negotiating the fabrication and physical environment for sanitation activities (Panda 2007; Singh 2006; Sultana 2011). Likewise, Hirve et al. (2015) document that the stresses caused among the seasonal migrant workers within slum dwellings practising open defecation were predominantly associated with lack of privacy due to vehicular and human traffic (Bhattacharyya 2019). Although public toilets have been made available for the slum dwellers in large numbers, it has been found that they happened to be even more dangerous as easy bacterial transmission led to more ailments (such as urinary or reproductive tract infection (UTI/RTI) among the women making them more vulnerable to poor health (Nagpal et al. 2019; Koner 2018).

According to the Millennium Development Goals (MDG) statistics of India, 2010, adequate water supply and improved drinking water supply sources were available to approximately 97% of the urban residents, but access to sanitation was limited to only about 30% (WHO/UNICEF 2012). The situation concerning access to sanitation is still far from better in the contemporary times as well wherein the women invariably raised concerns against poor sanitation in the slums. Likewise, the above estimates could be somewhat distorted as there lies severe speculation as to what can be constituted as "improved" (UN 2010). Water quality predominantly affects individuals' body mass index (BMI) (Yousefi et al. 2019; Yoshida et al. 2004). Seldom, the water supplied by the municipal water in slums is unfit for household chores and consumption (USAID 2016). Women have also complained of the presence of a high quantity of zinc in the water available to them in parts of Rajasthan that have severely exacerbated hair loss and damage to the skin if used frequently (Nagpal et al. 2019; Thakkar 2016).

The locus of gender-water discourse invariably lies in proper water usage within the domestic and productive spaces. Sultana (2009) argues that negotiation of water-related conflicts and disputes involved discourses that theorized, restructured and reshaped gendered identities, which inevitably strengthened and challenged traditional gendered access to water. Further, how it is often considered disgraceful for women in Bangladesh to venture out to get water from public sources. Chen (1995) identified how some households restricted women's mobility considering it a threat to the family's reputation. Thus, such households had a higher likelihood of equally shouldering the responsibility of water collection within the men and women or men taking this responsibility upon themselves altogether. Another significant problem is the alarming spread of diseases and health problems due to population congestion resulting in the growth of squatter settlements (Srivastava and Singh 1996). Further, sexually transmitted diseases (STDs), extramarital affairs, alcohol and drug abuse (among men especially) are some of the other perpetual challenges that slum dwellers are prone to and oftentimes, women and children are worst hit during such adverse circumstances (Barker and Das 2004). Consistent probing enabled the women to open up more during interviews wherein they mentioned how extramarital affairs, alcohol and drug abuse are rather considered mundane in the slums and women are habitually attuned to it.

The interviews with the participants are elucidated in the subsequent sections, and the following section discusses the issues linked to methodology.

8.3 Methodology

8.3.1 The Nature of Study

Through exploratory and qualitative fieldwork, the research attempted to capture the narratives and an in-depth understanding of the challenges of the migrant women working as part-time domestic worker/helper and residing in the slums and JJ (*Jhuggi Jhopdi*) colonies of Delhi. All the participants contributed to the household income as well as also shouldered responsibilities and chores of the household. The stories and experiences were captured through in-depth interviews, case studies and free-flowing conversations (Bhattacharyya 2016; Das et al. 2020; Kaur and Garg 2010). Amassing the complexity of social relationships of the researched is derived through the rich profusion of in-depth interviews (Das et al. 2020; Limb and Dwyer 2001; Bhattacharyya and Vauquline 2013), in this research, in-depth interviews were conducted to develop deeper insights into the complexity and density of contextual and multidimensional factors that impact the lives of slum dwellers invariably.

8.3.2 Data Collection

This research deployed qualitative fieldwork, which was conducted in Delhi during the months of March–April 2018. More significant assistance and guidance from two major local Non-governmental Organizations (NGOs) was solicited to gain access to the participants. The missions of both the NGOs aim at empowering women financially and economically within the informal sector through an integrated approach comprising of advocacy, organising, livelihood strategies and skillsets, education, social security and health and community microfinance in the form of training, vocational courses and classes, advocacy skills programmes, self-help groups (SHGs), and self-microfinance. It was during one of those training where potential encounters and connections with the participants were established wherein in-depth interviews were conducted in several sessions after taking the consents and following the WHO recommended ethical guidelines (WHO 1999). The groups comprised women who predominantly belonged to the JJ colony of Madanpur Khadar and slum of Sanjay colony in Delhi. Twenty in-depth interviews (Das et al. 2020) were conducted among the female domestic workers within the age group of 20 to 40. The interviews were audio recorded in Bengali and Hindi, which were later translated and transcribed into English following the nuances of translation and transcription (Das et al. 2015; 2020; Mukhtarova et al. 2019).

8.3.3 Data Analysis

The interview transcripts were carefully reviewed followed by a thematic analysis wherein specific categorizations enabled the understanding of crucial factors like cultural, familial and individual which tend to play a vital role in reshaping the lives of the women and their interplay in determining the subjective meanings of empowerment and the ability to exercise one's choice within their (women) contemporary livelihood.

Genealogies and life histories enabled the researchers to gather and trace a broader understanding of various migratory practices within the family of the participants that had been ongoing or prevalent since long. The names and ages of the respondents referred to in this research are through pseudonyms and identities are kept strictly confidential to maintain anonymity and privacy of the participants.

8.4 Everyday Struggles of Migrant Female Slum-Dwellers of Delhi

The interviews and fieldwork evoked four specific themes and individual responses concerning gendered violence and psychosocial stress among the female slum dwellers of Delhi: first, how migration has impacted and continuing to impact the lives of the women and their respective families. Second, it elicits the ongoing struggles and stresses emerging from inadequate sanitation and prolonged utilization of public toilets leading to bad health conditions and chronic illnesses; third, the ongoing battling of domestic violence at home as well as shouldering the responsibilities of domestic and public spaces, including employment and income; and ultimately, the participants share their customary challenges with water and electricity within their homes.

8.4.1 Migratory Experiences

Most of the women dwelling in the slums with their families seemed to have migrated from their respective villages in the hope of better employment and other opportunities. However, some were among the next generation, wherein their parents had migrated for (more or less) similar reasons and decided to settle in the city. Although the narratives of some of those women were quite overwhelming, wherein they discussed their paths to migration and the expansive transformations that followed right after their movement to the city, others stated how the process of migration was rather gradual for their families, and they have conveniently accustomed to their lives within the slums of the city.

Rita (25) described that she was relatively young when her parents migrated to the city as they were deluged in debt and migration seemed like the solitary refuge which could lift the burden of debt off their shoulders.

> our village is in the district of Moradabad, Uttar Pradesh and we visit our village every now and then. My father had taken an amount of INR 2 lakhs (2500 USD) to start a small business in the village from one of the moneylenders at a very high rate of interest. For unexpected reasons, the business could not flourish, and thus, we were left with a hefty loan that kept accumulating over the days. Soon after, demands, warnings and threats from the moneylender and his goons followed, and we lost our night's sleep over it. There was not enough work in the village and constant threats made our living almost impossible, forcing us to leave the village in search of work and repay the loan amount at the earliest. We have been living in the slums of Sanjay Colony ever since, where my father and two younger brothers work as contractual laborers and I work as a domestic worker. We have been able to successfully pay back the loan amount to the moneylender and wish to go back to our village sometimes

Rita mentioned that their joint incomes are enough to sustain their livelihood in the city as well as contribute to their savings. She explains that her parents have been planning for their return since long now and they keep reminding themselves that they would leave the city in the next year or two, however, that never really materialized. Back in their village, they have extended families and relatives as well as their small house which is presently being inhabited by her grandparents.

Where Rita seemed to have spent most of her life within the lanes of the slums, Renuka (30) stated that she moved here about six years ago. Some family debts accompanied by an urge to live the 'city life' motivated them to migrate to Delhi from the district of South 24 Parganas, West Bengal. The process of migration was rather easy and smooth for Renuka and her family as they received continuous support and help from a close friend (Ramesh) who also assisted them in getting a small room within the slums of Sanjay colony. However, the transition from her village life to the so-called urban city life was extremely difficult and draining. This is how Renuka recalls it:

> the transition from my home till here was a difficult and stressful one. Back home, we have some land, an ancestral house, family and facilities as such. There was not much to do though except helping out in the field and managing the household chores; nevertheless, life was simple. However, being relatively young, we found it monotonous and unrewarding

> whereas, Ramesh, on the other hand, did very well in the city. He earned enough to sustain himself as well as send money to his family each month. We approached Ramesh, and he promised to help us out – had it not been for him, we might never have been able to get ourselves out. We were advised against our decision by our families, but we still decided to take a chance. The slum life, congestion, lack of space, hygiene, pollution and filth was enormously overwhelming at first as compared to what we had back in our villages. Getting accustomed to this lifestyle and place was rather gradual, and we thought of giving up and returning to our village many times during our initial days, but that never happened, and today, I feel fortunate that we decided to stay back and fight.

Renuka works as a part-time domestic worker, and her husband works as a full-time driver for one of the private companies around the area. They have been able to save up enough, which would have otherwise never been possible if they were to stay in their village. Although Renuka misses the comfort of her home and village habitually, she believes that their struggles, busy lives and incomes are quite appealing and enough to keep them going in the city. She intends to purchase some land back home and eventually return to her village in some years.

Albeit the causes of migration essentially remain rooted to factors such as steady income and employment, opportunities, experiences of *city life,* temporary space of freedom and escape, independence (Shah 2006) and familial debts/feuds; however, the transition of spaces and the associated individual experiences tend to be highly differential and subjective. Where on the one hand, there were women like Renuka who left their house and its comforts for the struggle in the city or fled to the city to earn and repay off the debts back home and, on the other, there were women like Rita who were raised within the slums as an upshot of early migration initiated by their parents. Herein, lived experiences of the women and families who have been long term migrants in the city have been accustomed to life within slums. In contrast, recent migrants tend to reveal characteristics of withdrawal marked by potential reverse migration.

8.4.2 Health and Sanitation

Lack of sanitation and poor hygiene appeared to be one of the causes of tremendous concern among women within the slums. While women belonging to the JJ colony complained less about it, conditions in the Sanjay Colony slum were deplorable. There was a total of three community toilet complexes (CTC) in the area and some other public toilets (PTs) which are paid toilets run by a private contractor. Savitri (38) explained that the inhabitants prefer the neighbouring forest areas as the CTCs are filthy and extremely foul-smelling with no windows and a lack of proper ventilation. She further continues:

> Along with my daughter, I have contracted UTI (Urinary Tract Infection) multiple times, and each woman I know have contracted it at least sometime or the other - it is pervasive among the females in this slum. The previous time my daughter suffered from UTI, it became very serious as it went untreated during the beginning and the medicines prescribed by the local doctors could not cure it—we ended up taking her to a private doctor and spent a huge sum

> of money for her treatment. These are added expenses, and almost every household ends up spending a lot of money on medical bills and health treatments each month. Toilets lack water and people do not clean the toilets after their use, thus, if one wishes to use these toilets, one must clean up the latrines first and then use it for themselves. But who would wish for such a sight early morning? We have demarcated areas for men and women in the forest, but safety is the foremost concern when it comes to women. Unlike men, we do not have the luxury to go anytime and thus, we usually divide ourselves into groups and go together very early in the morning. Previous incidents of sexual harassment and rape have also been reported in the past, and thus, we usually avoid the forest after dusk.

According to Savitri, the challenges faced by women are often shadowed, and very few people have been vocal about it. She also discussed the case of her friend's daughter who was harassed by one of the local boys who managed to record a video while she was in the shower of the CTC with a broken door. People had earlier complained of the broken latches and doors of the showers and toilets (especially for the ladies), but no actions were initiated to fix them properly.

> We have also heard that the boys break it deliberately, however, there is nothing much that we can do unless we have a strong support from (at least few) the men which we clearly do not.

Anita (31) mentioned that congestion and insanitary conditions have led to multiple health problems among the slum dwellers, and further, it becomes incredibly detrimental if there is a prevalence of a disease that is contagious. She adds:

> We attempt to distance ourselves from people with illnesses and contagious diseases as it is very common in the slums to contract diseases from one another. A family member from one of the nearby rooms suffered from tuberculosis (TB), and they visited private hospitals for his treatment. Since then, his family contracted it from him, and now, many families in the slum have TB, some have even lost their lives to TB. During the monsoon, the poor drainage system and water-logging within the slums call it a breeding ground for mosquitoes and dengue fever is quite common, especially among the children. Thus, health and illnesses tend to be the primary concern for the inhabitants living here, and we have no choice but to deal with them due to the living conditions that we endure. Life is not easy for us at all.

The participants expressed their disappointment over the lack of care by the state to ensure the good health and wellbeing of the slum dwellers. Further, they complained of the lack of proper toilet facilities for girls in the local schools as well which often tend to be an excuse for the girls for skipping schools; thus, the challenges encountered by women and children in the slums are multiple as compared to men. The women articulated the need for basic sanitation and hygiene for themselves or an action plan to delegate the responsibilities, authority and finances to women which would enable them to form councils and undertake the control of ensuring safer public health facilities to fellow women.

8.4.3 Domestic Violence and Employment

Herein, women were yet again, working twice as hard compared to the men wherein they were responsible for the smooth functioning of the family and household as well as stepping out to earn for the family income.

Niti (38) is a domestic worker who currently works for four families with a monthly income of INR 7000 (USD 100) believes that life is traditionally more challenging for women in slums. She describes the typical state of affairs within every household in and around where women have been survivors or victims of domestic violence. She further adds:

> We women must work very hard to keep the family going. The customary course of action after migration into urban cities is often followed by alcoholism among the men, which I believe, is the primary cause of domestic violence. A portion of their monthly salary goes into alcohol, and women end up bearing the consequences of this newfound practice. My husband never raised his hand on me earlier; however, he has started doing so very recently—I believe this is an impact of his new circle of friends whom he socializes with. When he is sober, he is a very different person as opposed to when he is under the influence of alcohol—it seems he is influenced by his friends who are wife beaters as they consider it a symbol of masculinity.

Likewise, Rekha (34) explained her ongoing struggles as a mother and breadwinner for her family as well as enduring domestic abuse and molestation by her husband at one point in time. However, she states that she has been separated from her husband lately.

> It all started with my husband's alcoholism which soon turned into a nightmare for the family. I was beaten up almost every night by him, and eventually, he stopped providing for the family as well. Later, he confessed that he was in love with some other woman who lives down the lane from here and he abandoned us to live with her. I have been single-handedly taking care of my two children till then, lately, he has also stopped visiting and cut all ties with us.

Rekha mentions how such extramarital affairs are also common in certain households wherein few of the men have married women in the cities despite having a wife and family back home. In such cases, the wives residing in the villages do not have the slightest inkling about their husbands' extramarital affairs or marriages.

Such cases of domestic violence which is rife within the slums often stem from congestion and the confines that people are restricted to, which include large families occupying a small room. Lack of space limits the privacy of individuals which leads instances of frustration and aggression among men, ultimately leading to domestic abuse among women.

8.4.4 Experiences of Water and Electricity in Slums

When it comes to water and its inadequate supply, the slum dwellers have a harrowing time coping with it. One of the respondents, Renu (29) mentioned that the circumstances are worse in the slums as compared to her village concerning the lack of water. She further adds:

> Back in my village, we had two tube wells and one borewell at a distance from our house, and the women would often gather to fetch water twice a day. The women longed for that time of the day as the activity of fetching water was observed as a moment of recreation and socialization with fellow mates wherein everyone discussed their problems or daily routine activities to each other. However, the situation is not so similar in the slums as there is never enough water in our homes, and we end up purchasing it from the tankers by Delhi Jal Board (DJB). Further, the inhabitants push and fight their way ahead in the queue to receive water, and it is never an easy task. It is almost a struggle.

Renu delineates that there are issues relating to the distribution of water as well. Some of the lanes (*galis*) receive more water as compared to other lanes. Influential people manage to receive water regularly as opposed to others. Lately, her husband has restricted her mobility outside the house for fetching water as he feels responsible for undertaking the chore of fetching water for the household. Moreover, he feels very protective of Renu and he believes that fetching water from the tankers in the slums involves verbal spats which might even lead to possible argumentations with the fellow slum dwellers and thus, as a "man" of the household, it should be rather him who should tackle such feuds rather than Renu. As Renu remarked:

> Such role reversals in contemporary times have been also observed in some other households. For instance, my friends have also lately complained of similar concerns being raised by the men of their households and lately, we observe more men than women who queue up to fetch water from the tankers regularly

Niti mentions that electricity and water have explicitly been challenging for the inhabitants of JJ and Sanjay colony. She explains that there is a regular power cut every day for about 3 h and there are no individual meters owned by households, instead, a common meter for the aggregate of all households which is installed by the contractor.

> We have common contractors who take care of electricity and power supply within different parts of the slum. Since most of us are migrants, we tend to go home (*desh/gaon*) for a long vacation at least twice a year. However, we still have to pay a minimum electricity bill during our absence. This is because we do not own individual meters, and thus, we are forced to pay electricity even if we are using it. We have complained and protested multiple times, but all our appeals have gone into complete vain as nobody is ever willing to listen to us. The contractors are always attempting to make money from the poor, and we are unfortunately suffering.

Other respondents shared similar concerns wherein they mentioned the disparities that exist when it comes to power supply between different lanes (*galis*) within the slums. Some complained about the lack of consistency in receiving bills, whereas others mentioned higher monthly charges and regular power cuts. When prompted if the above issues have been ever addressed on a wider platform, the women respond by stating that they are invariably given false hopes and claims by multiple NGOs and welfare associations who often visit the slum all year round to gather interviews and case studies, however, there has been no substantial change. Lately, they have lost all hopes for a change.

8.5 Discussion

According to an estimation, there are approximately 1,080 slum clusters which occupy the sprawling megalopolis that Delhi is gradually on its way to becoming (Siddiqui and Pandey 2003). These slums are responsible for providing shelter to 1.5 million people and another 1.5 million have made their homes in 46 resettlement colonies (Kumar 1996). Despite multiple development interventions undertaken by the government, perpetual efforts by the United Nation (UN) to uplift the global sanitation crises through an established framework such as Millennium Development Goals (MDGs) and Sustainable Development Goals (SDGs), the current condition and setting in urban slums seem far from better. A new report suggests that 732 million people do not have access to toilets which is, evidently, the highest number of people with India having the second largest population in the world (Salve 2017). In another report by WaterAid, it is estimated that 355 million women and girls have been survived with no access to toilet facilities. Contemporarily, India has been fighting hard to reduce open defecation and improving basic sanitation and thus, some amelioration can be observed in the statistics wherein population without access to basic sanitation fell from 78.3% in 2000 to 56% in 2015 (WaterAid 2017). Lately, according to the claims made by the Indian government, household toilet coverage across India has risen from <30% in 2014–15 to 91–100% in 2019–20 under the Swachh Bharat Mission.

However, the narratives and individual cases foreground the ongoing struggles which slum dwellers (especially women) are continually grappling with and, essentially, points to a critical notion of sheer delinquency when it comes to addressing the needs of women. The participants stress their concerns on the failure of various initiatives such as Clean India Mission in their respective villages and typically how many such programmes fail to account for the cases of women. McFarlane (2008) conducted her research on sanitation within the informal settlements of Mumbai and discussed the varied preferences of men for the use of toilet blocks in the morning, whereas women were given no such benefits. On the other hand, women were expected to finish all their household chores and wait until evening to be able to use the toilet facilities. The plight, however, does not end as sexual harassment and psychosocial stresses further exacerbate the situation for women. One of the participants explained how a basic human need such as sanitation and hygiene could be identified as a state of fortune for some women, which the privileged few would often never understand. Further, younger women are invariably more prone to harassments and rapes as opposed to older women (detailing one of the horrific rape incidents that occurred some years ago wherein a young migrant woman in the slum was raped at night while she was out in the forest to relieve herself). The above finding has been supported by other scholars and research studies which claim concerns relating to the vulnerability of sanitation-related stress among the newly married women in India who are specifically targeted by men (Johnson and Johnson 2001; Koner 2018; Sahoo et al. 2015).

While policy recommendations on sanitation have underlined the growing implications of health among children under the age of five, the focus has been lately shifted to substantial threats posed by unique health risks among women and girls due to causes relating to inadequate sanitation and inferior hygiene, including poor infection control and unsatisfactory birthing practices (Cheng et al. 2012); urogenital tract infections (Mudey et al. 2010) and chronic constipation (Sahoo et al. 2015; Fisher 2006). Moreover, as discussed earlier, poor access to sanitation and hygiene among women has prompted critical discourses and its linking with psychosocial stress in other low- and middle-income countries such as Bolivia and Ethiopia (Stevenson et al. 2012) coupled with a heightened risk of sexual violence in schools of South Africa (Abrahams et al. 2006), and fear of harassment as well as sexual assaults within the slums of Delhi and other parts of India (Bhattacharyya 2016; Lennon 2011; Koner 2018) and Kampala, Uganda (Massey 2011a, b). Thus, crucial initiatives have been undertaken by the UN to address the issue of gender-responsive sanitation approaches which foregrounds the urgency of physical, social, and environmental needs of women (Tilley et al. 2013), development and planning of gender-sensitive programming to improve dignity and promote equality (Diczfalusy 1997; Gender and Water Alliance and UNDP 2006b, a), as well as an immediate demand for gender-disaggregated data (UN 2008; Sahoo et al. 2015). Identifying the perennial repercussions of inadequate sanitation and resources for women and children which stretches beyond the physical and social need to emotional-mental health and wellbeing is a crucial step towards empowering women and calls for a timely, contextualized understanding of sanitation-related psychosocial stress within countries with inadequate sanitation facilities.

The current research does not simply underscore the lived experiences of women within an urban slum, but, at the same time, opens up discussions on their newly defined roles and duties as a migrant worker, wife, mother and slum-dweller—essentially how do these women balance their lives and grapple with the challenges that surround their homes and employment. (Thakkar 2024, 2021; Thakkar and Smiley 2024) As discussed earlier, migration has invariably been associated to successes and remittances, source of livelihood, better quality of life, education, reduce risk exposure and more; which typically had discourses associate migration with 'greener pastures' (Poeze 2010; Lu 2002; Uggla 2015; Sward 2016). However, less has been explored about the vulnerabilities associated with migration. There remains a dearth of literature when it comes to discourses concerning how migratory patterns have led to changing household structures and dynamics, shaped peoples' adaptive capacities and abilities and most crucially, how have women adjusted themselves to the changing environment (Singh 2019; Thakkar 2021). Migration has been equally responsible for exacerbating vulnerability as it often pushes migrants into unwanted precarious and irregular livelihoods (Deshpande and Bazaz 2017; Chandrasekhar and Mitra 2018; Nguyen and Locke 2014; Samaddar 2016), entailing higher costs of living (De Haas 2007) and placing insurmountable pressures on those left behind or moving with them (Bhatta et al. 2015; Bhagat 2017). The cases in the current research depict women who were passive followers in the process of migration presented a mixed response detailing how migration served more of a bane than boon wherein the shifting household dynamics, structures and responsibilities impacted women more than the men. Along with an improved standard of living and fixed source of income in the slums, comes the various risks and psychosocial stresses associated with poor hygiene, congested spaces, inadequate sanitation, poor water and power supply, and consequentially, the vulnerable status of women who are more prone to infections and illnesses within the slums. For instance, TB, which is known to have higher rates of transmission within urban areas of India as opposed to the rural, and, within urban areas, is known to have high transmission rates in slums (Suganthi et al. 2008). Khan (2012) studied the aspects related to TB through a gendered lens and how the social constructions and discourses surrounding TB stigmatized women more than men in the slums of Delhi. She explains how this stigma is highly manifested through everyday interactions that reinforce discrimination and prejudices against women wherein the parents are particularly unwilling of their son's marriage to a girl who is suffering or has suffered TB in the past for fear of passing the disease to future progeny.

Furthermore, studies reveal that women residing in slums are invariably more prone to other infections such as UTI and RTI which can be attributed to causes relating to lower socioeconomic conditions, poor utilization of health care and inadequate personal hygiene (Dasgupta and Sarkar 2008). Often, such infections are never even taken too seriously by the women, as it has been observed from the responses gathered from the participants because of the mistaken belief that RTIs are not fatal and it is often neglected due to higher costs for its treatment. Cheap home remedies are utilized to treat these infections with no expert supervision, and women tend to conceal it from the men to escape the embarrassment and humiliation.

Amidst all this, women complained of domestic violence which is broadly prosaic among the slum-dwelling population. Pandey et al. (2009) note how majority of the population residing in slums develop advanced level of physiological and psychosocial stress due to low income, unsanitary conditions, insufficient dietary conditions, social instability and insecurity, deficient hygiene and sanitation. The effect impacts women making them prone to domestic violence and abuse due to promiscuity, alcohol and drug abuse, child labor, prostitution, and a mix of other conditions (Holmes and Rahe 1967). According to Sinha et al. (2012), psychological violence was more prevalent among slum-dwelling women of Kolkata as opposed to physical violence. Donta et al. (2016), in their study on the association of domestic violence from husbands and women empowerment in the slums of Mumbai, remarked that domestic violence is independent of women empowerment and there seems to be no correlation between the two. Likewise, the current research highlights cases of those women who are survivors of domestic abuse despite emerging as the breadwinner of their households (Thakkar 2018). Instead, aberrant behaviors among men such as alcoholism, extramarital affairs, temperamental attitude, erratic contribution to household expenses, concealment of facts can be correlated to heightened feelings of low self-esteem and insecurity due to adverse socioeconomic conditions borne amidst poverty (Babcock et al. 1993). Women are, unfortunately, scapegoats in the system where they tend to endure abuse and violence met as part of the compensatory behavior of men for the fulfilment of their lack of power in other areas of marriage.

8.6 Conclusion

The present study is a modest attempt to address the everyday struggles of migrant slum dwellers through a gendered lens. It identifies the plethora of challenges and duties that eventuate upon women as a result of migration to urban cities and how these women exceptionally choose their actions to deal with them. The common ground within the women's narratives about health and sanitation can enable the recentering of Gender and Development (GAD) discourses through a critical lens of gender sensitization as well as assist policymakers to develop frameworks prioritizing sustainable sanitation through a bottom-up approach. The complexities of out-migration are carefully explored, which sets the tone for vulnerabilities, insecurities and perils of precarity, which will be a meaningful contribution to the existing literature. Furthermore, the findings of this research could lay crucial groundwork for policy researchers, human rights activists and practitioners to raise awareness programmes against domestic violence, develop crucial strategies and sensitize the youth through a holistic approach.

References

Abrahams N, Mathews S, Ramela P (2006) Intersections of 'sanitation, sexual coercion and girls' safety in schools. Trop Med Int Health 11(5):751–756. https://doi.org/10.1111/j.1365-3156.2006.01600.x

Agarwal S, Taneja S (2005) All slums are not equal: child health conditions among the urban poor. Indian Pediatr 42:233–44. https://ssrn.com/abstract=2753358

Agarwal S, Satyavada A, Kaushik S, Kumar R (2007) Urbanization, urban poverty and health of the urban poor: status, challenges and the way forward. Demogr India 36(1). https://ssrn.com/abstract=3133050

Babcock JC, Waltz J, Jacobson NS, Gottman JM (1993) Power and violence: the relation between communication patterns, power discrepancies, and domestic violence. J Consult Clin Psychol 61:40–50. https://doi.org/10.1037/0022-006X.61.1.40

Babu BV, Swain BK, Mishra S, Kar SK (2010) Primary healthcare services among a migrant indigenous population living in an eastern Indian city. J Immigr Minor Health 12(1):53. https://doi.org/10.1007/s10903-008-9181-z

Bansal S (2016) 45.36 Crore Indians are internal migrants. The Hindu. https://www.thehindu.com/data/45.36-crore-Indians-are-internal-migrants/article16748716.ece
Barker G, Das A (2004) Men and sexual reproductive health: the social revolution. Int J Men's Health 3(3):147–155
Bhagat R (2017) Migration, gender and right to the city. Econ Polit Wkly LII:35–40
Bhagat RB, Mohanty S (2009) Emerging pattern of urbanization and the contribution of migration in urban growth in India. Asian Popul Stud 5(1):5–20. https://doi.org/10.1080/17441730902790024
Bhatta GD, Aggarwal PK, Poudel S, Belgrave DA (2015) Climate-induced migration in South Asia: migration decisions and the gender dimensions of adverse climatic events. J Rural Community Dev 10:1–23
Bhattacharyya R (2016) Street violence against women in India: mapping prevention strategies. Asian Soc Work Policy Rev 10(3):311–325. https://doi.org/10.1111/aswp.12099
Bhattacharyya R (2019) Child sponsorship: exploring pathways to a brighter future. Prog Dev Stud 19(1):83–84. https://doi.org/10.1177/1464993418799895
Bhattacharyya R, Vauquline P (2013) A mirage or a rural lifeline? Analysing the impact of Mahatma Gandhi rural employment guarantee act on women beneficiaries of Assam. Space Cult India 1(1):83–101. https://doi.org/10.20896/saci.v1i1.10
Bhattacharyya R, Sarma PK, Nath M (2020) COVID-19 and India's labour migrant crisis. Int J Innov Creat Change. Special Edition: COVID-19 Life Beyond 14(6):243–258. www.ijicc.net
Breman J (1996) Footloose labour. Cambridge University Press, Cambridge
Chandrasekhar S, Mitra A (2018) Migration, caste and livelihood: evidence from Indian city-slums. Urban Res Pract:1–17. https://doi.org/10.1080/17535069.2018.1426781
Chandrasekhar S, Sharma A (2015) Urbanization and spatial patterns of internal migration in India. Spat Demogr 3(2):63–89. https://doi.org/10.1007/s40980-015-0006-0
Chandrasekhar S, Montgomery MR (2009) Paper prepared for the international institute for environment and development (IIED). IIED, London. Broadening Poverty Definitions in India: Basic needs and urban housing
Chant S (1998) Households, gender and rural-urban migration: reflections on linkages and considerations for policy. Environ Urban 10(1):5–22. https://doi.org/10.1177/095624789801000117
Chant S, McIlwaine C (2015) Cities, slums and gender in the global south: towards a feminised urban future. Routledge
Chen M (1995) A matter of survival: women's right to employment in India and Bangladesh. In: Nussbaum MC, Glover J (eds) Women, culture, and development: a study of human capabilities. Oxford University Press, World Institute for Development Economics Research, New York, NY, pp 37–56
Cheng JJ, Schuster-Wallace CJ, Watt S, Newbold BK, Mente A (2012) An ecological quantification of the relationships between water, sanitation and infant, child, and maternal mortality. Environ Health 11:1–8. https://doi.org/10.1186/1476-069X-11-4
Chimankar DA (2016) Urbanization and condition of urban slums in India. Indones J Geogr 48(1):28. https://doi.org/10.22146/Ijg.12466
Dasgupta A, Sarkar M (2008) A study on reproductive tract infections among married women in the reproductive age group (15–45 years) in a slum of Kolkata. J Obstet Gynecol India 58(6):518–522
Das T, Bhattacharyya R, Alam F, Parvin A (2020) In-depth semi-structured interviewing: researching domestic violence as a public health issue in Bangladesh. SAGE Research Methods Cases Medicine & Health, Disciplines: Public Health. ISBN: 9781529719840, https://doi.org/10.4135/9781529719840
Das T, Alam F, Bhattacharyya R, Parvin A (2015) Causes and contexts of domestic violence: tales of the help seeking married women in Sylhet, Bangladesh. Asian Soc Work Policy Rev 9(2):163–176. https://doi.org/10.1111/aswp.12055
Datta P (2007) Urbanization in India. Indian J Reg Sci 39(1):124–133

De Haas H (2007) Remittances, migration and social development. United Nations Research Institute for Social Development, Geneva, Switzerland, p 46. https://www.imi.ox.ac.uk/pdfs/unrisd-remittances-mig-dev

Deshingkar P (2006) Internal migration, poverty and development in Asia. ODI Briefing Paper 11

Diczfalusy E (1997) In search of human dignity: gender equity, reproductive health and healthy aging. Int J Gynecol Obstet 59:195–206. https://doi.org/10.1016/S0020-7292(97)00230-0

Donta B, Nair S, Begum S, Prakasam CP (2016) Association of domestic violence from husband and women empowerment in slum community, Mumbai. J Interpers Violence 31(12):2227–2239. https://doi.org/10.1177/0886260515573574

Fisher J (2006) For her it's the big issue: putting women at the centre of water supply, sanitation and hygiene

Floro M, Swain R (2013) Food security, gender, and occupational choice among urban low-income households. World Dev 42:89–99. https://doi.org/10.1016/j.worlddev.2012.08.005

Gender and Water Alliance, UNDP (2006) Mainstreaming gender in water management. Version

Goli S, Arokiasamy P, Chattopadhayay A (2011) Living and health conditions of selected cities in India: setting priorities for the national urban health mission. Cities 28:461–469. https://doi.org/10.1016/j.cities.2011.05.006

Grover A, Parthasarathy D (2012) Relocation and changing nature of vulnerability—a case study of Krantinagar Slum, Mumbai. Horizon J Soc Sci 3:1–19

Gulyani S, Talukdar D (2008) Slum real estate: the low-quality high-price puzzle in Nairobi's slum rental market and its implications for theory and practice. World Dev 36:1916–1937. https://doi.org/10.1016/j.worlddev.2008.02.010

Gupta I, Mitra A (2002) Rural migrants and labour segmentation: micro-level evidence from Delhi slums. Econ Polit Wkly:163–168. https://www.jstor.org/stable/4411601

Haan AD (1997) Rural-urban migration and poverty: the case of India. IDS Bull 28(2):35–47

Habitat UN (2006) State of the world's cities 2006/7. United Nations, New York

Hirve S, Lele P, Sundaram N, Chavan U, Weiss M, Steinmann P, Juvekar S (2015) Psychosocial stress associated with sanitation practices: experiences of women in a rural community in India. J Water Sanit Hyg Dev 5(1):115–126. https://doi.org/10.2166/washdev.2014.110

Holmes TH, Rahe RH (1967) The social readjustment rating scale. J Psychosom Res 2:213–218. https://doi.org/10.1016/0022-3999(67)90010-4

Hossain S (2008) Rapid urban growth and poverty in Dhaka City. Bangladesh e-journal Sociol 5(1)

Jaysawal D, Saha S (2014) Urbanization in India: an impact assessment. Int J Appl Sociol 4(2):60–65. https://doi.org/10.5923/j.ijas.20140402.04

Johnson PS, Johnson JA (2001) The oppression of women in India. Violence against Women 7(9):1051–1068

Kaur R, Garg S (2010) Domestic violence against women: a qualitative study in a rural community. Asia Pac J Public Health 22(2):242–251. https://doi.org/10.1177/1010539509343949

Kapoor A, Thakur B (2020). Slum upgradation, redevelopment and relocation through slum vulnerability assessment in Delhi. In: Thakur R, Dutt A, Thakur S, Pomeroy G. (eds) Urban and Regional planning and development. Springer, Cham. https://doi.org/10.1007/978-3-030-31776-8_14

Khan KB (2012) Understanding the gender aspects of tuberculosis: a narrative analysis of the lived experiences of women with TB in slums of Delhi, India. Health Care Women Int 33(1):3–18. https://doi.org/10.1080/07399332.2011.610541

Khoo SE, Smith PC, Fawcett JT (1984) Migration of women to cities: the Asian situation in comparative perspective. Int Migr Rev 18(4):1247–1263. https://doi.org/10.1177/019791838401800417

Khosla R (2000) Women and sanitation: urban reality experiences of government programmes, Ngos and CBOs. Soc Change 30(1–2):192–207. https://doi.org/10.1177/004908570003000213

Koner K (2018) Sanitation and hygiene of Darjeeling city: a crisis for women and adolescent girls. Space Cult India 5(3):89–105. https://doi.org/10.20896/saci.v5i3.292

Kone ZL, Liu MY, Mattoo A, Ozden C, Sharma S (2018) Internal borders and migration in India. J Econ Geogr 18(4):729–759. https://doi.org/10.1093/jeg/lbx045

Kumar V (1996) Warning signs. Hindu Surv Environ 1996:67–71
Kundu A, Lalitha N, Arora SL (2001) Growth dynamics of informal manufacturing sector in urban India: an analysis of interdependence. In: Kundu A, Sharma AN (eds) Informal sector in India. Institute of Human Development, New Delhi
Lauermann J (2018) Geographies of mega-urbanization. Geogr Compass 12(8):e12396. https://doi.org/10.1111/gec3.12396
Lennon S (2011) Fear and anger: perceptions of risks related to sexual violence against women linked to water and sanitation in Delhi. LSHTM, India
Lu M (2002) 'Are pastures greener?' Residential consequences of migration. Int J Popul Geogr 8(3):201–216. https://doi.org/10.1002/ijpg.244
Limb M, Dwyer C (eds) (2001) Qualitative methodologies for geographers: issues and debates. Arnold, London
Massey K (2011a) Insecurity and shame: exploration of the impact of the lack of sanitation on women in the slums of Kampala, Uganda. LSHTM
Massey K (2011b) Insecurity and shame: exploration of the impact of the lack of sanitation on women in the slums of Kampala, Uganda. SHARE Briefing Note. SHARE Research, London, UK. http://www.shareresearch.org/LocalResources/VAW_Uganda.pdf
Mazumdar I, Neetha N, Agnihotri I (2013) Migration and gender in India. Econ Polit Wkly:54–64. https://www.jstor.org/stable/23391360
McFarlane C (2008) Sanitation in Mumbai's informal settlements: state, 'slum', and infrastructure. Environ Plan A 40(1):88–107. https://doi.org/10.1068/a39221
Ministry of Housing and Urban Poverty Alleviation (2008) Report of the committee on slum statistics/census. Government of India, Ministry of Housing and Urban Poverty Alleviation, New Delhi, India
Mitra A, Murayama M (2009) Rural to urban migration: a district-level analysis for India. Int J Migr Health Soc Care 5(2):35–52. https://doi.org/10.1108/17479894200900011
Mosse D, Gupta S, Mehta M, Shah V, Rees JF, Team KP (2002) Brokered livelihoods: debt, labour migration and development in tribal western India. J Dev Stud 38(5):59–88. https://doi.org/10.1080/00220380412331322511
Mudey A, Kesharwani N, Mudey G, Goyal R (2010) A cross-sectional study on awareness regarding safe and hygenic practices amongst school going adolescent girls in rural area of Wardha district, India. Glob J Health Sci 2:225–231. https://doi.org/10.5539/gjhs.v2n2p225
Mukherji S (1993) Poverty induced migration and urban involution in India: cause and consequences. International Institute for Population Sciences, Mumbai, pp 1–91
Mukherji S (2002) Urbanization and migration in India: a different scene. Chapters
Mukhtarova S, Karagulova B, Sideshova Z, Mukhtarov S, Yergazina A, Nasipov I (2019) Translating the untranslatable: challenges, sticking points and struggles. Space Cult India 7:250–263. https://doi.org/10.20896/saci.v7i1.421
Murthy S (2012) Land security and the challenges of realizing the human right to water and sanitation in the slums of Mumbai, India. Health Human Rights 14(2). https://doi.org/10.2307/healhumarigh.14.2.61
Nagpal A, Hassan M, Siddiqui MA, Tajdar A, Hashim M, Singh A, Gaur S (2019) Missing basics: a study on sanitation and women's health in urban slums in Lucknow, India. Geojournal:1–13. https://doi.org/10.1007/s10708-019-10088-0
Nguyen MT, Locke C (2014) Rural-urban migration in Vietnam and China: gendered house-holding, production of space and the state. J Peasant Stud 41:855–876. https://doi.org/10.1080/03066150.2014.925884
Nijman J (2012) India's urban challenge. Eurasian Geogr Econ 53(1):7–20. https://doi.org/10.2747/1539-7216.53.1.7
Nolan LB (2015) Slum definitions in urban India: implications for the measurement of health inequalities. Popul Dev Rev 41(1):59–84. https://doi.org/10.1111/j.1728-4457.2015.00026.x
O'hare G, Abbott D, Barke M (1998) A review of slum housing policies in Mumbai. Cities 15:269–283. https://doi.org/10.1016/S0264-2751(98)00018-3

Panda SM (2007) Mainstreaming gender in water management: a critical view. Gend Technol Dev 11(3):321–338. https://doi.org/10.1177/097185240701100302
Pandey GK, Dutt D, Banerjee B (2009) Partner and relationship factors in domestic violence: perspectives of women from a slum in Calcutta, India. J Interpers Violence 24(7):1175–1191. https://doi.org/10.1177/0886260508322186
Parikh P, Fu K, Parikh H, McRobie A, George G (2015) Infrastructure provision, gender, and poverty in Indian slums. World Dev 66:468–486. https://doi.org/10.1016/j.worlddev.2014.09.014
Prakash A, Kumar N, Chhatre A, Thakkar S, Dar A (2024) Navigating India's Groundwater Crisis: Legal and Institutional Perspectives on Regulation and Conservation. Water Policy, 26(8):835–855. https://doi.org/10.2166/wp.2024.123
Poeze M (2010) In search of greener pastures? Boat-migrants from Senegal to the Canary Islands (p 97). African Studies Centre, Leiden
Pryer JA (2017) Poverty and vulnerability in Dhaka slums: the urban livelihoods study. Routledge
Rashid SF (2006) Marriage practices in an urban slum: vulnerability, challenges to traditional arrangements and resistance by adolescent women in Dhaka, Bangladesh. Asia Research Institute Working Paper No. 72, https://ssrn.com/abstract=1317159 or https://doi.org/10.2139/ssrn.1317159
Shah A (2006) The labour of love: Seasonal migration from Jharkhand to the brick kilns of other states in India. Contrib Indian Sociol 40(1):91–118. https://doi.org/10.1177/006996670504000104
Sahoo KC, Hulland KR, Caruso BA, Swain R, Freeman MC, Panigrahi P, Dreibelbis R (2015) Sanitation-related psychosocial stress: a grounded theory study of women across the life-course in Odisha, India. Soc Sci Med 139:80–89. https://doi.org/10.1016/j.socscimed.2015.06.031
Salve P (2017) At 732 Million, India tops list on number of people without access to toilets: report. The Wire. https://thewire.in/health/732-million-india-tops-list-number-people-without-access-toilets-report
Samaddar R (2016) Migrant and the neo-liberal city: an introduction. Econ Polit Wkly II:52–54
Siddiqui RN, Pandey J (2003) Coping with environmental stressors by urban slum dwellers. Environ Behav 35(5):589–604. https://doi.org/10.1177/0013916503254748
Singh N (2006) Women women's s participation in local water governance: understanding institutional contradictions. Gend Technol Dev 10(1):61–76. https://doi.org/10.1177/097185240501000104
Singh S (2014) Women, environment and sustainable development: a case study of Khul Gad Micro watershed of Kumoun Himalaya. Space Cult India 1(3):53–64. https://doi.org/10.20896/saci.v1i3.45
Singh SB (2015) Women as milieu managers in integrated watershed management: perspectives from the hilly areas of Uttarakhand. Space Cult India 2(4):71–79. https://doi.org/10.20896/saci.v2i4.130
Singh C (2019) Migration as a driver of changing household structures: implications for local livelihoods and adaptation. Migr Dev 8(3):301–319. https://doi.org/10.1080/21632324.2019.1589073
Singh N, Keshri K, Bhagat RB (2016) Gender dimensions of migration in urban India. Routledge, New Delhi, pp 176–190
Sinha A, Mallik S, Sanyal D, Dasgupta S, Pal D, Mukherjee A (2012) Domestic violence among ever married women of reproductive age group in a slum area of Kolkata. Indian J Public Health 56(1):31. https://doi.org/10.4103/0019-557X.96955
Srivastava A, Singh RC (1996) Slums and associated problems: a case study of Bhilai, an industrial city, India. Int J Environ Stud 50(1):51–60. https://doi.org/10.1080/00207239608711038
Stevenson EG, Greene LE, Maes KC, Ambelu A, Tesfaye YA, Rheingans R et al (2012) Water insecurity in 3 dimensions: an anthropological perspective on water and women's psychosocial distress in Ethiopia. Soc Sci Med 75:392–400. https://doi.org/10.1016/j.socscimed.2012.03.022

Suganthi P, Chadha V, Ahmed J, Umadevi G, Kumar P, Srivastava R, Sharda MA (2008) Health seeking and knowledge about tuberculosis among persons with pulmonary symptoms and tuberculosis cases in Bangalore slums. Int J Tuberc Lung Dis 12:1268–1273

Sultana F (2011) Suffering for water, suffering from water: emotional geographies of resource access, control and conflict. Geoforum 42(2):163–172. https://doi.org/10.1016/j.geoforum.2010.12.002

Sultana F (2009) Community and participation in water resources management: gendering and naturing development debates from Bangladesh. Trans e Ins Br Geograph 34(3):346–363. https://doi.org/10.1111/j.1475-5661.2009.00345.x

Sward J (2016) Moving to 'greener pastures'? The complex relationship between internal migration, land tenure and poverty in mid-Ghana. https://opendocs.ids.ac.uk/opendocs/handle/20.500.12413/14847

Thakkar S (2018) Art in everyday resistance: A case study of the pink vigilantes of India. Junctions: Graduate J Humanit 3(1):8–17. http://doi.org/10.33391/jgjh.5

Thakkar S (2016) Is India really transforming? Narratives from Dabok, Rajasthan. Econ Polit Wkly 51(47):17–23. https://www.jstor.org/stable/44165872

Thakkar S (2021) Moving towards empowerment: migrant domestic workers in India. Asian J Women Stud 27(3):1–16. https://doi.org/10.1080/12259276.2021.1957195

Thakkar S (2024) Exploitation, harassment and violence: lived experiences of women paid domestic workers in India. J South Asian Develop 19(1): 44–60. https://doi.org/10.1177/09731741231164872

Thakkar S, Smiley KT (2024) Overlooked realities: Reimagining "Home" and "House" among women domestic workers in India. Violence Against Women 30(8):2053–2072. https://doi.org/10.1177/10778012241238245

Tilley E, Bieri S, Kohler P (2013) Sanitation in developing countries: a review through a gender lens. J Water Sanit Hyg Dev 3:298–314. https://doi.org/10.2166/washdev.2013.090

Uggla F (2015) Looking for a greener pasture: exploring the narratives of Gambian clandestine migrant. urn:nbn:se:su:diva-118574

UN Human Rights Council (2010) Report of the independent expert on the issue of human rights obligations related to access to safe drinking water and sanitation, C. de Albuquerque, UN Doc. No. A/65/254

UN (2008) Gender-disaggregated data on water and sanitation

UNDP (2006) Human development report (HDR): beyond scarcity: power, poverty and the global water crisis. New York

United Nations (2014) Slum population as percentage of urban. http://mdgs.un.org/unsd/mdg/SeriesDetail.aspx?srid=710

USAID (2016) Drinking water supply for urban poor: role of urban small water enterprises. https://www.safewaternetwork.org/sites/default/files/2016_India_Urban_Sector_Review.pdf

World Health Organization (1999) Putting women's safety first: ethical and safety recommendations for research on domestic violence against women (Report No. WHO/EIP/GPE/99.2). Global Programme on Evidence for Health Policy, World Health Organization. Geneva, Switzerland. https://www.who.int/gender/violence/womenfirtseng.pdf

WHO/UNICEF (2012) Joint monitoring program for water supply and sanitation. In Progress on drinking water and sanitation, pp 44–45

Wood G, Salway S (2000) Introduction: securing livelihoods in Dhaka slums. J Int Dev 12(5):669. https://search.proquest.com/docview/214887817?accountid=26511

Xu D, Hou G (2019) The spatiotemporal coupling characteristics of regional urbanization and its influencing factors: taking the Yangtze River delta as an example. Sustainability 11(3):822. https://doi.org/10.3390/su11030822

Yew CP (2012) Pseudo-urbanization? Competitive government behavior and urban sprawl in China. J Contemp China 21(74):281–298. https://doi.org/10.1080/10670564.2012.635931

Yoshida T, Yamauchi H, Sun GF (2004) Chronic health effects in people exposed to arsenic via the drinking water: dose–response relationships in review. Toxicol Appl Pharmacol 198(3):243–252. https://doi.org/10.1016/j.taap.2003.10.022

Yousefi M, Saleh HN, Yaseri M, Jalilzadeh M, Mohammadi AA (2019) Association of consumption of excess hard water, body mass index and waist circumference with risk of hypertension in individuals living in hard and soft water areas. Environ Geochem Health 41(3):1213–1221. https://doi.org/10.1007/s10653-018-0206-9

Chapter 9
COVID-19 Induced Family Violence in an Urban Setting in Bangladesh: Prospects for Social Work Interventions

Atia Binta Abdul Wahid and Tulshi Kumar Das

Abstract Family violence is a common scenario in South Asia including Bangladesh. Women in Bangladesh encounter some forms of family violence, particularly after marriage due to the strong prevalence of patriarchal norms. Such violence encompasses physical, psychological, and sexual abuse inflicted by male members upon their female counterparts. COVID-19 has led to a significant increase in family violence worldwide and Bangladesh is no exception. This study adopts a qualitative approach to understanding family violence and explores social work interventions to address it in an urban setting. Two organizations located in Sylhet city of Bangladesh, namely the Women Support Programme (WSP) and Bangladesh Legal Aid Services Trust (BLAST) dealing with family violence have been selected for this research. A total of 35 female respondents were surveyed using a close-ended structured interview schedule. Fifteen abused married women, and 5 perpetrators have been interviewed using a semi-structured interview guideline. Additionally, 2 FGDs with male respondents and 3 key informants' interviews have been carried out to extract their experience of dealing with family violence. The findings show the nature of family violence and the actual reasons behind it that have specifically occurred during COVID-19. It has portrayed a picture of the coping mechanisms adopted by the victims and the perpetrators, and explored social work interventions to resolve the crises so that a sustainable and violence-free community could be built.

Keywords COVID-19 · Family violence · Married women · Coping mechanisms · Social work interventions

---

A. B. A. Wahid
Apasen International, Sylhet, Bangladesh

T. K. Das (✉)
Department of Social work, Shahjalal University of Science and Technology, Sylhet, Bangladesh
e-mail: dastulshi-scw@sust.edu

S. Anand et al. (eds.), *Sustainability in South Asian Cities*, Advances in Geographical and Environmental Sciences, https://doi.org/10.1007/978-981-97-7455-5_9

9.1 Introduction

The COVID-19 pandemic has not just affected the health of people but more or less affected every sphere of life. The outbreak of coronavirus has intensified family violence (Usher et al. 2021). In today's society, family violence is a pernicious phenomenon like an infectious disease (Gelles 1985). Many studies (Kearney 2001; Sharma 2005; Wolfe et al. 1985) have focused on family violence where women are often abused within their own family structure. In most cases, the perpetrator of family violence is always a man (Solotaroff and Pande 2014). World Health Organization (WHO) (2013) revealed that 35% of women encountered family violence in one form or another in their lifetime, and WHO (2021a, b) noted that 1 (one) in 3 women is exposed to some form of family violence either physical, psychological or sexual during their lifespan. In South-East Asia, 33% of married women aged 15–49 years are estimated to have been subjected to intimate partner violence in their lifetime. Around 50% of women generally experience family violence in Bangladesh (WHO 2021a, b).

A number of studies (Adams and Adams 1984; Nishat and Rahman 2018; Parkinson 2019; Schumacher et al. 2010; Thomas et al. 2019; Weitzman and Behrman 2016) have shown that family violence took a new turn after the occurrence of natural disasters, namely eruptions, hurricanes, tsunamis, earthquakes and so forth. The outbreak of COVID-19 appears to be playing a role in increasing family violence (UNW 2020a, b; Singh and Bhattacharyya 2020). Likewise, the rate of family violence has increased in the aftermath of the coronavirus crisis across the globe (Campbell 2020; Singh and Bhattacharyya 2020). Although coronavirus is apparently a health issue, it has also affected our family, social, cultural, and economic life (Boserup et al. 2020).

Anybody can be a victim of family violence, but women are at increased risk of being exposed to violence amid the COVID-19 crisis (Ertan et al. 2020). UN Women (2020a, b) in a policy brief disclosed a notable escalation in the rate of violence against women as well as children in many countries throughout the COVID-19 lockdown. The confinement measure during COVID-19 is one of the underlying reasons behind this increased rate of violence (Ertan et al. 2020).

Lockdown was one of the main protocols to slow down the transmission of the spread of the infection across the world. Due to the lockdown, people were confined to their houses and the close proximity of family members for a long time increased the rate of family violence (Campbell 2020; Kagi 2020; Peterman et al. 2020; Wagers 2020). Zahran et al. (2009) identified a number of risk factors for family violence including financial distress, loss of property, and traumatic experiences of adversities which were the by-products of different natural disasters. Women are faced with an increased rate of violence in their own family usually by the male members after a disaster, mostly in developing countries (Parkinson 2019). Likewise, COVID-19 as a global disaster (Sheek-Hussein et al. 2021) cast its evil spell on families which induced family violence more than it used to be (Humphreys et al. 2020). The COVID-19 pandemic has led to a rise in family violence against women and girls in

Bangladesh because of lockdown, financial strain, and limited access to livelihood opportunities (Jahid 2022; Rayhan and Akter 2021).

In this pandemic, many people have lost their jobs and are facing economic hardships due to loss of income. Feelings of uncertainty have caused an increase in violent behaviour among family members (Capaldi et al. 2012; Fulu et al. 2013; Thomas et al. 2019; UN 2020). Women and children from poor families are the most affected by this crisis of COVID-19. In addition, the pandemic has enormously deteriorated the psychological health of people (Brooks et al. 2020). It has induced stress among them, playing a significant role in escalating family violence.

Much of the literature has highlighted the COVID-19-related factors that are believed to be playing a significant role in intensifying family violence. However, none of them explored the possibility of social work interventions to reduce family violence in an urban setting in Bangladesh. This chapter discusses the nature of family violence and the reasons behind it. It digs out the coping mechanisms adopted by the victims of family violence and also suggests possible social work interventions to address it among the lower economic strata in an urban setting during the time of COVID-19 in Bangladesh.

9.2 Methodology

This study adopts a qualitative approach, which is exploratory in nature. In-depth interviews with 15 females and 5 males were undertaken. Data were also collected from 2 FGDs (FGDs were conducted with male respondents) and 3 key informant interviews (KII). Moreover, data were gleaned from 35 female respondents using a survey questionnaire. Two organizations working in Sylhet city, namely WSP—a government organization, and BLAST—a non-government organization dealing with family violence and related issues were purposively selected for data collection. Separate checklists were developed for in-depth Interviews, FGDs, and KII for collecting relevant data.

Collected data were transcribed and translated from Bengali into English. Transcribed data were read again and again to develop different categories. Several themes were generated out of these categories. A thematic analysis technique was followed to analyse and interpret the transcribed data taking different themes into consideration.

9.3 Ethical Considerations

This study carefully followed different ethical principles. At the beginning of data collection, each respondent was clearly explained the aims and objectives of this research. Their oral consent was obtained as they looked convinced about the study. Moreover, respondents were given full freedom to withdraw their participation from the interview at any time if needed and were allowed to skip any question they wished not to answer. Since family violence is a highly sensitive issue in the socio-cultural context of Bangladesh, the participants were assured of the privacy and confidentiality of the data they shared. Pseudonyms have been used to hide the identity of the participants and collected data have only been used for the purpose of this research. Two females (the first author along with one social work graduate) were trained to ensure no physical or mental harm to the participants during the interviews.

9.4 Findings of the Study

Findings of Survey Conducted on Female Respondents

Socioeconomic and Demographic Background

The findings of this study disclosed that almost all the families from lower economic strata have had family violence during COVID-19. The survey was conducted on 35 female respondents who experienced family violence. The age range of the respondents was between 19 and 45 years. Except for 5 Hindu respondents, all other 30 respondents were from Muslim families. Among them, 27 respondents were homemakers, 7 respondents worked as maid servants and 1 was a tailor. They were of lower socioeconomic strata as their average monthly family income was TK. 7400 (1 USD = BDT 107.02) before separation from their husbands. Most of the respondents had little or no access to education. Out of 35 respondents, 13 respondents studied up to primary level, 7 up to secondary, 2 completed higher secondary, and 13 were illiterate. Except 1 widow, the other 30 respondents were married but remained separated. Among the separated, 29 were living in their paternal house, 3 were living alone in a rented house, 2 were living as boarders in WSP and only 1 widow respondent was living with her 25-year-old son in a rented house. We found 18 respondents had married for 0–3 years, 4 respondents for 3–6 years, 8 respondents for 6–9 years, 2 respondents for 9–12 years, 1 respondent for 12–15 years and 1 respondent for 15–18 years. Among them, 11 respondents did not have children, 16 had 1 child, 4 had 2 children and the other 2 had 3 and 5 children respectively.

Nature of Family Violence

The study found that the respondents experienced one or more than one form of violence within their family at the time of COVID-19. All the respondents went through either psychological abuse or physical, psychological, and financial abuse. In most of the cases, perpetrators were husbands; 25 respondents reported that they were experiencing violence by their husband, 9 respondents stated that violence was perpetrated by their in-laws along with the husband and only one respondent said that her son was the perpetrator. Even 6 respondents reported that they were seriously injured due to physical assault carried out by the husband. One respondent had to have an abortion due to physical torture and the rest had serious injuries on different parts of the body. Twenty five respondents went through these experiences after the outbreak of COVID-19 and during the lockdown, 6 respondents were experiencing violence a few days/months before the outbreak of COVID-19 and the lockdown exacerbated the conflicts, only 5 respondents mentioned that it started soon after their marriage but COVID-19 crisis made it worse.

Reasons behind Family Violence

Several reasons for committing violence were identified. Many of the reasons were directly or indirectly related to the COVID-19 crisis. The economic crisis caused by the pandemic played a major role in increasing domestic violence. The number of respondents who shared such reasons was 21. Here, 20 respondents blamed the lockdown as their husbands were becoming increasingly frustrated at home due to a lack of social interaction, occupational uncertainty, and economic crisis. More than half of the respondents (22) said that their husbands and in-laws wanted to have the dowry and considered it as a way out from their current economic constraints. Respondents were under severe pressure for bringing the dowry from their parents or even in some extreme cases they had experienced violence committed by the husband for dowry. Some other reasons were also identified by the respondents such as arguments with the husband (3 respondents), instigations of others to abuse the wife (9 respondents), the extramarital affair of the husband (5 respondents), the wife being tortured for not allowing the husband for second marriage (3 respondents), the husband was losing interest in their wife (2 respondents), the husband was feeling inferior to the wife (1 respondent), the husband was short-tempered, holding a patriarchal mindset (4 respondents), husband suspected the wife of infidelity (1 respondent).

Coping strategies

Of 35, 21 respondents were seeking help from BLAST to get a divorce with alimony and child support but 13 respondents wanted to make up with their husband, and 1 widow respondent wanted to patch up with her son. Here, 30 respondents from BLAST and the rest from WSP were interviewed. Among them, 24 respondents wanted to resolve their problems through arbitration and 11 respondents filed a lawsuit to resolve the issue. All the respondents were getting help from their parents or other family members from their paternal house.

Twenty-four respondents stated that they tolerated everything that happened to them for various reasons like society would blame them; they were women, so they had to endure all this; they had to accept that men would do such things; they were married and could not go back to their paternal house; they had to tolerate everything thinking of their parents and also the future of their children. Only 11 respondents claimed that they protested the violence.

9.5 Findings of In-Depth Interviews with the Females

Collected data are sorted into 5 broad categories. Simultaneously, varied themes have been generated under different categories (Fig. 9.1).

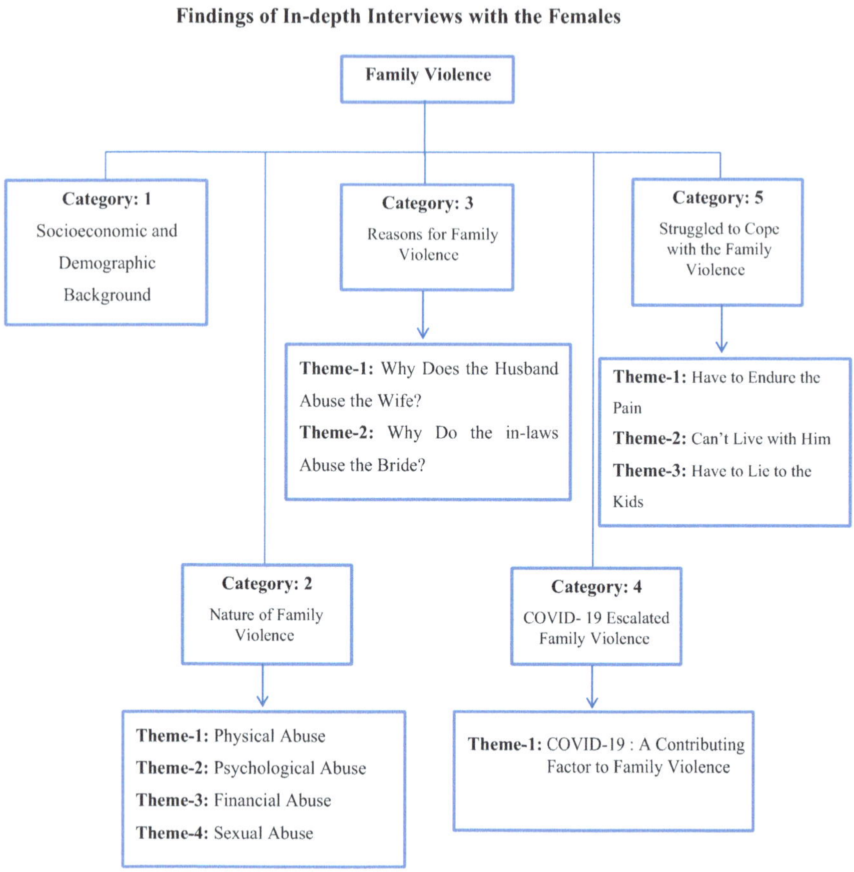

Fig. 9.1 Source: Self-created

9.6 Socioeconomic and Demographic Background

Fifteen in-depth interviews were conducted with female respondents who had experienced family violence. The age of the respondents was from 18 to 35 years. All of them were Muslims. Among them, 14 respondents were homemakers, one of them was a homemaker as well as a teacher providing Arabic (Quran) tuition to the children in her neighborhood. Only 1 respondent worked as a maidservant. They all belonged to the lower socio-economic strata as their average monthly family income was TK. 6500 (1 USD = 107.02 BDT). Most of the respondents had little or no access to education. Six respondents studied up to the primary level, 4 respondents up to the secondary, and 5 respondents were illiterate. Here, 14 respondents were found separated from their husbands. There was an exceptional case where the victim of family violence died during the arbitration initiated by BLAST (data of this respondent was collected from her sister as well as from the official documents of BLAST). Among the separated respondents, 13 were living in their paternal house, and 1 respondent was living in the shelter of WSP. There were 4 respondents who had been married for 0–3 years, 5 respondents married for 3–6 years, 2 respondents married for 6–9 years, 3 respondents married for 9–12 years, and 1 respondent was married for 15–18 years. Eight respondents had no children, 3 had 1 child, 3 had 2 children and 1 had 3 children.

9.7 Nature of Family Violence

Physical Abuse

Physical abuse involves physical assault directed at a person. Physical abuse is a common form of family violence. Mostly, women face violence by their close family members (Hegarty et al, 2000). In many cases husband and in-laws are responsible for perpetrating violence against women (Hossain 2016). During the COVID-19 pandemic, people spent more time at home due to lockdowns. It was challenging for families who might have already experienced conflicts. The stress and uncertainty caused by pandemic had intensified existing conflicts and created fresh ones, even it led to physical abuse. In this study, victims disclosed different sorts of physical abuse they experienced.

Alia Begum stated,

> When my husband gets angry, he beats me very badly. He hits me with whatever he gets around his hands. Beating, kicking, and hitting by different objects were common during the lockdown period. He pulled my hair and slapped me on the cheek on small pretexts. Once I was seriously injured and had to be hospitalized for three days.

This study found different types of physical abuse including kicking, slapping, punching on sensitive body parts, throwing objects (like shoes, utensils, etc.) at the wife, grabbing and pulling her hair, etc. It revealed that battered women incurred serious injuries like miscarriage, cuts, losing eyesight, bruising, and so on. Amina painfully shared how she lost her unborn baby in the time of the pandemic,

> I never imagined that my life after marriage would turn into such a nightmare. For a few months after the marriage, I was in a delusion and that did not last long. My husband and in-laws showed up their real face. We went through a financial crisis during COVID-19. At that time, my husband and in-laws started harassing me for dowry which later turned into beating. I had a miscarriage when I was expecting my first baby. One day my husband kicked me on my lower abdomen and I lost my baby.

Psychological Abuse

People generally do not recognize psychological abuse as abuse. But it is as harmful as physical one or sometimes far more harmful than physical abuse (Sackett and Saunders 1999). Its effects are also more long-lasting than physical torture. Some common forms of psychological abuse are: verbal abuse, humiliation, neglect, yelling, bullying, threatening, name-calling, disrespecting the parents of the wife, stop talking to the wife, kicking her out of home, etc. However, the COVID-19 pandemic has contributed to an increase in psychological abuse within the family (McNeil et al. 2023). One respondent named Jahanara said,

> I am 18 now. I lost my father at a young age. My uncle is my guardian. About a year ago my uncle arranged my marriage. I did not want to get married as I wanted to continue my study. But I could not refuse my uncle's decision. I agreed to the marriage against my will, hoping that I may get a supportive life partner. But the reality is, "I am not happy". He did not abuse me physically, but he is constantly torturing me mentally and using the financial crisis due to the pandemic as a justification for his abusive behavior. Now I am at my parent's home and he did not talk to me in the last month.

Rina's sister added,

> We made a mistake by marrying my sister to this guy. Within a few days of marriage, we learned that he had an affair with his sister-in-law. During lockdown, the man and his sister-in-law became closer to each other, which made my sister even more difficult to cope with that situation. My sister emotionally broke down. She suffered from this mental agony till her death.

Financial Abuse

In addition to physical and psychological abuse, financial abuse is also a form of family violence (Singh 2020; Fitzgerald et al. 2022)). Financial abuse happens to women alongside physical and psychological abuse. It makes the victims more vulnerable. This study's findings revealed many ways of financial abuse like the husband limiting the wife's access to resources, exploiting her possessions, forcing her to borrow money from different organizations that provide microcredit to the women and so forth. However, during the pandemic, some of the key factors contributed to financial abuse, such as many people lost their jobs, faced financial stress, economic uncertainty, etc. Farzana commented,

> My husband and in-laws tortured me for dowry. During the pandemic, they took the advantage of the financial distress and they forcibly sold my gold jewellery. They also verbally abused me as my father could not fulfil their demands for dowry. In every Eid, my husband buys clothes for everyone else, but he does not buy a single *saree* for me.

Jahanara added,

> My husband used to live in Dubai. He came to Bangladesh to get married in early 2020. Later, he lost his job in Dubai due to COVID-19 outbreak. And my husband wanted to get microcredit from an organization through me, forced me for my consent to get the microcredit. After the arguments about this, he forcefully sent me to my mother's house. My husband threatened me of divorce if I do not take microcredit for his business.

Sexual Abuse

It is true that many women may not be willing to share openly about their experiences of sexual abuse. Likewise, it was not possible to gather much information about sexual abuse through this study. There are two main reasons behind that: sexual abuse is not that much frequent as physical and psychological abuse; and another reason was that the victim refused to talk about this topic. However, some respondents agreed to provide information regarding sexual abuse. They highlighted that the husband neglects their opinions, rejects their desire, and disrespects them during physical relationship. Few of them also mentioned that after the outbreak of COVID-19 pandemic, their husbands become more aggressive during intimate moments due to stressful life circumstances. Israt shared,

> My husband is obsessed with sex. He has to do it even if I do not want to. He does not try to understand my physical illness even during my menstruation he could not control his impulsion. Being confined to home, he became more impulsive for sex during lockdown and I was feeling uncomfortable with his level of impulsivity. He became aggressive during sex. He made the thing very painful for me but he had no sympathy for me.

Jesmin shared,

> My husband was impotent. But he did not want me to tell anyone about his complications. My mother-in-law, on the other hand, was pressuring me to have a baby. During the COVID-19 crisis, my life became more vulnerable and I had to move back to my parent's house.

9.8 Reasons for Family Violence

Why Does the Husband Abuse the Wife?

In our patriarchal society, mainly women are the victims of family violence. This study demonstrates that the husband perpetrates violence to the wife than other family members. This study reports multiple factors that provoke men to commit intimate partner violence. Dominant mentality of a male is a significant factor behind committing violence against women. Economic insecurity, scarcity of resources, and unemployment of male members are some other major reasons that contribute to the family violence (Berger 2005; Bowlus and Seitz 2006; Nagassaar et al. 2010). Studies show that the husband tortures his wife due to her parents or brothers have not met the husband's demand for dowry. During the financial crisis because of COVID-19, many husbands started asking their wives to bring money from their parents. Failing to do so, they abused and tortured them. In some cases, the husband pressurized the wife to give permission for the second marriage. Rupali stated,

> We fell in love with each other and later got married. We have been married for 3 years and we have a daughter. During the first COVID-19 lockdown, I realized that my husband was involved in an extramarital affair. He started physically abusing me because I forbade him to do so. He wanted me to allow him to remarry. I was tortured and sent to my father's house because I did not give him permission.

Rasheda shared,

> I lost my left eyesight due to physical abuse committed by my husband. He lost his job during COVID-19 and we were in a severe financial crisis. He blamed me and taunted me, labelling as a woman of misfortune!

Why Do the In-Laws Abuse the Bride?

Women are not only abused by their husbands, most of them are also abused by their in-laws (Hossain 2016). Some of the respondents shared that they were physically abused by their in-laws, especially by mother-in-law as well as sister-in-law. They mentioned that their in-laws provoked their husbands against them and it led to the conflict between them. They generally did this because of the negative attitude towards the bride as they could not accept her as their daughter. Their demands for dowry also played a prominent role here. Because of COVID-19 health protocol everyone had to remain confined at home and such conflicts occurred more frequently. Ruma disclosed,

> Both my mother-in-law and sister-in-law used to taunt me by calling me "*fakirnir baccha*" (daughter of a beggar). They verbally humiliated my father for dowry. After the onset of the COVID -19 pandemic, it became regular and every day they taunted me for dowry money. My mother-in-law used to say that his son was struggling with financial crisis; so my father had to give dowry money. My in-laws instigated their son to torture me. At their instigation, my husband stopped talking to me several times.

Fahima mentioned,

> My first four children died shortly after their birth. During the fifth time of my pregnancy I had a son. But the irony of fate is that when he was nine years old, he was diagnosed with heart disease. My son died within a year of diagnosis. During the pandemic, my husband lost his job and could not afford my son's treatment cost. After his death, everyone from my in-laws' family even my husband started blaming me because I could not manage money from my brother. They abused me mentally. Even they wanted to get their son remarried. My mother-in-law wanted me to divorce my husband. Now I am living in my brother's house and do not want to go back to my husband.

9.9 COVID-19 Escalated Family Violence

COVID-19: A Contributing Factor to the Family Violence

COVID-19 was a hot topic all over the world. Life became stagnant during this crisis. COVID-19-induced family violence has been reported in different countries (UNW 2020a, b). Bangladesh was no exception to that. Family violence against women was on the rise in our society during the COVID-19 crisis (Das 2020). Families of lower economic strata suffered from family violence more than those of others. The COVID-19 pandemic was a global health issue, but it had far-reaching impacts on almost every sphere of life. To control the infection caused by COVID-19, health and safety protocols were imposed on the people by the government. The COVID-19 situation contributed to the family violence in multiple ways.

Alia Begum pointed out,

> I got married 15 years ago. For a few years after the marriage, there was no conflict between me and my husband, everything was going well. Gradually, the financial crisis in the family increased after the birth of my children. At the same time, family conflicts began to grow.

> Poor people like us cannot even manage their food regularly, so it is normal to have fights among family members. On top of all that, corona virus became a curse on us. Financial crisis erupted due to unemployment during COVID-19 and my husband struggled a lot to meet our daily basic needs. This stressful and uncertain situation exacerbated our pre-existing issues and led to new instances of violence.

9.10 Struggled to Cope with the Family Violence

Had to Endure the Pain

A large number of married women have been frequently abused by their family members in Bangladesh (Das et al. 2016). Most of them tolerate all abuses that happen to them. For multiple reasons, the married women have to endure all the pains or in many cases they are forced to endure all sorts of abuse. Some of the reasons for tolerating abuses are: they remain silent for the fear of society as our society considers family violence as a normal matter in the life of a woman; the victims do not get support from their parents, neighbours or even from other close ones; women often digest tortures just thinking of the future of their children. These types of abusive behaviour multiplied during the COVID-19 lockdown. Although some respondents were currently separated from their husband, they wanted to return to their husband. Shamima stated,

> I am the second wife of my husband. He married me after the death of his first wife. But his first wife's elder sister instigated him against me. As a result, my husband tortured me in every possible way. He wanted me to do some unethical thing which I cannot even share now. But I refused to do that. He tortured me. During the lockdown he stayed at home and I was beaten up by him every single day. Later, he sent me back to my parents' house. My family members along with the community leaders tried to solve this issue but they failed. Then I sought help from BLAST to resolve this matter. Despite all that happened, I want to go back to my husband thinking of my child's future.

Can't Live with Him

Violence against women is a very common phenomenon in Bangladeshi society. In our social system, women are considered subordinate to men. They are often subjected to abuse (Das et al. 2016). Although a large number of women accept this as a reality of their life (García-Moreno et al. 2005), there are some women who do not tolerate this abuse. A few battered women sought help from the community leaders, but they failed to reconcile with their husband and suggested the victims to take legal action against the perpetrator(s). These conflicts became so unbearable during COVID-19 that more than half of the respondents wanted to end their marriage and therefore asked for a divorce. They sought legal assistance either from BLAST or WSP. Naju recalled,

> Because of the COVID-19 pandemic, my husband went through a significant loss in business. So, my husband and my in-laws tortured me for dowry and sent me back to my parents' house. They sold my gold jewellery without my permission. After the first lockdown, my

husband demanded TK. 20000 from my father to start a new business. My father borrowed money and gave it to my husband on his demand. Even then he did not take me back to his house. Recently, he married again without my permission. Now I do not want to go back to him anymore after all that had happened.

Lima commented,

We did not have any issues between us till six years after the marriage. Suddenly, I noticed changes in my husband's behavior. During the lockdown, he started spending so much time on his phone. Gradually I found out that he got engaged in a virtual extramarital affair and he started neglecting me. Even right after the lockdown lifted, he secretly married that girl. I cannot live with my husband's second wife. So I asked for a divorce from my husband.

Had to Lie to the Kids

Family violence interrupts the psychological development of children. It causes negative impacts on their psychology. The most important factor in the development process of children's psychology is family relationships, especially the relationship between their parents. If the child repeatedly witnesses abuse and violence between the parents, it disturbs their healthy development (Mueller & Tronick, 2019). Unfortunately, many children had to face conflicts with their parents on a daily basis during the COVID-19 lockdown.

One respondent named Alia Begum pointed out,

When my children witnessed violence between me and their father, they became anxious. I understand that it affected them emotionally. Not only that, it had an adverse effect on their study. They saw their father treating me so badly since they were kid. And that took a worst turn during the COVID-19 lockdown as we had fights almost every day.

Similarly, Nesarun shared,

Our conflicts caused serious harm to our children's psychological health. During the first lockdown, my husband lost his job. We were in a great financial crisis. The conflict between us became a part of our daily life. At one point I moved to my father's home with my children. My husband never inquired about the kids. In the beginning, they used to ask about their father but now they do not.

Most of the time children have a closer relationship with their mothers than their fathers. When they observe their father treating their mother badly and abusing her, they cannot take it easily. It may trigger issues relating to mental health like depression, anxiety, stubbornness, anger, etc. among the children. It was a tough time during COVID-19 even for children who experienced unbearable site of domestic violence that definitely made them mentally more vulnerable. One of the respondents named Shaheda mentioned,

My children were terrified seeing their father beating me. They started crying in fear. After those incidents, they were scared to go to their father. Sometimes, my younger son asked me, "Mom, will Dad beat us too?" Then I assured him by saying, "No, your father is a good person, he will not hurt you all". As my husband was at home all the time because of the lockdown, my children were in constant fear of their father's anger.

9.11 Male Perspectives on Family Violence

Five themes are developed on the basis of data gathered from in-depth interview as well as FGD conducted on male respondents.

Husband should have control over Wife

Men think that women should always be subordinate to them. They want women not to do anything without men's permission. Men are the master of the family and women should be under the control of men. These men want women to be submissive towards them. During COVID-19 they were almost isolated from the outside world and spent most of the time at home with their family, which made them bored and violent. Conflicts often broke out because of that. The data collected from almost all male respondents reflected such perspectives. They think that women need to be controlled by the husband in order to ensure happiness in the family. Abdul Aziz stated,

> As the leader of my family, what I believe is that wife and children should follow the instructions of their husband and father respectively. It is totally unacceptable when they do not do such thing. Unfortunately, I had to face such behaviour during the lockdown period of COVID-19. Violent behaviour brings nothing but only creates chaos in the family.

One FGD respondent commented,

> As a man and the only breadwinner of my family I do not like it when my wife and children do not obey me or do not do things the way I want them to do. Man is the leader and the rest of the family members should follow him or should do as he asks them to do. During COVID-19 lockdown, I spent more time together with my family members than usual. I often found myself in a situation where conflicts arose as they did not obey my order.

Another FGD respondent shared,

> Male member should be strict as the head of the family; otherwise he cannot control the family members. It is our tradition that man leads the family matters and women do the household works. In a harsh condition of COVID-19, there were many fights between me and my partner on almost every trifle issue. She should have mind her business maintaining the kitchen instead of fighting with me.

She Hardly Cooked

Both men and women have different perspectives about family violence. Just as women blame men, so is that men blame the women for family conflicts. A common misconception in the society is that women will do the cooking in the house. In this study, many male respondents complained that during lockdown period women did not do their household chores properly which prolonged the conflict between them. It made the stressful situation worse. Belal stated,

> I think close proximity of family members due to COVID-19 lockdown increased internal conflicts and violence within the family. Usually when we stay closer to someone all day long, we will find ourselves in situations where things our partner say or do will seriously offend us or we might feel angry which will end up in conflicts. During this lockdown, I had

frequent conflicts with my wife. Due to the conflicts between us, my wife stopped cooking several times for me and my children.

She Argued

Women are considered inferior to men. It is considered that women's duty is to obey the orders of men in the society and this has been going on for ages. Therefore, women have no right to argue with men (Bhattacharyya 2015; Dalal et al. 2009). Because of COVID-19 lockdown, couple had to stay in close proximity for a long time which led to conflict among them. Men blamed their wife for such conflicts. They did not want women to have any opinion or protest against their orders, prohibitions or decisions. Tarek argued,

> Already life became hell because of COVID-19 and after reaching home when you found your wife nagging, it was very hard to control anger. The moment you lost control over your anger you would find yourself in the middle of conflict.

Ali commented,

> I do not like staying at home all the time, but for COVID-19 I had to stay at home for 24 hours. No job, no money, not going outside; this was enough to make me mad. In such a mood when I saw my children or wife doing something that I won't want them to do, I could not control my anger and, as a result, started beating.

She was Demanding

In this study, male respondents complained that even in the midst of the unfavorable situation of COVID-19, their wives taunted their husbands for losing their job and for not having enough savings. Financial insecurity during this crisis was a major cause of this kind of family conflict. One respondent shared,

> We all went through the fear of COVID-19 along with tons of uncertainties. During this pandemic-related crisis the earning almost became zero and staying mentally calm was very hard. On top of that, my wife taunted me all the time for not earning enough. These things irritated me and it led to family conflicts which were unavoidable.

One FGD respondent, Rahman said,

> We never had a major conflict before the coronavirus outbreak. But during the lockdown, my wife and I had a conflict over economic issues which later escalated. So we had to resolve it through arbitration. We sought help from the BLAST for arbitration.

She Talked to a Male Stranger

It's not just men who get involved in extramarital affairs. The male informants in this study mentioned that many of them were suspicious of their wives. They doubted that their wives talked with other men when they were out of home for work. Some informants said that they found it out while they were at home during the lockdown. They pointed out this issue as the reason behind their aggressive behaviour towards their wives. Alamgir disclosed,

> My wife was having an extra-marital affair and I could not take proper steps against her as I truly loved her, and we had children together. But during this lockdown period it became almost impossible for me to stay under the same roof with her. My wife always fought with me on small pretexts. Sometimes I felt really helpless because of frequent conflicts that I had with my wife during this pandemic.

9.12 Findings of Key Informant Interview

"No Income" Increased Family Violence

Family income plays a pivotal role in maintaining the stability of the family relationships. If the income of a family decreases, problems and disputes arise within the family. During the COVID-19 lockdown period, the income of the people decreased and conflicts broke out within the family (Parvin et al. 2021). Mostly, the families of the lower economic strata were exposed to family violence due to the financial crisis.

The Assistant Director of WSP stated,

> People with a low income struggle all the time in their lives. Those who live hand to mouth became vulnerable when they could not get out for work due to an uncertain lockdown. In such a situation, it is natural that their mental state will also be adversely affected. Thus, they got involved in arguments with the wives even on small issues, which contributed to the family violence during the time of COVID-19.

One of the Key Informants (Social Welfare Officer of WSP) pointed out,

> Due to COVID-19, the earnings of low-income people have been reduced. Many of them lost their job. The business of small traders had to be stopped due to the lockdown. COVID-19 caused a severe financial crisis mostly among the people from lower economic strata. It has negatively affected their mental health and also their relationships with their family members. Even there are a number of cases where the husband kicked his wife out of the home. Their miserable financial condition due to the COVID-19 pandemic is largely responsible for the family conflicts.

The Unit Coordinator of BLAST added,

> Coronavirus not only affected our physical health but also our psychological state. We all went through different uncertainties. This pandemic made the life of the poor people miserable. COVID-19 caused their financial condition more insecure than before. It is one of the reasons which triggered family conflicts and violence.

Decreased Organizational Activities

The first outbreak of COVID-19 pandemic in Bangladesh occurred on 7th March, 2020. To reduce the transmission of COVID-19, the Bangladesh government declared lockdown throughout the country and ordered closure of public and private offices too (Das 2020). Thus, the BLAST and WSP had hardly any activities at the field levels during that period.

One of the Key Informants, (Assistant Director of WSP) stated,

> The functions of the organization are to provide legal assistance to the abused women and to provide accommodation facilities for the most vulnerable women. Although the official work of the organization continued during COVID-19, no new cases of domestic violence were taken. There was no boarder staying at WSP's shelter during the lockdown period. After the Government relaxed COVID-19 lockdown, this organization immediately restarted its activities. We noticed that COVID-19 had contributed a lot to the family violence. But there is no official statistics of family violence with us during COVID-19.

Social Welfare Officer of WSP commented,

> WSP provides different types of assistance to the women who are the victims of family violence. Some of these services provided by WSP are: arbitration at free of cost, counseling, and providing accommodation and vocational training to the victims, etc. This organization tries to resolve conflicts between spouses through arbitration. If the female victims want to stay at the shelter of WSP, the organization provides them with accommodation. They can stay here for six months with a maximum of two children under 12 years of age. Sewing training is provided to the boarders for rehabilitation. We also follow up on the cases. Those who come to us for arbitration, we try to counsel both parties to resolve the issues. However, the activities of this organization have been slowed down for a long time during the COVID-19 pandemic.

The Unit Coordinator of BLAST added,

> BLAST seeks to reach out to vulnerable women who lack access to the judicial system. The core activities of this organization are providing legal aid to marginalized people like women and children. We also provide support to the victims of family violence through arbitration. However, the activities of BLAST were limited during the COVID-19 crisis.

9.13 Prospects for Social Work Interventions

Family violence, especially violence against women, has reportedly increased in Bangladesh during the period of lockdown imposed by the government due to higher rate of infection caused by COVID-19 (Das 2020, p. 29). Research participants of the current study who belong to the low income category identified different reasons, like income crisis, close proximity between the husband and wife for longer time, dowry demands of the husband, arguments of wife with the husband, wife instigating husband, extramarital affairs of the husband, wife not allowing the husband for another marriage, husband losing interest in wife, infertility of wife, husband is short-tempered having patriarchal mind-set, husband suspects the wife of infidelity, etc. for increasing family violence during the pandemic. Social work interventions at micro-, mezzo- and macro-levels in the context of increasing violence against women during the time of COVID-19 could mitigate the intensity of the problems caused through the aggressive behavior of the husband towards their wife. Social work practitioners have the scope of applying social work knowledge and skills in order to address the issues that eventually create cracks in the relationship between husband and wife. Moreover, social work knowledge may also be applied to renovate the relationship between the wife and her in-laws which often turns out to be bitter, especially in a situation of a crisis like the COVID-19 pandemic. Micro social work practices like working with individuals, mezzo social work practices like working with the family, and macro social work practices like working with the community and organizations (Zastrow 2010) in the context of increasing family violence during COVID-19 in Bangladesh may produce positive results. Social casework or family casework may be practiced with married women who were psychosocially affected due to the frequent abuse meted out by the husband at the time of COVID-19. A casework practitioner could provide psychosocial guidance and counseling to the violence-affected women as well as their husbands during COVID-19. Social workers have the scope to have direct practice with individuals, couples, families, and groups by applying social casework, family casework, and group work methods in the face of increasing family violence because of COVID-19 (Midgley 2001). Those who are traumatized due to the physical and mental torture of their husband and in-laws in the COVID-19 situation may seek counseling, sympathy, empathy, motivation, mediation, negotiation, etc. from micro-social work practitioners (Das and Alam 2013). Social workers can extend required help to abused women in the face of COVID-19 using strengths based and ecosystem perspectives, applying role theory, cognitive behavior theory as well and learning theory (Das and Alam 2013; Kirst-Ashman, 2010). At the community level, social workers can go for an 'awareness campaign' to make the common people aware of the increasing violence committed by men while passing through the crisis of COVID-19, using the social action method of social work, which could bring down the rate of family violence (Das 2012). The aforementioned examples may be considered important while exploring the prospects of social work interventions to deal with COVID-19-induced family violence in Bangladesh.

9.14 Recommendations Proposed by Different Organizations

There are some recommendations proposed by the officials of BLAST and WSP. These are hereunder:

- Our society is a patriarchal one. Most of the men look down upon the women. When a woman who is a victim of family violence seeks help from the chairman and members of the union council, most of the time they take men's side and they are biased against women (Ashrafun 2018; Das et al. 2016; UNDP 2020). There is a need to change the perspective of these responsible authorities.
- Women's empowerment is very much needed to get them out of this miserable situation.
- Government and other authorities should review and evaluate the current laws and policies addressing family violence.
- Researchers should conduct research on family violence more and more for renewed policy formulation that can effectively address increasing family violence in Bangladesh.

- Public and private organizations working for women's safety should have mutual collaboration among themselves to deal with family violence.
- Social media can play a key role in awareness-raising campaigns against family violence.
- The government should take the initiative to strengthen local judicial systems in rural areas of Bangladesh so that the victim women can get justice.

9.15 Conclusion

The COVID-19 pandemic has contributed much to the situation of family violence around the world. Family violence against women is a common phenomenon in Bangladesh but it has been exacerbated by the strike of coronavirus pandemic, especially in urban settings. The current study exposed a grim picture of family violence during the pandemic leaving the married women of lower economic strata in Sylhet city in a state of massive suffering. The victims were confronted with several forms of family violence such as physical, psychological, financial, and sexual violence committed by the husband and their in-laws during COVID-19. The study explores the possibility of social work interventions that may be efficacious in combating increased family violence in a situation that emerged in an urban setting due to the spread of COVID-19 triggering huge uncertainties in the life of the poor.

References

Adams PR, Adams GR (1984) Mount Saint Helens's ashfall: evidence for a disaster stress reaction. Am Psychol 39(3):252

Ashrafun L (2018) Women and domestic violence in Bangladesh: seeking a way out of the cage. Routledge, London. https://doi.org/10.4324/9781351256643

Bhattacharyya R (2015) Understanding the spatialities of sexual assault against Indian women in India. Gend Place Cult 22(9):1340–1356. https://doi.org/10.1080/0966369X.2014.969684

Berger LM (2005) Income, family characteristics, and physical violence toward children. Child Abuse Negl 29(2):107–133

Boserup B, McKenney M, Elkbuli A (2020) Alarming trends in US domestic violence during the COVID-19 pandemic. Am J Emerg Med 38(12):2753–2755

Bowlus AJ, Seitz S (2006) Domestic violence, employment, and divorce. Int Econ Rev 47(4):1113–1149

Brooks SK, Webster RK, Smith LE, Woodland L, Wessely S, Greenberg N, Rubin GJ (2020) The psychological impact of quarantine and how to reduce it: rapid review of the evidence. The Lancet 395(10227):912–920

Campbell AM (2020) An increasing risk of family violence during the COVID-19 pandemic: strengthening community collaborations to save lives. Forensic Sci Int Rep 2:100089

Capaldi DM, Knoble NB, Shortt JW, Kim HK (2012) A systematic review of risk factors for intimate partner violence. Partn Abus 3(2):231–280

Dalal K, Rahman F, Jansson B (2009) Wife abuse in rural Bangladesh. J Biosoc Sci 41(5):561–573

Das TK (2012) Applicability and relevance of social work knowledge and skills in the context of Bangladesh. SUST Stud 15(1):45–52

Das TK, Alam MF (2013) Micro social work practice in the context of Bangladesh. J Soc Work Commun Soc Work (Special Issue) 3(8):105–122

Das TK, Bhattacharyya R, Alam MF, Pervin A (2016) Domestic violence in Sylhet, Bangladesh: analysing the experiences of abused women. Soc Change 46(1):106–123. https://doi.org/10.1177/0049085715618561

Das TK (2020) Bangladesh. In: Dominelli L, Harrikari T, Mooney J, Leskošek V, Tsunoda EK (eds) COVID-19 and social work: a collection of country reports. COVID-19Social Work Research Forum, IASSW, pp 24–30. https://www.iassw-aiets.org/wp-content/uploads/2020/07/IASSW-COVID-19-and-Social-Work-Country-Reports-Final-1.pdf

Ertan D, El-Hage W, Thierrée S, Javelot H, Hingray C (2020) COVID-19: urgency for distancing from domestic violence. Eur J Psychotraumatol 11(1):1800245

Fitzgerald AJ, Barrett BJ, Gray A, Cheung CH (2022) The connection between animal abuse, emotional abuse, and financial abuse in intimate relationships: evidence from a nationally representative sample of the general public. J Interpers Violence 37(5–6):2331–2353

Fulu E, Jewkes R, Roselli T, Garcia-Moreno C (2013) Prevalence of and factors associated with male perpetration of intimate partner violence: findings from the UN multi-country cross-sectional study on men and violence in Asia and the Pacific. Lancet Glob Health 1(4):187–207

García-Moreno C, Jansen HA, Ellsberg M, Heise L, Watts C (2005) WHO multi-country study on women's health and domestic violence against women. World Health Organization

Gelles RJ (1985) Family violence. Ann Rev Sociol 11(1):347–367

Hegarty K, Hindmarsh ED, Gilles MT (2000) Domestic violence in Australia: definition, prevalence and nature of presentation in clinical practice. Med J Aust 173(7):363–367

Hossain A (2016) The impact of domestic violence on women: a case study of rural Bangladesh. Soc Crimonol 4(1):135–142

Humphreys KL, Myint MT, Zeanah CH (2020) Increased risk for family violence during the COVID-19pandemic. Pediatrics 146(1)

Jahid A (2022) Unsafe at home: the increased trend of domestic violence during the COVID-19 pandemic in Bangladesh. J Adult Prot 24(1):15–21

Kagi J (2020) Crime rate in WA plunges amid coronavirus social distancing lockdown measures. ABC News Aust

Kearney MH (2001) Enduring love: a grounded formal theory of women's experience of domestic violence. Res Nurs Health 24(4):270–282

Kirst-Ashman KK (2010) Introduction to social work and social welfare: critical thinking perspectives. Brooks/Cole, Cengage Learning, Belmont CA

McNeil A, Hicks L, Yalcinoz-Ucan B, Browne DT (2023) Prevalence & correlates of intimate partner violence during COVID-19: a rapid review. J Family Violence 38(2):241–261

Midgley J (2001) Issues in international social work: resolving critical debates in the profession. J Soc Work 1(1):21–35. https://doi.org/10.1177/146801730100100103

Mueller I, Tronick E (2019) Early life exposure to violence: developmental consequences on brain and behaviour. Front Behav Neurosci. https://doi.org/10.3389/fnbeh.2019.00156

Nagassar RP, Rawlins JM, Sampson NR, Zackerali J, Chankadyal K, Ramasir C, Boodram R (2010) The prevalence of domestic violence within different socio-economic classes in Central Trinidad. West Indian Med J 59(1):20–25

Nishat KJ, Rahman MS (2018) Disaster, vulnerability, and violence against women: global findings and a research agenda for Bangladesh. In: Handbook of research on women's issues and rights in the developing world. IGI Global, pp 235–250. https://doi.org/10.4018/978-1-5225-3018-3.ch014

Parkinson D (2019) Investigating the increase in domestic violence post disaster: an Australian case study. J Interpers Violence 34(11):2333–2362

Parvin RA, Islam MS, Akter MS, Mohammod SS, Hossain MB, Rashid S (2021) Psychosocial anxiety from lockdown due to COVID-19 to income earner of the family: an evidence from Northern Bangladesh. Sociol Mind 11(3):65–80

Peterman A, Potts A, O'Donnell M, Thompson K, Shah N, Oertelt-Prigione S, Van Gelder N (2020) Pandemics and violence against women and children, vol 528. Center for Global Development, Washington, DC
Rayhan I, Akter K (2021) Prevalence and associated factors of intimate partner violence (IPV) against women in Bangladesh amid COVID-19 pandemic. Heliyon 7(3):e06619
Sackett LA, Saunders DG (1999) The impact of different forms of psychological abuse on battered women. Violence Vict 14(1):105–117
Schumacher JA, Coffey SF, Norris FH, Tracy M, Clements K, Galea S (2010) Intimate partner violence and hurricane Katrina: predictors and associated mental health outcomes. Violence Vict 25(5):588–603
Sharma S (2005) The needs of pacific women when they are victims of family violence. Soc Policy J N Z 84:1–13
Sheek-Hussein M, Abu-Zidan FM, Stip E (2021) Disaster management of the psychological impact of the COVID-19 pandemic. Int J Emerg Med 14(1):1–10
Singh S (2020) Economic abuse and family violence across cultures: gendering money and assets through coercive control. Crim Coercive Control Family Violence Crim Law:51–72
Singh S, Bhattacharyya R (2020) A review of domestic violence against women in India during lockdown. Int J Innov Creat Change. Special Edn: COVID-19 Life Beyond 14(6):230–242. https://www.ijicc.net/images/Vol_14/Iss_6/PUL015_Singh_2020_R.pdf
Solotaroff JL, Pande RP (2014) Patterns of violence against women and girls in South Asia.
Thomas B, Trivedi H, Subhash R, Pathak SN (2019) Understanding proximate factors associated with perpetration of intimate partner violence by men in India. Space Cult India 7(3):29–39. https://doi.org/10.20896/saci.v7i3.574
United Nations (UN) (2020) Monthly briefing: covid-19 disrupting lives, economies and societies
United Nations Development Programme (UNDP) (2020) Almost 90% of men/women globally are biased against women. https://www.undp.org/press-releases/almost-90-men/women-globally-are-biased-against-women
United Nations Women (UNW) (2020) COVID-19 and ending violence against women and girls
United Nations Women (UNW) (2020) Violence against women and girls: the shadow pandemic. https://www.unwomen.org/en/news/stories/2020/4/statement-ed-phumzile-violence-against-women-during-pandemic
Usher K, Bradbury Jones C, Bhullar N, Durkin DJ, Gyamfi N, Fatema SR, Jackson D (2021) COVID-19 and family violence: is this a perfect storm? Int J Ment Health Nurs 30(4):1022–1032
Wagers S (2020) Domestic violence growing in wake of coronavirus outbreak. The Conversation 23
Weitzman A, Behrman JA (2016) Disaster, disruption to family life, and intimate partner violence: the case of the 2010 earthquake in Haiti. Sociol Sci 3:167–189
Wolfe DA, Jaffe P, Wilson SK, Zak L (1985) Children of battered women: the relation of child behavior to family violence and maternal stress. J Consult Clin Psychol 53(5):657
World Health Organization (2013) Global and regional estimates of violence against women: prevalence and health effects of intimate partner violence and non-partner sexual violence. World Health Organization
World Health Organization (2021a) Devastatingly pervasive: 1 in 3 women globally experience violence. New York
World Health Organization (2021b) Violence against women prevalence estimates, 2018: global, regional and national prevalence estimates for intimate partner violence against women and global and regional prevalence estimates for non-partner sexual violence against women
Zahran S, Shelley TOC, Peek L, Brody SD (2009) Natural disasters and social order: modeling crime outcomes in Florida. Int J Mass Emerg Disasters 27(1):26–52
Zastrow C (2010) Introduction to social work and social welfare, 10th edn. Brooks/Cole, Belmont CA

Chapter 10
Going Beyond the Traditional Microcredit Approach for Socioeconomic Development: A Case of a Slum from Bangladesh

Md Mahmud Hasan and Tulshi Kumar Das

Abstract Concomitantly, microcredit receives hails and condemnation because of its complex facets in the development discourse. However, its role in reducing poverty and ensuring women empowerment from the downtrodden has often been appreciated. Criticisms such as high-interest rates, forcible collection of instalments, pushing the poor into a debt cycle and exploiting the poor by transferring their capital to multinational financial corporations are well known. The study shows an alternative microcredit approach different from those of leading MFIs since 2000 for poverty alleviation and women empowerment. *Pathgudam Duldul Camp Landless Cooperative Society* (PDCLCS) has been successfully operating a microcredit program in a slum located in the Mymensingh district with almost 100 per cent credit recovery without forcible collection of instalments. Slum dwellers seem to have benefited from the microcredit program of PDCLCS adopting a community development approach. The study follows in-depth interviews, focus group discussions, and key informants' interviews to explore the facts of success. The findings show that the slum dwellers feel confident about paying their instalments as they own their capital through a cooperative system managed and operated by themselves, which other NGOs generally do not practice. Moreover, the practice of democratic norms, quick response to the crisis of slum dwellers, and borrowers' socioeconomic development and empowerment have triggered the success of an alternative microcredit program. Instead of the neoliberal practice of traditional microcredit programs, the community-owned capital approach of PDCLCS could be an alternative, aiming at ensuring the socioeconomic development of the downtrodden.

M. M. Hasan
Department of Public Administration, Shahjalal University of Science and Technology, Sylhet, Bangladesh
e-mail: mahmud-pad@sust.edu

T. K. Das (✉)
Department of Social Work, Shahjalal University of Science and Technology, Sylhet, Bangladesh
e-mail: dastulshi-scw@sust.edu

S. Anand et al. (eds.), *Sustainability in South Asian Cities*, Advances in Geographical and Environmental Sciences, https://doi.org/10.1007/978-981-97-7455-5_10

Keywords Microcredit · Neoliberalism · Social capital · Slum eviction · Urban poverty · Bangladesh

10.1 Introduction

The instances of improving the living standards of the poor through offering credit by the government and other institutions are not very atypical in history. Coupled with the longstanding examples of usury laws and Islamic prohibition on interest, various historical money-lending institutions directly supplied credit to the poor (Banerjee 2013; Guinnane 2011). However, the modern microcredit system, introduced by Bangladeshi Professor Dr Muhammad Yunus in the 1970s in a village next to the Chittagong University campus, drew attention globally and occupied one of the central places in development policy (Morduch 1999). From the beginning, with social and profit missions, NGOs like BRAC, Grameen Bank (GB) and many other MFIs provide credit to the poor to whom mainstream banks are unwilling to lend (Banerjee and Duflo 2013). Because of its impressive success on both missions, Dr Yunus and Grameen Bank were awarded Noble Prize for Peace in 2006, and this has been portrayed as a successful model of poverty alleviation in the global media and academic arena (Banerjee and Duflo 2013; Bloomberg 2022). According to Convergences (2019), about 140 million borrowers received $124.1 billion from Microfinance Institutions (MFIs), where 80% were women, and 65% were rural inhabitants.

Since the beginning, it has been considered an excellent opportunity for the poor to receive credit from NGOs instead of exploitative local money lenders by avoiding outrageously high-interest rates (Banerjee and Duflo 2013; Bateman 2010). Initially, microcredit programs were appreciated for their role in poverty alleviation, women empowerment, self-employment creation, and about 95 per cent of loan recovery (McKernan 2002; Morduch 1999). However, the rise of neoliberal economic policy in the 1980s stimulated the microcredit system targeting the rural poor by the donors. Consequently, from the 1980s to the end of the twentieth century, the number of NGOs increased rapidly with the assistance of foreign funds from donor agencies (Karim 2008; Muhammad 2009). In addition, GB and BRAC extended their activities globally, having joint ventures with the World Bank and multinational corporations, which turned into a kind of group of companies, ignoring their initial promises to provide social well-being to the poor and reduce inequalities (Muhammad 2015).

At this stage, many other philanthropists and global commercial institutions got involved with microcredit, transforming it into microfinance with various micro offers like ensuring savings and introducing insurance and so on (Bateman 2010; Karnani 2007; Mia et al. 2019; Rankin 2001). In the following years, microfinance became the most instrumental global phenomenon, and the United Nations declared 2005 as the International Year of Microcredit, Muhammad Yunus and GB Bangladesh received the Nobel peace prize in 2006, describing microcredit as an ever more effective instrument in fight for against poverty (Yunus 2006). These created a powerful narrative regarding the effectiveness of microfinance in alleviating poverty, as Yunus declared with confidence that the poverty will be put into the museum by 2030 in Bangladesh (Karim 2017; Karnani 2007; The Daily Star 2007). Correspondingly, many scholars find the positive role of microfinance in bringing up the socioeconomic development of Bangladesh, which has been achieved in the last decades (Drèze and Sen 2013; Khandker and Samad 2017).

After the first decade of the twenty-first century, microfinance began to experience a period of crisis, leaving behind its golden age (Haldar and Stiglitz 2016). Since then, many scholars have become sceptical about the contribution of microcredit to women's empowerment, poverty reduction, education, healthcare etc. (Banerjee and Duflo 2013; Das and Pulla 2014; Haldar and Stiglitz 2016; Karnani 2007; Sinclair 2012). Moreover, several scandals like borrowers committing suicide after failing repayment, selling organs for paying instalments, siphoning off funds and indulging in corruption etc., had been revealed associating the leading MFIs, including GB, which flooded global media (Associate Press (AP) 2012; BBC News 2013; Biswas 2010; Karim 2017; Swapan 2022). Moreover, many studies conducted in the last two decades have highly criticised the neoliberal approach of MFIs (Bateman et al. 2019; Chowdhury and Willmott 2019; Mia et al. 2019; Weber 2002). Negative aspects of microfinance, like high-interest rates, creating debt traps, causing agrarian dispossession, public shaming technique, forceful loan renewal, developing poverty traps etc., have frequently been mentioned in recent studies (Chowdhury and Willmott 2019; Haldar and Stiglitz 2016; Karim 2008; Maîtrot 2019; Paprocki 2016; Sandberg 2012). Moreover, the existing microfinance system is widely depicted as a neoliberal tool, which is being used as a megaprofit-making technique of MFIs of Multinational corporations (Bateman 2017; Karim 2017; Muhammad 2015). Therefore, the potential of microfinance appears to be in severe crisis now, which Dr Muhammd Yuns also echoed in his article published in the New York Times, where he urged to save microfinance from the megaprofit motive of MFIs (Yunus 2011).

Within this context, the study explores an alternative microcredit approach that has been practised by the dwellers of a slum in Bangladesh since 2000. It explains the mechanisms and features of this system and investigates its potential for being an alternative aimed at replacing the existing neoliberal microcredit system.

10.2 Traditional Microcredit Approach and Its Limitations

10.2.1 Features of the Traditional System

Dr Muhammad Yunus primarily designed the microcredit system to protect rural women from the excessive interest rate of local moneylenders (Wahid and Hsu 2000; Yunus and Jolis 2007). In the 1980s, the MFIs in Bangladesh followed the approach, and later it became global through profit-making neoliberal mechanisms

(Bateman 2017; Muhammad 2015). Because of geographical locations and organisational mottos, rules and regulations of microfinance differ to an extent, but in most cases, MFIs follow the same strategies around the world and have become a multibillion-dollar industry (Banerjee 2013; Haldar and Stiglitz 2016; Mia et al. 2019). Traditional microfinance introduced in a community starts with a motto of making savings for all group members. Saving is mandatory for a borrower in the microcredit system, as it is a precondition for receiving a loan from an MFI (Yunus and Jolis 2007). This practice created a positive mindset among the borrowers as they could use it during their crisis (Banerjee and Duflo 2013; Choudhury Haque et al. 2017). However, after the 1990s, flexible savings replaced mandatory savings without following a proper legal framework (Dowla and Alamgir 2003).

Group lending and group liability are essential characteristics of the traditional microcredit system, which makes tracking, monitoring, and ensuring repayment of the instalments easier for MFIs (Haldar and Stiglitz 2016). In this process, group members are jointly responsible for the failure of loan repayment (Suzuki et al. 2013). In the case of most MFIs, the borrowers sit at a particular spot every week, and the group leader collects the instalment from every member, which is handed over to the credit officer (Banerjee 2013). Moreover, women are given priority in MFIs lending, aiming to empower them in their families and societies (Das and Pulla 2014). However, achievements relating to gender characteristics of microcredit are questionable, as various studies show intense male dominance in the families of Bangladeshi society (Haldar and Stiglitz 2016).

Noncredit programs like vocational and income-generating training for the borrowers are integrated aspects of the microcredit system (Alam 2013; McKernan 2002). Before receiving the credit from an NGO, the group members should receive vocational training to invest the loan in income-generating activities, which helps the borrowers directly by increasing their profit (Jansen and Dsw 1998). Alam (2013) and McKernan (2002) identified more profits from microcredit loans than that of commercial banks. Moreover, Nikkhah and Redzuan (2010) explain how NGOs' noncredit programs make the borrowers aware of attaining sustainable community development. Other key characteristics of the traditional microfinance system include small repayment, weekly payment and disbursement of a new credit after repayment of the previous loan (Banerjee 2013).

10.3 Limitations of the Existing Microfinance System

The microcredit system is often praised as an effective strategy for alleviating poverty. McKernan (2002) and Khandker and Samad (2017) found that both the credit and noncredit approaches of microcredit contribute to poverty reduction in Bangladesh by providing credit to obtain capital inputs, transforming human beings into human resources. Drèze and Sen (2013) appreciated the contributions of the NGOs and their microcredit program to the recent progress of Bangladesh reflected in its various social and economic indicators, especially for women. However, Huda

(2020) disagreed with Drèze and Sen (2013), arguing that the relationship between NGOs and rural women in Bangladesh is always problematic and distrustful. He has criticised the NGOs for their approaches, which have engineered various neoliberal mechanisms, making the rural poor socioeconomically more vulnerable. Moreover, Banerjee and Jackson (2017) have strongly criticised the market-oriented neoliberal microfinance approach. Their study found that this approach has increased the borrowers' indebtedness, throwing them into a situation of helplessness.

The impacts of microfinance on the lives of the poorest and the overall economy are debatable. Raihan et al. (2017) assessed the impacts of microfinance on the macro economy of Bangladesh, showing its positive effects on the labour market, which add 9 to 12 per cent to the national GDP and between 13 and 17 per cent to the rural economy. However, Banerjee (2013) was sceptical about the impacts of microfinance on the lives of the poor. He identified insignificant implications on their daily consumption, education and health improvement. On the other hand, Gehlich-Shillabeer (2008) showed the adverse impacts of microcredits on the lives of the poorest in Bangladesh, often causing indebtedness and a potential poverty trap.

The borrowers get the opportunity to improve their economic well-being by using microfinance. Because of the group lending approach, many studies (Field et al. 2010; Ito 2003; Tahmasebi and Askaribezayeh 2021) explained that microfinance triggers social interactions that significantly contribute to building social capital among the borrowers. However, Hossain (2022) has argued that the borrowers use their already established social capital to repay the instalment, which helps increase the recovery rate of the loan significantly, but that does not improve the economic status of the poorest borrowers.

Apart from the economic and social contributions, it has been argued that microfinance makes borrowers politically aware, which helps them understand their political, social and economic rights. Consequently, this awareness enables them to ensure their rights (Bayulgen 2008). Conversely, several studies (Bateman 2017; Bateman et al. 2019; Karim 2011; Muhammad 2009) show how the poorest borrowers become the victims of MFIs as their funds being drained to multinational corporations, but they failed to raise their voices against this sort of injustice meted out by the field officials. For example, Muhammad (2009, 2015) explained how the leading MFIs, including GB, siphoned the borrowers' profits to create multiple independent companies. Bateman (2010, 2017) described the process of using MFIs to expand global capitalism to the grassroots and made them the profit-making clients of Wallstreet. As a former employee of the MFIs, Sinclair (2012) has explained the involvement of the USA and Europe-based corporations in controlling the global MFIs.

Consequently, profits made from the poor by imposing excessive interest rates are enjoyed by corporate investors, ignoring their social commitments. It pushed capitalist globalisation to the grassroots of Bangladesh, causing women vulnerable to the interest of neoliberal economic policies. The microcredit programs have made the rural people dependent on the NGOs, turning them into consumers of multinational corporations' products like cell phones, seeds, fertiliser, etc. (Karim 2008, 2017; Weber 2002).

A higher percentage of loan recovery and flexible repayment schemes are widely recognised as success tales of microfinance. McKernan (2002) and Jain and Moore (2003) showed that the monthly repayment rate of microcredit ranges from 90 to 95 per cent, much higher than commercial banks. The borrowers are trained from noncredit services of MFIs in different trades and involved in varied income-generating activities, which generally play the most critical role in this higher repayment rate. Moreover, the group lending system and the program's obligations contribute to the repayment success of MFIs (Al-Amin and Islam 2020; Jain and Moore 2003). However, social pressure from peer group members has been considered an instrument for loan repayment, which has been shown a success. Karim (2008, 2011) described this process as a 'shame economy'. Moreover, Banerjee (2013) criticised the existing repayment system introduced by the MFIs, which compelled the borrowers to repay after the first week of loan receiving.

10.4 Methods and Study Area

The institutional ethnography approach of qualitative methods has been used in this study to understand the dwellers' experience regarding the microcredit programs operated in the *Patgudam Duldul Camp* Landless Cooperative Society (PDCLCS) since the 1990s. This approach is considered a method of inquiry to understand people's attitudes and reactions in a particular setting or definite local conditions, describing an individual's experiences within the institutional relationships and outside the institution that creates an understanding which pushes them to act a particular way (McCoy 2006; Rankin 2017; Smith 2005). The data of this study were collected from September to December of 2021 to understand the borrowers' experience and action toward MFIs operating microfinance programs in this slum and their knowledge gained beyond that MFIs that helped them act in different ways. During this period, 27 in-depth interviews, 8 Key-Informent interviews, and 3 FGDs were conducted among the slum dwellers, general members, the executive committee members of the PDCLCS and field officials of the MFIs working in Duldul Camp Patgudam slum. All the participants in the study voluntarily participated and gave their oral consent before the interview, and the researcher thanked them for sharing their experiences.

All the participants of this study were selected using judgemental sampling. The participants for the in-depth interviews were 27 (21 female and 6 male), selected based on three criteria: gender, membership tenure of the PDCLCS, and loan borrowing status. Among them, the PDCLCS members were 20, and the rest 7 were non-members but slum inhabitants. The members of PDCLCS were selected considering their tenure of membership: less than three years (6 participants), 4 to 10 years (6 participants), and over ten years (8 participants). The variations of these categories facilitated understanding the borrowers' perceptions towards PDCLCS and how its activities evolved over the decades. All non-members of PDCLCS selected for in-depth interviews were borrowers of MFIs and relatively new dwellers or preparing for migration elsewhere. They were selected for interviewing to understand their perspectives towards the PDCLCS as not being members of it. Two FGDs, with the members of the PDCLCS members, were conducted inside the slum. The president, secretary, vice-president, cashier and manager of the PDCLCS, and three NGO officials were the key informants who were interviewed to develop a better understanding of the role of the PDCLCS and MFIs in this slum.

All the interviews were conducted in Bangla, the participants' mother tongue, with each interviewee's approval and consent. The proceedings were recorded using a digital audio recorder for preservation and analysis. Subsequently, the recordings were transcribed, translated into English, and subjected to a process of coding in order to identify and categorise recurring themes.

Data were collected from different sources to crosscheck them so that the validity and reliability of the data could be ensured. The researcher stayed a long time with the slum dwellers and interacted with many segments of its population, which helped them extract genuine information to analyse and conclude the participants' perceptions. It may be noted that a substantial number of the slum dwellers have been interviewed and interacted in different manners, but many of them have not been reached out. Moreover, the current study was conducted on a microcredit system practised in a particular slum, which has not been studied in other communities. Therefore, it becomes difficult to conclude that this microcredit system could benefit others. There is an opportunity to undertake further studies on this microcredit system practice for different areas in the country.

10.5 Case of Duldul Camp and Its Alternative Practices of Microcredit

The emergence of Patgudam Duldul Camp Slum and their Cooperative Society

The *PDCLCS* is situated at the *Patgudam Duldul Camp* slum of Mymensingh division, which is about 120 km away from the North of Dhaka, Bangladesh. Around 2300 people of 315 families live on approximately 2 acres of land. Most of them have come here from different regions of the country, seeking a job and pursuing a better living standard. An average of 4 persons live in a 120 square-feet house in this slum. Most of the working people (adult men and children) in this slum are hawkers who sell fruits, cakes, chutney, pickles, boiled eggs etc., made by their female family members at home. Near this slum, the Bangladesh-China friendship bridge over the Brahmaputra river is the source of different earning activities for the dwellers, as many of them sell homemade products around the bridge. Other notable occupations of the dwellers are small businesses, rikshaw pulling and daily labour, whose monthly income ranges from about TK. 10 thousand to TK. 15 thousand. A primary school and a madrassa have been run inside the slum by the Duldul Camp Landless Society. Many students from this slum have been admitted to college after completing their primary education in the last few years.

After the partition of India in 1947, some Bihari[1] families started living in the *Pathgudam Duldul* camp, which was government-owned but abandoned land. From 1947 to 1971, the Duldul camp was populated by refugee Bihari migrants from India and some Bengalis. However, after the liberation war of Bangladesh in 1971, only the Bengalis settled and gradually expanded there. In 1999, slum dwellers applied to lease the land but were denied because of an ongoing case associated with the land. Thus, they created a landless cooperative society aiming to secure the possession of the land from two local influential persons who strongly claimed land ownership. In 2000, the cooperative society got registration from the government and started operating its activities in the slum. Besides the *Patgudam Duldul Capm Landless Cooperative Society* (PDCLCS), many international NGOs and development agencies, including UNDP, WorldVision, CARE, GB, BRAC, ASA, and some local NGOs, have also operated microcredit and development activities here. However, no leading NGOs currently work except ASA, which now operates the microfinance program amongst very few inhabitants in the slum.

There are 315 general members of PDCLCS, along with an executive committee of nine members, which is the top decision-making authority. The general members directly elect the executive committee members for three years. The election process is very competitive, and only the inhabitants of the *Pathgudam Duldul Camp* slum can be a candidate and vote in the election. The committee appoints a manager to maintain the PDCLCS's transaction account, which is regularly audited by the government authority. The PDCLCS runs a primary school, a madrassa and a mosque in the slum with its funding. Moreover, it also contributes to different cultural, educational and religious events held in the slum. The government has given it the best cooperative society award multiple times at the national level for its extraordinary achievements.

10.6 Microcredit of PDCLCS

In the beginning, the PDCLCS tried to secure the possession of their land as they collected small subscriptions from the dwellers per week to maintain the cost required for fighting the ongoing case lodged with the court concerning the land of the slum. Later, when their savings increased, they started lending money to their members, following other NGOs operating microcredit programs in the slum. In 2020, the PDCLCS distributed about Tk 30 hundred thousand (about $30,000) of loans among the general members. It follows the same interest rate as most MFIs charge for providing a loan. According to the rules of PDCLCS, if a borrower borrows Tk. 1000 ($10), he has to repay it with interest Tk 25 (25 cents) per week, and the entire repayment has to be completed in 45 instalments. However, the PDCLCS is least bothered about where borrowers invest their loans, and it distributes 60 per cent of its annual gross profits amongst the members based on the amount of savings. The

[1] Non-Bengali Muslims migrated to East Pakistan, now Bangladesh, from the Bihar state of India after the partition of India in 1947.

remaining 40 per cent of the profit is spent on various development activities in the slum, including PDCLCS's day-to-day activities and the costs of ongoing litigation pending in the court. The borrowers repay their instalments every Friday, which is redistributed among the members who have applied for a loan. Without any collateral or nominee, any member with minimum savings of Tk 2000 ($20) can ask for credit. However, the previous repayment history is considered while releasing a new loan. The PDCLCS approves a maximum of Tk one hundred thousand ($1000) loan at a time.

10.6.1 Positive Impacts of the PDCLCS

Tenure security

All the respondents feel that the PDCLCS plays an influential role in ensuring the tenure security of the slum dwellers. Most respondents argue that the microcredit program of PDCLCS helps them survive and to remain united. The secretary of the cooperative society stated:

> Initially, the savings started through collecting subscriptions from the dwellers to bear the expenses for fighting the case in court regarding the ownership of this land. Later, the savings turned into a microfinance program, and now it has become the fuel of our every initiative. Suppose we explain the achievements of this cooperative society; we should mention its crucial role in ensuring tenure security, financial benefits and unity among us.

The importance of the microfinance program in keeping dwellers united against the encroachment of outsiders has also been echoed among the participants of the FGD. According to the participants, the microfinance program has been helping them to secure their tenure directly and indirectly. One of the participants shared,

> Primarily, other NGOs like ASA and CARE made us aware of our fundamental rights, but we were not united then. After the formation of this cooperative society, our unity has become stronger. Consequently, several times, we boldly encountered evacuation attempts by outside encroachers.

No group liabilities

Group liability is a common practice undertaken by the leading MFIs, but PDCLCS does not follow this procedure while offering microcredit. Since they have lived together for many years, they are familiar with each other; consequently, they do not need to make anybody liable for microcredit repayment. Moreover, the long-standing relationship helps them select the right borrowers to provide with credits. The president of the cooperative society explained,

> Our cooperative society members are known to each other as we have been living together for many years. Therefore, we are aware of everyone's pains, sorrows, weaknesses, and strengths. We are like a family, and we all are generally liable to each other's loans. Thus, no one should be separately held responsible for it.

One of the members of the cooperative society, who took microcredit from some other NGO, shared,

> I took a loan from an NGO and was made one of the group leaders along with two others. Suddenly, one of our borrowers left this area without repaying the loan. Consequently, as group leaders, we three had to repay her loans. That was a huge burden on my family and me. Afterwards, I had to go through mental and physical abuse from my husband. I, therefore, left that NGO after repaying the credit.

Politically empowered

Political parties and local politicians now count the slum dwellers because of the existence of PDCLCS. Solidarity among them has caused them to be politically empowered. They can effectively present their rights and demands to the local government and administration. As a result, many development activities like; water, sanitation and social safety net programs have been introduced in the slum. The secretary of the PDCLCS argued,

> Due to the PDCLCS, the slum dwellers have become united and are now in a position to draw the attention of local politicians and political parties. They are currently concerned about our problems. The local government and the field administration have also come forward to introduce development projects for our socioeconomic improvement. We're now strongly in possession of this area.

Reliable and least formalities

The dwellers of the slum stated that they face more formalities in receiving a loan from MFIs. Even they do not understand their actual interest rates, conditions and other issues. However, they feel that obtaining a loan from their cooperative society is a more straightforward process. A borrower who received loans from an NGO stated,

> We go through a complex procedure to receive a loan from some NGOs, which is tedious. You have to fill up and sign in various forms. I am not educated enough to understand everything, and I trust their promise without understanding enough. I received a lesser amount of loan than I applied for, which I didn't understand why. But, taking a loan from our society is much easier and more reliable because I know the people and can charge them If cheated.

Opportunity for Savings

The PDCLCS offered savings opportunities for its members without taking loans, which is impossible for MFIs. The members of this cooperative society can deposit money and receive profit based on their deposited amount. On the other hand, NGOs do not allow deposits without taking a loan, discouraging the slum dwellers from continuing with MFIs. Unconditional deposit opportunity provided by the PDCLCS encourages slum dwellers to save money, which helps them deal with future crisis. One of the female members of the *Duldul Camp Cooperative Society* shared,

> I was a member of various NGOs operated here at different times. I haven't found one among them which would allow me to deposit without taking a loan. Their rule is that you cannot save without a loan. They never care about us, considering only their interest. But,

> the microcredit program of our cooperative society allows savings with interest even without taking a loan. I deposited little money weekly with our society, which I later used to bear the education expenses of my elder son.

10.7 No economic exploitation

Most respondents perceived that they were economically exploited by the MFIs, which had been operating in the slum over the last four decades. Besides, many of them feel that PDCLCS has saved them from economic exploitation. According to them, none of the MFIs has ever shared any part of their profit with the locals; instead, the money was drained into the NGOs' funds. However, the PDCLCS distributes their profit among the shareholders regularly, making this program popular in the community. This opportunity has developed a sense among the dwellers that the PDCLCS has freed them from the MFIs economic exploitation. The President of the PDCLCS explained,

> Since the end of the 1980s, many NGOs like Grameen Bank, BRAC, ASA, RDRS etc., have been operating the microfinance program in this slum, but we never received any part of the profit they made from us. Their attitude was different as we never thought of them as our own people. They treated us like customers of their loans as other money lenders did. After the success of our own credit program, we can now understand how much money they have transferred from our community to their bank accounts over the years.

A female participant of the FGD stated,

> NGOs started operations in this slum, promising to do good for us. Initially, they did some excellent activities, like making us aware of our rights and providing sanitation facilities. Later, their attitude changed, and they focused on just selling the loan to us. Their loan is easier to get than the traditional banks, but the interest rate is high. The credit program of our society follows the same interest rate, but we are not worried about that because we know we'll get a return from this. But the NGOs exploited our situation caused by poverty. On the other hand, PDCLCS has reduced economic exploitation.

Development activities

The PDCLCS has been playing a significant role in the infrastructural development of the slum. The City Corporation implemented most development activities under the Urban Partnership for Poverty Reduction (UPPR) project in this slum, which UNDP has funded. Because of this project, the road, supply of drinking water, sanitation and drainage systems in this slum have substantially been improved. The slum dwellers spontaneously participated in all these development activities. Moreover, the PDCLCS has been involved in varied activities like supporting children with sports equipment and financial support to the local school, mosque, madrassa and other cultural events. The field organiser of the UPPR project, who worked here from 2010–2012 , explained the role of the PDCLCS as a success of their program. He stated,

> As a community organiser of UPPR, I was assigned to organise four communities, including the Pathgudam Duldul Camp slum, to improve the living standards of the poor slum dwellers. During the work, I got the most positive response from the dwellers of the Pathdudam Duldul Camp slum. I think the previous cooperative society had a role in facilitating positive responses from the dwellers. It made the work easier for us, leading to accomplishing development and other work. Because of the PDCLCS, they were already organised, having privileged and received a prompt response from City Corporation and other government bodies. As far as I know, the project has been running successfully in Pathgudam Duldul Camp Slum.

The vice president of the PDCLCS shared;

> The PDCLCS has united us, and our national recognition increased our importance to the City Corporation, government offices and political parties. We had development works because of our reputation. Basic necessities are fulfilled in our slum now as we pursue development works with the City Corporation, UNDP and other government offices. Currently, all of them treat us with respect because of the reputation of PDCLCS.

Law and order maintained

Criminal activities like drug dealing and robbery are widespread in the slums of Bangladesh (Atkinson-Sheppard 2016; Khan 2013). However, the situation in the slum under this study is different because the role being played by PDCLCS in maintaining law and order seems effective. It arbitrates within the slum to address any crime peacefully committed by the dwellers. The cooperative society also punishes criminals or handover them to the police if needed. The cashier of the PDCLCS stated,

> If you look at the nearby slums, you will find many complaints of drug use and robbery. But that doesn't happen here. We are concerned about the reputation of our slum. If any complaint is lodged against anyone, our association act immediately.

One of the participants of the FGD explains,

> The PDCLCS regularly warns us not to be engaged in any criminal activities. We are very strict in this regard. Besides, most of the children in this slum are involved in the study or income-generating activities. Thus, there is a least chance for them to be involved is such illegal activities.

Social-Capital Increasing

One of the most crucial achievements of the PDCLCS is increased social capital among dwellers. Because of improving their social capital, these slum dwellers have benefited in multiple ways. It has strengthened their unity, ensuring tenure security, economic prosperity, and infrastructural development. Now the slum dwellers easily draw the attention of government officials and others. A participant of an FGD revealed,

> Because of the PDCLCS, our love and affection for each other have increased. It has taught us that our possession of this slum land will surely be lost if we do not remain united. Our unity and integrity benefit us in many ways.

Profit Sharing

One of the significant aspects of the PDCLCS's microcredit program is its profit-sharing policy. Sixty per cent of its annual profits are distributed amongst the members according to their deposits and share, which is generally uncommon in other MFIs. The dwellers appreciate it very much, and many respondents shared that they have left other MFIs because of the high profits available in this society. One of the dwellers shared,

> I have left ASA as it always puts me under pressure. I calculated that they took the entire profit from the money I deposited. But, I am getting part of the profit of what I am paying to the PDCLCS. So, why should I take their loan?

The cashier of the PDCLCS argued,

> An NGO profits Tk 150 in 11 months by providing Tk 1000 to a poor woman. This profit excludes service charges, insurance and other deductibles. If Tk 3 million loans are distributed, imagine how much profit they can gain in a year. It is about Tk 0.45 million!! We now understand their exploitation and have stopped taking loans from them. The profits that are made are shared or reinvested for our community, but most NGOs take away money from us.

Sense of ownership and pride

The slum dwellers have developed a kind of ownership towards the *PDCLCS.* They always consider the *PDCLCS* as their own, taking pride in its success and achievements at the national level. On the other hand, they have grown negative attitudes toward microcredit-providing NGOs, describing them as profit-making organisations involved in the business by taking advantage of their poverty. One of the participants of an FGD shared,

> We created the PDCLCS for our protection and development. It has been protecting us from eviction and illegal possession of the slum land by outsiders. Later, the PDCLCS introduced the microcredit program, contributing to our socioeconomic development. Because of its achievements and contributions *to the slum dwellers, it has received the gold medal award* twice from the prime minister. All the dwellers of this slum are proud of it.

A participant in an FGD echos,

> The PDCLCS is our organisation, and everyone in this slum owns it from their heart. But the NGOs are outsiders doing their business and making profits on our hard-earned money. You won't find any sympathy for us from the microcredit operating NGOs as they forcefully collect instalments during our crisis. They treated us inhumanly when we failed to repay an instalment. These NGOs should not be compared with the PDCLCS.

Easy repayment

The PDCLCS's flexible attitude in instalment repayment has made it more popular than other MFIs working in this slum. The dwellers are critical of the insulting behaviour of the MFIs' officials while collecting the instalments. The officials of PDCLCS never force or insult their members if they fail to repay one or even more instalments. One of the members of the PDCLCS who has also taken a loan from an NGO stated,

> The NGO officials are very aggressive. They never try to understand our crisis. We're bound to pay our instalment weekly by any means, and no excuse will be accepted by them. If I fail to repay an instalment, the NGO field officials come and sit at my house until the instalment is paid. They insult us before the other dwellers, which is humiliating, but the PDCLCS never does it with anyone. The management of PDCLCS never forces me to repay. If I failed to repay an instalment, I just let the cashier know about my problem. But, usually, I don't try to be late repeatedly. If I do it, I may not be considered for a loan next time.

The vice president of the PDCLCS explained their flexible payment system,

> If a member of our society fails to repay his/her instalment, we never force to repay immediately. If he doesn't inform us of the problem, we visit his house to know the situation. If a few weeks go by without repayment, then we sit with him/her and try to make him/her understand to repay the instalments. We have been living here for many years. So we know everyone's strengths and limitations. It is not required to force or insult anyone like the NGOs do, and it is not our strategy.

Hundred per cent recovery

The credit repayment rate of the PDCLCS is about a hundred per cent. The loan default case of it is rare with its microcredit program. Some PDCLCS members do not pay their instalments on time, but running away without repaying or not repaying at all does not happen. When a member leaves the slum permanently, his/her borrowed money is adjusted with his/her savings. The feelings of ownership for the PDCLCS motivate its members to repay. Moreover, the PDCLCS's members have been living in the slum for many years, contributing to this high recovery rate.

> The recovery rate of our society is about a hundred per cent, and it is rare to find any loan defaulter in our credit history. As we don't force anyone to repay, so late repayment is common among some of our members. Some can't repay the total loan in 25 instalments. In such cases, the borrowers pay off the entire loan in delayed instalments; and we accept it. Everyone repays the loan amount because she knows that the PDCLCS is his/her organisation.

An FGD female participant explains why she does not flee away without paying the loan,

> Why should I flee away without repaying? We created this society for our betterment, which is helping us in many ways. Our livelihood has been revolving around this slum. No one ever thinks of being a loan defaulter in our society.

Savings brings out more facilities

Different facilities based on savings have made the PDCLCS popular among slum dwellers. For example, if any member fails to repay a loan instalment, the MFIs deduct the instalment amount from his savings. Conversely, the PDCLCS never adjusts the pending instalment with the savings of the borrower. A borrower can easily withdraw the entire or any portion of deposits without interest in any crisis, which is impossible in other MFIs. These facilities generally encourage the members to save as they can use it during emergencies. Many members of the PDCLCS deposit a certain amount of money every week because of these facilities without taking a loan. The President of the PDCLCS shared,

> We never deduct an instalment from a member's savings. A member deposit this money for his future. If we deduct or adjust instalments from their savings, then it will reduce their annual return with interest. The NGOs deduct their members' savings if someone fails to repay an instalment. The main purpose of the NGOs' savings is to secure their instalment. They follow the business attitudes with us that we never do it. We have formed the PDCLCS to serve our community members.

Another female interviewee shared,

> I have been a member of PDCLCS for the past eighteen years. I have taken credit from this society three times so far and paid on time. Though I have no loan now, I regularly deposit money there for savings. When I fall into a crisis like a major illness, then I withdraw any amount of money from my savings without any hassle. It pays no interest; It is my money. If these opportunities were not there, I would have borrowed money with interest from someone else. It is a huge opportunity for us. I know many members who do not take a loan but deposit money here to use it during a crisis.

No group shaming

The dwellers are very critical regarding the attitudes and behaviour of the MFIs in collecting their instalments. According to the dwellers, shaming techniques, including insulting and abusing in front of all, are widespread practices followed by the MFIs' field officials. Conversely, the PDCLCS receives high praise for not following shaming techniques in its instalment collection process. Instead, it follows flexibility and other motivating methods, like consultations in the instalment collection process. Moreover, the group liabilities credit system plays a central role in the shaming technique of the loan collection procedure of the MFIs, which is absent in the PDCLCS's credit system.

One of the female participants of an FGD said,

> You're bound to repay an NGO's instalment in any circumstance, no matter what danger you face. In any case, if you fail to repay an instalment, the field officer will hunt you and finally treat you like a criminal. Later, the other members of your group will also ask for money as the officer pressures them. Insulting and using foul language is a common practice followed by the officials and sometimes by the group members in NGOs loan. It is totally absent in the PDCLCS's loan. We have never seen the PDCLCS insults any borrower for collecting instalments.

The manager of PDCLCS explained;

> We never insult or abuse any of our borrowers to collect instalments. It is out of our practice. The NGOs practice it because they are doing business with us; we aren't. Why should I use foul language or insult others for what is not my personal property?

Role in decision making

An executive committee consisting of nine members take the day-to-day management-related decisions. They also decide whom to lend to and other relevant expenses. However, it is found that the general members participate in the meeting where important decisions are made, increasing their sense of ownership of the PDCLCS. Besides regular auditing and a democratic leadership selection process, it also helps to ensure accountability and transparency of the PDCLCS.

10.8 Other NGOs Cannot Survive Here

It is found that the number of microcredit-based NGOs that have been working in the *Pathgudam Duldul Camp* slum has decreased over the decades. According to the slum dwellers, an MFI began microcredit operations in this slum at the end of the 1980s. Within two decades, many leading MFIs, like GB, BRAC, and ASA, started operating their microfinance activities here. Moreover, many local NGOs also launched their microfinance program. Consequently, at the beginning of 2010, about ten microfinance-operating NGOs were active in this slum, and most of the dwellers turned into their borrowers. However, with the success of PDCLCS, the borrowers of the MFIs started declining. Consequently, only the ASA currently operates its group-based microcredit program among a limited number of dwellers. Moreover, some dwellers have borrowed money from NGOs working in neighbouring areas. However, it is evident that most of the MFIs could not survive here after increasing the credit volume of the PDCLCS.

The vice president of the PDCLCS shared,

> After the devastating flood of 1988, some NGOs started operating microcredit programs here, and their number increased within two or three decades. As a result, most dwellers received loans from them, hoping to improve their economic conditions. But the reality was different as many fell into a debt trap. Many families failed to repay their loans and were forced to sell their possessions. They had to leave the slum forever. But the success of the PDCLCS changed the situation, and most borrowers stopped taking loans from NGOs. Now, you'll find only one or two NGOs operating microcredit programs among the limited borrowers.

In this regard, one of the borrowers of ASA explained why it survived here. He argued,

> ASA has changed its program strategies now. A borrower can get loans from ASA without having any savings. He receives a loan by paying the cost of a credit form. But you must have a minimum of Tk 2000 ($20) in saving to be eligible to apply for a loan from PDCLCS. Moreover, ASA can provide a larger loan than the PDCLCS can. That is why the ASA is still surviving here.

10.9 Discussion

Microcredit programs were first introduced in Bangladesh by some NGOs with social and profit missions to reduce poverty and empower rural women (Banerjee and Duflo 2013; Yunus and Jolis 2007). Later, it received recognition worldwide as an effective tool for reducing poverty and was awarded the Nobel peace prize in 2006 (Banerjee and Duflo 2013; Mia et al. 2019; Sinclair 2012). However, in recent decades, microfinance operated by the MFIs has been receiving massive criticism on many limitations relevant to its role in the expansion of the neoliberal capitalist agenda at the grassroots by using the capital of multinational corporations (Bateman et al. 2019; Karim 2011; Muhammad 2009; Sinclair 2012). Nevertheless, the microcredit approach followed by the PDCLCS at the *Pathgudam Dulcul Camp* slum in Bangladesh has provided some positive dimensions leading to a success story.

Muhammad (2009, 2015), Sinclair (2012), and Bateman (2010, 2017) have criticised the existing microcredit system because of its connections with multinational financial corporations, which extract the profits from the money of the poor community, transferring them into their Wallstreet accounts. The PDCLCS does not have any chance to transfer the community funds to an outside account. The study finds that sixty per cent of the profits earned from the loan are distributed among the members, and the rest forty per cent is spent to bear the operation costs and community development activities. Thus a poor community hardly has a chance to be used as a profit-making machine.

The study shows a positive impact on the savings tendency of the members of PDCLCS for use in a future crisis. This outcome is contrary to that of Banerjee and Duflo (2013) and Choudhury Haque et al. (2017), who found a positive tendency towards savings among the borrowers of MFIs for meeting emergencies. However, the saving tendency of the borrowers of MFIs in the *Pathgudam Duldul* camp does not seem to be linked to addressing uncertainties. On the contrary, meagre interest rates and deducted instalment amounts from savings for the loan defaulters appear to have made the MFIs borrowers disinterested in savings. Moreover, the flexible saving policy replaced by a mandatory one, as described by Dowla and Alamgir (2003), could also be a reason.

The study supports the findings relating to the high recovery rates of the MFIs found in previous studies (Jain and Moore 2003; McKernan 2002). Group lending, forceful approach to defaulters, social pressure, etc., play vital roles in this regard (Al-Amin and Islam 2020; Field et al. 2010; Karim 2008, 2011). Interestingly, the PDCLCS has ensured a hundred per cent loan recovery without using force and social pressure. The sense of ownership among the members towards the PDCLCS has perhaps made it possible.

The findings of this study are closer to Hossain (2022) than to Ito (2003) and Tahmascbi and Askaribezayeh (2021) in the case of increasing social capital through microfinance provided by the MFIs. This study shows that the PDCLCS has been playing a significant role by uniting the slum dwellers, which helps secure the possession of slum land and facilitates other development activities. However, the MFIs cannot play such a role among their borrowers in this slum. Hossain (2022) argues that already-established social capital among the borrowers of MFIs facilitates the repayment of instalments without substantial socioeconomic improvement, which supports the current study.

The findings of the study revealed that the PDCLCS has politically empowered the slum dweller, which is helping them to ensure their socioeconomic rights, as Bayulgen (2008) expected from the borrowers of MFIs. The study participants feel that the MFIs-operated program does not secure their political rights. However, the study findings support the argument of Huda (2020), which shows the tense relationship between the borrowers and the MFIs. The activities of the PDCLCS have made the dwellers conscious of the exploitation made by the MFIs. Consequently, the number of MFIs in this slum has been gradually decreasing.

Although most of the slum dwellers feel the microcredit program of PDLCS has improved their socioeconomic conditions, it may not be alone responsible for it.

The UPP project and some NGOs have also been working in the slum to improve their conditions. In addition, this is to be mentioned that the Bangladesh-China friendship bridge situated nearby has generated different kinds of employment for the slum dwellers, which may be one of the primary reasons for their overall improved socioeconomic status.

The PDLCS microcredit program seems to be running successfully because of the strong unity among the slum dwellers. The slum dwellers appear to have developed a strong unity among themselves to prevent the illegal encroacher from taking possession of the slum land. Thus, the microcredit program of the PDLCS needs to be experimented with in some other settings in order to understand its effectiveness in a real sense. However, innovative ideas applied in the PDLCS microcredit program may help address many of the limitations of traditional microcredit programs of MFIs.

10.10 Conclusion

The microcredit programs initiated by NGOs in the mid-1970s, spearheaded by Dr Muhammad Yunus, aimed to eliminate poverty among marginalised women. In the first few decades, apart from poverty alleviation and women empowerment, the role of microfinance in bringing about socioeconomic development for the marginalised had been globally appreciated. Consequently, many global financial institutions, directly and indirectly, launched microfinance programs, considering it an effective tool for making profits. This microfinance program of MFIs has often been criticised for implementing neoliberal agendas at the grassroots. Some of the negative impacts of microfinance programs on borrowers, like economic exploitation, high-interest rates, capital draining, creating poverty traps etc. have been identified as significant limitations of MFIs. In this context, the microcredit program run by the PDCLCS in the slum addresses the limitations of the leading MFIs, socioeconomically benefiting dwellers. Moreover, PDCLCS has developed a sense of ownership among the dwellers for its profit-sharing practices of microcredit programs, which is generally not found among the borrowers of MFIs.

References

Alam S (2013) The impact of credit and non-credit aspects on self-employment profit : a comparison of microcredit programs and commercial lenders in rural Bangladesh. J Dev Areas 47(1):23–45

Al-Amin M, Islam MN (2020) The application of Foucault's disciplinary power to microcredit programmes in two villages of Sylhet, Bangladesh. Dev Pract 30(1):101–112. https://doi.org/10.1080/09614524.2019.1635568

Atkinson-Sheppard S (2016) The gangs of Bangladesh: exploring organised crime, street gangs and 'illicit child labourers' in Dhaka. Criminol Crim Just 16(2):233–249. https://doi.org/10.1177/1748895815616445

Associated Press (AP) (2012) Hundreds of suicides in India linked to microfinance organizations. Business Insider. https://www.businessinsider.com/hundreds-of-suicides-in-india-linked-to-microfinance-organizations-2012-2

Banerjee AV (2013) Microcredit under the microscope: What have we learned in the past two decades, and what do we need to know? Annu Rev Econ 5:487–519. https://doi.org/10.1146/annurev-economics-082912-110220

Banerjee SB, Jackson L (2017) Microfinance and the business of poverty reduction: critical perspectives from rural Bangladesh. Human Relat 70(1):63–91. https://doi.org/10.1177/0018726716640865

Banerjee AV, Duflo E (2013) Poor economics: rethinking poverty & the ways to end it (Paperback). Penguin Random House India.

Bateman M (2010) Why dosen't microfinance work? The destructive rise of local neoliberalism. In: Zed Books. Zed Books

Bateman M, Blankenburg S, Kozul-Wright R (eds) (2019) The rise and fall of global microcredit: development. Routledge, Debt and Disiilustion

Bateman M (2017) The political economy of microfinance. In: Bateman M, Maclean K (eds) Seduced and betrayed: exposing the contemporary microfinance phenomenon. School for Advanced Research Press, Univesity of New Mexico Press, pp 17–31. https://doi.org/10.1057/9781137364210

Bayulgen O (2008) Muhammad yunus, grameen bank and the nobel peace prize: what political science can contribute to and learn from the study of microcredit. Int Stud Rev 10(3):525–547. https://doi.org/10.1111/j.1468-2486.2008.00803.x

BBC News (2013) The Bangladesh poor selling organs to pay debts. BBC News. https://www.bbc.com/news/world-asia-24128096

Biswas S (2010) India's micro-finance suicide epidemic. BBC News. https://www.bbc.com/news/world-south-asia-11997571

Bloomberg (2022) The dark side of microfinance. https://www.bloomberg.com/news/videos/2022-05-02/the-dark-side-of-microfinance-video

Choudhury Haque A, Das A, Rahman A (2017) The effectiveness of micro-credit programmes focusing on household income, expenditure and savings: evidence from Bangladesh. J Compet 9(2):34–44. https://doi.org/10.7441/joc.2017.02.03

Chowdhury R, Willmott H (2019) Microcredit, the corporatisation of nongovernmental organisations, and academic activism: the example of Professor Anu Muhammad. Organisation 26(1):122–140. https://doi.org/10.1177/1350508418768002

Convergences (2019) Microfinance barometer 2019. https://www.convergences.org/wp-content/uploads/2019/09/Microfinance-Barometer-2019_web-1.pdf

Das TK, Pulla V (2014) Is micro credit in urban poverty making a difference? A review from Bangladesh. Int J Soc Work Human Serv Pract Horizon Res Publish 2(2):73–81. http://www.hrpub.org/download/20140405/IJRH10-19200088.pdf

Dowla A, Alamgir D (2003) From microcredit to microfinance: evolution of savings products by MFIS in Bangladesh. J Int Dev 15(8):969–988. https://doi.org/10.1002/jid.1032

Drèze J, Sen A (2013) An uncertain glory: India and its contradictions (Course Boo). Princeton University Press. https://doi.org/10.23943/9781400848775

Field EM, Ambrus A, Banerjee A, Besley T, Chandra A, Duflo E, Fernandez R, Leggett D, Niederle M, Sioux A, Shapiro J, Somani A (2010) Feigenberg et al 2010 building SC through microfinance (No. 16018; NBER working paper series). https://www.nber.org/system/files/working_papers/w16018/w16018.pdf

Gehlich-Shillabeer M (2008) Poverty alleviation or poverty traps? Microcredits and vulnerability in Bangladesh. Disaster Prev Manag Int J 17(3):396–409. https://doi.org/10.1108/09653560810887310

Guinnane TW (2011) The early German credit cooperatives and microfinance organisations today: similarities and differences. In: The handbook of microfinance, pp 77–100. https://doi.org/10.1142/9789814295666_0004

Haldar A, Stiglitz JE (2016) Group lending, joint liability, and social capital: insights from the Indian microfinance crisis. Polit Soc 44(4):459–497. https://doi.org/10.1177/0032329216674001
Hossain MA (2022) Misplacing social capital in women's economic wellbeing—the experience of microcredit from Bangladesh. J Asia Pac Econ. https://doi.org/10.1080/13547860.2021.2024381
Huda KAN (2020) Micro-credit, trust, and social solidarity in Bangladesh: a socio-philosophical analysis. Cult Dyn 32(4):282–306. https://doi.org/10.1177/0921374020920152
Ito S (2003) Microfinance and social capital: does social capital help create good practice? Dev Pract 13(4):322–332. https://doi.org/10.1080/0961452032000112383
Jain P, Moore M (2003) What makes microcredit programmes effective? Fashionable fallacies and workable realities. In: IDS working paper 177. https://opendocs.ids.ac.uk/opendocs/bitstream/handle/20.500.12413/3968/Wp177.pdf
Jansen GG, Dsw JLP (1998) The Grameen bank in Bangladesh : helping poor women with credit for self- employment. J Community Pract 5(1–2):103–123. https://doi.org/10.1300/J125v05n01
Karim L (2008) Demystifying micro-credit: the Grameen Bank, NGOs, and neoliberalism in Bangladesh. Cult Dyn 20(1):5–29. https://doi.org/10.1177/0921374007088053
Karim L (2017) The "scandal" of Grameen: the nobel prize, the bank, and the state in Bangladesh. In: Bateman M, Maclean K (eds) Seduced and betrayed: exposing the contemporary microfinance phenomenon. School for advanced research press, Univesity of New Mexico Press, pp 203–217
Karim L (2011) Microfinance and its discontents: women in debt in Bangladesh, 1st edn. The University of Minnesota Press
Karnani A (2007) Microfinance misses its mark. Stanford Soc Innov Rev:34–40. http://www.ssireview.org/images/articles/2007SU_feature_karnani.pdf
Khan MJ (2013) City slums den for crimes. Dhaka Tribune. https://archive.dhakatribune.com/uncategorized/2013/07/27/city-slums-den-for-crimes
Khandker SR, Samad HA (2017) Bangladesh's achievement in poverty reduction: the role of microfinance revisited. In Sawada Y, Mahmud M, Kitano N (eds) Economic and social development of Bangladesh: miracle and challenges. Palgrave Macmillan, pp 177–198. https://doi.org/10.1007/978-3-319-63838-6
Maîtrot M (2019) Understanding social performance: a 'practice drift' at the frontline of microfinance institutions in Bangladesh. Dev Chang 50(3):623–654. https://doi.org/10.1111/dech.12398
McCoy L (2006) Keeping the institution in view: working with interview accounts of everyday experience. In: Smith DE (ed) Institutional ethnography as practice. Rowman & Littlefield Publishers, Lanham, pp 109–125
McKernan SM (2002) The impact of microcredit programs on self-employment profits: do noncredit program aspects matter? Rev Econ Stat 84(1):93–115. https://doi.org/10.1162/003465302317331946
Mia MA, Lee HA, Chandran VGR, Rasiah R, Rahman M (2019) History of microfinance in Bangladesh: a life cycle theory approach. Bus Hist 61(4):703–733. https://doi.org/10.1080/00076791.2017.1413096
Morduch J (1999) The microfinance promise. J Econ Lit 37(4):1569
Muhammad A (2009) Grameen and microcredit: a tale of corporate success. Econ Polit Wkly 44(35):35–42. https://www.jstor.org/stable/25663491
Muhammad A (2015) Bangladesh-a model of neoliberalism: the case of microfinance and NGOs. Mon Rev 66(10):35. https://doi.org/10.14452/MR-066-10-2015-03_3
Nikkhah HA, Redzuan MB (2010) The role of NGOs in promoting empowerment for sustainable community development. J Hum Ecol 30(2):85–92. https://doi.org/10.1080/09709274.2010.11906276
Paprocki K (2016) "Selling our own skin:" social dispossession through microcredit in rural Bangladesh. Geoforum 74:29–38. https://doi.org/10.1016/j.geoforum.2016.05.008
Raihan S, Osmani SR, Khalily MAB (2017) The macro impact of microfinance in Bangladesh: a CGE analysis. Econ Model 62(January):1–15. https://doi.org/10.1016/j.econmod.2017.01.002

Rankin KN (2001) Governing development : neoliberalism, microcredit, and rational economic woman. Econ Soc 30(1):37–41. https://doi.org/10.1080/03085140020019070

Rankin J (2017) Conducting analysis in institutional ethnography: analytical work prior to commencing data collection. Int J Qual Methods 16(1):1–9. https://doi.org/10.1177/1609406917734484

Sandberg J (2012) Mega-interest on Microcredit: are lenders exploiting the poor? J Appl Philos 29(3):169–185. https://doi.org/10.1111/j.1468-5930.2012.00560.x

Sinclair H (2012) Confessions of a microfinance heretic: how microlending lost its way and betrayed the poor. Berrett-Koehler Publishers, Inc. http://books.google.com/books?id=x3lqCbyAKKgC&pgis=1

Smith DE (2005) Institutional ethnography: a sociology for people. In Suparyanto dan Rosad, vol 5(3). Rowman & Littlefield Publishers Inc

Suzuki Y, Barai MK, Uddin SMS (2013) Islamic banking and the Grameen mode of microcredit in Bangladesh: an institutional comparison. Contemp South Asia 21(4):413–428. https://doi.org/10.1080/09584935.2013.856378

Swapan HUR (2022) অভাবে অসহায় মানুষ তীব্র হচ্ছে ঋণের ফাঁদ—–DW—29.08.2022. Deutsche Welle (DW). https://www.dw.com/bn/অভাবে-অসহায়-মানুষ-তীব্র-হচ্ছে-ঋণের-ফাঁদ/a-62961968

Tahmasebi A, Askaribezayeh F (2021) Microfinance and social capital formation—a social network analysis approach. Socioecon Plann Sci 76:100978. https://doi.org/10.1016/j.seps.2020.100978

The Daily Star (2007) Poverty to be put in museums by 2030: Prof Yunus Says. The Daily Star. https://archive.thedailystar.net/2007/02/18/d7021801116.htm

Wahid A, Hsu M (2000) The Grameen bank of Bangladesh: history procedures, effects and challenges. Asian Aff 31(2):160–169. https://doi.org/10.1080/0002889708506224

Weber H (2002) The imposition of a global development architecture: the example of microcredit. Rev Int Stud 28(3):537–555. https://doi.org/10.1017/S0260210502005375

Yunus M, Jolis A (2007) Banker to the poor: micro-lending and the battle against world poverty. Public Affairs

Yunus M (2006) Nobel lecture. https://www.nobelprize.org/prizes/peace/2006/yunus/lecture/

Yunus M (2011) Sacrificing microcredit for megaprofits. The New York Times. https://www.nytimes.com/2011/01/15/opinion/15yunus.html

Chapter 11
Growth and Development of Squatter Settlements: A Challenge to Urban Sustainability in Nepal: the Case of Kathmandu City

Uttam Sagar Shrestha

Abstract Squatter settlement is related to ugly urban landscape in Nepal. Over the years, the number of squatter settlements in large cities like Kathmandu has increased, occupying public areas usually on the riverbanks due to migration of people in search of jobs and other opportunities. These people are mostly poor who usually get work in informal sector. Most of them do not hire rooms due to expensive rent. To fulfil the SDG goal 11, Nepalese government is committed to provide shelters to all squatters by the year 2030. This paper chapter reviews Nepal's urban squatter settlements in general and then explores the environment condition of squatter settlements in Kathmandu city. The study draws on information from text-mining method, interview with the key informants, and observation of the city's squatter settlements. The findings show that the environment condition and facilities of the squatter settlements are pathetic, and the government efforts so far initiated seem to be impractical and inadequate. At the end, the paper offers suggestions towards fulfilling the SDG and achieving liveable cities for the poor dwellers—the squatters.

Keywords Living condition of squatters · River corridor · Informal sector · Liveable cities

11.1 Introduction

Squatter settlement is defined as a low and poor residential area, which is developed without legal right to the land or permission from the concerned authorities to build, and as a result, due to their illegal status, infrastructure and services are usually inadequate (UN-Habitat 2003). Numerous squatter settlements have emerged in many large cities across Nepal including Kathmandu, the capital city being the epicentre over the last two decades (Gupta and Sharma 2006). Number of squatter settlements is

---

U. S. Shrestha (✉)
Tribhuvan University, Kathmandu, Nepal
e-mail: usashrestha@gmail.com

S. Anand et al. (eds.), *Sustainability in South Asian Cities*, Advances in Geographical and Environmental Sciences, https://doi.org/10.1007/978-981-97-7455-5_11

growing in the country's other fast-growing cities such as Pokharā, Dharān, Birganj, Bharatpur, Mechinagar, etc. (MoUD 2015).

Urbanization is very rapid and haphazard in Kathmandu valley due mainly to immigrants from different parts of the country. Such migration is due to comparably low socio-economic growth, poor land administration system, inadequate capacity to cope with housing needs and poor governance (Shrestha 2013). In the cities of Kathmandu valley, the housing problem is severe (UN-Habitat 2010a, b). In the urban valley, land and housing price and rent have risen extremely and as a result, problem of finding space and housing is intensified particularly to the poor people who are at extreme poverty level. This is one of major factors to cause to the growth of squatter settlements. Urban poverty in Kathmandu valley is at 11.5% as compared to 8.7% in the urban hills and 22% in the urban Tarai (CBS 2012). The people of squatter settlements are very poor who survive on less than US$ 1.9 (NRs 222) per day or NRs. 6,600 per month.

In 2012, there were 8,000 squatter families in Kathmandu valley. Currently, Kathmandu city alone has squatter settlements in 73 places and three places each in the valley's two other cities such as Lalitpur and Bhaktapur. Most of the squatter settlements have encroached riverbanks and other public lands since over four decades (The Himalayan Times 2017). In 2012, the government tried to evacuate the squatters from the riverbanks in the valley but failed. Recently, the valley has received 105 Rohingya Muslim families and their 500 members from Myanmar who now are living as squatters in Kapan of Kathmandu city (The Kathmandu Post 2020). Large and growing number of squatter settlements in Kathmandu is a huge challenge to the SDG. These settlements are very vulnerable to river floods, polluted environment, poor structured houses, and fire hazards and face lacking basic services such as drinking water, electricity, etc. These pose to a main question: how they are living and maintaining their livelihood in the context of environmental sustainability of the cities. In this backdrop, this study attempts to explore the environment and sanitation conditions of squatter settlements and their physical facilities in Kathmandu city. Policy options are also provided to regulate the squatter settlements.

11.2 Study Area

Kathmandu valley has three large cities comprising Kathmandu, Bhaktapur and Lalitpur (Fig. 11.1) The valley with an area of 570 km^2 is the largest urban conglomeration and oldest among the cities in Nepal. The valley floor is situated at an altitude of 1,350-m average sea level surrounded by the mountain ranges with an average elevation at 2,740 m. Of these, Kathmandu is the largest metropolitan city in Nepal.

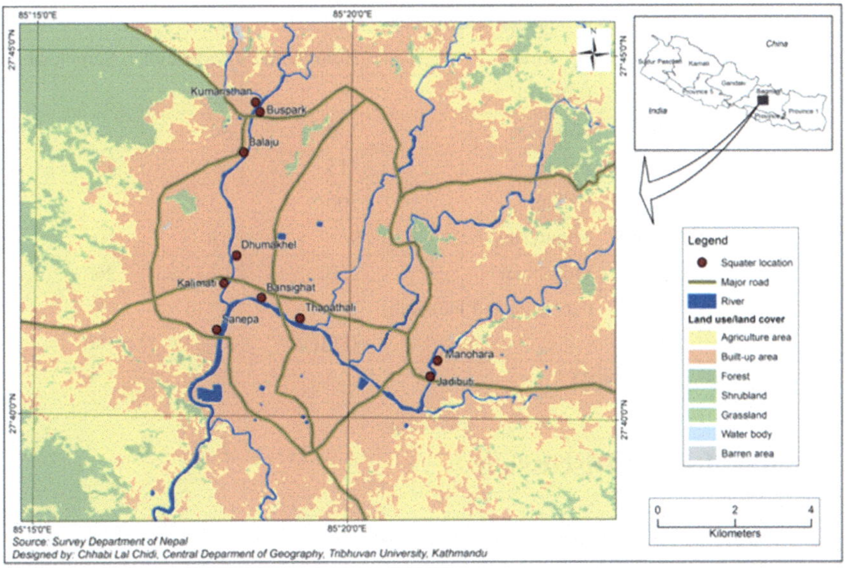

Fig. 11.1 Location of Kathmandu city, Nepal

11.3 Methodology

For survey, the study area was divided into three river corridors such as the Manoharā, Bāgmati, and Bishnumati representing the east, centre, and west for the study purpose and seven out of 73 squatter settlements and their 180 households (about 10.45%) were selected randomly. Field surveys comprising observation, household survey, FDGs, and key informant interview was carried out for gathering data during two weeks from December 25, 2020, through January 2, 2021.

Table 11.1 shows seven sample squatter settlement places (Fig. 11.1) and their households, population, and sample households. Observations were conducted at seven squatter settlements to record their environment and sanitation condition. FDGs comprising the number of participants ranged from 4 at Jāgriti tole to maximum 10 at Balkhu, both males and females, and different ethnic groups. Other surveys such as household survey (HH) and Key Informant Interview (KII), case studies and informal discussions were held for the data collection. Chairperson of *Tole Sudhār Samiti* (Locality Improvement Committee), promoters and aged persons of the squatter settlements were taken as key informants. The secondary information was also gathered.

Table 11.1 Sample locations of squatter settlements

Rivers	Sample site	Localities	Households	Population	Sample hhs
Bishnumati	Dhumakhel	Kālimāti	80	520	8
	Jāgriti tole	Bālāju	130	845	14
	Kumāristhan	Bālāju	115	745	12
Bagmati	Balkhu	Balkhu	300	2,145	31
	Banshighāt	Tripureshwar	96	650	10
	Kuriyagāu	Thāpāthali	250	1,712	26
Manoharā	Manoharā	Pepsicola and Jadibudhi	750	4,915	78
Total			1,721	11,532	180

Source Field Survey, 2020–2021

11.4 Results and Discussions

11.4.1 Growth of Squatter settlements

The squatter is someone who settles unlawfully on the public land with an intent to acquire title to it. Such people are called as squatters and their occupied land for living is called squatter settlement. In Nepal it is called *Sukumbāsi*[1] referring to the people without legal land right (Ministry of Land Reform 1964).

The process of urbanization in Kathmandu city is a major cause of growth of squatter settlements. In Nepal, the process of urbanization is not due to an economic reason, but due to the extension of existing municipal boundary, designation of new municipalities, high urban population growth rate, and high urban bound migration (Pradhan 2003).

The rate of urbanization in Kathmandu city is at 4.5%. With this rapid urbanization rate, the number of squatter settlements has also increased significantly. Table 11.1 shows that the number of squatter settlements grew from five (Khadapakha, Ramhiti, Pathivara, Kangeshwori and Bijeshwori) in 1970 to 73 in 2020 in Kathmandu, signifying the growth by nearly 15 times. During the same duration of five decades, the number of households and their population also grew substantially at 201 and 80.5 times respectively (Table 11.2).

The squatters arrived for the first time in 1970 at those five localities were due to washed away their houses and properties by landslide and flood hazard (Lumanti 2008; Thapa 2006). Besides, political crisis and insecurity and increasing unemployment in rural areas during the Maoist insurgency (1996–2006) are major reasons for increasing squatter settlements. Further, increasing squatter settlements are also due

[1] The people living in Sikkim was called Sukhim people. The term was for the first time defined in Nepal and India Sugauli Treaty on 2 December 1816 for residents of Darjeeling, part of west Bengal of India (Ministry of Land Reform, 1964).

Table 11.2 Growth of squatter settlements in Kathmandu city

Years	Number of settlements	Number of households	Total Population
1970	5	36	360
1980	13	38	30
1990	19	859	4,294
2000	61	2,031	11,862
2010	66	2,031	11,862
2020	73	7,250	29,000

Sources Shrestha (2015); Field Survey, 2020/21

to poverty, hardship life, remoteness, and deprive of basic facilities (health and education) services in the migrants' origin areas and availability of employment opportunities and amenities in the destination place—the Kathmandu city. In terms of relative share, unemployment is the largest cause (70%) of migration. Next includes reasons, such as landlessness, poverty, and conflict together represent 20%. Lastly, lack of economic opportunities and education facilities, internal conflict, and natural calamities (flood, landslide, and earthquake) share 10%, causing migration of people to Kathmandu city. These reasons are somewhat similar to the findings of the studies carried out in Afghanistan and other countries (Schutte 2006). They show that chronic conflict and political instability decisively influence livelihood strategies of urban poor and increasing level of risk and constant uncertainty most probably lead to the people to leave their living places to move to other relatively better places.

A total of 40 districts out of 77 country's total districts representing all three ecological regions—Mountain, Hill, and Tarai are the sources of squatter settlers in Kathmandu city. The hill districts share the largest, followed by Kathmandu district alone (Table 11.3). Three northern neighbouring districts of Kathmandu, namely Nuwākot, Rasuwā, and Dhāding are major sources of squatters. In terms of total population, Tarai region shares over half of the country's total population, while the mountain region the least, but in terms of relative share of the squatters, the mountain region is larger than the Tarai region. Noted that the mountain region is composed of rugged terrain, drudgery of movement, harsh climate, and limited agricultural land, whereas the Tarai is the plain region having fertile and extensive agricultural land, and better condition of other facilities. However, despite relatively well-off people in terms of adequate agricultural land for self-sufficient production of food, people from the latter region found to be tended to migrate to Kathmandu and occupy public land there with the hope that the occupied land would be granted to them by the government free of cost.

The squatters appear to be encouraged even by local leaders by initiating them to live as squatters and get the occupied land free of cost after some years. This is the case with some squatters of Jāgriti, Banshighāt, Thāpāthali, and Balkhu. In case of Banshighāt, Jāgriti, and Manoharā, there are some squatters who found to be staying at the squatted homes during the daytime and return to their own home (legal) in the evening.

Table 11.3 Distribution of migrants by origin places in the sample squatter settlements

Squatter settlements	Migrants' origin places (in %)				
	Kathmandu	Mountain districts	Hill districts	Tarai district	Not stated
Dhumakhel	60.0	5.0	10.0	20.0	5.0
Jāgriti tole	10.0	15.0	60.0	7.0	8.0
Kumāristhan	7.0	15.0	68.0	5.0	5.0
Balkhu	55.0	10.0	20.0	15.0	0.0
Banshighāt	40.0	10.0	25.0	15.0	10.0
Kuriyagāu	6.0	11.1	62.7	19.6	0.0
Manoharā	1.2	21.1	63.4	14.3	0.0
Distribution (%)	25.6	12.5	44.2	13.7	4.0
Total sample hhs	46	22	80	25	7

Source Field Survey, 2020/21

There exists a close relationship of geographical contiguity with the origin of squatters and squatted lands. For instance, majority of squatters of Thāpāthali are migrated from surrounding districts such as Sindhuli (south-east), Kābhre (east) and Sindhupalchowk (north). Road facility is a factor to make them move to the current occupied land from those districts. The same situation also occurs in case of Manoharā, which is located in the eastern part of Kathmandu, where majority of the squatters are from Kābhre, the eastern adjoining district of Kathmandu valley, followed by the squatters from the districts of Morang, Rāmechhāp and Khotāng. These findings follow the migration pattern and direction, according to the population census of 2011 (CBS, 2014). During the 1990s, over 40% of the urban growth in Kathmandu was contributed by migration, which was the largest net inflow of migration among the country's urban areas. The net inflow of migrants in Kathmandu valley accounted for 36% of its urban population and the migrants prefer Kathmandu due to employment opportunities (40%), education (30%), health (20%) and other reasons (10%) (CBS 2012).

Lastly, the study's finding is that most squatters have no interest to go back to their previous (original) own home places. For instance, most squatters (90%) of Manoharā, Siddheswar, Balkhu, Thāpāthali, and Jāgriti responded 'unwillingness' to return to their previous own home places due to better income generating opportunities available here in the city of destination.

11.4.2 Features of Squatter settlements

The features of the squatter settlements are described in terms of area, land use change, social groups, and education status. The distribution of occupied land area among the squatter settlements varies greatly ranging from 0.31 ha in Dhumakhel to 5.7 ha in Manoharā with an average area of 1.4 ha per settlement (Table 11.4). The total area of seven sample squatter settlements shares approximately 20% of the total occupied land of 86.43 ha in Kathmandu valley (KC 2010).

There is a rapid growth in squatted areas. Though information for all sample squatter areas is not available, the information available for Manoharā shows that it grew from 0.05 hectare in 2001 to 3.57 hectare in 2009, whereas Kuriyagāu grew from 0.009 hectare in 2001 to 1.36 hectare in 2009 (KC 2010).

All the squatter settlements are located on the riverbanks. Before being occupied by the squatters, those public lands were found to be covered by grassland, other vegetation, and roads. The channels of the rivers have been narrowed down. At some sites along the river course, under the river training project retaining walls found to be built to protect the banks from erosion and mud cutting by the river, as well as from further encroachment by the squatters. One example is Kumāristhan, which is situated on the bank of the Bishnumati River, where Siddheswar temple is located. Under this project, road corridors have been built along the banks of this river.

The population of the squatter settlements consists of different castes and ethnics, which are grouped into four major types (Table 11.5). All squatter settlements have all these four social groups, but their relative size varies largely. The Janajātis (ethnic groups primarily from the hills and mountains) are the largest in all six squatter settlements except Dhumakhel, where Dalits are the largest. Bahun/Chhetri and Dalits belonging to Hindu castes occupy all squatters with sizeable numbers.

The field survey shows that the literacy level of the people of the squatter settlements was 39%. The people having school level education was highest at over 90% and the rest was college level education. Majority of the children with over 70% found to be enrolled in the public schools/colleges due to affordability, and very few children of Dhumakhel also found to be enrolled in the private boarding schools. The students enrolled in the public colleges also found to have worked in a day to support their families. Due to extreme poor, many of the squatter households found to have chosen between two options: either send their kids to school or to the street as beggars to earn money. Therefore, some of the school going age children found

Table 11.4 Distribution of land area of squatter settlements

Name	Area in hectare	Percentage
Dhumakhel	0.31	1.86
Jāgriti Tole	3.31	19.95
Kumāristhan	2.47	14.88
Balkhu	1.88	11.33
Banshighāt	1.26	7.59
Kuriyagāu	1.68	10.12
Manoharā	5.68	34.23
Total	16.59	100.0
Average	1.37	

Source Google earth image retrieved on 1/30/2021

Table 11.5 Distribution of squatters (%) by castes/ethnics

Settlements	Bahun/Chhetri	Dalits	Janajātis	Others
Dhumakhel	5.0	62.0	26.0	7.0
Jāgriti tole	30.0	15.0	40.0	20.0
Kumāristhan	25.0	20.0	30.0	25.0
Balkhu	22.0	2.5	63.0	2.5
Banshighāt	15.0	30.0	40.0	10.0
Kuriyagāu	22.2	6.9	69.4	1.4
Manoharā	22.0	17.1	61.0	0
Total (%)	20.3	22.1	48.1	9.5

Source Field Survey, 2020–21

to be not going to school but instead to streets for begging. The view expressed by Nagma Khatun, a 20-year-old woman squatter of Balkhu is:

> There is no chance to send our children to school. From the early morning to the evening the whole family members must work in collecting recyclable waste goods from the garbage dumping sites and sell them at nearby *Kawādi* (scraps dealers). If we do not do this work, we all will be without food and remain to be hungry.

Thus, the squatter settlements pose obstacle to the conservation of river ecology. Increasing number and size of the squatter areas is mainly due to weak or lack of implementation of river conservation legislation and so, no demarcation of the rivers channels is made yet (Shrestha 2015). In addition, scarcity of land, high land value, and concentration of most of the central government activities in Kathmandu city are other reasons. Above all, lack of urban land use zonation is a key factor for haphazard urbanization and reckless use of the river corridors.

11.4.3 Informal sector

Urbanization is considered as one of the main drivers of poverty reduction through creating employment opportunity in informal activities in the cities. Large cities often witness increased vulnerable populations like slum dwellers, squatters, rage pickers, etc. Open spaces in such cities are often occupied by squatters, foot–loose sellers, rag pickers, etc. (Pradhan 2013). ILO (2004) found that workers in the informal economy in Nepal were most likely to involve in low skilled, manual, and dirty jobs which lie at the bottom of the work hierarchy due to low level skills, education, etc. Since they get under paid wage, most of the people of the squatter settlements must engage in multiple works to sustain their living.

The findings of this study are that all squatter settlements under study have Kawādi. Daily wage labour is the largest employment source (nearly 59%), followed by butcher. Majority of the squatter households earn NRs 3,000–4,500 per month, which is very low to sustain their living. Some of the families are engaged in retail outlets of groceries, electric goods, tailoring, carpenter, insane stick making, etc. Number of such vending shops range from 35 in Banshighāt to 130 in Manoharā.

So, the economic condition of most of the people living in squatter settlements is miserable due to under paid wage and lack of regular jobs. There is no provision of institutional financial support to them for operating any sort of economic activities. However, few of them found to have received supports from international and non-governmental organizations like Action Aid, Water Aid, and Lumanti for operating skill-based works.

11.4.4 Housing Condition and Basic Facilities

Most of the houses (over 70%) in the squatter settlements have one storey having one or two rooms. The walls are made up of brick mud materials. Most of the houses are of temporary structure made up of plastic sheets, sacks, and bamboo. Few houses have roofs of galvanized iron and cement wall and two storeys with steel trust pipe. All houses are constructed with inadequate lighting and ventilation. The discussions with the people of Balkhu show that more than 35% houses have been built on less than 2 anna (63.53m^2) land area, 25% have land with 2–2.5 anna (79.47m^3), and 5 percent have 3 anna (95.37m^3). Thirty per cent squatters did not want to state the area they occupy.

Each house had a temporary toilet. About two-thirds of them used to discharge sewage directly into the nearby river and one-third houses had safety tank. The sources of drinking water of the squatter settlements were covered well, stone spout, and public tap. The water quality of these water sources was poor and not potable. Few of the households of Kuriyagāu and Manoharā were connected with the pipe system for which they had to pay water tariff.

11.4.5 Nepal Government's Policy Efforts Towards SDG

The main objective of SDG is to develop human settlements with inclusiveness, security, and sustainability. The SDG's 11th goal is related to urban poor or squatter settlement. To this context, the Kathmandu Valley Authority yielded a slogan "Squatter Free Kathmandu Valley" in 2010 under which the project was supposed to relocate the squatters living along the bank of the Bagmati River to other suitable places within the valley, but the project remained unimplemented until now. There is a problem with the government that no policy exists to prevent unnecessary evacuation and violation of housing rights (UN-Habitat 2010a, b).

The Kathmandu Municipal Government together with Lumanti, a non-governmental organization and UN Habitat initiated to establish "Urban Poor Community Donation Fund" with an investment of NRs.200 million. Under this joint project, a colony named "Kirtipur Āwās Chhetra" (Kirtipur Residence Area) with 44 houses having facilities of water supply, open space, garden, and incineration yard were constructed in Ward 3 of Kirtipur municipality located in southern part of the valley. Two types of houses with the costs of NRs. 330,000 (US$ 2950) and NRs. 350,000 (US$ 3125) were constructed in the colony. The settlers required to deposit NRs. 10,000 (US$ 90) as the first instalment and the rest to be paid on instalment basis within 15 years and after complete payment, they would become landlords of their houses. This scheme was beyond the capacity of most of the squatters. Another project named as "Ichangu Narayan Housing Project" located at Ichangu Narayan, of Nagarjun municipality was initiated by the Department of Housing and Physical Planning, Government of Nepal to relocate the squatters of Kuriyagāu to that project site in two phases. The project constructed 6 storey buildings with rooms containing bedroom, kitchen, dining room, drawing room, and balcony for five families in 0.12 ha area with the cost of around US$ 115 million. In the first phase, 127 families were supposed to be relocated and another 100 families to be shifted to the site in the second phase. A Committee of Integrated Development of the Bagmati Civilization (HPCIDBEC) having high-power was formed to relocate and manage the Kuriyagāu squatters to the project site, but unfortunately it remained futile due to unsolved issues like renting of the rooms, maintenance, management, ownership to generations raised by the squatters. One of the main problems was to identify genuine squatters. The main reason to reject the site for living by the squatters was that the site is being far from the working place. Most of them were daily workers and there was tough competition to get loading and unloading works from the trucks for which they wanted to live close to such working place.

11.5 Conclusion

The squatter settlement is a major problem of Kathmandu city in viewpoint of conservation of urban environment. The number of squatter settlements has not only grown rapidly because of political instability, slow economic growth, limited budget, weak rules and regulations, and high aspiration by politicians to grant free land/houses to the squatters, but also, they are responsible to cause unpleasant city landscape.

Though the central and local governments have made several attempts to solve the problem of squatters by constructing buildings in places within Kathmandu valley, they got failure due to the inability of recognizing the real squatters, importance of close distance relationship between living and working places on which the livelihood of the squatters rely, limited budget, etc. In fact, there are numerous successful works on solving squatter settlements problem elsewhere in several developing countries. For instance, the government of Nepal should learn from the Colombo city experience regarding the way to solve the squatters' problem. Since there is lack of city land use zonation policy and act and likewise lack of road network master plan, not only urbanization has taken place haphazardly, but informal settlers, mostly back by major political parties tend to occupy the public lands.

The Fifteen Five-Year Plan (2019/20–2023/24) has formulated objective of constructing a total of 35,000 safe citizen housing units and 5,000 housing units for urban poor in the city area (NPC 2019). This is in line with achieving the SDG goal by the Nepalese government by 2030. It requires however a strong and sincere willpower, and commitment of the government to implement the plan honestly.

References

CBS (2003) Population monograph of Nepal, vol II. Central Bureau of Statistics, Kathmandu

CBS (2012) Nepal population and housing census 2011: national report. Central Bureau of Statistics, Government of Nepal, Kathmandu

CBS (2014) Population monograph of Nepal census 2011 (National report). Central Bureau of Statistics, Government of Nepal, Kathmandu

Diana M (2004) Understanding urban poverty: what the poverty reduction strategy papers tell us. IIED, London. http://www.sdinet.co.za/static/pdf/understanding_urban_poverty_mitlin.pdf

Gupta US, Sharma S (2006) The challenge of squatter settlements in Kathmandu: addressing a policy vacuum. Int Dev Plan Rev 28(1):105–126. 10.38281Idpr.28.1.5

Gyawali GP (2017) Urbanization in Nepal. In: Gautam S, Karki IB, Dhamala BP, Bhandari NB, Dahal BN (eds) Patanpragya. Tribhuvan University Teachers' Association, Patan Multiple Campus 3(1)

ILO (2004) Global employment trends, International Labour Office ILO), Switzerland

KC K (2010) A study of squatter settlements of Kathmandu using GIS, aerial photography, remote sensing and household survey. Unpublished MSc Dissertation, Pokhara University, Pokhara

Lumanti (2008) Status of squatter community along Bagmati River and its tributaries in Kathmandu. Kathmandu

Ministry of Finance (2018) Economic survey2018. Ministry of Finance, Kathmandu

Ministry of Land Reforms (1964) Land revenue records (Unpublished Official Document). Ministry of Land Reform, Kathmandu

MOUD (2015) National urban development strategy (Draft). GoN/Ministry of Urban Development (MOUD), Kathmandu

NLSS (2004) Nepal living standard survey, 2003–2004. National Planning Commission/Central Bureau of Statistics, Kathmandu

NPC (2019) The fifteenth five-year plan (2019/20–2023/2024), National Planning Commission (NPC), Kathmandu

Pradhan PK (2013) Urban infrastructure planning and investment. Asian Development Bank, Manila

Pradhan PK (2003) Population growth, migration and urbanization, and environmental consequences in Kathmandu valley Nepal, Nepal. In: Unruh J, Krol M, Kliot N (eds) Environmental changes and implications for population migration, Kluwer Academic Publishers B.V. Dordrecht (The Netherlands)

Schutte S (2006) Poor, poor's, poorest: urban livelihoods and vulnerability in Mazar-i-Shrif (case study series). Afghanistan Research and Evaluation Unit (AREU), Kabul. Retrieved from www.geo.fuberlin.de/geog/facinhrichutungen/anthrogeog/elf.median/downloadedschuettee

Shrestha BK (2013) Squatter settlements in Kathmandu valley: looking through the prism of land rights and tenure security. Urban forum, 24 (119–135)

Shrestha S (2015) Assessment of the Bagmati river encroachment through application of GIS and RS. Unpublished Master's Thesis, Pokhara University, Pokhara

Shrestha N (2020) Covid and macro economy. Karobar

Thapa G (2006) A study on livelihood strategies of urban poor in Ramhiti, Kathmandu, Nepal. Unpublished M.A. Thesis, Tribhuvan University, Kathmandu Nepal

The Kathmandu post (2020)

The Himalayan Times (2017)
UN (2007) World urbanization prospects: 2007 revision population database. Population Division, Department of Economic and Social Affairs, United Nations, New York
UN (2013) A country analysis with a human face. United Nations Country Team in Nepal. Kathmandu. http://un.org.np/sites/default/files/Nepal_CountryAanalysis_2011_Feb2013.pdf
UN-Habitat (2003) The challenges of slums. Global report on human settlements. Earthscan, London
UN-Habitat (2010a) Nepal urban housing sector profile. United Nations Human Settlements Program, Nairobi. https://unhabitat.org/nepal-urban-housing-sector-profile
UN-Habitat (2010b) Nepal urban housing sector profile. United Nations Human Settlements, Nairobi
UN-Habitat (2012) State of world's cities 2012/2013 prosperity of cities. UN Human Settlement Program, Nairobi
UNSD (n.d.) Millennium development goals indicators database. http://mdgs.un.org/unsd/mdg/Data.aspx

Chapter 12
The COVID-19 Crisis and Indian Cities: Managing Urban Contours in Times of Pandemic

Anuradha Banerjee

Abstract Historically, cities have been the driving force of growth and development and are termed 'engines of growth' that have helped national economies forge ahead. We envision a situation when the world is becoming progressively urban. However, the COVID-19 crisis has created a situation that necessitates revisiting the urban scenario, the population base and communities concerned, urban densities, infrastructure and amenities, governance, management and planning. Many Indian cities are modern impositions on a colonial framework, in many cases growing haphazardly, dominated by heightened mobility and in-migration. Cities like Mumbai, Delhi, Pune, Ahmedabad, Surat and Chennai are interconnected with the global economy arising out of globalisation and the international division of labour and have evolved as prime centres for global capital investments. This has reinforced the migration pattern to these cities, with many migrant workers working in the informal sectors and living in congested habitats. The COVID-19 crisis has reinforced the already existing vulnerabilities in urban areas. These include overcrowding, insanitary conditions and deprivations in large parts of the cities, like in the older portions and in the slums and low-income residentials. The urban poor face deep-seated structural and economic inequalities, many of whom are informal migrant workers engaged in low-income activities like daily wage earners and casual labourers in construction, transport, petty trade, etc. With research intensifying on the socio-economic aspects of COVID-19, it should also be of major concern to address our cities, to reorganise them, improving health and hygiene, particularly in the congested parts and thus increase the factor of urban resilience. Against this backdrop, this chapter focuses on Indian cities during the COVID-19 crisis. Using data from the Census of India; health data from the National Family Health Survey; and crowd-sourced data for Covid-19, the aim is to analyse the reinforced vulnerabilities in Indian Cities during the Coronavirus Pandemic. The aim is also to critically evaluate the urban policies and programmes in the face of extenuating Institutionalised Inequalities that stand in the way of instilling the factor of resilience in making the urban systems sustainable.

A. Banerjee (✉)
CSRD, SSS Jawaharlal Nehru University, New Delhi 110067, India
e-mail: banerjee.anb@gmail.com

S. Anand et al. (eds.), *Sustainability in South Asian Cities*, Advances in Geographical and Environmental Sciences, https://doi.org/10.1007/978-981-97-7455-5_12

The specific issues that need to be addressed at this hour also constitute the scope of the present chapter.

Keywords Pandemic · Vulnerabilities · Inequalities · Globalisation · Resilience

12.1 Introduction

The COVID-19 outbreak that was first reported in Wuhan, a city in the Hubei province in China, had spread thenceforth very rapidly to other cities of the world through person-to-person contact arising out of human mobility and migration.

COVID-19 pandemic as declared by WHO much later than its appearance, initially by its character, had proved to be an urban phenomenon, both in the developed and the developing countries. In this respect, the pandemic has proved to be favouring urban areas, with the world cities being the hotspots of the disease spread, be it New York, Sao Paulo or Mumbai.

Cities have been historically the driving force of growth and development and are often termed as 'engines of growth' that have helped national economies to forge ahead, and we are envisioning a situation when the world is becoming progressively urban. However, the COVID-19 crisis has created a situation that necessitates revisiting the urban scenario, the population base and communities concerned, urban densities, infrastructure/amenities, governance, management and planning.

According to the urban theorist Richard Florida, (Professor, University of Toronto) cities have evolved as hotspots of COVID-19 due to a multitude of reasons and no single reason can suffice to explain effectively the urban scenario. Florida states that multiple factors are responsible for the virus spread that includes population, age distribution, education levels, inequality, religiosity, occupation and level of social capital. Steven Goodman, an epidemiologist at Stanford University, views that population density, together with very good intra-city connectivity have been the enemies in cities like New York, Chicago and New Jersey. New York's high COVID-19 mortality seems to support this argument, although it had been found that mortality is lower in many of Asia's hyper-dense cities.

India is no exception to this; though the mortality regimes are lower, but the COVID hotspots have seemed to favour our megacities and the million-plus cities initially, and at present, in addition to these, the pandemic has extended its tentacles to other sizes classes as well. To begin with, we had six cities that had accounted for around half of all the reported cases in the country till May 2020. These include Mumbai, Delhi, Ahmedabad, Chennai, Pune and Kolkata.[1]Later, there had been apprehensions that community transmissions were well underway in the largest Indian cities. Several other cities like Thane, Indore, Jaipur, Surat and Agra, sooner or later also became the prominent centres of the pandemic outbreak. Almost ten

[1] *The Week*. Retrieved 20 May 2020. "Infections over 1 lakh, five cities with half the cases: India's coronavirus story so far".

months into the pandemic, as of March 2021, these cities still keep adding the highest number of cases to India's COVID-19 cases. To make things worse, these densely populated cities have the highest positivity ratios in the country and contribute to the highest fatality rates. For instance, Maharashtrathe worst affected state in India owes its misery to its urban centres like Pune, Nagpur, Mumbai, Thane and Nashik. It has a Confirmed Per Million (CPM) rate of 21, 291.6 as of March 2021, much higher than that of the national average (8887.5 CPM) (Table 12.1).

The percentage of death of COVID-19 patients is highest in Delhi, followed by Gujarat, Maharashtra, Tamil Nadu, Madhya Pradesh and West Bengal (above 10%); these states are highly urbanised (except Madhya Pradesh). Urbanised districts of these states show that the rate of recovery is also highest in Delhi, 90.2% of the patients recovered, the death rate due to COVID-19 is 2.7%. The rate of COVID-19 death is highest in Ahmedabad (5.6%), Mumbai (5.4%), Agra and Solapur (above 4%). The recovery rate is highest in Delhi and second highest in Chennai (87.5%). Delhi also records 14.4 Lakh people tested for COVID-19 (Table 12.2).

Many Indian cities are modern impositions on a colonial framework, in many cases growing haphazardly, dominated by heightened mobility and in-migration. Many cities like Mumbai, Delhi, Pune, Ahmedabad, Surat and Chennai are interconnected with the global economy arising from globalisation and the international division of labour. They have evolved as prime centres for investments of global capital. This has reinforced the migration pattern to these cities, with many migrant workers working in the informal sectors and living in congested habitats.

Indian cities particularly the largest ones are immensely diverse in their attributes. They are reflective of a highly variegated landuse with dense built-up areas in certain pockets reflective of congestion and low-income group residentials. In other parts spacious affluent habitats can also be seen, associated with other types of landuse. This is the physical expression of prevailing socio-economic disparities within these megacities. Sometimes the transition from affluence to poverty is very stark. For example, in Mumbai, the slums of Dharavi, Madanpura, Elphinstone and Grant Road coexist with the posh enclaves of Malabar Hill, Chruchgate, Worli and Juhu in close proximity.[2] Dharavi, which is Asia's largest slum and has a population density upwards of 2.5 lakh people per sq. km, and the informal settlement of Shivaji Nagar are Mumbai's bigger COVID-19 hotspots.

According to **Souvanic Roy**, *professor of urban planning at the Indian Institute of Engineering Science and Technology (IIEST), Shibpur* due to acute land scarcity, six to eight persons share a room of 100 sq. ft, rendering physical distancing meaningless in congested localities of Mumbai. Environmental sanitation is at the lowest levels and about 40% of households here do not have toilets, forcing several families to share one community latrine. Often, it has been noticed that the poor have to pay or purchase drinking water, particularly during the summer months and hence frequent

[2] *Souvanic Roy, professor of urban planning at the Indian Institute of Engineering Science and Technology (IIEST), Shibpur.* **How Our Cities Turned the National COVID-19 Crisis From Bad to Worse Science the WIRE.**

Table 12.1 Percentage of COVID-19 death and urbanisation across Indian States/UTs, 2020

State/UT	Percentage covid 19 deaths	Rate of Urbanization
Delhi	37.1	97.5
Gujarat	20.0	42.6
Maharashtra	13.3	45.22
Tamil Nadu	12.4	48.4
Madhya Pradesh	10.4	27.63
West Bengal	10.3	31.87
Jammu and Kashmir	8.6	27.38
Punjab	8.2	37.48
Rajasthan	6.7	24.87
Haryana	6.5	34.88
Uttar Pradesh	6.1	22.27
Karnataka	5.9	38.67
Goa	4.8	62.17
Uttarakhand	4.7	30.23
Andaman and Nicobar Islands	4.5	37.7
Puducherry	4.4	68.33
Andhra Pradesh	3.8	33.36
Jharkhand	3.4	24.05
Telangana	3.2	38.9
Ladakh	2.9	
Tripura	2.9	26.17
Chandigarh	2.7	97.25
Chhattisgarh	2.4	23.24
Bihar	2.4	11.29
Himachal Pradesh	1.9	10.03
Odisha	1.7	16.69
Manipur	1.4	29.21
Assam	1.3	14.1
Kerala	1.1	47.7
Nagaland	0.7	28.86
Meghalaya	0.7	20.07
Sikkim	0.6	25.15
Arunachal Pradesh	0.6	22.94
Dadra and Nagar Haveli and Daman and Diu	0.5	46.72
Mizoram	0.0	52.11

Source Data taken from Ministry of Health and family Welfare website https://www.mohfw.gov.in/ & Urbanization data is taken from Census 2011

Table 12.2 COVID-19 Status of India's urbanised districts, (data on 25/08/2020)

Urban districts	Confirmed	Active	Recovered	Deceased	Persons tested	Percent recovered	Percent deceased
Pune	152,511	43,065	105,681	3765	6,300,000	69.3	2.5
Mumbai	137,096	18,267	111,082	7442	7,000,000	81.0	5.4
Thane	123,499	19,335	100,570	3593	4,500,000	81.4	2.9
Nashik	32,778	10,486	21,536	756	1,200,000	65.7	2.3
Raigad	26,812	5091	21,031	688	8,420,000	78.4	2.6
Phalghar	23,659	6648	16,455	556	1,300,000	69.6	2.4
Jalgaon	22,698	6554	15,376	768	1,000,000	67.7	3.4
Nagpur	20,324	8769	11,032	522	1,400,000	54.3	2.6
Solapur	17,200	4310	12,196	693	1,300,000	70.9	4.0
Chennai	126,677	13,258	110,819	2600	8,600,000	87.5	2.1
Bengaluru Urban	109,793	34,735	73,363	1694	6,900,000	66.8	1.5
Lucknow	22,585	6966	15,328	291	2,900,000	67.9	1.3
Kanpur Nagar	12,756	3591	8798	367	1,300,000	69.0	2.9
Ghaziabad	7341	1070	6204	67	1,600,000	84.5	0.9
Gautam Buddha Nagar	7216	927	6246	43	1,700,000	86.6	0.6
Agra	2532	293	2133	106	1,000,000	84.2	4.2
Delhi	162,527	11,626	146,588	4313	14,400,000	90.2	2.7
Kolkata	36,722	5691	29,844	1187	NA	81.3	3.2
Patna	19,112	2915	16,073	124	1,100,000	84.1	0.6
Ahmedabad	30,362	3406	25,268	1688	6,200,000	83.2	5.6
Surat	18,850	2498	15,756	596	3,000,000	83.6	3.2
Vadodara	7316	1614	5584	118	1,000,000	76.3	1.6

Source Ministry of Health and Family Welfare website https://www.mohfw.gov.in/

washing of hands to prevent contamination is a far-fetched dream as observed in Shivaji Nagar, Mumbai.

The colonial cities of India portray a variegated pattern, with many portions of the older cities being overcrowded and extremely vulnerable to an outbreak of contagious diseases. These are also the areas, with households dominated by large family size, economic and social deprivation, regressive beliefs, inadequate sanitation, deplorable living environment and lack of open spaces. Religious polarisation in these parts as well as in the slums and squatters also have a role to play. The infection rate is higher in Ahmedabad's slums and low-income settlements in Jamalpur, Danilimda, Dariyapur and Behrampura.[3]

Kolkata has the third highest mortality rate among India's cities, and its cases are concentrated in infrastructure-deficient north and central Kolkata.

In Delhi, Mangolpuri, Wazirpur and Shahdara are the hotspots. However, with the pandemic engulfing every nook and corner of these cities, containment strategies have evolved to designating smaller areas as hotspots rather than entire districts so as not to cause severe economic distress. Even in the planned city of Chandigarh, a large number of cases have been reported from the Bapudham slum rehabilitation colony.[4]

With researches intensifying on the socio-economic aspects of COVID-19, it should also be of major concern to address our cities, to reorganise them, by improving health and hygiene, particularly in the congested parts and thus increase the factor of urban resilience.

Against this backdrop, the present chapter focuses on Indian cities during the COVID-19 crisis. The primary objectives of the present chapter aim to delve into the following heads:

- Urban Profile and the geography of COVID-19 spread in India
- Reinforced vulnerabilities and Indian Cities
- Urban policies and programmes in the face of pandemic
- The urban future in the context of resilience in urban planning and development
- Conclusion.

12.2 Urban Profile and the Geography of COVID-19 Spread in India

India, which reported relatively very few cases during the beginning of the pandemic, has constantly remained among the top three countries for the past few months. India currently occupies the third position in a total number of cases reported worldwide, next only to the United States and Brazil. As of March 26, 2021, India has reported total cases of 11,787,534 (9.46% of global cases) and total deaths of 160,692 (5.86% of global cases) (WHO Covid-19 Dashboard). By October 2020, almost entire India had come under the influence of pandemic with urban areas accounting for disproportionate share of total cases.

India exemplifies a top-heavy structure of urbanisation. The disease spread had also exhibited a similar pattern with a maximum concentration of cases in India's biggest and most populated cities. These include the Municipal Corporation of Mumbai, Delhi, Ahmedabad, Chennai Pune, Indore, Jaipur, Kolkata, Surat, Hyderabad and Jodhpur. Incidentally, Bangalore and Hyderabad being on the top list of cities, had revealed a much lesser number of COVID-19 cases till the end of May.

---

[3] Ibid.

[4] Ibid.

Bangalore had initially stood out as an example in containing COVID-19-positive cases. However, with the progress in the pandemic and multiple waves occurring as a result of staggered unlocking, Bangalore has now been facing the rage of COVID-19 like any other Metropolitan city in the country, with Bangalore contributing more than 68% of all cases in Karnataka as of March 2021.

Mumbai Urban Agglomeration has consistently spread the pandemic, where the infection has shown little chance of abatement. Delhi had initially shown a rapid growth but exhibited a deceleration in the end of July and August, to which many epidemiologists believe that a flattening of the graph had begun here. However, these were short-lived, and Delhi is already experiencing its third wave and cases are again accelerating with a high positivity rate in March 2021. Ahmedabad had shown a considerable increase till May end and a decline thereafter. However, Chennai was able to contain the virus until April end, only to be followed by an explosion thereafter. This shows that the trajectory of the spread of the pandemic does not have simultaneity. The boom in virus spread in the leading cities has been transferred to other cities and towns of India. So, by the time some of the largest cities and municipal corporations have shown a decline, others have joined the queue.

12.3 Reinforced Vulnerabilities of Health Infrastructure in Indian Cities

The COVID-19 crisis has reinforced the already existing vulnerabilities existing in urban areas. Besides, already deep-rooted inequalities and social problems have been exposed, posing severe limitations to our fight against Covid-19 (Kihato and Landau 2020). These include congested and insanitary conditions in large parts of the cities like in the older portions and in the slums and low-income residentials (Biswas 2020). Deep-seated structural and economic inequalities faced by the urban poor, many of whom are informal migrant workers engaged in low-income activities of daily wage earners and casual labourers in construction, transport, petty trade and other such activities have rendered them much more vulnerable than many other better-off groups. This is in line with historical trends in the impact of epidemics and pandemics (Wade 2020).

However, the largest crisis of all has been the sudden excessive pressure on our inadequate health infrastructure. Though the Central Government has identified and set up adequate health infrastructure and health facilities for COVID-19 management in the country, yet many cities have shown paucity to cope with the significant emerging cases in terms of health infrastructure, medical and paramedical personnel, leave alone the quality of services provided and the existing stigma attached to the COVID-19 infected patients.

Public Health facilities dedicated to COVID-19 case management are categorised into:

Dedicated COVID Hospital (DCH),

Dedicated COVID Health Centre (DCHC) and
Dedicated COVID Care Centre (DCCC).

Dedicated COVID Hospitals constitute the first category that would provide comprehensive care particularly to clinically assigned severe cases. The Dedicated COVID Health Centre comprises the second category that is supposed to cater to clinically assigned moderate cases. The last category i.e. the Dedicated COVID Care Centre should cater to patients who have been clinically assigned as mild or very mild cases or COVID suspect cases.

It was recommended that the Dedicated COVID Care Centres be established as temporary facilities by the states/UTs in various venues such as hostels, hotels, schools, stadiums, and lodges, whether public or private, as well as in homes. There will be distinct spaces in these centers, areas for suspected and confirmed cases. The Government stated, "Each Dedicated COVID Care Centre is connected to one or more Dedicated COVID Health Centres and at least one Dedicated COVID Hospital for referral purposes."

The central Government made it a priority for all states and UTs to notify and upload information about the three assigned types of COVID-dedicated facilities on their websites for public awareness. According to Government sources, "32 states and union territories have already uploaded this information on their websites or public information platforms, while the remaining are in the process of doing so."

It is imperative that because of India's huge population base, testing capacity should be increased, The Government had realized this and hence the National Centre for Disease Control (NCDC), was involved in the procurement of a high powered machine as per the recommendations and approval of the Empowered Group 2 Committee. Hence the Cobas 6800 testing machine was successfully installed at NCDC. NCDC is providing support for testing samples from Delhi, NCR, Ladakh, J&K and various other states, as per need. With these efforts, India's testing capacity increased manifold, and India is testing one of the highest numbers per day even when the positivity rate remains much lower than the worst affected countries around the globe.

However, though these exist in paper and pen, ground realities report different stories altogether. In many instances, Government health facilities had not been able to cater to COVID patients efficiently, resulting in a lack of apathy of individuals for public facilities. Moreover, due to the surge in Coronavirus infections, patients have found it extremely difficult to get admitted to coronavirus-designated hospitals. On the other hand, the private facilities often charge exorbitantly, leaving the poor man out of its ambit. There has been ample media reporting regarding the struggle involved in getting patients hospitalised, and even after the long-drawn struggle, even if a bed could be managed, the patient had died due to delay in treatment. In some cases, deaths have occurred in ambulances carrying patients. These incidents highlight the fragility of our already overburdened health infrastructure of urban India. With the largest cities under severe pressure, imagine the plight of small and medium towns that are already deprived in the face of an existing urban bias towards megacities and metropolitan areas.

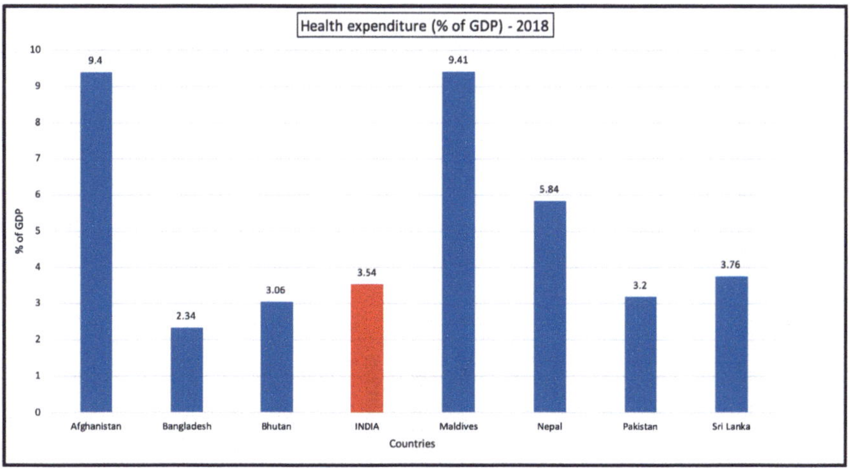

Graph 12.1 Health expenditure (as percent of GDP), in some developing countries of the world—2018. *Source WHO Global Health Expenditure Database, 2020 Note Health expenditure here includes public expenditure + out of pocket expenditure*

India has one of the lowest investments in health expenditure compared to not only developed countries like the US. UK and Italy, but also in terms of some of the developing countries. Low public investment in health has made the country even more vulnerable to the COVID-19 pandemic (Graph 12.1).

In addition, India also has a severe shortage of healthcare workers. According to a report by the Centre for Disease Dynamics, Economics and Policy (CDDEP) (2020), India has a shortage of around 600,00 doctors and 2 million nurses with a huge rural–urban divide. Moreover, per the Health Ministry data released in October 2019, there is one doctor for every 11,082 people, which almost falls short by 10 times the doctor-patient ratio that the World Health Organization (WHO) suggests. The WHO mandates that the doctor-to-population ratio should be 1:1,000, while India had a 1:1,404 ratio as of February 2020.

Graph 12.2, shows the data for the population served by one doctor in significant economies and South Asian countries vis-à-vis India:

In rural areas of India, this doctor-patient ratio is as low as 1:10,926 doctors as per National Health Profile 2019 (Central Bureau of Health Intelligence (CBHI) 2020).

According to data from the **Organization for Economic Co-operation and Development** available for India for 2017, India has 0.53 beds for 1,000 people compared with 0.87 in Bangladesh, 1.1 in Indonesia, 2.11 in Chile, 2.73 in Turkey, 1.38 in Mexico, 4.34 in China and 8.05 in Russia.

A recent study by the Center for Disease Dynamics, Economics & Policy (India) and Princeton University pointed out that the country currently has 713,986 beds, including 35,699 in intensive care units and 17,850 ventilators for 1.3 billion people.

However, India has progressed considerably in terms of its strategies for dealing with the pandemic. As we move towards the vaccination phase, the challenges posed

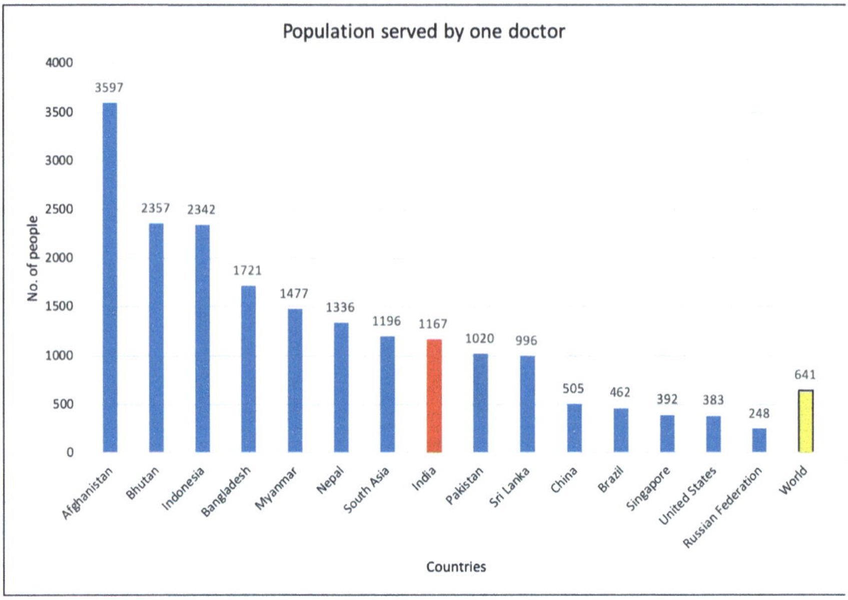

Graph 12.2 Population served by a doctor in selected countries of the world, 2018. *Source* Global Health Workforce Database, 2018

by inadequate healthcare infrastructure and human resources deepen the crisis and limit our response. This again poses threats of delayed recovery from the pandemic and enhanced burden on closely packed cities.

12.4 Urban Policies and Programmes in the Face of the Pandemic

The Coronavirus crisis has brought to the fore that we should revisit the urban policies and programmes and try to rejuvenate them according to the need of the hour. Policymakers and planners should take cognisance of the community-level diversities and try to make target-specific interventions.

The Jawaharlal Nehru Urban Renewal Mission (JNNURM), a significant urban revitalization initiative introduced in 2005,, features two associated schemes for Non-Mission Small and Medium Towns: (1) Urban Infrastructure Development Scheme for Small and Medium Towns (UIDSSMT), which integrates two existing Programmes of Integrated Development Scheme for Small and Medium Towns (IDSMT) and Accelerated Urban Water Supply Programme (AUWSP), and (2) the Integrated Housing and Slum Development Programme (IHSDP), which combines the existing Valmiki Ambedkar Awas Yojana (VAMBAY) and discontinued National Slum Development Programme (NSDP).

However, relatively more developed states such as Maharashtra, West Bengal, Tamil Nadu, Uttar Pradesh, Andhra Pradesh, and Gujarat have attracted a greater share of funds and projects, leaving smaller states behind. Out of the total JNNURM funding (approximately Rs. 18,462 crore), only 22.7% was allocated to the Basic Service for the Urban Poor (BSUP) component, while Urban Infrastructure and Governance (UIG) received 77.3%. There has been a systematic decline in the percentage of the population covered by JNNURM as the size-class of urban centers decreases (Kundu and Samanta 2011). Scholars have critically argued that by promoting a neo-liberal urban agenda, this scheme exhibits biases against the poor in several ways. The construction of flyovers and roads often results in slum demolitions, displacing the poor, and some of the prescribed reforms, such as full cost recovery, user charge policies, and the participation of private agencies, have adversely affected them.

Development of small and medium towns is handicapped because of their limited technical and financial capacity (Mahadevia 2006; Mukhopadhyay 2006). Rajiv Awas Yojana (RAY), launched in 2011, also aimed towards providing basic urban amenities to slum dwellers and finding suitable vacant land for rehabilitating them in affordable houses, finally reaching a goal of slum-free cities in the long run.

Various government-sponsored schemes since the late 1980s to early 2000s, (e.g. National Slum Development Programme, 1996; Valmiki Ambedkar Awas Yojana, i.e. VAMBAY, 2001) have attempted to improve the living environment of slums. However, the inadequacy of funds and the diversion of little available funds were some of the challenges that NSDP experienced in reality (India: National Report 2001).

It has been highlighted in the Eleventh Five-Year Plan document (2007–12) that the slum dwellers are marginalized, with inadequate basic urban facilities such as water supply and sanitation in these areas. Thus, the VAMBAY programs that were already in place as well as the NSDP program that was stopped were combined into the Integrated Housing and Slum Development Program (IHSDP). This initiative aimed to improve the living conditions of slum dwellers and to promote holistic slum development by providing better shelter and other physical amenities (Planning Commission 2007).

According to an approach paper to the Twelfth Five-Year Plan (2011) highlighted that the urban poor continue to "lack access to basic amenities such as water supply, sanitation, healthcare, education, social security, and decent housing" and are experiencing "multiple deprivations and vulnerabilities" (Planning Commission 2012).

Despite numerous government interventions and initiatives, a significant portion of the urban poor continues to reside in the slums and squatter settlements. It is important to note that, while the names and funding patterns of these schemes have changed, their focus on improving living conditions in slums and their components, such as physical infrastructure (water supply and drainage systems) and social infrastructure (health centers and schools), have remained largely unchanged.

Moreover, the norms and guidelines associated with many of these initiatives and schemes are often ambiguous in nature. For example, the "basic norms" of the

Environmental Improvement of Urban Slums, as defined by the Delhi Urban Shelter Improvement Board—the agency responsible for managing jhuggi-jhopdi clusters in Delhi—mandate "1 Tap/hydrant for 50 persons"/"1 WC Seat for 25 persons." While these standards may appear reasonable to the government, a one-hour water supply during the summer only allows for approximately three minutes of water use per five-member household. The long waiting times for toilet use can lead to chaos, prompting slum dwellers to resort to open defecation, especially in the mornings.

In order to enhance aesthetic appeal, economic growth and promote urban management of cities, the Central Government hopes to eliminate slums in the near future. Scholars have argued that the urban rich and poor coexist in a symbiotic socio-economic sphere and share common public spaces (Patel 2011). However, the urban poor have limited influence in political decision-making. For instance, the preparation for the Commonwealth Games in Delhi (2010), as part of the effort to create a 'world-class city,' led to the demolition of several slums. Additionally, Delhi's slum policy, which is biased against poor residents, tends to result in the proliferation of squatter settlements (Dupont 2008).

Different national and state-level planning agencies place considerable effort in preparing planning instruments in India. It is time to establish a coordinated effort in preparing a comprehensive and integrated urban plan. This should be implemented appropriately with systematic analysis and review from time to time. Also, rural development should not be de-linked from these efforts.

12.5 The Urban Future and Resilience in Urban Planning and Development

Against such a backdrop, it becomes imperative to locate Indian cities within the resilient framework since the concept of resilience is inextricably linked to sustainable development. 'Cities of resilience' as coined by Musacchio and Wu (2002), refers to complex urban forms arising out of the functional integration of several aspects that play specific roles in the spatial heterogeneity in both ecological and social functioning of urban areas.

It indicates how cities can endeavour across different sectors and creates situations enabling cities to bounce back and recover swiftly from sudden shocks or slow-onset disasters.

This ultimately translates into risk minimisation and vulnerability reduction. A comprehensive definition would perhaps read as "urban resilience refers to the ability of an urban system and all its constituent socio-ecological and socio-technical networks across temporal and spatial scales, to maintain or rapidly return to desired functions in the face of a disturbance, to adapt to change and to quickly transform systems that limit current and future adaptive capacity (Meerow et al. 2016).

Our urban future would depend on the mechanisms through which we can create resilient cities. Some of the issues, if not all, that should be addressed in moving along the line of resilience include mainstreaming health in urban planning and development. Some of the required interventions in this respect include:

1. A decentralised approach with the serious involvement of the urban local bodies is the need of the hour.
2. Moreover, resource mobilisation is fundamental to empower the ULBs for the development of health infrastructure and health delivery systems.
3. Access to adequate data and information systems. Digital technologies, including tracking mechanisms and the incorporation of big data, are the need of the hour, notably to empower the local Governments for their knowledge base and effective targeting.
4. Integrated planning of the physical and social environment in urban areas and efficacy in the health delivery system should be the prime agenda of the ULBs. Water, sanitation, and adequate nutrition for all should constitute the main areas in this effort.
5. Instilling eco-friendly practices in public health should also form priorities with adequate stress on healthy lifestyles and building immunity in individuals. Therefore, urban planning components should be environmentally friendly so as to boost public health. It is high time that urban planning and design is reimagined in the post-Covid scenario wherein concepts like the ones of ***"Garden Cities" and "Neighbourhoods"***– propounded by **Clarence Perry in 1929** are brought back and aligned to new realities. The idea is to make urban spaces more humane, self-sustained and functional with environmental sustainability at the heart of any such planning process.
6. The fact that Covid-19-led lockdowns have made our cities more livable when one considers the environmental problems of pollution and global warming re-emphasises that climate change is already upon us, and the onus of course correction lies squarely on humankind. Various research has shown that the near-complete halt in transportation had led to a drastic decrease in the concentration of NO_2, SO_2 and particulate matter in the environment (Saadat et al. 2020).

12.6 Conclusion

As Governments and cities manage their immediate response to the COVID-19 pandemic and are probing into mechanisms that would look ahead to plan for long-term recovery efforts, there is an opportunity to transform cities in meaningful ways that not only protect vulnerable people from immediate threats but also build resilience towards any emergency. The Sustainable Development Goals must guide this work, in particular, Goal No. 11 'Make cities and human settlements inclusive, safe, resilient and sustainable.' While planning for resilient cities it is imperative in this digital era that newer technologies like big data and artificial intelligence be increasingly used so that population mapping by disaggregated groups is possible. In addition, global inter-city collaboration regarding consultancy services and technical and administrative assistance is the need of the hour; as due to heightened mobility, an epidemic can take the turn of a pandemic within a very short period.

The current crisis can be seen as an opportunity to rethink the way we live in cities, the symbiosis between cities and their inhabitants, and the role of physical and virtual urban public spaces, and thus to design the cities of tomorrow. In addition, there is a felt need to reconfigure the workspace to allow physical distancing and the use of smart technology to address the new priorities arising from the pandemics.

It is time to redefine the identity and priorities of Indian cities. This should entail a rethinking of urban policies to strengthen risk preparedness and response capacity, and thereby making urban India more resilient by making cities smarter, greener, more inclusive; resilient and above all, livable.

References

Biswas PP (2020) Skewed Urbanisation and the Contagion. Econ Pol Wkly 55(16):13–15

CDDEP, India (2020) Covid-19 for India updates. New Delhi

Central Bureau of Health Intelligence (CBHI) (2020). *National Health Profile (NHP), 2019.* New Delhi: Ministry of Health & Family Welfare, GoI.

Planning Commission (2007) The 11th five year plan 2007–12. Government of India, New Delhi

Planning Commission (2012) The 12th five year plan 2012–17. Government of India, New Delhi

Dupont, V. (2008, July 12). Slum Demolitions in Delhi since 1990s: An Appraisal. *Economic & Political Weekly*, 79–87.

Kihato CW, Landau LB (2020) Coercion or the social contract? Covid-19 and spatial (in)justice in African Cities. City & Society

Kund D, Samanta D (2011) Redefining the inclusive urban agenda in India. Econ & Polit Wkly, 46(5)

Mahadevia D (2006) NURM and the poor in globalizing mega cities. Econ & Polit Wkly:3399–3406

Meerow S, Newell JP, Stults M (2016) Defining Urban Resilience: A Review. Landsc Urban Plan 147:38–49

Mukhopadhyay P (2006) Whither urban renewal? Econ & Polit Wkly, 879–884

Musacchio LR, Wu J (2002) Cities of resilience: Integrating ecology into urban design, planning, policy, and management. In: 87th annual meeting of the ecological society of America/14th annual conference of the society for ecological restoration. Ecological Society of America, Washington DC

Perry CA (1929) The Neighbourhood Unit: A scheme of arrangement for the family-life community regional plan of New York and its environs. Arno Press, New York

Saadat S, Rawtaani D, Hussain CM (2020) Environmental perspective of covid-19. Sci Total Environ

Wade L (2020) An unequal blow. sciencemag

Chapter 13
Representation of Dhaka in *The Book of Dhaka: A City in Short Fiction*

Mohammad Shafiqul Islam

Abstract Dhaka, the capital of Bangladesh, is often marked as one of the most densely populated and polluted cities in South Asia and the world. For various reasons, Dhaka is an important city, and historically, the city has a broader significance. Since the emergence of Bangladesh in 1971 through a nine-month-long war against Pakistan, Dhaka has risen to a metropolis despite its numerous challenges. Devastated by the war, the city has developed rapidly, but because of a lack of long-term planning and deep-rooted awareness among the people of the country to sustain the development, Dhaka is turning into an almost uninhabitable city day by day. Alarmingly, the city is on the verge of losing ecological balance. Environmental hazards, created mainly by people, are putting the city at risk of various calamities, including earthquakes, water crises, and droughts. This article intends to examine how Dhaka is placed in the glorious history of Bangladesh. The city has, no doubt, developed quite amazingly, but the people of the city have failed to warrant sustainability. The article also addresses how the city has been portrayed in different stories in *The Book of Dhaka: A City in Short Fiction*, with a special focus on the development and sustainability of the environment.

Keywords Dhaka · *The Book of Dhaka* · Liberation war of Bangladesh · Environment · Development · Sustainability

13.1 Introduction

The history of Dhaka is as old as it is of Bangladesh, which emerged as an independent and sovereign country in 1971. Until 26 March 1971, the country was known as East Pakistan, which was then part of West Pakistan. Through the 1947 Partition, the subcontinent had been divided into India and Pakistan, and pathetically religion had been one of the main issues taken for granted during the Partition. India (Hindustan) was for Hindus, and Pakistan was for Muslims. Even though Bangladesh, then East

M. S. Islam (✉)
Department of English, Shahjalal University of Science and Technology, Sylhet 3114, Bangladesh
e-mail: msislam-eng@sust.edu

S. Anand et al. (eds.), *Sustainability in South Asian Cities*, Advances in Geographical and Environmental Sciences, https://doi.org/10.1007/978-981-97-7455-5_13

Pakistan, has no land connection with Pakistan, then West Pakistan, this country was annexed to Pakistan only because Muslims were a majority here. It was illogical as well as unexpected because division based on religion sparked horrible conflicts between Hindus and Muslims throughout the whole subcontinent. But the reality of East Pakistan was beyond this conflict—West Pakistan began to exploit the people of East Pakistan immediately after the Partition had taken place. Soon after this major event in history, the people of East Pakistan, especially the political leaders, including Sheikh Mujibur Rahman, the Father of the Nation, realized how unusual and unreasonable it had been to merge this part of the region, currently Bangladesh, with Pakistan. Therefore, the people of East Pakistan had begun to dream of secession from Pakistan, sovereignty, and independence, and their dream came true in 1971.

The history of Bangladesh is a history of struggle, exploitation, conspiracies, war, and deaths, since the country was born out of a nine-month-long bloody war against Pakistan. At the cost of about three million lives of the Bengalis and the honour of two to four lakh women, Bangladesh earned independence. According to Das et al. (2022), "... this independence was gained after the country witnessed one of the worst genocides in world's history" (1113). As a result, unlike many other countries in the world, and more specifically in South Asia, the country had to struggle to flourish. A country devastated by war needs the time and support of the whole world to stand on its own feet. Roads, highways, bridges, schools, colleges, universities, mosques, temples, and buildings of different offices were destroyed by Pakistani forces. Jahan (2009), an accomplished scholar and critic, explains the war, "[o]n March 25, 1971, General Yahya abruptly broke off the negotiations and unleashed a massive armed strike against the population of Dhaka, the capital city. In two days of uninterrupted military operations, hundreds of ordinary citizens were killed, houses and property were destroyed, and the leader of the Awami League, Sheikh Mujibur Rahman, was arrested" (p. 247). Instead of handing over power or continuing discussions with the nationalists of Bangladesh, the Pakistan government unleashed Operation Searchlight to eliminate the freedom-loving people of East Pakistan. When Sheikh Mujibur Rahman became President of the independent country, there were debris, destruction, and signs of death everywhere throughout the country. Dhaka being one of the most important cities in the country was more affected during the war.

13.2 Theoretical Underpinning

Main cities of various countries draw the attention of writers and poets who write about cities taking their history, heritage, transformation, and transition into consideration. In this respect, the environment remains one of the critical factors in the representation of cities. With the advent of ecocritical studies in literature, there have been tendencies to explore fictional works, nonfiction, and poetry through the lens of ecocriticism. Huggan and Helen (2010b) argue that "there is no social justice without ecological justice" (p. 35). Huggan and Helen suggest, social justice and ecological justice are interconnected, as both human and nonhuman entities deserve justice equally. Doing damage to the environment means causing harm to human societies. Huggan and Helen (2010a) further observe that "human liberation will never be fully achieved without challenging the historical conditions under which human societies have constructed themselves in hierarchical relation to other societies, *both* human *and* nonhuman" (p. 22). This statement is key to understanding ecosystems that help various species on earth survive and environmental justice. To build a society free from discrimination and injustice, human beings need to challenge their asymmetrical relations to other societies, including ecosystems.

Maxwell (2009) speculates that among scholars and critics of postcolonialism, there is a "widespread concern about [the] destruction of the finely balanced ecosystems that keep the planet's various life forms alive" (p. 17). The critics, poets, and writers of postcolonial spaces are concerned about the consequences of ecological imbalance. Global capitalism agents are responsible for creating this kind of imbalance, as they promote urbanization at the cost of green terrains. Maxwell also notes that postcolonial poets and writers show a "remarkable commitment to exposing the neo-colonial realities helping to propel globalization" (2009, p. 17). Poets and writers of formerly colonized countries bring to light the systems designed and developed by agents of capitalist societies, giving ample space to globalization. In the name of development, if people destroy nature, intruding on ecological balance, they simply restrict the route to human liberation. Much damage has also been done to Dhaka's ecological balance, and it has been done in the name of development. Influenced by its multilayered history and rich tradition, many writers and poets have depicted Dhaka in their seminal works. They also keep the celebration of nature and lamentation for the loss of eco-friendly territories at the backdrop. *The Book of Dhaka: A City in Short Fiction* is such an attempt to present Dhaka with its diversity and transition to modernity. The book is, however, a literary representation of the city. With ten stories well chosen from the literary oeuvre of Bengali literature from Bangladesh, *The Book of Dhaka: A City in Short Fiction* is a work of translation, which is published by Bengal Lights Books, a publishing house based in Dhaka that promotes Bangladeshi literature in English translation or English writing from Bangladesh across the globe. Comma Press, UK, has simultaneously published the book.

The literary representation of the city is common, and the leading cities of different countries throughout the world have appeared in formative works of the literature. Writers and poets have presented the cities in a great deal of ways. Bird (2017) quotes James Donald stating that the "living space of the city exists as representation and projection and experience as much as it exists as bricks and mortar" (p. 3). It sometimes appears that a city is a living entity in a work of fiction or a single poem. A city of a particular country often works as a mirror to map the country's overall state. In fact, a city sometimes represents a nation to which it belongs. Ashcroft (2011), one of the prominent postcolonial theorists, observes:

> Cities... appear to be, if not similar, at least similarly messy irruptions of global modernity. Since medieval times, cities have been locations of health and justice, freedom and enterprise on one hand, and dingy dystopias of class inequality on the other. But from a postcolonial

> perspective the crucial feature of all great cities in history is that they have always been imperial; they have always been the centres of one form of empire or another (p. 497).

Ashcroft thinks that cities are the incursions of global modernity, because in cities people are provided with better health facilities and administrative services, and city-dwellers seem to enjoy more personal freedom. On a different note, cities are a hub of corruption, illegal activities, human trafficking, drug dealing, incommensurable power relations, and wide discrimination between the rich and the poor. People in cities, especially the ones who are politically powerful or affluent, behave as though they are superior to the country's people. The country-city dichotomy, which is deep-rooted, exists and cannot be denied. And treatment of the power to the underprivileged demonstrates that they are imperialists—Ashcroft compares them with the agents of imperialism.

Varma (2012b) observes that the city is "a cacophonic, polyvocal space of difference and 'excessive' postcoloniality" (p. 138). Varma's observation is apt as cities are doubtless home to people with multifaceted identities, speaking different languages in varied tones, creating chaos. And there is, above all, a widening gap between the well-off and the destitute. Although Varma states this about Bombay, the same is true about Dhaka, and *The Book of Dhaka* sketches the reality existing in many megacities of the world. Varma (2012a) further notes, "The postcolonial city... is constituted by the tensions and contradictions between the global, national and the local concepts and practices of urban space; hence the idea of the postcolonial city as a conjunctural space" (p. 14). Dhaka, along with many other cities of Asian, African, and Latin American countries, is undoubtedly a postcolonial city, which, as Varma asserts, is at the juncture of complex relations between the cities of the Global North. As an urban space, a city like Dhaka cannot be isolated from other metropolises of Western countries, but the city is intrinsically connected to its own history and culture. It is, however, undeniable that an uneven economic condition is prevailing in the cities of the Global South.

13.2.1 City in Literature and The Book of Dhaka

Edited by Arunvava Sinha and Pushpita Alam, *The Book of Dhaka: A City in Short Fiction* has received rave reviews in several well-known journals and magazines. Ten stories by ten established and prominent writers of Bengali literature were selected for translation, and ten translators were selected from the country. The translations were discussed, analyzed, and edited in a workshop organized by Dhaka Translation Center in collaboration with Comma Press, the British Centre for Literary Translation, and Commonwealth Writers. Ra Page, a writer and editor from the UK, was, along with Arunva Sinha, the workshop leader. It is understandable how rigorous the processes were to bring out the book. Several critics have commented positively on the book in renowned publications, including *World Literature Today* and *Kitaab*. Readers have also received the stories in the book warmly.

With a foreword by Kaiser Haq and an introduction by Kazi Anis Ahmed, *The Book of Dhaka: A City in Short Fiction* contains ten well-chosen stories in which Dhaka appears in multiple forms, state, and realities. The city concerning different times of history has been portrayed in the stories. Dhaka seems to play a role as a character in the book. In his introduction to the book, Ahmed (2016) observes:

> Dhaka is arguably the world's most densely populated city. The largest cities in the region—Delhi, Kolkata, Mumbai, Karachi—count an average of about 40,000 people per square mile. The figure for Dhaka is an astounding 130,000. And without the infrastructure to match the population, it is regularly ranked among the most unliveable cities in the world. Many of the basic amenities expected by city dwellers elsewhere are scarce or completely missing in Dhaka: there are few nice parks, clean movie theatres, quality museums, or indeed—being officially dry—bars or nightclubs (pp. ix-x).

It is not an overstatement when Dhaka is called the "most densely populated city" in the world, as Ahmed offers population figures per square mile in other remarkable megacities. A witness to a great deal of events in history, Dhaka has a rich tradition and heritage, but the city, compared to various capitals and metropolises in the world, has failed to provide "basic amenities" to the people living here. A lack of long-term planning, proper management, honest purposes of the politicians, thoughtful decisions of the concerned people in authoritative positions, and, above all, noble intentions of the governments has prevented the city from progressing as expected in terms of habitability.

But Ahmed (2016) calls the city home to millions of people, "Yet, tens of millions still live here, and, improbably enough, love this maddening mess they call home" (p. x). People from around the country swarm into the capital city for various purposes, especially for livelihood, as people from rural areas look for work with better pay in the city. But ultimately, they make the city a "maddening mess" in the form of shanties and slums. With people's tendencies to migrate from country to city, Ashcroft's argument is thought-provoking. Ashcroft (2011) articulates that "postcolonial cities are a particularly intense demonstration of the diasporic movement of populations, microcosms of the global flow of peoples that intensifies during and after the period of European imperialism" (p. 498). Ashcroft reflects on the movement of the country's people to the city from a deeper perspective, connecting it to the global phenomenon of displacement through mass migration, especially from the Third World to the West. After European imperialism, Ashcroft thinks, people have tended to move to diasporic locations, and he considers the massive flow of the country people to the city as a microcosm of the global flow.

Despite numerous problems that the people living in Dhaka face in their everyday lives, the city owns some establishments, which are historically significant, as well as some great places that attract visitors around the year. These are what add to the significance of Dhaka as a metropolis. In regard to what the city contributes, Ahmed (2016) remarks, "It is a city of contrasts: Louis Kahn's monumental parliament building stands in the middle of the city—a totemic reproach to Dhaka's lack of space and grandeur. An incomplete red-brick fort in the Old Town marks Dhaka's earlier history as a regional trade entrepôt. The tiny but charming British-era enclave similarly bears testimony to faltering starts as a capital city" (p. x). Here K. Ahmed mentions and then gives brief information about a few monuments that bear historical significance. Dhaka is proud of its parliament building that the world-famous American architect Louis Kahn designed—it is one of the most beautiful architectural designs in the country. To add to the glorious side of the city, Ahmed (2016) further articulates, "Yet Dhaka's concrete regime is softened, it has to be said, by a surprising preponderance of trees, which stay green year-round. A few lucky neighbourhoods enjoy expanses of man-made lakes, which serve as sanctuaries in winter to millions of 'guest birds.' In spring, krishnachura trees set entire roads ablaze with their blooming red flowers" (p. xi). Although Dhaka is called the city of concretes, rickshaws, mosquitoes, and mosques, it has hundreds of trees by various roads and streets making the city green. For example, krishnachura trees proliferate around the city, and when krishnachura flowers bloom, Dhaka adorns itself with a distinct look. Besides, some lakes attract thousands of migratory birds from Siberia during winter. But this brighter part of the city is getting lost to the greed of some people who are destroying natural sites in the name of urbanization.

This article addresses, among many other things, the issue of environmental justice and the problems created out of a lack of proper planning about Dhaka. There is a huge gap between the proper planning and the visible developments. Land grabbers and terrorists are destroying rivers, canals, lakes, and by doing this, they are causing damage to the ecological balance not only to the city but to the whole country as well. Huq (2020) remarks:

> The whole of the Dhaka region sits on a very substantial series of connected wetlands of rivers, canals and big and small water bodies. The current part of Dhaka city is built over those water bodies and has meant that we have to suffer flooding due to drainage congestion whenever we have a heavy downpour.... The most important point is that the Dhaka Plan is aware of this and we have laws against destroying our wetlands, but that does not stop dishonest developers from breaking the laws and getting away with it. This cannot be allowed to happen (para. 8).

It is already mentioned that Dhaka is now an overcrowded city, which is expanding, no doubt, at the cost of rivers, lakes, trees, and playing grounds. For obstruction of the natural flow of water, flood occurs, and waterlogging is a common concern in the rains. Some parts of the city are submerged if it rains for some time. Because of the poor drainage system, rainwater sometimes blends with sewage, polluting the city to an alarming extent. As a result, people are affected by various kinds of diseases. They go through sad experiences because of all these problems created by a few dishonest developers—their sufferings sometimes become so severe that they cannot but attempt to protest.

The problems such as traffic jam, widespread density, exponential population growth, air pollution, land grabbing, dust, mosquitoes, garbage, waterlogging, and extreme heat have made life in Dhaka difficult. To rid the city-dwellers of these crises, the government at times tries to chalk out a long-term plan and take measures, but in most cases, they do not come to fruition. Some development works, including the elevated expressway, traffic system, eviction of illegal establishments, and so on, are visible, but the crises are so wide-ranging and the situation is so complex that the concerned authority, especially the city corporation and RAJUK,[1] should be more active and dynamic in implementing the decisions to turn the city into an ideal one, making urban facilities possible for the people. Unplanned housing is another obstacle to the progress of Dhaka as a civic modern city. People are exposed to health hazards because of their living in unsafe areas where there are few facilities of untroubled life free from pollution. Ahmed (2019), an urban planner, rightly points out the problems of the city:

> Generally, the discharge of untreated industrial effluent, urban waste water agrochemicals, sewage water, storm run-off, solid waste, oil spillage, sedimentation and concurrent encroachment are major reasons responsible for the pollution of Dhaka's watershed. Industrial effluent discharged from and domestic sewerage from a large part of the city through storm sewers are responsible for the pollution of the Buriganga-Turag-Bangshi Dhaleshwari river system. Besides, people living on river banks throw their household wastes into the rivers (para. 3).

There has not been a well-organized plan to manage industrial wastes in Dhaka, although many industries and factories, small and big, have been set up from time to time. Hundreds of garment factories are concentrated in or around Dhaka, and their sewage and solid wastes are poorly managed. Most of the factories dump their wastes into the Buriganga and other rivers. As a consequence, the river water is so badly polluted that no one can use it. Earlier people used to bathe in the river and utilize the water for their family work, but it is poisonous now. If anyone touches the water, they are surely exposed to deadly diseases.

13.2.2 Representation of Dhaka in The Book of Dhaka

Once, many parts of Dhaka, especially the city's adjacent areas, were replete with "swamps and rice paddies" (Elias 2016, p. 5), but in the name of development, the agricultural land has been turned into part of the city. Many natural swamplands have been made in residential areas where only concrete buildings are visible. The green city comprising croplands and trees is turning into a drab, dry, and colourless place for city dwellers. Once there was "a chill to the rains of late autumn" (Elias 2016, p. 9), but now it rarely rains to assuage unbearable heat in the city. People have cut down trees to cause damage to the natural beauty as well as the ecological balance of the city. As a result, heat has risen high, and life has become more difficult in summer. There is a wonderful description of rain soaking Dhaka now and then in "The Raincoat":

> That was in the thick of the rainy season, and it had rained a lot in Dhaka this time. He had said something about the endless rain, to which the labourer had muttered, 'The rainy season is the most fitting.' He had said it twice. What did that mean? Someone in the staff room had

[1] RAJUK is a portmanteau word that is extended as Rajdhani Unnoyon Kortripoksha, meaning Capital City Development Authority. This authority is entitled to make plans for the development of the city and solve the crises that the city-dwellers face.

whispered once or twice, 'The bastards don't know the rains of Bengal. Russia had General Winter, we have General Monsoon' (Elias 2016, p. 17).

It is worth mentioning that Akhteruzzaman Elias's story "The Raincoat" is based on the 1971 Liberation War of Bangladesh. In this story, Nurul Huda, the central character, puts on his brother-in-law Mintu's raincoat while going out to meet Principal of his college in the rain. The Pakistani military men catch him in the college office as some freedom fighters come to the campus in the guise of labourers with Nurul Huda's help. A military man informs him that one of the labourers described the rain. It is interesting to see how Elias relates rain with the city and how he shows the difference between Dhaka and other cities, especially the ones inhabited by the Pakistani Army. Bengal is famous for monsoon and the land of Bengal is fertile because it rains profusely here during the rainy season. There is a special mention of Dhaka where people regularly met with rain during the monsoon.

The stories in *The Book of Dhaka: A City in Short Fiction* depict the city in multidimensional ways, some of them offering its details of roads and streets, beautiful trees and flowers, orchards and parks, monuments and historical buildings, while others featuring pollution, environmental injustice, traffic jam, extortion, water-logging, unplanned construction, political violence, and so on. Syed Manzoorul Islam's story "The Weapon" (2016) depicts the dilapidated condition of Old Dhaka, including the idiosyncrasies of the people of this part of the city. What is striking about the story is that it gives details of each street and alley in a creditable manner. As the writer informs, "My story has been progressing in a straight line so far—from Gulistan along Bongshal Road, for instance. But it will become more complicated now, turning towards Nawabpur and negotiating the congested roads of Adalat Para leading to Patuatuli. It might even get lost up a blind alley. Or down a manhole" (Islam 2016, pp. 26–27). The writer refers to different roads that most of the time of the day remain overcrowded and traffic jams are common scenes in these places. It is also worth noting that the roads are narrow, and people living in surrounding areas, most of whom are either poor or belong to the middle class, appear complacent, although many of the members of the young generation form gangs to exhibit muscle power, eventually killing their counterparts just to dominate in the locality. These narrow streets remind readers of the alleys of twentieth century London described by T S Eliot in "The Love Song of J. Alfred Prufrock" and *The Waste Land*.

Parvez Hossain's story "The Decision" (2016) presents Dhaka with its both outstanding exuberances and snags. Although it is a powerful love story, the writer offers a beautiful description of the city, putting the Ekushey Book Fair at the centre of the narrative. In the story's first paragraph, Hossain (2016) vividly describes how the city is soaked in rain, "Yesterday's rain had cleansed the air but not the trees. There were no puddles of mud on the ground. Soft clouds hung in a clear blue sky, and a gentle breeze signalled spring. Radhachura flowers blossomed by the road. Such splendour didn't normally flourish in this city oppressed by brick and mortar, advertisement hoardings, and the glare of electric lights" (p. 41). Everything looks and sounds perfectly in shape and splendour if nature does not act bizarrely. Trees, lakes, and orchards, among others, make a city ideal for human beings to inhabit.

Seasons behave properly when these natural entities are given due care and are not destroyed. In the above description, the writer has placed the opposite pictures side by side—rain and concrete jungles—to portray the city as credibly as people see it and undergo the experience of inhabiting the city amidst pain and pleasure.

Slums and shanties, which have burgeoned like mushrooms in Dhaka, have added to the derelict state of the city. People in these shanties struggle to live hand to mouth, and most of them live below the poverty line. As a result, the slum-dwellers, especially adolescents and youths in these places, sometimes take recourse to corruption, terrorism, and drug addiction, eventually engaging in horrible acts like killing, raping, and smuggling. Sultana (2016) successfully captures the reality of slum life in the story titled "Mother." As Sultana (2016) writes, "[t]he four of them lived with their uncle in a house in the Methor Slum, occupied by sweepers and sewage workers, in Hajaribagh" (p. 51). The story recounts the life of a poor family living in the Methor Slum, which the lower class people of the society inhabit. As the story progresses, a woman, the mother of a little child, is compelled to trade drugs, especially marijuana and heroin. Moinul Ahsan Saber's story "The Circle" (2016) also portrays the life of a middle-class family in Dhaka. The people in the family cannot come out of the circle they have been dawdling around for a long time. The statement "The entire city must have changed" (Saber 2016, p. 69) suggests how Dhaka has undergone a tremendous change all these years.

Bipradash Barua's story "The Princess and the Father," (2016) which is based on the Bangladesh War of Independence, is the story of Bokul, a young man, who skates around the city. The writer describes Dhaka's roads and streets while informing the readers of Bokul's skating. Barua (2016) writes, "Bokul skated noisily down the street, slowing down where the surface was uneven. He wanted to get to North South Road, either through Bongshal or Alubazar. He chose the first option. The North South Road was fairly smooth, but it was shrouded in fog. It would be wonderful to get to Ramna Park, he told himself" (p. 91). Against the backdrop of Bokul's skating, the condition of Dhaka's roads and streets is described here. Some names of familiar roads and their condition are described as well. Since the story is about the Liberation War, Dhaka's state during 1971 is given with a little context. In Anwara Syed Haq's story "Helal Was on His Way to Meet Reshma," (2016) readers are presented with a love story that also portrays Dhaka's past and present condition. Thus Haq (2016) offers a comparison:

> The road was incredibly crowded. It was the same story every Friday. There was a time when Dhaka would be a quiet city on the weekend. Not any more. Every day was a busy one now. Millions of people migrated to the city every day, people who had no idea of the lay of the land or of what money was worth. New slums rose up all the time, with new faces in them. The funny thing was, you wouldn't know from the people on the roads that hundreds of thousands, even millions, of them were unemployed (pp. 107–108).

This is indeed a true picture of Dhaka where millions of "unemployed" people wander aimlessly, maybe in search of jobs, or, in many cases, for food. Most of them undergo existential crisis as far as poverty and hunger are concerned. As a result, vehicles, including rickshaws, remain stuck on roads for hours. Earlier, weekends were a bit better in terms of traffic conditions, but the situation, in this case, has also worsened now. Thus Dhaka is gradually turning into almost an uninhabitable city. Coming all the way from remote villages, the poor people have made the city a mess of crowds. In many cases, they have sheltered on pavements, blocking walkways, and, above all, ruining beauty and balance.

13.3 Conclusion

Despite multifarious crises and concerns, Dhaka can be taken for granted as a modern and cosmopolitan city. Ahmed (2016) rightly observes, "Against all odds, three factors have conspired to render Dhaka one of the most colourful and chaotically jolly cities: a rich interplay between folk and high culture, a fecund natural climate and the city-dwellers own irrepressible instincts for creative entrepreneurship" (p. xii). On different occasions, especially during the celebration of Bengali New Year and the celebration of Victory Day, Dhaka wears a festive look. On special days, the city reverberates with the jubilant mood of its dwellers and is adorned with colourful designs and decorations. It is interesting to note that the city is the habitat of the people living in both skyscrapers and slums. Culturally rich, Dhaka celebrates and accommodates diversity. The people of Bangladesh, as a whole, have an instinctive tendency to promote arts and letters and music, and the dwellers of Dhaka exhibit their creativity in diverse fields on special occasions. And Dhaka has been home to many great poets, writers, revolutionaries, cultural activists, architects, artists, and musicians. If environmental justice can be ensured, if the city-dwellers are provided with urban facilities, if wastes can be managed properly, if city-based corruption and terrorism can be contained, if sound and air pollution can be controlled, if natural sites and resources, including rivers, lakes, orchards, and trees, are saved from destruction or extinction, and if a long-term people-friendly urban plans are implemented, Dhaka must be an ideal, progressive, and peaceful, cosmopolitan city.

References

Ahmed KA (2016) Introduction. In: Sinha A, Alam P (eds) The book of Dhaka: A city in short fiction. Bengal Lights Books, pp ix–xii

Ahmed T (2019) Protecting environment of Dhaka City. The New Age. https://www.newagebd.net/article/65995/protecting-environment-of-dhaka-city

Ashcroft B (2011) Urbanism, mobility and Bombay: Reading the postcolonial city. J Postcolonial Writ 47(5):497–509. https://doi.org/10.1080/17449855.2011.614774

Barua B (2016) The princess and the father. In: Sinha A, Alam P (eds) The book of dhaka: A city in short fiction (M. S. Islam, Trans.). Bengal Lights Books, pp 91–104

Bird E (2017) 'The things not in the picture': Bombay's poets and the re-representation of the city. J Postcolonial Writ 53(6):644–458. https://doi.org/10.1080/17449855.2017.1391797

Das TK, Bhattacharyya R, Sarma PK (2022) Revisiting geographies of nationalism and national identity in Bangladesh. GeoJournal 87:1099–1120. https://doi.org/10.1007/s10708-020-10305-1

Elias A (2016) The raincoat. In: Sinha A, Alam P (eds) The book of dhaka: A city in short fiction (P. Alam, Trans.). Bengal Lights Books, pp 1–18

Haq AS (2016) Helal was on his way to meet Reshma. In: Sinha A, Alam P (eds) The book of dhaka: A city in short fiction (M. Rahman, Trans.). Bengal Lights Books, pp 105–126

Hossain P (2016) The decision. In: Sinha A, Alam P (eds) The book of dhaka: A city in short fiction (P. Alam, Trans.). Bengal Lights Books, pp 41–49

Huggan G, Helen T (2010a) Introduction. Postcolonial ecocriticism: Literature, animals, environment. Routledge, pp 1–24

Huggan G, Helen T (2010b) Postcolonial ecocriticism: Literature, animals. Routledge, Environment

Huq S (2020) Entering the anthropocene era in a befitting manner. The Daily Star. https://www.thedailystar.net/opinion/politics-climate-change/news/entering-the-anthropocene-era-befitting-manner-2015669

Islam SM (2016) The weapon. In: Sinha A, Alam P (eds) The book of dhaka: A city in short fiction (A. Sinha, Trans.). Bengal Lights Books, pp 19–40

Jahan R (2009) Genocide in Bangladesh. In: Totten S, Parsons WS (eds) Centuries of genocide: Essays and eyewitness accounts. Routledge, pp 245–265

Maxwell A (2009) Postcolonial criticism, ecocriticism and climate change: A tale of Melbourne under water in 2035. J Postcolonial Writ 45(1):15–26. https://doi.org/10.1080/17449850802636499

Saber MA (2016) The circle. In: Sinha A, Alam P (eds) The book of dhaka: A city in short fiction (M. A. Nusrat, Trans.). Bengal Lights Books, pp 59–72

Sultana R (2016) Mother. In: Sinha A, Alam P (eds) The book of Dhaka: A city in short fiction (S. N. Royhan, Trans.). Bengal Lights Books, pp 51–58

Varma R (2012b) The postcolonial city and its subjects: London, Nairobi. Routledge, Bombay

Varma R (2012a) Introduction. The postcolonial city and its subjects: London, Nairobi, Bombay. Routledge, pp 1–35

Chapter 14
Vulnerability and Coping Mechanism of the City Dwellers: A Study of Old Dhaka

Md. Anwar Hossain and Tulshi Kumar Das

Abstract Urbanization is one of the most significant phenomena closely connected with economic growth and development. Although the level of urbanization is low in South Asia, the people of Bangladesh experienced a rapid growth of urbanization in the recent past. About 29.7% of people lived in urban areas in 2009 while it increased up to 38.2% in 2020. Along with facilities like job opportunities, health care, education, business, etc. city dwellers also face multiple socioeconomic problems including poverty, lack of proper housing, pure drinking water and sanitation facilities and many more. On the other hand, water, air and noise pollution; traffic jam, and different types of disasters emerged as by-products of urbanization. Considering the gravity of the situation, the United Nations included SDG-11 (Sustainable Cities and Communities) in its Sustainable Development Agenda 2030. The government of Bangladesh has already taken measures for achieving SDG-11 and improving the overall conditions of urban dwellers, especially in Dhaka city. It is important to know to what extent the residents of this city are resilient to vulnerability, and how they cope with the situation. A study is carried out at the Lalbag area of Dhaka South City Corporation using a qualitative approach for a better understanding of the situation. Data are collected by employing face-to-face interviews with a semi-structured interview schedule to address the research questions. Data are also collected through focus group discussion (FGD) and non-participant observation to supplement the findings placed on the basis of face-to-face interviews. Thematic analysis technique is used to analyze the collected data. Verbatim of the research participants is also presented for a greater understanding of the situation that has affected them most. The findings show that the people of Old Dhaka experience some problems such as income insecurity, problem with drainage, disruption of gas supply, water logging,

Md. A. Hossain
Department of Social Work, Jagannath University, Dhaka 1100, Bangladesh
e-mail: ahossainsw@yahoo.com; anwarh@sw.jnu.ac.bd

T. K. Das (✉)
Department of Social Work, Shahjalal University of Science and Technology, Sylhet 3114, Bangladesh
e-mail: dastulshi-scw@sust.edu

S. Anand et al. (eds.), *Sustainability in South Asian Cities*, Advances in Geographical and Environmental Sciences, https://doi.org/10.1007/978-981-97-7455-5_14

fire outbreak, problems of movement, traffic congestion, water, air and noise pollution, and so on that make them vulnerable to lead their daily lives. Accordingly, they adopt some indigenous mechanisms such as looking for alternative jobs, cooking early in the morning or night, abstaining from going outside during water logging, being alert for fire outbreak and so on to cope with the situation.

14.1 Introduction

Bangladesh is historically known as a rural-agrarian country. But the people are experiencing a rapid growth of urbanization. About 29.7% of people lived in urban areas in 2009, it increased to 38.2% in 2020 (https://knoema.com/atlas/Bangladesh/Urban-population) and it is estimated that by 2035, 50% of people will live in urban areas in Bangladesh (Ministry of Planning 2020: 155). The rapid growth of the urban population in Bangladesh has been remarkable in the past four decades. Urbanization is the by-product of industrialization and development. Basically, industrialization, development and urbanization—this trio is closely connected with each other. Although the existence of the city was found about 550 years ago, modern urbanization came into light after industrialization in the 1850s. It is found that humankind crossed a milestone in 2008 having half of the population of the world living in urban areas (BBS 2015). It is also estimated that the urban population will be 66% of the total population of the world by 2050 (UN 2014). That is why the twenty-first century is called an urban century.

Dhaka, the capital city of Bangladesh, is the eleventh mega city in the world. It is projected that Dhaka will be the 6th largest mega city in the world in 2030 with a population of 27.37 million (UN 2014). It will also transform the socio-economic and infrastructural situation of Dhaka city seriously. However, the rapid growth of urbanization in Dhaka city has already created a variety of problems leading to hampering the daily lives of the city dwellers. It has been found that civic services like electricity, water and gas supply, roads and transport facilities have become severely inadequate.

The greater portion of the urban population is concentrated in Dhaka which is more than one-third of the total urban population of Bangladesh. It is estimated that the density of the population in Dhaka city is 47,400 persons per square kilometer (Dhaka Tribune 2018a, b) which indicates over over-density of the population. As it is the land of business, economic, trade and administrative activities people of many parts of the country migrate here. As a result, number of population of Dhaka city is increasing day by day. Despite being the center of all types of activities, sometimes it poses serious challenges to the people that include poverty, scarcity of utility services, lack of transportation facilities, environmental degradation and so on (Islam 2018).

However, the growth of population as well as economic progress increases both the demand and supply of transport facility, job opportunity, civic facility and housing facility. Considering the gravity of the situation, the United Nations focused on sustainable management of these issues and included SDG-11 (Sustainable Cities and Communities) in its Sustainable Development Agenda 2030 of which the main theme is to ensure access to adequate, safe and durable housing, transport, other basic services, waste management, green and public spaces and so on (UN n.d.). The government of Bangladesh has already taken measures for achieving SDG-11 and improving the overall condition of urban dwellers (Ministry of Planning 2020), especially in Dhaka city. It is important to know to what extent the residents of this city are resilient to vulnerability, and how they cope with the situation. Accordingly, a study has been carried out at Lalbag area of Dhaka South City Corporation using a qualitative approach for better understanding the above-mentioned situation and the findings have been presented in this paper.

14.2 Understanding the Meaning of Concepts Used in This Chapter

Vulnerability

The term 'vulnerability' is a central concept that states the condition of the people. Vulnerability is used more frequently in the field of disaster management. However, it is a dominant approach to explain people's condition and is widely used in different disciplines of social science. The concept 'vulnerability' has been defined by a wide range of authors. According to Canon (1994 cited in Islam et al. 2020:354), vulnerability is a characteristic of individuals and groups of people who inhabit a given natural, social and economic space. Vulnerability refers to the characteristics and circumstances of a community that make it susceptible to risk. But it has no universal definition and scholars of different disciplines use it in different ways. Scholars of disaster study use the term vulnerability to explain the susceptibility of people to hazards (Alexander 1993; Gabor and Grifith 1980; Timmerman 1981). Some scholars focused on political, social and economic conditions (Liverman 1990; Smith 1992; Dow and Downing 1995). It is the lack of ability of a system to resist against the concerns of external stressors. Social scientists consider a community's inability to cope with a variety of socioeconomic factors to define vulnerability (Paul 2013). However, social vulnerability focuses on the susceptibility of social groups to potential losses from extreme events and to absorb the impact (Emrich and Cutter 2011). In this paper, vulnerability has been conceived as the susceptibility of any system or people (city dwellers) to cope with the challenges of daily life in urban area. It is a complex characteristic produced by a combination of factors derived especially (but not entirely) from class, gender and ethnicity. Factors related to vulnerability have been divided into three categories (Table 14.1).

Coping Mechanism

Cope means to deal successfully with some difficulties. On the other hand, mechanism refers to a system for achieving something (Hornby 2000). Coping is the approach in which the people act within existing resources and the range of expectations of a situation to achieve various goals. Baqutayan (2015: 481) defined coping as "any effort, healthy or unhealthy, conscious or unconscious, to prevent, eliminate, or weaken stressors, or to tolerate their effects in the least hurtful manner". The notion of 'coping mechanisms' is strongly connected with the idea of survival. Coping mechanisms are remedial actions adopted by vulnerable people whose survival and livelihood are threatened (WHO 1998). It includes the range and combination of activities and choices that households choose for survival in any crisis (Palmino-Raganit 2005). These strategic options may include short-term considerations such as ways of earning a living, coping with shocks and managing risks (Aasoglenang and Boney 2013). These strategies have human and social costs that include community self-help initiatives (World Health Organization 1998). It is an adaptation to environmental stress that is based on conscious or unconscious choice and that enhances control over behavior or gives psychological comfort (Dictionary.com). Taking the above-mentioned views into consideration, the term 'coping mechanism' has been defined in this paper as such activities that are based on existing resources and conscious or unconscious choices by the city dwellers of Old Dhaka to prevent, eliminate, or weaken stressors, or to tolerate the impact of daily challenges of respondents' lives.

14.3 Methodology

The paper is qualitative in nature and based on empirical data. A purposive sampling procedure was followed in this study as it allows us to choose a case that illustrates some feature or process in which we are interested (Silverman 2001). Accordingly, the Lalbag area of Dhaka South City Corporation was selected purposively. Then 10 males and 10 females were selected purposively as a sample who were living for more than five years in the study area. However, the paper is based on both primary and secondary data. Primary data were collected by employing interviews with semi-structured interview schedules to address the research questions (for a better understanding of the situation), where open-ended questions were included. The interview schedule was designed to cover the vulnerabilities and coping mechanisms of the respondents. Data were also collected through focus group discussion (FGD) and non-participant observation to supplement the findings placed on the basis of face-to-face interviews. On the other hand, secondary data were collected by consulting research articles, research reports and books. Thematic analysis technique has been used to analyze the collected data. Verbatim of the research participants have also been presented for a greater understanding of the situation that has affected them most.

14.4 Findings

14.4.1 Socioeconomic Profile of the Respondents

The respondents were from various socioeconomic groups. With regards to age, the lowest age of the respondents was 25 years and the highest age was 55 years. Some of them were highly educated and a few of them were educated only up to grade 5. They were from different occupations that included rickshaw-pulling, small business, driving, teaching, studying (student), household works and housemaid. They were from different income groups. Their monthly household income ranged from BDT 15,000 to 55,000. All of the respondents, except one, lived in rented house. Their housing condition was not good. Most of them lived in dilapidated building, two of them lived in tin-shed house and only two of them lived in multi-storied building with modern facilities.

14.4.2 Vulnerability and Coping Mechanism of the City Dwellers

Physical Vulnerability and Coping Mechanism

In this study, physical vulnerability has been assessed on the basis of some indicators that include housing, availability of water for drinking and other household works, sanitation facility, supply of gas and electricity, roads and communication facility and drainage system. The targeted respondents were asked about the availability of above mentioned structural facilities. The respondents opined that the locality was composed of old buildings although some new multi-storied buildings have been constructed in the very recent past. Some of the respondents were found to live in modern buildings. On the other hand, some were found to still live in old dilapidated buildings. In modern buildings, sunlight and wind pass easily but in old buildings the environment is not healthy. The people were found to live in these old buildings with unhygienic conditions. They had to use electric lights in the daytime too. In new buildings, modern sanitation facility was found available. But in old buildings, only one toilet was found available for 5–7 persons. Participants of FGD also mentioned that in the study area some buildings were not furnished with modern toilet, washroom and bathroom facilities. Even some old buildings had no such facilities at all. Residents of these houses used a common toilet and bathroom with other residents. Even some people used common toilets constructed beside the road. They take bath on the road using water from roadside water tap.

With regard to the supply of water, most of the respondents were found to have a connection with a pipeline of water supplied by WASA. But the quality of water was not good for drinking directly. They used to drink it after boiling. Some of the respondents mentioned that they had no water connection at their house. They were found to collect water from roadside water. They also used to take bath at roadside bathrooms. They were to maintain a long queue and spent more time as many people were in the queue for this purpose. But it was a matter of regret that although the water was available at other times, they had to face a water crisis in the dry season (March–May). Then they could not take bath for some days. One of the respondents, Madina, explained her vulnerability and coping mechanism in the following way:

> I have no water connection at my house. I collect water from roadside tap for drinking and other household works. But in summer I am to face water crisis. Then I am to maintain long queue and spend more time for collecting water. Sometimes I collect water from deep tube-well of neighbor's house by spending TK. 10. per gallon

Sohana mentioned, "*In summer I am to face water crisis. I spend more time for collecting water because of a long queue. In my locality water is available from 9:00 am to 4:00 pm. But it is difficult to maintain time as I have other household works.*"

The respondents also faced a scarcity of gas supply. Usually from 9:00 am to 4:00 pm, gas supply was unavailable in some parts of the study area. The problem became acute during winter. Although the gas supply used to resume after 4:00 pm the respondents faced low pressure of gas supply. They had to spend more time for cooking. It hampered the daily lives and other works of the respondents. Sometimes the respondents could not have breakfast and reach the office on time. Some of the respondents were found to have become habituated to this situation. They used to cook their food in previous night or early in the morning and ate it in the morning and at noon. On the other hand, they have had their dinner at late night. Nadir Ali explained his family's coping mechanism with the scarcity of gas supply in the following way:

> In our area gas supply is very unstable. Very often we are to face scarcity of gas. The problem becomes more severe in winter. In the early morning (7:00 am) gas supply is stopped and we get it in the evening (9:00 pm). Although gas supply is resumed at 9:00 pm pressure remains low. So my wife starts to cook at 10:00 pm and we eat our dinner at 12:00 midnight. Not only this, we also cook our food for next morning and noon in the previous night.

It was also found that some of the respondents bought cooked foods from restaurants during the gas crisis. Again, some of them were found to use LPG cylinders to cope with the gas crisis. Even they used firewood as the fuel for *kancha* oven for cooking food during the gas crisis. They also used kerosene stove or electric oven for cooking during the disruption of the gas supply.

Supply of electricity was almost satisfactory during winter. Although now and then they suffered from power disruption, it was not unbearable in winter. But the problem became acute during summer. Almost every day they had to suffer from load shedding for one or two hours more than once a day. Some of the respondents were found using IPS during load shedding and they were economically in good position. On the other hand, those who were not in a good position economically used candles for light at night and used hand-fan to protect themselves from the stifling hot.

Road and communication system in old Dhaka was not up to the desired level. The study area was no exception in this regard. The roads were found narrow. Lanes and by-lanes were also very narrow. It was tough to move from here to there. Most of the lanes were congested and overcrowded area which made it difficult for the people to walk on the road and lanes. Very often the residents throw dirty water and waste from multi-storied buildings, causing the pedestrian annoyed. The following statement reflects the situation:

> The old Dhaka is very over-crowded. Roads, lanes and bye-lanes are very narrow. It is very difficult to walk on the road. Some people throw dirty rubbish or spit from upper floor of multi-storied building on you. Some people throw away wastage from the roof of the building. The wastage is rotted here until comes the cleaner and that pollutes the environment. So we are to prepare to meet/face such awkward situation. It is true that we are habituated with this situation.

However, traffic jam was found daily feature of Old Dhaka. Dwellers of Old Dhaka were compelled to spend more time on movement. Actually, rickshaw, rickshaw-van, pick-up van, mini-truck, and push-cart ply on the roads of the study area. As a result, people experience traffic jams every day. Rickshaw or rickshaw-van pushed from behind the pedestrians. It created quarrel between the pedestrians and rickshaw-pullers. Since it was a regular feature, they settled it very soon.

Drainage system of the study area (Old Dhaka) was not good enough. Although drains of some lanes were found working properly, drains of other lanes were not working properly. Many of the drains were blocked. Thus, wastewater could not pass away properly. During the rainy season, due to heavy rain water-logging started in the study area. Some parts of the study area went under 3/4 feet of water. Then small boats were found plying on the roads. Some drains remained open. There were no covers/lids over drains. Therefore, wastes (stool, dead bodies of insects and worms or birds) were seen in the drain. In the study area, some manholes were found open. As a result, many people fell down in manholes. During waterlogging many rickshaw pullers fell down into manholes with their passengers. The water logging stayed for 2/3 days during the rainy season. Then it became very difficult to move from here to there. One of the respondents, Sohana, stated it in the following way:

> Drainage system of our area is not good. Most of the drains are very narrow. Drains are not covered with lid. Dirty wastes are seen. On the other hand, many of the drains are blocked. As a result, water logging is created because of moderate rainfall. Then it becomes very difficult for movement. We pass our time inside the room.

Actually, because of the defective and inadequate drainage system the people of the study area had to suffer a lot. Some of the respondents mentioned that they had been habituated to this situation. They had been living for a long time in this area. So, the situation was not unusual for them. If most of the roads were submerged due to rain, they did not go outside of their house. If they had to go outside for urgent business they used rickshaw. Even sometimes they used small boats for movement during water-logging.

Socioeconomic Vulnerability and Coping Mechanism

Vulnerability is the state of susceptibility to untoward situations. It is related to features of a community or groups susceptible to any harmful social phenomenon. Social vulnerability indicates the lack of ability (incapability) of people to resist undesirable situations. It is connected with the absence of well-being of the people. It is related to lack of education, healthcare, communication, etc. On the other hand, economic vulnerability indicates lack of resources (unemployment, lack of income) (Islam et al. 2020; Choirunnisa and Giyarsih 2018). The present study focuses on the education of the respondents, schooling of children, security of income, social network, health care and life insurance to understand the socioeconomic vulnerability of the respondents. Data show that the respondents were from a variety of educational groups. Some of them did not cross the boundary of primary school while some of them obtained bachelor's and post-graduate degrees. Accordingly, they were involved in various occupations that included teacher, student, rickshaw-pulling, driving, salesman, household works (maidservant), etc. Most of these jobs were temporary. There was no security of their jobs. Anytime they might lose their jobs. As a result, they were to face a financial crisis. Lack of security of income led them to frustration reflected in the following statement:

> I have read up to grade-3, I do household works in various houses. But there is no security of my work. Anytime I may lose my job. So, every now and then I suffer from frustration, (Modina, a maidservant of 28 years old).

The businessmen were also found suffering from income insecurity. For example, after the fire outbreak in *Churihatta* in 2019, the income of many households had been decreased. The household of Sohana was one of the households whose income had decreased after the fire outbreak in *Churihatta.* Not only the income of many households and businessmen had been decreased but also many workers had lost their jobs. The same incident happened during covid-19 pandemic. Many workers in the non-formal sector had lost their jobs. On the other hand, the income and jobs of employees of the government and autonomous bodies were secured. But the number of government job holders was very poor. The findings indicate that income security was connected with the nature of the job and workers in the non-formal sector were found suffering from income vulnerability.

However, the respondents were found to adopt a variety of mechanisms to cope with income loss/insecurity. Some respondents took loan from relatives or friends to cope with financial crisis. Somebody took loan from bank and other financial institutions. Somebody was found to reduce daily expenses. For example, some respondents reduced expenditure on buying meat, fish and fruits. On the other hand, respondents from the poor income groups reduced their daily meals from thrice to twice a day. When they lost their temporary jobs they tried to find another one. They were found to start new jobs even though the salary was a little bit lower than the previous one. Sometimes they changed their business, especially those who were involved in small business. Some respondents changed their occupations (from salesman to rickshaw-pulling).

Children of the study area were found sincere to schooling as per the opinion of the respondents. The respondents from the better off families did not feel any problem to continue the study of their siblings. But the respondents from low-income groups faced difficulties in bearing the expenses. It was difficult for them to pay tuition fee for school/college and salary of private tutor. One of the respondents, Modina, shared,

> Tuition fee of school of my daughter is very high. It is difficult for me to pay the tuition fee. Not only this, I am to pay BDT 3000 per month as salary of private tutor. It is a burden on me.

It was really difficult for the respondents of low-income groups to bear the expenses related to education. To cope with the situation some respondents took loans from different sources (relatives, friends and neighbors). Sometimes they received an advanced salary from the employer. It has been found that some respondents had changed the school of their children as they were not in a position to bear the expenses of school. They admitted their children into the government-run primary school where expenditure was a little bit low. Some parents sent their children to *madrasa* (religious school) where education was provided almost free of cost (as opined by participants of FGD).

Social relation was found strong among the people in the study area. The respondents used to come forward to help each other in case of danger. The owner of the house helped the tenant of a rented house in their danger. Again, the exceptional situation was not rare. As the buildings were constructed side by side sometimes they were involved in quarrels. Then the *panchayet* (local committee composed of community leaders) took the initiative to solve it peacefully. Sometimes it was beyond the control of *panchayet,* then the people went to court to solve that problem.

Many hospitals (both government and private) and clinics were situated near the study area. For example, Dhaka Medical College Hospital, Mitford Hospital, Azimpur Maternity Center, K.G. Hospital and Fair Health Medical center, etc. were noteworthy among them. They had consultations with the physicians and some medicines were available in these health facilities. But they had to buy expensive medicine from outside the hospital. They were compelled to go to a private hospital for better consultation if needed. It was very expensive and difficult for people of low-income groups to bear the expenses. Affluent families could make quick arrangements but the poor respondents had to suffer a lot. They had to depend on loans or financial assistance from the better-off families. Sometimes they are required to sell their valuable goods for treatment purposes.

Overall insurance penetration stood at a meager scenario (0.57% only in 2018), the lowest in emerging Asia. Due to lack of awareness, confidence and understanding about life insurance, coverage was very poor in Bangladesh (Dhaka Tribune, 16 September 2019). The situation of the respondents was gloomier. None of the respondents had an insurance policy. Actually, due to lack of awareness as well as poor financial condition, they were not under the facility of insurance.

Environmental Vulnerability, Natural Disaster and Coping Mechanism

Very often environmental sphere cannot be separated from the socioeconomic spheres because of the mutual interaction of the components of these two phenomena. In this study, some variables such as water logging, waste management, air pollution, sound pollution, water pollution and natural disasters such as earthquake, fire outbreak have been incorporated to understand the environmental vulnerability of the city dwellers of Old Dhaka.

Waste Management

The data show that waste management was a critical problem in Old Dhaka as it was a densely populated area. The number of dustbins was poor compared to the need. Although residents of some lanes and bylanes appointed a person who collected waste from houses, still some people were found throwing waste from their houses through windows. They did not give their waste to the waste collector on time, rather they threw it through windows according to their will. On the other hand, in some other regions, the residents did not appoint such a waste collector. But there were some dustbins kept beside the road. Despite this, still some people did not use these dustbins rather they threw their household waste on the road, polluted the environment and hindered the free movement of the people. The residents of this area and pedestrians used these roads very cautiously by closing their noses and mouth to protect them from ill-smelling.

Water Logging

Though the study area was located inside the flood control embankment (Dhaka-Narayanganj-Demra Dam), short-term water logging during the rainy season happened because of heavy and incessant rain. It has been mentioned earlier that the drainage system was not good enough in the study area, which made the inhabitants vulnerable to water logging. The drains were too narrow to pass out the rainwater. Not only in the rainy season, but in other seasons too, dirty water from the drains overflowed, causing water logging because of the blocked drains. The people used rickshaw-van to cross the water-logged area. Some people took the initiative to clean the drains by the cleaners. Few people did not go outside of their homes as far as possible during this time. Some of them went to their work, crossing the dirty water despite the threat of health hazards. People who had to go outside spent more money on rickshaw or rickshaw-van fare.

Air Pollution

Dhaka is one of the most vulnerable cities in the world with respect to air pollution with an Air Quality Index (AQI) score of 188 while an AQI value between 151 and 200 signifies that residents of this area may begin to experience health effects. Air Visual's World Air Quality 2019 ranked Bangladesh as the most contaminated country in the world for exposure to PM 2.5 (Dhaka Tribune 2020). It is found that exposure to household air pollution is from burning fuels that include coal, wood or biomass for cooking. According to Real-time Air Quality Index (AQI), PM 2.5 AQI of Dhaka was 227 as on 30 January 2020 while 0–50 is good and 50–100 is moderate that indicates air of Dhaka city is very unhealthy (www.Aqicn.org/city/Bangladesh/Dhaka/us-consulate). The respondents expressed their opinion about the air pollution of Dhaka city from their own perception. They opined that air of Dhaka city was very much polluted and unhealthy. The excessive use of biomass, construction work, use of chemical, cutting of tiles, crushing of bricks and stone were responsible for air pollution. Diseases occurring from air pollution include respiratory problems such as chronic obstructive pulmonary disease (COPD), asthma, bronchiolitis, lung cancer, cardiovascular events, and central nervous system dysfunction (Manisalidis et al. 2020). People in the study area were also found to suffer from the above-mentioned diseases. It was very difficult for the general people to avoid air pollution. They had to live with polluted air. But they tried to avoid dust. Sometimes they used facemasks to protect themselves from dust. But before the COVID-19 pandemic, the rate of using masks was very poor. Although the government tried to reduce air pollution by imposing restriction on running brickfield, stone and brick-crushing machine, it was not found at the desired level.

Sound Pollution

One of the environmental hazards, the city dwellers face in Old Dhaka, is noise pollution. It is a great concern like other Megacities of developing countries. The sources of noise pollution are manifolds that include industries, indiscriminate use of loudspeakers, construction works, motorized traffic and public gatherings. Noise pollution is generally defined as regular exposure to elevated sound levels that may lead to severe effects on human beings and other living organisms. Sound pollution is unwanted or excessive sound that may have harmful effects on human health, wildlife and environmental quality. It is a complex sound wave and hazardous to public health and hearing. It has also psychological and social implications, and affects the well-being and quality of our lives (Chowdhury et al. 2010). If we very often expose ourselves to high levels of noise, it will hamper our physical and mental peace and may cause damage to our health. According to World Health Organization, sound levels less than 70 dB are not harmful to living organisms regardless of how long the exposure is., while exposure for more than 8 h to constant noise beyond 85 dB may be hazardous to human beings and living organisms (environmentalpollution-center.org/noise-pollution). According to the Department of Environment, the lowest and highest level of noise pollution in Dhaka City is 47 dB and 132 dB respectively, while 75 dB is comfortable for human hearing (Dhaka Tribune 2018a, b). The study area was no exception to this situation. The respondents had to face noise pollution almost every moment because of traffic congestion, construction work, and industrial sound. Very often high voice of loudspeakers added additional sound to the existing one. One of the respondents, Probir, explained the situation in the following way:

> People of Dhaka city are to face severe sound pollution. Because of construction work and brick crushing, very high level of sound exists in our locality. When we go to main road we can't listen to others because of hydrochloric horn of buses, pick-up vans and other traffics. Not only this, people of Old Dhaka play music with high sound on different occasions that include Durgapuja (worship of Durga), Swaraswatipuja (worship of Swaraswati), Sakrain (kite flying), 31st night and happy new year, Bangla nabobarsho (Bangla Happy New Year), observing holud program (ceremony of smearing the bride and bridegroom with turmeric preparatory to marriage) etc.

Another respondent, Sohana, stated:

> Because of high level of sound we can't do our regular activities properly. Sounds of tiles cutting, brick crushing, and sounds of different types of vehicle make us annoyed. During

night, because of loud noise of songs I can't concentrate on my study. Even I can't sleep properly.

Although the government has adopted the Noise Pollution (Control) Rule, 2006 it could not prevent noise pollution up to the desired level. In such a situation, the city dwellers seem to have quietly accepted the high level of sound. Sometimes they close the doors and windows of their houses to prevent acute sound that causes much irritation. Sometimes they request the organizers of the party to play the music at a low sound. But most of the people feel helpless as they can do very little to prevent noise pollution.

Water Pollution

Water pollution is another concern of the city dwellers. Inadequate sewage and inefficient waste management contribute to the pollution of water in Dhaka city. These elements are especially responsible for the pollution of surface water (Bashar and Fung 2020). Drinking water supplied by WASA is also contaminated with bacteria (The Daily Star 2019, Barta24 2019). The respondents mentioned that the quality of water in their area was not bad. But sometimes water becomes polluted because of broken supply line. Since the supply line is broken dirt penetrates into the supply line of WASA and the people can't drink this water. The respondents generally adopt some mechanisms to make this polluted water drinkable. First, they filter the water and then boil it before drinking. Some of the respondents use electronic filters for water purification. Sometimes they collect pure water from deep tube wells of neighbors' houses by giving BDT 10 per bucket/jar *(kolos)* that comply with the statement of Shadhona:

> Every now and then dirty water is supplied by WASA. It is not drinkable. First we boil and then filter it for drinking. Sometimes we collect it from deep tube-well of our neighbor's house. Even we buy bottled water during water crisis

Earthquake

The topological setting and geographical location of Dhaka city make it vulnerable to earthquakes. Unplanned urbanization (indiscriminate use of land and unplanned construction of buildings) as well as the high density of population added additional vulnerability to earthquake. Narrow road networks, lack of preparedness and low level of awareness among city dwellers are really a matter of concern (Ferdous and Rahman 2015; Rahman et al. 2017). However, there are some dilapidated buildings in the study area built more than 75 years ago causing high level of vulnerability. Besides, narrow lanes and bylanes have made the respondents more vulnerable to earthquake. The concerned government and non-governmental organizations have taken some measures as preparedness and response to earthquake. Awareness among the city dwellers has also been created by different organizations and mass media. Training on earthquake has been provided to the people, especially to the students of academic institutions of the study area. Some of the respondents reported that they had training about what to do if earthquake occurred. One of the respondents explained:

> If earthquake occurs I shall take shelter on an open place. If it is not possible I shall stand under the beam of the house or beside the column, (Shadhona).

Opposite view was also found. Some of the respondents mentioned that they had nothing to do if earthquake happened. What Allah (the creator) wills no frost can kill. What Allah wills is for good. Because of their belief in fatalism, they did not take any measures of preparedness. One of the key informants Mr. Mohanta explained the view of the residents of Old Dhaka in the following way:

> Most of the people of Old Dhaka irrespective of religion bear fatalistic attitude. They have no headache about earthquake. They are God-fearing. They think what is lotted cannot be blotted. They do not accept that we have to take some measures to reduce loss and damages.

Another respondent, Modina, stated her opinion regarding earthquake:

> If earthquake happens I will remain silent inside my house. I depend on Allah. What Allah wills no frost can kill.

Fire Outbreak

Fire outbreak is a devastating hazard for mankind. It destroys the lives and livelihood of the people. Old part of Dhaka city is very much vulnerable to fire outbreaks because of the density of population, preserving chemicals in storehouse, and lack of maintaining safety code (especially in using electricity). The people of Old Dhaka had experienced some major fire outbreaks, among which *Nimtoli* fire outbreak (3 June 2010) and *Churihatta* fire outbreak (20 February 2019) were devastating, killing 124 and 71 people, respectively. The situation of the firebreak sites worsened very quickly because of connecting flammable substances deposited nearby. But these locations are still vulnerable as actions required for improvements have not been taken yet in spite of imposing some restrictions by the government.

However, the respondents were asked about their vulnerability to fire outbreaks and their coping mechanisms. As per the opinion of all the respondents, participants in focus group discussion and key informants, the study area was very much vulnerable to fire hazards. The Old Dhaka is densely populated; roads and lanes are narrow, buildings are constructed side by side (closely connected with each other). Thus, fire may break out from nearby sources very easily. Moreover, a good number of buildings had limited and in some cases had no access to roads causing more vulnerability of the people. One of the respondents, Boishakhi, reported:

> We are very vulnerable to fire outbreaks. We noticed severe fire outbreak at Nimtoli and Churihatta. Many people were killed in those fire outbreaks. Because of flammable substances fire spread out very quickly. On the other hand, due to narrow passes the victims could not come out from their houses. It took more time for fire service and rescue team to reach out the spots. As a result, many casualties occurred.

Another respondent, Shimul, replied:

> We are scared. If conflagration occurs we won't be able to come out from our house. Because stair of our house is very narrow. Only Allah knows, what will happen!

In spite of vulnerability to fire hazard, preparedness was not found satisfactory. It has already been mentioned that houses are constructed side by side and very closely. There is small space left between and among the buildings. Some buildings are used both as stores of chemicals as well as residences of the families. It is highly risky for the inhabitants. However, at the individual level, some respondents were found watchful about fire breakouts. They preserve the contact number of the fire service so that they can call it if there is any such incident occurs. These conscious people use electric goods and electricity cautiously. They neither preserve combustible substances in their house nor do they rent out their houses for the same purposes.

14.5 Discussion

The study examines the vulnerabilities and coping mechanisms of the city dwellers of Old Dhaka. It shows that the residents of Old Dhaka are vulnerable in many respects. It finds physical/structural, socioeconomic and environmental vulnerabilities. But vulnerabilities vary according to education, profession, income and other resources that are in possession of the residents of Old Dhaka. Most of the respondents live in dilapidated buildings which make them remain in constant fear about the possible collapse of the building. They also suffer from disruption of gas and power supply. There is a lack of sanitation facilities, and drainage system in the study area is not good enough. Thus people suffer from water-logging during the rainy season. However, the respondents have been found to adopt a variety of coping mechanisms to cope with the emerging susceptibilities (to overcome the vulnerable situation). However, their coping strategies vary from family to family or person to person, depending upon their economic condition. For example, poor people go to their workplace in spite of water-logging through knee-deep water, but the members of better-off families use rickshaws or other transports to reach their workplace. Even some people do not go outside of their house during water-logging.

The findings of our study are consistent with a body of previous studies that show a high prevalence of vulnerabilities of the city dwellers of Old Dhaka to natural disasters (Rahman et al. 2017; Huq et al. 2020). Though these studies identified the vulnerabilities and sufferings of the city dwellers because of natural disasters, the current study investigates the vulnerabilities of the people and also explores their coping strategies. The findings show that the respondents are socioeconomically poor, and they often suffer from income insecurity. They struggle to bear expenses for the education of their children and the health care of the family members. Therefore, they adopt some strategies to cope with the situation. For example, suffering people often switch over their jobs for better income, change the school of their children, and go to government hospitals instead of private ones for treatment.

However, previous studies assessed the environmental vulnerabilities of the city dwellers that included flood, water-logging, earthquakes, fire outbreaks, etc. (Rahman et al. 2017; Huq et al. 2020; and Ahmed 2014). But this study explores some additional hazards that include water, noise and air pollution. These hazards are responsible for further vulnerabilities that cause physical and mental annoyance in the lives of the respondents. Thus, the respondents were found to employ some extra measures to overcome these vulnerabilities. The findings indicate that the coping mechanisms may vary from person to person or family to family but nobody living in Old Dhaka is free from the effects of physical/structural, socioeconomic and environmental hazards.

14.6 Recommendations

Vulnerability of the city dwellers, especially poor residents of Old Dhaka, is evident in the study area. It is believed that addressing vulnerabilities is imperative for the betterment of city dwellers. Although males and females as well as poor and rich households could be victims of vulnerabilities, nature/extent of being exposed to vulnerabilities varies from person to person. Our findings show that some factors expose people to vulnerable conditions in general, while effects and coping mechanisms vary as per the socioeconomic condition of the people. Therefore, vulnerability prevention interventions should take these findings into account.

The study finds that some people in the study area have no toilet and bathroom at all. They use community/public toilet and bathroom constructed beside the road. But the condition of those toilets and bathrooms is deplorable. Therefore, the study recommends rebuilding toilets and bathroom with modern facilities at the doorsteps of the city residents. Disruption in power and gas supply is also found in the study site makes the lives of the respondents vulnerable. The people, especially the poor ones, suffer a lot because of the disruption in the power and gas supply sectors. The authorities should take appropriate measures to ensure uninterrupted gas and power supply.

Roads, lanes and bylanes are very congested in the area. Moreover, some people often throw dirty water and waste from multi-storied buildings that make the movement of the people very difficult. In addition, because of throwing waste here and there, some drains have been blocked, which creates water-logging in the study area. It causes a lot of sufferings that hinder movement and hamper the normal life of the people. Therefore, public awareness should be raised so that people abstain from such activities that pollute the entire environment. Social work methods, especially community organization and social action, could be effective tools in this regard.

This study unveils that the respondents from the poor groups, especially those who are involved in informal sectors for livelihood, suffer from income insecurity. Therefore, insurance coverage should be expanded as well as the government should take legislative measures for job security.

The government of Bangladesh is very keen to provide the people with health services from the government hospitals. Our study reveals that the respondents receive consultation services and medicine from the healthcare facilities introduced by the government, but they are compelled to go to private hospitals for the treatment of complex and chronic diseases and buy expensive medicine from outside of the government hospitals. This is an extra burden for the poor and middle-class households. Apart from that, the COVID-19 pandemic has caused an additional burden for them. Therefore, the present study recommends increasing the healthcare facilities at the government hospitals.

The study further finds that water, air and noise pollution make the lives of the people vulnerable. City residents have been suffering from various diseases because of such pollution. The authorities concerned (the government, city corporation, WASA) have to take legal actions, structural measures, and steps to create awareness among the people to reduce and prevent pollution.

The literature shows that the old part of Dhaka city is critically vulnerable to earthquakes because of its geographical location and physical structure. If any medium-scale earthquake strikes in the study area major casualties will happen. The dilapidated buildings, narrow lanes and bylanes may delay rescue operations and add more vulnerability. Therefore, the study recommends that major renovation has to be done. It might be a new urban plan, abolishing dilapidated buildings and construction of new buildings with available space and wide roads, lanes and bylanes. In doing so, mass awareness has to be created and people's participation in different actions should be ensured. Community organization, one of the basic methods of social work, could be an effective tool to bring about desired changes.

Fire outbreak is one of the devastating natural hazards which is a major concern for the residents of Old Dhaka. It should be noted that one of the causes of the quick spread of fire in Old Dhaka is connecting flammable substances. Although the government has imposed restrictions on preserving combustible goods/chemicals in Old Dhaka, the situation has not been improved yet. In this circumstance, the law should be enforced very strictly along with taking appropriate measures to transfer the flammable goods to other places that are comparatively safer than Old Dhaka.

14.7 Conclusion

People's exposure to social, economic, physical (structural) and environmental vulnerabilities is very high in the study area. This study reveals that some issues expose people to some vulnerabilities but they vary from person to person, family to family based on their socioeconomic condition. Accordingly, they take on a variety of coping mechanisms to cope with the situation created through multiple vulnerabilities. Keeping it in view, the study proposes a set of recommendations to curb the vulnerabilities of city dwellers. However, the study followed a qualitative approach because of which it did not show any cause-effect relationship among the independent, dependent and extraneous variables. Further studies may be carried out to fill up the gaps indicated in the current study.

Table 14.1 Factors related to Vulnerability Source: Complied by the authors

Physical factors	Socioeconomic factors	Environmental factors
Housing	Education	Waste management
Availability of water	Schooling of children	Water pollution
Sanitation facilities	Security of income	Air pollution
Supply of gas	Social relation	Sound pollution
Supply of electricity	Healthcare	Water logging
Roads and communication		Earthquake
Drainage system		Fire outbreak

References

Aasoglenang AT, Bonye SZ (2013) Rural livelihoods diversity: Coping strategies in WA West district in Northern Ghana. Eur Sci J 9(35):139–156

Ahmed I (2014) Factors in building resilience in urban slums of Dhaka, Bangladesh. Procedia Econ Financ 8:745–753

Alexander D (1993) Natural disasters. Chapman and Hall, New York, NY

BBS (Bangladesh Bureau of Statistics (2015) Changing pattern of urbanization in Bangladesh: An analysis of census data. Ministry of Planning, Government of the People's Republic of Bangladesh, Dhaka

Baqutayan SMS (2015) Stress and coping mechanisms: a historical overview. Mediterr J Soc Sci, 6(2S1):479–488

Barta 24 (2019) WASA water contains fecal wastes and harmful bacteria. Barta 24

Bashar T, Fung IWH (2020) Water pollution in a densely populated megapolis, Dhaka. Water, 12(8). https://doi.org/10.3390/w12082124

Choirunnisa AK, Giyarsih SR (2018) The socioeconomic vulnerability of coastal communities to abrasion in samas, bantul regency, Indonesia. Quaest Geogr 37(3):115–126

Chowdhury SC, Razzaque MM, Helali MM (2010) Assessment of noise pollution in Dhaka city. In: The paper presented in the 7th international congress on sound & vibration held at Cairo, Egypt on 18–22 July, 2010

Dhaka Tribune (2018) Noise pollution: A bane of Bangladeshi urban life. Dhaka Tribune

Dhaka Tribune (2018) Dhaka remains the world's most densely populated city. Dhaka Tribune

Dhaka Tribune (2019) Insurance coverage in Bangladesh lowest in emerging Asia. Dhaka Tribune

Dhaka Tribune (2020) Air visual ranks Dhaka as worst in air pollution. Dhaka Tribune

Dictionary.com. https://www.google.com/search?q=https%3A%2F%2Fwww.dictionary.com%2Fbrowse%2Fcopingmechanism%23&oq=https%3A%2F%2Fwww.dictionary.com%2Fbrowse%2Fcopingmechanism%23&aqs=chrome.69i58j69i57.8725j0j7&sourceid=chrome&ie=UTF-8 Retrieved on 07 June 2022

Dow K, Downing TE (1995) Vulnerability research: where things stand. Hum Dimens Q 1:3–5

Emrich CT, Cutter SL (2011) Social vulnerability to climatesensitive hazards in the southern United States. Weather, Clim, Soc 3(3):193–208

Ferdous I, Rahman MK (2015) Citizens at risk from earthquake hazard in Dhaka city: Scaling risk factors from household to city region level. J Geogr Inst Cvijic 65(1):91–106

Gabor T, Griffith TK (1980) The assessment of community vulnerability to acute hazardous materials incidents. J Hazard Mater 3(4):323–333

Hornby AS (2000) Oxford advanced learner's dictionary of current English, 6th edition. Edited by Sally Wehmeier. Oxford University Press, Oxford

Huq ME, Shoeb AZM, Javed A, Shao Z, Hossain MA, Sarven MS (2020) Measuring vulnerability for city dwellers exposed to flood hazard: a case study of Dhaka City, Bangladesh. In: Yuan

X, Elhosney M (eds) Urban intelligence and applications, studies in distributed intelligence. Springer Nature, Gewerbestrasse, Switzerland, pp 207–215

Islam MR, Khan MA, Reza MM, Rahman MM (2020) Vulnerabilities of river erosion-affected coastal communities in Bangladesh: a menu of alternative livelihood options. Global Social Welfare 7:353–366

Islam N (2018) Urbanisation in Bangladesh: Recent trends and challenges. Daily Sun

Liverman DM (1990) Drought impacts in Mexico: Climate, agriculture, technology, and land tenure in Sonora and Puebla. Ann Assoc Am Geogr 80(1):49–72

Manisalidis I, Stavropoulou E, Stavropoulou A, Bezirtzoglou E (2020) Environmental and health impacts of air pollution: A review. Front Public Health 8:1–13. https://doi.org/10.3389/pubh.2020.00014

Ministry of Planning (2020) Sustainable development goals. Bangladesh Progress Report 2020. Bangladesh Planning Commission. Ministry of Planning, Government of the People's Republic of Bangladesh, Dhaka

Palmiano-Raganit M (2005) Analysis of community's coping mechanisms in relation to floods: A case study in Naga City, Philippines. Unpublished Master's thesis, International Institute for Geo-Information Science and Earth Observation Enschede, the Netherlands

Paul SK (2013) Vulnerability concepts and its application in various fields: A review on geographical perspective. J Life Earth Sci 8:63–81

Rahman MM, Akter S, Ashikuzzaman M (2017) Multi-hazard vulnerability assessment of an urban area: a case study on ward 34 of Dhaka South City corporation. J Asiat Soc Bangladesh 43(3):181–195

Silverman D (2001) Interpreting qualitative data. Methods for analysing talk, text and interaction, 2nd edn. Sage Publications, London, UK

Smith K (1992) Environmental hazards: Assessing risk and reducing disaster. Routledge, London, UK

The Daily Star (2019) Harmful bacteria in Wasa water. The Daily Star

Timmerman P (1981) Vulnerability resilience and collapse of society. A review of models and possible climatic applications. Institute for Environmental Studies, University of Toronto, Toronto, Canada

UN (2014) World urbanization prospects: The 2014 revision, highlights (ST/ESA/SER.A/352). Department of Economic and Social Affairs, Population Division

UN (n.d) Transforming our world: The 2030 agenda for sustainable development. A/RES/70/1. Sustainabledevelopment.un.org. on 29 April 2021

WHO (World Health Organization) (1998) Emergency health training programme for Africa. Panafrican Training Centre, Addis Ababa

https://knoema.com/atlas/Bangladesh/Urban-population. Retrieved on 29 April 2021

https://knoema.com/atlas/Bangladesh/Urban-population. Retrieved on 1.08.2021

www.Aqicn.org/city/Bangladesh/Dhaka/us-consulate. Retrieved on 01.08.2021

Chapter 15
Integrating Accessibility: A Critical Review of the Transportation Policies and Plans in Dhaka

Paromita Nakshi and Anindya Kishore Debnath

Abstract Transportation planning has long been centered on mobility-oriented approaches, where level-of-service (LOS) measurements and new road construction have been prioritized. However, in the past few decades, there has been a gradual shift from a mobility-oriented to an accessibility-oriented approach in both research and policymaking processes worldwide. Instead of prioritizing LOS measurements, accessibility-based planning considers land use-transportation interactions and people's ability to reach desired destinations. Nevertheless, mobility-oriented approaches still frequently dominate policymaking and planning processes. In this context, this chapter aims to critically evaluate the extent to which Dhaka's policy and planning documents meaningfully integrate the concept of accessibility. Dhaka, the capital of Bangladesh, is one of the fastest-growing megacities globally. While Dhaka's policy and planning documents increasingly include accessibility, the relevance and usefulness of such integration remain underexplored. To fill this research gap, the five most extensively used policy and planning documents applicable to Dhaka have been critically examined using a mixed content-structuring analysis method. The findings revealed that accessibility had been defined ambiguously across the documents, including a clear lack of explicit inclusion of accessibility in the vision, goals and objectives. Furthermore, the documents also do not identify any measurable accessibility indicators. This critical evaluation is expected to offer insights into accessibility in Dhaka's context and aid policymakers and planners in the Global South in more substantial integration of accessibility in the planning processes.

Keywords Accessibility · Dhaka · Mixed content-structuring analysis · Performance indicators

P. Nakshi (✉) · A. K. Debnath
Department of Urban and Regional Planning, Bangladesh University of Engineering and Technology (BUET), Dhaka, Bangladesh
e-mail: paromita_nakshi@urp.buet.ac.bd

S. Anand et al. (eds.), *Sustainability in South Asian Cities*, Advances in Geographical and Environmental Sciences, https://doi.org/10.1007/978-981-97-7455-5_15

15.1 Introduction

The notion of accessibility is one of the most widely discussed yet elusively used concepts in the past few decades. Transportation planning practices have been traditionally shaped by a mobility-centred approach which is predominantly based on level-of-service (LOS) measurements. This approach has two significant shortcomings. First, it shifts the focus from travel being a derived demand. Second, it ignores the relationship between land use and transportation. Accessibility, on the other hand, considers both land use and transportation. This gives accessibility a substantial edge over mobility since land use-transportation interaction is considered important in sustainable development. Hence, a paradigm shift from mobility-geared to accessibility-driven transportation policy is increasingly gaining importance worldwide due to its far-reaching impacts on urban structure, the performance of transportation systems and various social and economic issues (Cui et al. 2019; Deboosere and El-Geneidy 2018; Grengs 2010; Kawabata 2003a; Kawabata and Shen 2006). An accessible and sustainable transportation system has been emphasised in Sustainable Development Goals (SDG) as well to ensure sustainable cities and communities. Yet, considerable challenges exist in this field.

This chapter is motivated by the observation that there is a significant lack of conceptualisation and integration of accessibility in research and policymaking processes, particularly in developing countries like Bangladesh. Dhaka, the capital of Bangladesh, is one of the most densely populated cities in the world. Dhaka's transportation sector has received increasing attention in recent years since the city is suffering severely due to rapid urbanisation, haphazard development and an increasingly worsening public transportation system. Interestingly, accessibility is gradually becoming a buzzword in the transportation planning process here. Several policy and planning documents of Bangladesh have recognised accessibility as an important aspect of investment criteria in transportation infrastructure projects and overall urban liveability (Dhaka Transport Coordination Board and Ministry of Communications 2010; Japan International Cooperation Agency and Dhaka Transport Coordination Authority 2015; Ministry of Communications and Dhaka Transport Coordination Board 2005; MinistryCommunicationations and Planning Commission 2008; Rajdhani Unnayan Katripakkha n.d.). In spite of that, ironically, accessibility has received little systematic attention and there is a significant lack of clarity among researchers, practitioners and policymakers in conceptualising accessibility. As a result, although accessibility has been highlighted in many ways in pen and paper, it could not be translated effectively in the implementation process.

15.2 The Concept and Measures of Accessibility

> Accessibility… is a slippery notion... one of those common terms that everyone uses until faced with the problem of defining and measuring it (Gould 1969, p. 64).

The classic yet simplest definition of accessibility probably dates back to 1959 when Hansen (1959) defined it as "the potential for interaction." However simple that might sound, conceptualisation and measurement of accessibility have been far from simple for urban planners and transportation experts. A more operational definition of accessibility refers to the number of opportunities available within a certain travel time or distance from a given location by a particular transport mode (Handy 1992; Kawabata and Takahashi 2005; Shen 1998).

Accessibility is a powerful concept because it incorporates the intricate interaction of four components- land use, transportation, temporal and individual (Geurs and Van Wee 2004). The land use component includes the demand and supply, the quality of various spatial opportunities and the resulting competition. The transportation component includes the trip's time, cost and effort and the demand and supply of various transport infrastructures. The temporal component highlights the availability of opportunities at different times of day and individual's availability to participate in relevant activities. Finally, the individual component includes various socioeconomic, demographic and attitudinal characteristics of individuals (Geurs and Van Wee 2004).

Accessibility can be measured in a number of ways- each having its own advantages and disadvantages (Table 15.1). Infrastructure-based accessibility measures concentrate on the service and performance of the transportation system and use travel time and speed, congestion, etc. as evaluation criteria (Ewing 1993; Geurs and Van Wee 2004; Linneker and Spence 1992). Location-based measures analyse the spatial distribution of opportunities at the macro-level (Geurs & Van Wee, 2004). Several such measures have been extensively used (Deboosere and El-Geneidy 2018; Grengs 2010; Guy 1983; Ingram 1971; Kawabata 2003a; Vickerman 1974). The isochronic or cumulative opportunity-based accessibility and gravity measures are the most used. Cumulative opportunity-based is the simplest and one of the earliest accessibility measures. It counts the number of relevant opportunities available within a certain generalised cost, usually travel time or distance and assumes all opportunities with the predetermined threshold as equally attractive (El-Geneidy and Levinson 2006; LaMondia et al. 2010). Unlike the cumulative measure, gravity measure imposes a 'friction factor' or impedance function that can take the form of power, Gaussian or logistic function, negative exponential function, etc. (Geurs and Van Wee 2004; LaMondia et al. 2010). This measure assumes that the farther an opportunity lies in terms of the generalised cost, its attractiveness decreases according to the impedance function (El-Geneidy and Levinson 2006). Compared to cumulative opportunity-based accessibility measures, this is more complex and difficult to interpret.

As the name suggests, the person-based accessibility measure analyses accessibility at the individual level. It is based on the 'space–time prism' concept of Hägerstrand and considers that an individual's movement is constrained or determined by a space–time budget (Hägerstrand 1970; Kwan 1998). This approach is quite robust in nature but complex and data-intensive and, therefore, is used less often.

Utility-based accessibility measure is another highly complex, data-intensive and difficult-to-interpret measure. This approach considers travel behaviour theories and

Table 15.1 Summary of Accessibility measures*

Accessibility measures		Components[a]					Operationalization[b]	Interpretation[b]
		transport	Land use		Temporal	Individual		
			demand	supply				
Infrastructure-based measures		±	–	–	±	–	+	+
Location-based measures	Cumulative opportunities/Contour measure/Isochronic measure	±	±	–	±	–	+	+
	Gravity-based measure	+	+	–	±	±	+	±
	Adapted potential measure	+	+	+	±	±	+	±
	Balancing factors	+	+	+	±	±	+	±
Person-based measures		+	+	–	+	+	–	–
Utility-based measures	Logsum benefit measure	+	+	–	–	±	+	±
	Space–time measure	+	+	–	+	+	–	±
	Balancing factor benefit measure	+	+	+	–	±	+	±

[a] Score: + = criterion satisfied; – = not satisfied; ± = partly satisfied [b] Score: + = easy to operationalize/interpret; – = difficult; ± = moderately difficult
* Adapted from Geurs and Van Wee (2004)

incorporates individual preferences from a set of potential transport choices (El-Geneidy and Levinson 2006; Geurs and Van Wee 2004; Handy and Niemeier 1997; LaMondia et al. 2010).

The extremely crucial role of transport modes in accessibility has been highlighted in several studies. Studies have been conducted on the variation of private automobile and public transport accessibility and led to the development of several concepts like a modal mismatch, transport poverty, etc. (Grengs 2010; Kawabata 2003a, 2009; Kawabata and Takahashi 2005; Kwok and Yeh 2004; Niedzielski and Śleszyński 2008; Preston and Rajé 2007). Accessibility has been one of the pivotal arguments in the conceptualisation of modal mismatch theory from the original concept of spatial mismatch. The spatial mismatch hypothesis examined the impact of housing-market segregation, crudely defined as geographical distance on employment level (Kain 1968). While the spatial mismatch hypothesis was grounded simply on the spatial aspect, several studies found that modal differences, performance of transportation system, land use patterns, etc. play a crucial role in employment level since accessibility to jobs varies significantly depending on travel modes, even more than residential location (Blumenberg and Ong 2001; Kawabata 2003a, b; Kawabata and Shen 2007; Sanchez 1999). Therefore, accessibility is far more decisive in this regard (Grengs 2010).

In the past few years, the dynamic nature and temporal variability of public transit accessibility have also been studied (Farber et al. 2014; Owen and Levinson 2015). Yet, such studies are more often than not based on developed countries, whereas cities of developing countries as well, are increasingly becoming auto-dependent. The multimodality of public transport has received significant attention in recent years, unlike earlier studies that used to ignore the walking, waiting and transfer times associated with public transport travel. The emerging capabilities of GIS have made it easier and almost imperative to capture the First Mile/Last Mile issue in contemporary studies (Boarnet et al. 2017; Farber et al. 2014; Mavoa et al. 2012).

While the study of accessibility has been primarily focused on public transport and private automobile, walking and bicycling have been getting attention in recent years. Studies have found that investment in bicycling infrastructure improves accessibility substantially and is a cost-effective alternative to providing automobile infrastructure (The Accessibility Observatory 2019; Wu and Levinson 2019). Although a large body of studies has been primarily dedicated to job accessibility, over time accessibility to various non-work facilities e.g. shopping, health care, public parks, etc. has also been widely studied indicating their important role in urban life (Farber et al. 2014; Handy 1992; Hass 2009; Luo and Wang 2003; Maat and Konings 2018). However, developing countries, especially South Asian ones, still considerably lack studies related to accessibility.

15.3 Mobility to Accessibility: the Paradigm Shift

Although the concept of accessibility is not new, it took a long time for urban planners and policymakers to embrace accessibility-based planning instead of traditional mobility-focused planning. Mobility, defined as 'the ease of movement' and often measured by the traditional level-of-service measures, dominated the twentieth-century transportation planning and engineering. The notion of accessibility-based planning arose from the widely accepted notion of transportation being a 'derived demand'. Although neither 'sustainable development' nor 'sustainable transportation' emerged during the twenty-first century, the importance of accessibility in planning has taken a major leap with the growing emphasis on 'sustainable development' during this century (Zegras 2005). Yet, for a long time, mobility and accessibility have been used interchangeably and enhanced mobility has been thought to directly result in improved accessibility. Unfortunately, this ambiguity can still be observed in transportation policies and plans worldwide (Boisjoly and El-Geneidy 2017). Although increased mobility can result in increased accessibility, this is not a simple and universal outcome. People can experience poor accessibility in spite of having good mobility and vice-versa (Handy 2002). The traditional mobility-based planning, while focusing on better and faster movement, loses the core idea of the need for travel: it becomes pointless to enhance movement unless people can avail the spatial opportunities they want. Another in-built characteristic of mobility-based planning is to focus primarily on the road network using the 'predict and provide' approach. However, the road network does not function in isolation from different land uses. More precisely, unlike other land uses, road network does not have an exclusive function of its own; rather, it exists to support the need to reach different land uses for a variety of purposes. Thus, this seeming separation of land use from transportation in a mobility-based approach does not work well in an era of rapid urbanisation and industrialisation. Accessibility, on the other hand, inherently integrates land use and transportation system and focuses on whether the ease of movement is being translated into the ease of reaching desired opportunities.

Another important aspect of accessibility-based planning is increased choices both in terms of destinations and modes of travel (Handy 2002). Thus, accessibility is directly related to social and environmental equity issues because it allows prioritising public transit and non-motorised modes which in turn, makes travel easier for poor, women, senior citizens, physically challenged people or simply anyone who does not own or cannot drive a car (El-Geneidy et al. 2016; Grengs 2010; Manaugh et al. 2015). Furthermore, prioritising public transit and non-motorised modes subsequently reduces congestion and environmental pollution. This is clearly a massive gain over the mobility-based approach since no matter how fast movement becomes or how much road space is provided above a threshold, movement without spatial opportunities will result in growing congestion and, consequently, slower movement and increased travel times. No wonder sprawl development and mobility-based planning have resulted in almost standstill conditions in cities (Handy 2002). Therefore,

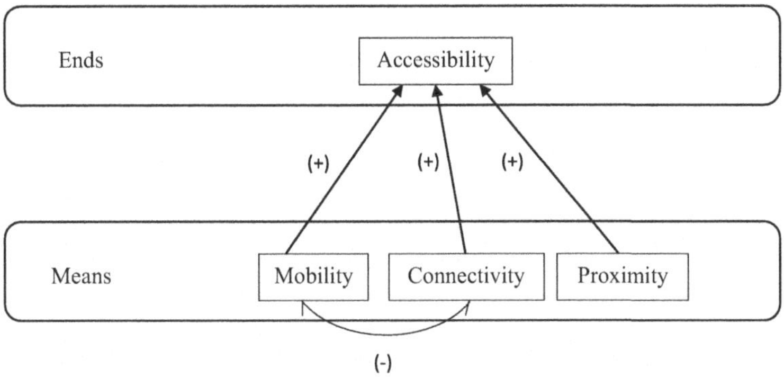

Fig. 15.1 Accessibility as the goal of transportation policy (Grengs et al. 2010)

the ultimate goal of transportation planning should be accessible with its means being mobility, proximity and connectivity (Fig. 15.1) (Grengs et al. 2010).

15.4 The Context of Dhaka

Dhaka is the capital of Bangladesh, a small country located in South Asia. The city, however, is one of the most densely populated (45,000 persons/sq.km.) and the fastest-growing megacities in the world. Dhaka alone accommodates nearly 10% of the country's population and 37% of the country's urban population (Rajdhani Unnayan Katripakkha n.d.). Rapid urbanisation in the core area of Dhaka has resulted in its functional relationship with several adjacent settlements. The current planning practices are increasingly focusing on this functional relationship, resulting in a large territory termed the Dhaka Metropolitan Region (DMR) (Fig. 15.2) (Rajdhani Unnayan Katripakkha n.d.).

On average, 21 million trips are generated in the city in a typical working day (Rajdhani Unnayan Katripakkha n.d.). Dhaka is a city with diverse motorised and non-motorised transport modes. Non-motorised modes, including walking and rickshaw (a human-paddled three-wheeler transport mode), contribute to nearly 20 and 38% of the modal share. Public buses constitute 30% of the modal share while private car constitutes around 5% (Rajdhani Unnayan Katripakkha n.d.). Apart from these, para-transit modes (known as Mishuk/CNG-compressed natural gas driven three-wheelers) and Human Hauler (known as Leguna, a motorised mode with a carrying capacity of 10–12 passengers) also play a crucial role in moving the city residents every day.

Severe and steadily increasing traffic congestion has been at the center of attention in the context of Dhaka's transportation sector for the past several years. Along with high population density, deteriorating public transportation system and increasing private car ownership and use are said to be primarily responsible for this status quo.

Out of all newly registered cars in 2018 in Bangladesh, nearly 90% have been registered only in Dhaka. The number is increasing steadily- in 2011, 11,423 cars were newly registered and by the first three months of 2019, 3,627 private cars had been newly registered (Bangladesh Road Transport Authority 2020, March 5). On the other hand, although the modal share of public buses is nearly six times higher than cars, the bus system is still semi-structured without any specific schedule and regulation (Rajdhani Unnayan Katripakkha n.d.). The bus routes with substantial overlapping mostly expand in the north–south direction and lack connectivity in the east–west direction (Fig. 15.2). This poses massive challenges since public transportation can hardly provide reliable service and reasonable coverage.

Several mega infrastructure projects, including a number of Mass Rapid Transit (MRT) and Bus Rapid Transit (BRT) lines and initiatives like Bus Route Rationalization, have been undertaken by the government in recent years to improve the public transportation system. However, their efficacy can be fully understood once they start operating.

15.5 Research Framework and Methodology

Over the years, a number of transportation policy and planning documents have been drafted to meet the latest challenges in the transportation sector of Dhaka. Some of these documents have been aimed solely at the transportation sector while some others encompassed all relevant sectors to provide a holistic guideline for the city's planning process. Some of these documents also drew direct motivation from the limitations or recommendations of some previous planning documents. This chapter aims to assess how meaningful accessibility has been incorporated in the relevant planning documents of Dhaka and how it can be integrated and operationalised to all its intents and purposes in the transportation planning practices in the city.

We have reviewed the five most relevant and widely cited policy and planning documents. These documents have been considered guidelines for urban transportation planning for the city. Documents like the Strategic Transport Plan (STP) for Dhaka, 2005 received the status of official transportation framework as well (Japan International Cooperation Agency and Dhaka Transport Coordination Authority 2015). To conceptualise its policies, the Dhaka Urban Transport Network Development Study (DHUTS) in Bangladesh 2010 was undertaken, and later on, the Revised Strategic Transport Plan (RSTP) for Dhaka, 2015 came into being to cater to the demand of changing times. The documents reviewed are as follows:

1. Strategic Transport Plan (STP) for Dhaka, 2005
2. Dhaka Urban Transport Network Development Study (DHUTS) in Bangladesh, 2010
3. National Integrated Multi-modal Transport Policy (NIMTP), 2013
4. Revised Strategic Transport Plan (RSTP) for Dhaka, 2015
5. Draft Dhaka Structure Plan (2016–2035).

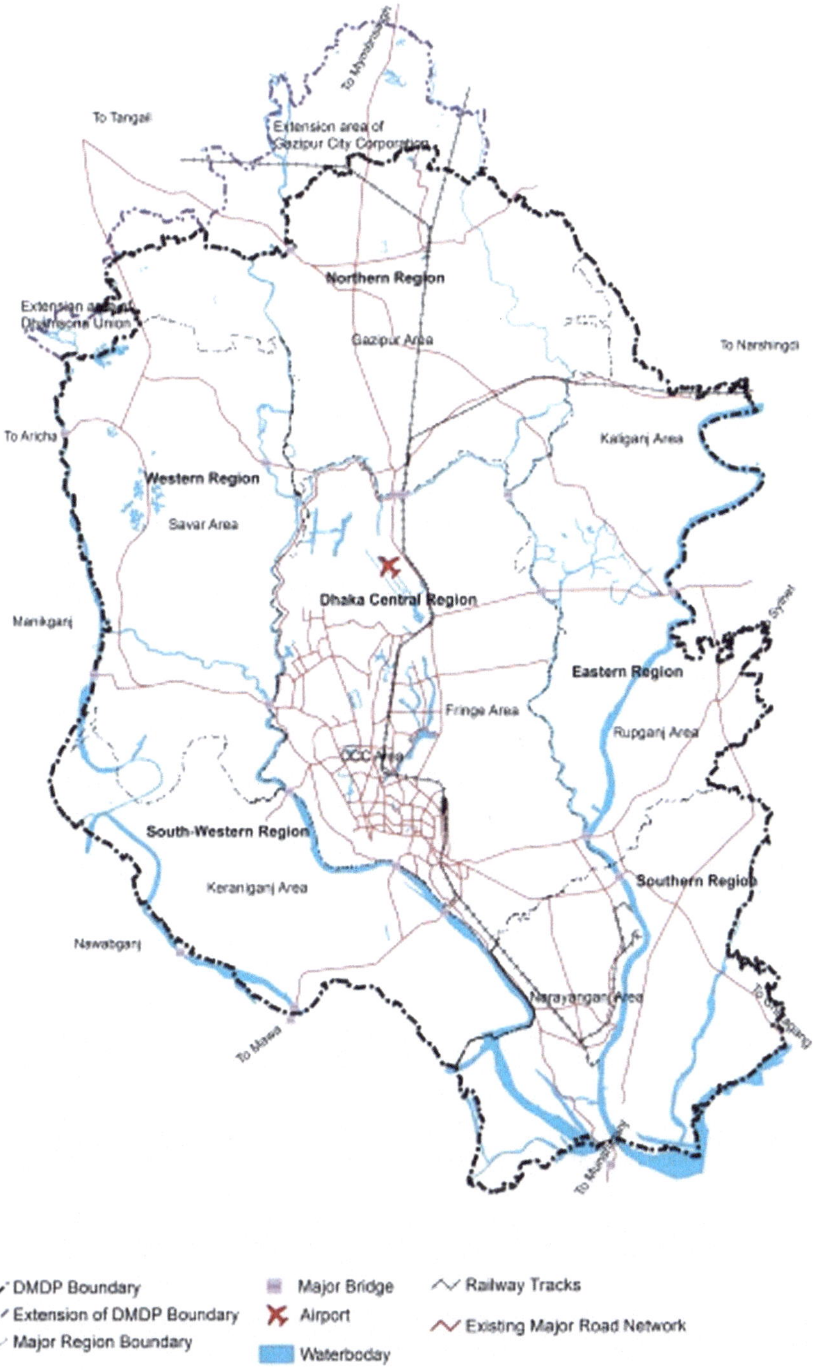

Fig. 15.2 Map of dhaka metropolitan region (Rajdhani Unnayan Katripakkha n.d.)

These policy and planning documents do not necessarily encompass the same scale of geography as the study area of this research. However, all of them have components relevant to the analysis and any component outside of the study area has been carefully excluded from the analysis.

To achieve the goal of this chapter, we address the following research questions:

- To what degree multimodal accessibility has been incorporated in the transportation plans of Dhaka over the years?
- What types of performance indicators have been used to translate accessibility-related visions, policies and goals into specific objectives and strategies?
- To what extent do the goals, objectives and strategies meaningfully reflect multimodal accessibility as the ease of reaching destinations and availing desired opportunities?

To answer the research questions, visions, policies, goals, objectives and strategies were selected from the documents that are related to accessibility either explicitly or implicitly. However, this is not a straightforward process. Accessibility, mobility, and connectivity are not isolated concepts. A measure oriented towards enhancing mobility may enhance accessibility as well and vice-versa. This is why several mobility-oriented objectives or strategies were also included as part of the synthesis process if they seemed to be linked with accessibility-oriented goals, objectives or strategies. In general, objectives or strategies aiming at increasing roadway capacity, reducing congestion, improving travel times or enhancing pedestrian or non-motorised movement without clear linkage between origin and destinations have been defined as mobility-oriented objectives or strategies. Accessibility-oriented objectives or strategies were defined as the ones aimed at reducing the need for travel, increasing ease of availing desired opportunities, improving the clear linkage between origin and destination, focusing the need for disadvantaged or marginalised population by increasing travel options and land use strategies (Handy 2005). They were then critically assessed to see how well they reflect the concept of accessibility and how meaningfully they have been translated into performance indicators or quantifiable measures.

A mixed content-structuring analysis (also known as 'keyword in context') was conducted to explore how accessibility has been included in the plans both explicitly and implicitly (Mayring 2014). Qualitative data analysis software ATLAS.ti was used in this process. At first, the plans were carefully skimmed through to find out accessibility-related vision, goals or objectives. A deductive summative content analysis approach was taken to explore how accessibility has been incorporated into the plans. Themes were deliberately focused on the unit of count instead of just words because accessibility is often expressed through different conceptual ideas (Lune and Berg 2016). Based on theory and literature review, some key themes were selected, e.g. accessibility, access, access, reach, ease of reaching destinations, get to, link, etc. This step was crucial for the study to get a good understanding of the degree to which the concept of accessibility has been able to manifest itself in its core form in the planning documents. Unfortunately, for some plans, this step did not provide much useful information.

This gave the impression that there is a possibility of implicit or latent manifestation of the notion of accessibility in those plans in the form of theoretical concepts or strategies. Compared to the deductive approach, exploring implicit manifestation from the planning documents, however, is more challenging. Moreover, different documents being drafted in years of gaps and by different authorities, have different choices of words or themes to express similar ideas. To deal with this, the words and themes that might implicitly reflect accessibility were sorted out using ATLAS.ti. A wider range of themes was considered in this inductive approach, including but not limited to multimodal transport, intermodal transport, public transport, mass transit, reducing travel demand, increasing choice of travel modes, addressing the needs of a specific population, connectivity, etc. Here, it is to be noted that the choice of words plays an important role in this approach because the same idea could be conveyed in a number of different ways as well as the same word could be used in different contexts. For instance- while Draft Dhaka Structure Plan (2016–2035) mentions 'Providing a greater choice of travel modes', STP, 2005 speaks of 'Provide alternative transportation modes and options'. Therefore, specific words from the themes were also picked as keywords, counted each instance the words appeared, and carefully assessed the context in which the words or themes were used.

15.6 A Critical Evaluation of Accessibility in Transportation Policies and Plans

The mixed content-structuring analysis shows that all the planning documents incorporate accessibility- either explicitly or implicitly (Table 15.2). However the extent to which accessibility is explicitly incorporated remains questionable. However, the complexity arises because none of the plans have defined accessibility. Moreover, the term 'accessibility' has been used to convey different ideas in the same document without any clear distinction. For example- RSTP, 2015 used the same term 'accessibility' in conceptualising the functional classification of roads, referring to the percentage of households with different infrastructure services, accessibility to bus stops and accessibility to urban services. As a result, its translation from vision, mission or policy to objectives, strategies or implementation tools remains highly ambiguous.

Mobility and accessibility have often been used interchangeably. Interestingly, sometimes even when some issues, measures or statements explicitly or implicitly reflect accessibility, they are included under the heading of mobility. This observation indicates that mobility has potentially received more direct emphasis than accessibility. Among the documents, NIMTP, 2013 has done a relatively better job of articulating accessibility while also focusing on social inclusion and non-work activities. However, the other plans have suffered from more ambiguity.

The concept of multimodality and first-mile/last-mile connectivity issues are largely absent in the plans. The issue of walking or non-motorised transports has

Table 15.2 Accessibility-related policies, goals, objectives and strategies in the selected policy and planning documents

Transportation plan	Policies, goals, objectives and strategies	Focus (accessibility or mobility oriented)	Accessibility-related performance indicator	Comments
Strategic Transport Plan (STP) for Dhaka (2005)	Strategic goal 2: Mobility and accessibility	Accessibility-oriented	None	Acknowledges walking as feeder mode, thereby indicating the importance of multimodality. Different groups of disadvantaged or marginalised population have been focused. Different work and non-work facilities have been considered
	Provide a basic level of mobility and accessibility for all segments of society to ensure reasonable access to employment, education, health, social and other programs, services and opportunities			
	Sub-Goal 2A: Meet mobility requirements of all segments of society	Accessibility-oriented	None	
	• Supply all areas with good transport services, Provide a basic level of transport services to the poorest people, Serve transportation needs of the disabled and handicapped, Provide good access to educational institutions and employment opportunities, Provide good access to educational institutions and employment opportunities, Provide alternative transportation modes and options within major travel corridors	Accessibility-oriented	None	
Strategic Transport Plan (STP) for Dhaka, 2005	• Emphasise schemes that make walking easier, less hazardous and less stressful, so that walking (as a complete trip, or as access to/from other modes of travel) becomes a preferred choice rather than a necessity	Mobility-oriented	None	

(continued)

Table 15.2 (continued)

Transportation plan	Policies, goals, objectives and strategies	Focus (accessibility or mobility oriented)	Accessibility-related performance indicator	Comments
	Strategic goal 7: Social development	Both		
	Sub-goal 7A: Develop affordable transport options for low-income groups to provide access to additional employment opportunities and other services	Accessibility-oriented	None	
	• Provide good transport services to areas with concentrations of low-income groups	Accessibility-oriented	None	
	• Provide good pedestrian facilities to encourage and facilitate walking	Mobility-oriented	None	
	Sub-goal 7D: Ensure that all transport services are suitable for use by all persons, particularly including women and selective societal groups (children, disabled, elderly, etc.)	Accessibility-oriented	None	
Dhaka Urban Transport Network Development Study (DHUTS) in Bangladesh, 2010	Mission 2: Equitable of people mobility	Accessibility-oriented	Accessibility-related Performance Indicator	
			None	Different work and non-work facilities have been considered. Focus revolves around mobility in general. Accessibility-related strategies remain vague
	The transport sector is generally responsive for assuring social equity, providing all people with equitable accessibility to places for their employments, schools, commercial activities, social services, medical care, etc			
	Strategy 1: Improvement of people's mobility			
	• Introduction of Traffic Demand Management (TDM)	Mobility-oriented	None	

(continued)

Table 15.2 (continued)

Transportation plan	Policies, goals, objectives and strategies	Focus (accessibility or mobility oriented)	Accessibility-related performance indicator	Comments
	Strategy 3: Safe and secure transport			
	• Priority to pedestrian traffic	Mobility-oriented	None	
	Strategy 4: Accessible transport for all people			
	• Accessible transport for all the poor	Accessibility-oriented	None	
	• Accessibility to a new satellite community			
National Integrated Multi-modal Transport Policy (NIMTP), 2013	Objectives:	Accessibility-oriented	None	'Social inclusion' gives an idea of mainstreaming disadvantaged groups of population. Different work and non-work facilities have been considered. Instead of only poor people, it specifically mentions people without a car. Objectives are supported by investment criteria
	• Ensure that transport meets social needs—in terms of its cost and accessibility to all sectors of society			
	• Reduce the need for travel by better land use planning			
	Investment criteria:	Accessibility-oriented	None	
	• Accessibility—improving access to everyday facilities for the poor and those without a car and reducing community severance			
	• Social inclusion—meeting the needs of women, the elderly and the disabled			
	• Providing accessibility to basic health and education facilities			

(continued)

Table 15.2 (continued)

Transportation plan	Policies, goals, objectives and strategies	Focus (accessibility or mobility oriented)	Accessibility-related performance indicator	Comments
Revised Strategic Transport Plan (RSTP) for Dhaka, 2015	Vision and goal:			No clear consideration of disadvantaged groups of the population in objectives and strategies although the vision explicitly speaks about equity
	Ensure mobility and accessibility to urban services that are vital for the people and the society, by providing a transport system characterised by safety, amenity and equity and sustained by an efficient public transport system	Accessibility-oriented		Accessibility-oriented strategies are mostly vague
	Objectives and strategies:			
	B. Effective management of urban growth and development			
	Promotion of integrated urban and transport development, particularly Transit-Oriented Development (TOD)	Accessibility-oriented	None	
	C. Promotion and development of attractive public transport			
	Early introduction of an integrated public transport system in the effort to maintain public transport share, Promotion of public transport use and expansion of services, Providing an affordable public transport system	Accessibility-oriented	None	

(continued)

Table 15.2 (continued)

Transportation plan	Policies, goals, objectives and strategies	Focus (accessibility or mobility oriented)	Accessibility-related performance indicator	Comments
Revised Strategic Transport Plan (RSTP) for Dhaka, 2015	Exploitation of para-transit and NMVs	Mobility-oriented	None	
	E. Effective Transport Demand Management (TDM)			
	Integrating urban development and transport (TOD), Providing efficient public transport alternatives	Accessibility-oriented	None	
	Regulating motorised vehicle access and proper charging of road use and parking	Mobility-oriented	None	
Draft Dhaka Structure Plan (2016–2035)	Long-term policy framework: Transport for efficient connectivity			Explicit focus on some innovative ideas e.g. staggered office hours and IT and several mobility-limiting strategies as well as reducing SOV use
	Goal: Safe, affordable, sustainable and connected communities			'Development that caters to the transit-dependent population should be located on or within walking distance of a bus route.' is unclear and confusing given there is no explanation of how some developments would cater only to these people and why this is focused instead of their universal access
	Objective 1: To prepare long term transport network plan			

(continued)

Table 15.2 (continued)

Transportation plan	Policies, goals, objectives and strategies	Focus (accessibility or mobility oriented)	Accessibility-related performance indicator	Comments
	Policy 1.1: Enhancing the linkage between land use and transport network			
	Implementation tools:			
	Development that caters to the transit-dependent population (e.g., elderly, disabled, or disadvantaged persons) should be located within walking distance of a bus route	Accessibility-oriented	None	
Draft Dhaka Structure Plan (2016–2035)	Policy 1.5: Encourage the development of sidewalk and bicycle route for both mobility and recreation purposes			
	Implementation tools:			
	Provide pedestrian/cycle accessibility to neighboring facilities including, schools, shops, parks and recreation facilities, community facilities, bus stops, train stations, etc	Accessibility-oriented	None	
	Objective 2: To Make Public Transport Efficient and Sustainable			
	Policy 2.1: Introduction of Mass Rapid Transit (BRT and MRT)			
	Implementation tools:			

(continued)

Table 15.2 (continued)

Transportation plan	Policies, goals, objectives and strategies	Focus (accessibility or mobility oriented)	Accessibility-related performance indicator	Comments
	Promote Transit-Oriented Development (TOD) along BRT/MRT Stations in close cooperation with transport corridor and urban development	Accessibility-oriented	None	
Draft Dhaka Structure Plan (2016–2035)	Policy 2.2: Promote improved bus transport system, Network restructuring and route franchising			
	Implementation tools:	Accessibility-oriented	None	
	Establish a subsidy system for the poor/vulnerable transport users			
	Objective 3: To ensure effective traffic management			
	Policy 3.1: Integration of Travel Demand Management (TDM) in planning process			
	Implementation tools:	Mobility-oriented	None	
	Provision of incentives to use alternative modes and reduce driving like road and parking pricing, road space allocation (dedicated bus lanes, bike lanes and transit-only lanes)			
	Introduction of Parking Management, like Parking cash-out programs, Priority parking for carpools, vanpools and short-term parkers	Mobility-oriented	None	

(continued)

Table 15.2 (continued)

Transportation plan	Policies, goals, objectives and strategies	Focus (accessibility or mobility oriented)	Accessibility-related performance indicator	Comments
Draft Dhaka Structure Plan (2016–2035)	Provision of high-capacity transit service in reducing travel demand and shifting travel away from single-occupant vehicles (SOV) to more efficient modes	Mobility-oriented	None	
	Alternative work arrangements (such as compressed work weeks and telecommuting) to eliminate commute trips one or more days per week	Mobility-oriented	None	
	Sharing of time and space by staggering office hours (different timing for private and public office) and school and university time can greatly reduce the peak hour traffic volume	Mobility-oriented	None	
	Using information technology, video conferencing between corporate head office and branch office of commercial establishments for important meetings would minimise the need for travel	Mobility-oriented	None	
	Land use management, Provision of more mixed-use development and increased densities in transit corridors	Accessibility-oriented	None	

(continued)

Table 15.2 (continued)

Transportation plan	Policies, goals, objectives and strategies	Focus (accessibility or mobility oriented)	Accessibility-related performance indicator	Comments
Draft Dhaka Structure Plan (2016–2035)	Build compact communities with sufficient density to support high-frequency transit service	Accessibility-oriented	None	
	Policy 3.2: Management of Rickshaw-based Transport	Mobility-oriented		Introduction of affordable taxi cabs as an alternative to rickshaws remains questionable
	Implementation Tools:		None	
	Expedite mass transit development, introduce affordable taxi cab, and limit movement of NMT within neighborhoods and on major roads. But all the measures should be taken up simultaneously so that NMT users do not suffer for want of an alternative mode of transport			

come as independent modes. Except for STP, 2005, the objectives or strategies have not adequately emphasised their role as feeder modes. Additionally, when the Draft Dhaka Structure Plan (2016–2035) talked about the role of rickshaws as a feeder mode of public transport, ironically, it proposed affordable taxi cabs as an alternative to rickshaws. However, taxi cabs involve fuel costs and environmental impacts.

Moreover, accommodating automobiles like taxi cabs or cars requires considerable road and parking space. Considering these impacts, it remains unclear from the document how taxi cabs would offer a sustainable solution. This indicates the underlying premeditated preference for mobility in the planning process. Therefore, the goals, objectives or strategies do not meaningfully reflect multimodal accessibility as the ease of reaching destinations and availing desired opportunities.

Unfortunately, none of the planning documents has incorporated any quantifiable metric or performance indicator of accessibility. This, however, is not surprising given the ambiguity in the concept of accessibility reflected in the plans. Therefore, the absence of any performance indicator in the plans drafted over all these years makes even the noblest goals and objectives fail because they hardly have anything to contribute in the subsequent implementation stages.

In contrast to the other reviewed documents, the Draft Dhaka Structure Plan (2016–2035), the latest one, surprisingly does not mention accessibility in any of its policy, goal or objective statements. Although it highlights 'Accessibility to services & facilities' as one of the parameters of liveability (liveability, functionality and resilience are included as the three pillars of its vision), the focus revolves around connectivity and mobility. The mobility-oriented planning is also evident in RSTP, 2015, which highlights road network performance through volume-capacity ratio (VCR). However, the entire plan does not have any sort of performance measure on changes in accessibility from proposed road construction or rapid transit projects.

We do not underestimate the importance of enhancing mobility or connectivity as along with proximity; they are the means of achieving accessibility (Levine et al. 2012). It is crucial here to understand that mixing these concepts and staying vague in determining what they actually want to achieve is misleading and even dangerous. These reveal a significant lack of clarity in conceptualising accessibility at the policy and planning level.

Another striking feature in the Draft Dhaka Structure Plan (2016–2035) is the lack of integration among different sectoral plans and the spatial development plan in general. For instance, it speaks about enhanced accessibility by concentrating facilities and indicates polycentricity to 'reduce long-distance commute'. However, the transportation planning section barely reflects these spatial development concepts. Similarly, an objective under the goal 'Making Dhaka Increasingly Functional and Productive' is about increasing economic and employment densities in urban areas. However, it misses the opportunity to integrate the transportation sector with it and develop any quantifiable performance indicator.

DHUTS, 2010, Draft Dhaka Structure Plan (2016–2035) and RSTP, 2015 have repeatedly focused on enhancing connectivity and minimising missing links in road networks. RSTP, 2015 mentions explicitly MRT Line 5: "In order to provide a high-capacity, high-speed and frequent public transport system to the city for trips which do not commence or end in the CBD and which will be served by the radial MRT Line 1, MRT Line 2, BRT Line 3, MRT Line 6 and BRT Line 7, the Study Team has identified the need for a circumferential MRT Line 5 that would provide a "bypass" public transportation service for the city's suburban areas and provide good connectivity between suburban zones." Nevertheless, MRT Line 5, the only MRT/BRT line that is aligned in east–west direction in the proposed plan, gets little focus for the near future (Fig. 15.3). It appears that the relatively low estimated ridership of MRT Line 5 and the urgency to improve connectivity of the city with CBD through other lines are the reasons behind this. This would have sounded more reasonable if new bus routes with east–west connectivity could have been developed in the meantime. A map of the basic network of the primary bus service has been included in RSTP but no plausible plan is outlined on how and when these will be developed. Focus has been given on integrating primary and secondary bus routes with the MRT/BRT system which is indeed required. But until 2035, when all the MRT/BRT lines are expected to be operational, the public transport system in the city would possibly suffer from the highly criticised east–west connectivity issue.

MRT Line 4 and Dhaka Elevated Expressway essentially have the same alignment in most parts while BRT Line 3 also overlaps with them in a considerable segment. The point here is not to simply criticise it but rather emphasise that there should be adequate and reasonable justification for this overlapping in the proposed plan currently absent in these documents. Without a sound, rational explanation, such massive investments would remain questionable.

15.7 The Way Forward

The ambiguity in defining accessibility, the complete absence of performance indicators and both explicit and implicit emphasis on mobility-oriented planning leave little room for any critical assessment of the goals and objectives outlined in the reviewed documents. But, again, this reveals that indeed a lot is to be done for an effective decision-making process for the city.

First, a clear distinction has to be made among concepts like mobility, accessibility and connectivity. Second, the vision, goals, objectives and strategies should explicitly incorporate accessibility. While enhanced mobility can lead to enhanced accessibility and vice-versa, they are not substitutable and should not be used interchangeably.

Third, the accessibility-related goals and objectives should be translated to quantifiable measures or performance indicators. A simple ridership count estimation or financial feasibility analysis for new infrastructure investment is barely enough to justify its implementation. Such planning initiatives and investments should be validated by examining the extent to which they bring changes in accessibility level and enhance social equity. This is not possible to assess without clearly stated indicators in the guiding planning documents. Would the proposed planning initiatives improve accessibility and reduce transport-related social exclusion? If yes, to which extent would the initiatives bring (positive and more equitable) changes? How would these

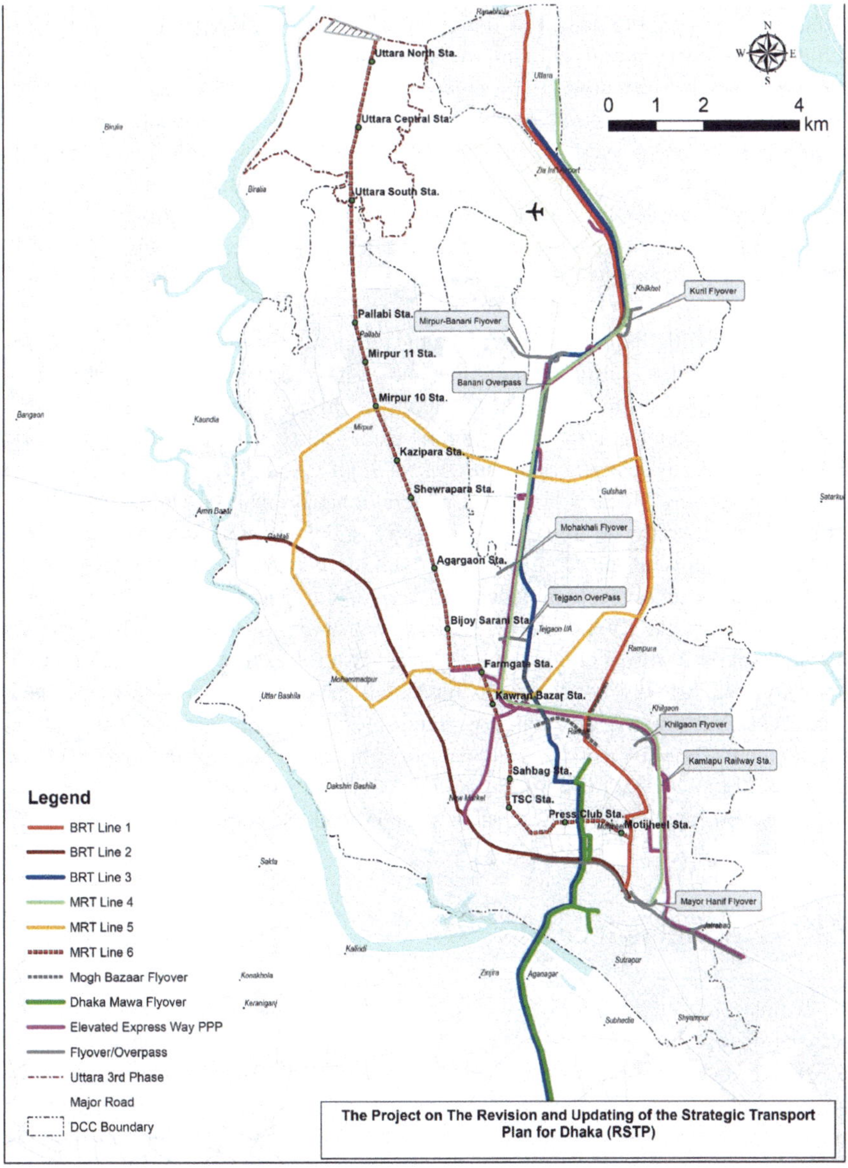

Fig. 15.3 Urban transport projects in Dhaka (*Source* Japan International Cooperation Agency, & Dhaka Transport Coordination Authority, 2015

be linked and contribute to the policies and goals? The existing planning documents include no assessment criteria to answer such vital questions.

Fourth, accessibility needs to be measured for different modes and for different times of the day to reflect the diversity of choices and needs. Down that line, first mile/ last mile connectivity issues of public transport should be considered. The importance of non-motorised transportation and walking has to be reflected in the planning process. For example- instead of proposing taxi cabs as feeder modes of public transport, promoting bicycle use through a park-and-ride system and separate bicycle lanes is far more sustainable and equitable from social, environmental and economic perspectives.

Fifth, along with the spatial dimension, the temporal dimension should be considered. Non-work trips constitute more than half of the total trips in the city (Rajdhani Unnayan Katripakkha n.d.). Therefore, along with CBDs, non-work trip destinations should also have a satisfactory level of accessibility. It is possible to have good accessibility during peak hours because of the high transit level of service but poor accessibility during off-peak hours when more non-work trips are made due to lack of transit service to the desired destinations. It is again possible to have good public transit accessibility all day long but poor walk or bicycle accessibility due to the absence of dedicated bicycle or pedestrian lanes. Aggregated accessibility measures would not be able to capture these temporal and modal differences. The reviewed documents have hardly addressed the temporal issue- on a few occasions, temporal dimension was considered; the focus has solely been on peak hour. At the same time, accessibility to various non-work facilities has been incorporated in most of the documents. However, no clear linkage could be found between the two given that accessibility to non-work activities would also require a focus on the off-peak hour. The documents appeared to disregard this spatial and temporal link completely.

15.8 Concluding Remarks

The evaluation sketched in this chapter is based on reviewing the documents. This study could not incorporate the complex decision-making process that works in the background before and during the documents are drafted. Understanding the decision-making process, including social and political influences, if any, would possibly shed more light on this area. That is, undoubtedly, a highly challenging task.

Nevertheless, this chapter reveals some of the striking pitfalls in incorporating the concept of accessibility as well as addressing accessibility-enhancing or mobility-limiting strategies in our transportation planning. The clearest drawback of the reviewed documents is their ambiguity or vagueness in defining crucial concepts like accessibility, mobility or connectivity and the resultant absence of any performance indicator to evaluate the various planning initiatives. Unless the concepts are defined clearly, and the policy and goals are translated to meaningful objectives and strategies, the documents will not guide the city's development process.

Funding This research is supported by the Social Science Research Council (SSRC) of Bangladesh.

References

Bangladesh Road Transport Authority (2020) Number of registered motor vehicles in Dhaka (Yearwise). Retrieved from http://www.brta.gov.bd/

Blumenberg E, Ong P (2001) Cars, buses, and jobs: Welfare participants and employment access in Los Angeles. Transp Res Rec: J Transp Res Board 1756(1):22–31

Boarnet MG, Giuliano G, Hou Y, Shin EJ (2017) First/Last mile transit access as an equity planning issue. Transp Res Part A: Policy Pract 103:296–310

Boisjoly G, El-Geneidy AM (2017) How to get there? A critical assessment of accessibility objectives and indicators in metropolitan transportation plans. Transp Policy 55:38–50

Cui B, Boisjoly G, El-Geneidy A, Levinson D (2019) Accessibility and the journey to work through the lens of equity. J Transp Geogr 74:269–277

Deboosere R, El-Geneidy A (2018) Evaluating equity and accessibility to jobs by public transport across Canada. J Transp Geogr 73:54–63

Dhaka Transport Coordination Board and Ministry of Communications (2010) Preparatory Survey Report on Dhaka Urban Transport Network Development Study (DHUTS) in Bangladesh: Final Report. Retrieved from http://open_jicareport.jica.go.jp/710/710/710_101_11996774.html

El-Geneidy A, Levinson D, Diab E, Boisjoly G, Verbich D, Loong C (2016) The cost of equity: Assessing transit accessibility and social disparity using total travel cost. Transp Res Part A: Policy Pract 91:302–316

El-Geneidy AM, Levinson DM (2006) Access to destinations: Development of accessibility measures. Retrieved from University of Minnesota Digital Conservancy: https://conservancy.umn.edu/handle/11299/638

Ewing R (1993) Transportation service standards-as if people matter. Transp Res Rec: J Transp Res Board 1400:10–17

Farber S, Morang MZ, Widener MJ (2014) Temporal variability in transit-based accessibility to supermarkets. Appl Geogr 53:149–159. https://doi.org/10.1016/j.apgeog.2014.06.012

Geurs KT, Van Wee B (2004) Accessibility evaluation of land-use and transport strategies: Review and research directions. J Transp Geogr 12(2):127–140

Gould PR (1969) Spatial diffusion, Resource Paper No. 4. Retrieved from Washington, D.C. https://files.eric.ed.gov/fulltext/ED120029.pdf

Grengs J (2010) Job accessibility and the modal mismatch in Detroit. J Transp Geogr 18(1):42–54. https://doi.org/10.1016/j.jtrangeo.2009.01.012

Grengs J, Levine J, Shen Q, Shen Q (2010) Intermetropolitan comparison of transportation accessibility: Sorting out mobility and proximity in San Francisco and Washington, D.C. J Plan Educ Res, 29(4):427–443. https://doi.org/10.1177/0739456x10363278

Guy CM (1983) The assessment of access to local shopping opportunities: A comparison of accessibility measures. Environ Plann B Plann Des 10(2):219–237

Hägerstrand T (1970) What about people in regional science? Pap Reg Sci 24(1):6–21

Handy S (2005) Planning for accessibility: In theory and in practice. In: Levinson DM, Krizek K (eds) Access to destinations. Emerald Group Publishing Limited, pp 131–147

Handy SL (1992) Regional versus local accessibility: Variations in suburban form and the effects on non-work travel. Doctoral dissertation, University of California Transportation Center, California, the USA

Handy SL (2002) Accessibility-versus mobility-enhancing strategies for addressing automobile dependence in the US: Institute of Transportation Studies, UC Davis

Handy SL, Niemeier DA (1997) Measuring accessibility: An exploration of issues and alternatives. Environ Plan A: Econ Space 29(7):1175–1194

Hansen WG (1959) Accessibility and residential growth. (Doctoral dissertation), Massachusetts Institute of Technology, Massachusetts, The USA

Hass K (2009) Measuring accessibility of regional parks: A comparison of three GIS techniques (Master's thesis)

Ingram DR (1971) The concept of accessibility: A search for an operational form. Reg Stud 5(2):101–107

Japan International Cooperation Agency, & Dhaka Transport Coordination Authority (2015) The project on the revision and updating of the strategic transport plan for Dhaka. Retrieved from http://www.rthd.gov.bd/docs/Institution%20Development%20(RSTP).pdf

Kain JF (1968) Housing segregation, Negro employment, and metropolitan decentralisation. Q J Econ 82(2):175–197. https://doi.org/10.2307/1885893

Kawabata M (2003a) Job access and employment among low-skilled autoless workers in US metropolitan areas. Environ Plan A: Econ Space 35(9):1651–1668

Kawabata M (2003b) Job accessibility by travel mode in US metropolitan areas. Theory Appl GIS 11(2):43–50

Kawabata M (2009) Spatiotemporal dimensions of modal accessibility disparity in Boston and San Francisco. Environ Plan A: Econ Space 41(1):183–198

Kawabata M, Shen Q (2006) Job accessibility as an indicator of auto-oriented urban structure- A comparison of Boston and Los Angeles with Tokyo. Environ Plann B Plann Des 33(1):115–130. https://doi.org/10.1068/b31144

Kawabata M, Shen Q (2007) Commuting inequality between cars and public transit: The case of the San Francisco Bay Area, 1990–2000. Urban Studies 44(9):1759–1780

Kawabata M, Takahashi A (2005) Spatial dimensions of job accessibility by commuting time and mode in Tokyo metropolitan area. Theory Appl GIS 13(2):31–40

Kwan MP (1998) Space-time and integral measures of individual accessibility: A comparative analysis using a point-based framework. Geogr Anal 30(3):191–216

Kwok RC, Yeh AG (2004) The use of modal accessibility gap as an indicator for sustainable transport development. Environ Plan A: Econ Space 36(5):921–936

LaMondia JJ, Blackmar CE, Bhat CR (2010) Comparing transit accessibility measures: A case study of access to healthcare facilities. In: Paper presented at the Proceedings of the 89rd Annual Meeting of Transportation Research Board, Washington, DC, USA

Levine J, Grengs J, Shen Q, Shen Q (2012) Does accessibility require density or speed? J Am Plann Assoc 78(2):157–172. https://doi.org/10.1080/01944363.2012.677119

Linneker BJ, Spence NA (1992) Accessibility measures compared in an analysis of the impact of the M25 London Orbital motorway on Britain. Environ Plan A: Econ Space 24(8):1137–1154

Lune H, Berg BL (2016) Qualitative research methods for the social sciences, 9th edn. Pearson Education Limited, Essex, England

Luo W, Wang F (2003) Measures of spatial accessibility to health care in a GIS environment: Synthesis and a case study in the Chicago region. Environ Plann B Plann Des 30(6):865–884

Maat K, Konings R (2018) Accessibility or innovation? Store shopping trips versus online shopping. Transp Res Rec: J Transp Res Board 2672(50):1–10

Manaugh K, Badami MG, El-Geneidy AM (2015) Integrating social equity into urban transportation planning: A critical evaluation of equity objectives and measures in transportation plans in North America. Transp Policy 37:167–176

Mavoa S, Witten K, McCreanor T, O'Sullivan D (2012) GIS based destination accessibility via public transit and walking in Auckland, New Zealand. J Transp Geogr 20(1):15–22. https://doi.org/10.1016/j.jtrangeo.2011.10.001

Mayring P (2014) Qualitative content analysis: Theoretical foundation, basic procedures and software solution Retrieved from https://www.ssoar.info/ssoar/bitstream/handle/document/39517/ssoar-2014-mayring-Qualitative_content_analysis_theoretical_foundation.pdf

Ministry of Communications and Dhaka Transport Coordination Board (2005) Strategic Transport Plan (STP) for Dhaka. Retrieved from http://fpd-bd.com/wp-content/uploads/2015/07/Strategic-Transport-Plan-Draft-2005.pdf.

Ministry of Communications and Planning Commission. (2008) National Integrated Multi-modal Transport Policy (NIMTP). Retrieved from http://fpd-bd.com/wp-content/uploads/2015/07/Integrated-Multi-Modal-Transport-Model-Draft-Final-2008.pdf

Niedzielski M, Śleszyński P (2008) Analyzing accessibility by transport mode in Warsaw, Poland. Geographia Polonica 81(2):61–78

Owen A, Levinson DM (2015) Modeling the commute mode share of transit using continuous accessibility to jobs. Transp Res Part A: Policy Pract 74:110–122

Preston J, Rajé F (2007) Accessibility, mobility and transport-related social exclusion. J Transp Geogr 15(3):151–160

Rajdhani Unnayan Katripakkha (n.d.) Draft dhaka structure plan 2016–2035. Retrieved from http://www.rajukdhaka.gov.bd/rajuk/image/slideshow/Dhaka_Structural_Plan.html

Sanchez TW (1999) The connection between public transit and employment: The cases of Portland and Atlanta. J Am Plann Assoc 65(3):284–296

Shen Q (1998) Location characteristics of inner-city neighborhoods and employment accessibility of Low-wage workers. Environ Plann B Plann Des 25(3):345–365

The Accessibility Observatory (2019) Access across America. Retrieved from http://access.umn.edu/research/america/index.html

Vickerman RW (1974) Accessibility, attraction, and potential: A review of some concepts and their use in determining mobility. Environ Plan A: Econ Space 6(6):675–691

Wu H, Levinson D (2019) Access across Australia. Retrieved from http://hdl.handle.net/2123/20509

Zegras PC (2005) Sustainable urban mobility: Exploring the role of the built environment. Doctoral dissertation, Massachusetts Institute of Technology, Massachusetts, the USA

Chapter 16
Informal Sector Towards Urban Sustainability in Kathmandu Valley, Nepal

Puspa Sharma

Abstract Over the past decades, informal enterprising activities have been an important sector towards urban sustainability in Kathmandu valley, Nepal. Informal marketing is a traditional system and has been operating in the valley for a long. Other informal activities including urban waste management for recycling and reuse recently emerged in the valley. The contribution of these informal activities towards urban sustainability in the valley is discussed in terms of employment generation and income among the poor, migrants, and women, which they take on to meet their basic needs (food, health, children's education, amenities, etc.) and environmental conservation. These informal activities show unique features in terms of marketing, spatial location, and transformation. Policy measures are also provided for better managing of the informal marketplaces and marketing systems, and environmental conservation towards urban sustainability in Kathmandu valley.

Keywords Informal sector · Marketplace · Employment and income generation · Urban wastes recycling

16.1 Introduction

Nepal's large cities are characterized by the unprecedented growth of population due mainly to the influx of migrants from different parts across the country, as well as from neighbouring countries. Kathmandu Valley, with its five large cities and other 15 small towns, is the largest urban agglomeration in Nepal.[1] According to the 2011 population census, there were nearly 1.5 million urban population in the valley, accounting for nearly three-fifths of the country's total urban population (CBS 2012). Over the past six decades from 1952 to 2011, a great transformation occurred

[1] The valley occupying 475 km^2 comprises five large cities, viz. Kathmandu and Lalitpur metropolis, Bhaktapur, Kirtipur, and Madhyapurthimi and 15 recently inducted urban municipalities after 2011.

P. Sharma (✉)
Central Department of Geography, Tribhuvan University, Kathmandu, Nepal
e-mail: puspa.sharma@cdg.tu.edu.np

S. Anand et al. (eds.), *Sustainability in South Asian Cities*, Advances in Geographical and Environmental Sciences, https://doi.org/10.1007/978-981-97-7455-5_16

in the valley's built-up area. One key factor to this change has been the growth of population due mainly to the influx of migrants into the valley. For instance, the valley's urban population grew by over 8 times (181,082–1,465,000) during the same decades, to which migrants contributed approximately 70%. This overwhelming flow of migrants included both Nepalese and foreigners, particularly Indians (Kansakar 1985; Skeldon 1985; Pradhan 2003; CBS 2003, 2012; Pradhan et al. 2020). In addition to the population growth, the valley also has received relatively large but increased investments in the utility facilities, and expansion of road networks, health, education, banking, tourism, industries, and government administration functions (Pradhan 1988). Along with these, there has also been an increase in the number of informal or squatter settlements from 17 to 45 and their population from 3,000 to 13,300 in the valley during the years 1985–2008 (Rabenau 1990; CARE Nepal 2008). The recent trend is that when people migrate, they choose larger cities because they have relatively better facilities and opportunities than small towns. This is because, small towns often suffer from the minimal public investments in physical infrastructure (MoUD 2017). Thus, bigger cities continue to grow at a faster rate than small towns, creating more challenges to employment in large cities (ADB 2010). Particularly in the Kathmandu Valley's cities, the magnitude of the influx of migrants has been so huge that has outnumbered the creation of employment and absorption of workers by the formal sector, resulting in a large surplus of labour force. Despite this, the flow of people into the valley continued, making its estimated population over four million. Thus, some of the formal sector's surplus labour force have no way but to engage in a variety of jobs related to the informal sector.

Informal activities are an important sector to provide employment and income to the urban poor to keep up their livelihood, which appears to be essential for urban sustainability in developing countries. By virtue of the small investment required, easy entry, among many other features, informal activities often deal with locally available resources, including urban wastes for production, utilization or trade. These are related to SDG's three goals (UNDP 2020), such as capacity building and eradication of extreme poverty (SDG 1), women empowerment and gender equality (SDG 5), and making cities inclusive, safe, resilient and sustainable (SDG 11).

Against these backdrops, this chapter (paper) intends to portray the features and roles of informal enterprising activities including trading and waste-scrap collection and composting particularly in the fields of employment and income among the urban poor in the Kathmandu Valley. Considering the importance and scope of the informal activities, measures are suggested for better planning of the informal activities in the valley's cities.

16.2 Database and Methods, and Concepts

The data and information gathered for this paper have been based on the field surveys carried out in Kathmandu Valley in the years 2015–16 and the surveys included three types of informal enterprising activities such as trading, wastes or scraps dealing outlets, and organic wastes-based manure production (Sharma 2018). Two survey methods such as observation and informal interview were adopted and prior to conducting these methods, a reconnaissance survey was carried out to acquaint the author herself with the locational attributes of those activities across the valley. In the observation survey, standard observation protocol sheet was used to gather information on the features such as site, area occupancy, number and types, and sanitation facilities of the localities that had marketplaces containing at least 20 petty traders within the cities' ring road. This was accompanied and verified by toposheet and Google Maps, GPS, and camera. This was done because of two assumptions. First, it was assumed that new growth of informal activities should occur within the rind road, which includes parts of three municipalities, viz. Kathmandu, Lalitpur, and Madhyapurthimi. Next, it was assumed that there were certain conditions such as urban expansion and density, and new roads that should lead to locating the informal activities along those development activities.

The informal interview was carried out with the sample enterprises, including both stationary and itinerant firms engaged particularly in petty trading at the marketplaces. The interview was based on the checklist that acquired information on their social and economic features,[2] problems and prospects. The informal interview was also carried out with some scrap dealers and one wastes-based manure producing enterprise as a case study and gathered information on their activities including volume, supply sources or backward and forward linkages, and employment. Due to their homogeneity nature in most cases, the consulted scrap dealers and manure producing enterprise were assumed to represent all others that existed in the valley. Additional information was also gathered from the existing informal sector studies carried out in the valley, as well as from the institutional public sources such as policies and programmes related to the informal enterprising activities.

The meaning of informal activities has evolved over time in the cities of the developing world. The term 'informal sector' was first used by Lewis in the academic literature in 1954 to describe the employment among the poor for livelihood in the developing world (Lewis 1955). Later, it came into the limelight as 'the informal economy' by the work of Keith Hart in Ghana in 1971 (Hart 1973) and referred to by the International Labour Organization in Kenya in 1972 (ILO 2002). Since then, several scholars have used it with different names, such as traditional sector, unorganized sector, invisible sector, shadow economy, and so on. Further, others have described it with these features such as marginal but productive, small-scale, and self-employed activities (Sethuraman 1976, 1981; Souza and Tokman 1976; Harris 1978; Moser 1978; Bromley and Gerry 1979; Harper and Soon 1979; Hussmanns 2004). By virtue of its small size and self-employment natures, the informal sector is widely considered to serve as a source for promoting growth, employment, equity, and inclusive objectives. Thus, the sector is one of the means for creating a somewhat equitable distribution of income among the poor communities in the cities of the developing world. In Kathmandu Valley, the informal enterprising activities are

[2] The social aspects included family size, migration, and economic aspects comprised expenditure patterns, employment, income, enterprise history and so on.

characterised by tradition and have existed since the historic time that continues till today.

In this chapter, an analysis of the informal activities related mainly to marketing and marketplaces has been carried out in the relevant context. Broadly, urban informal enterprises are defined to deal with the trading, production, or distribution activities in the cities. In specific terms, they include individuals, households, or communities engaged in the activities related to production, distribution and delivery of goods and services such as: (i) trading (both stationary and mobile in terms of space), (ii) manufacturing (cottage or craft related like blacksmith, goldsmith, silversmith, iron/aluminium, straw, bamboo, cotton, wool, herbs, agro and dairy produce), (iii) construction (mason, contractor, carpenter, painter, plumber, electrician), (iv) transport works (pushcart, rickshaw, porter), and (v) personnel services (hairdresser, laundry, tailoring, repairing, wage labour, rag picketing, etc.). These varieties of informal activities are related to both traditional and modern systems in the valley's cities.

The rapid urban growth has given rise to several informal activities associated with all five categories described above. The workers employed by these activities should be substantial, but due to lacking information on all four categories including manufacturing, construction, transport, and personnel service works except marketing was extremely lacking, employment has been omitted in the analysis. Further, though small in terms of size and volume, the contribution of informal activities to reduce the urban wastes at sources and their productive use is estimated. Lastly, being dynamic over time, the relative importance of the informal activities related to the marketing and marketplace is expressed in terms of growth, spatial location, employment, and the type of goods and services.

Assumed that these analyses can be useful to decision-makers and planners for further planning and management of the informal activities in the context of continuing growth and expansion of the valley's cities, as well as of their environmental sustainability.

16.3 Findings and Analysis

16.3.1 Features and Roles of Informal Marketing and Marketplaces

The characteristics and developments of the informal marketing and marketplaces in Kathmandu Valley, which lies in the central hills of Nepal (Fig. 16.1), can be described by a variety of factors, such as historical account, growth and urbanization of the cities, migration pattern, growth and spatial location, and initiatives of the municipal authorities.

(i) *Growth and development*

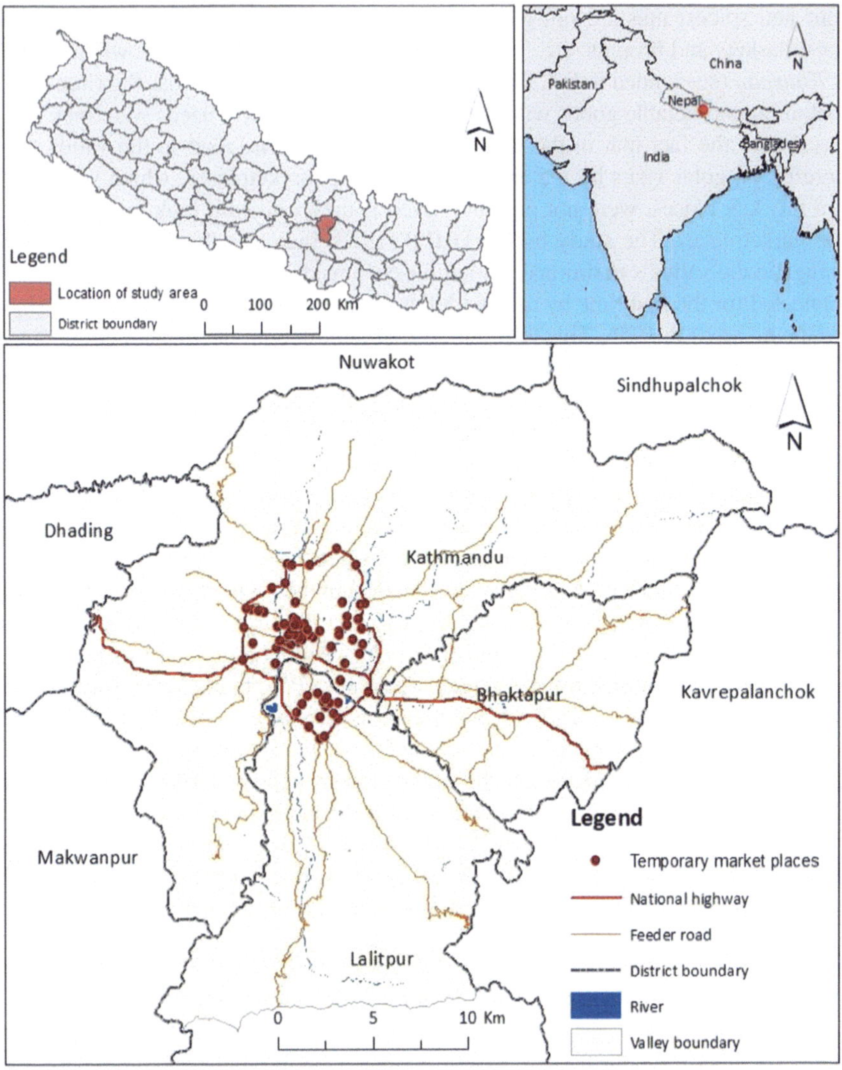

Fig. 16.1 Location of Kathmandu Valley, Nepal (*Source* Sharma 2021)

The valley's informal marketing systems dealing with local products seem to have evolved remarkably along with the development of roads and the growth of population. According to the studies by HMG/UN/UNESCO (1975) and Regmi (1978), the traditional marketing activities in the earlier times across the valley's cities and towns were conducted through a few scattered marketplaces and largely by the mobile vendors. The local mobile vendors-cum-producers were locally known as *Banjaras*. They often used to move on foot from locality (*Tol*) to locality within the cities for selling their own products such as fresh vegetables and fruits; grains, rice flake, flour and spices; mustard oil, ghee, milk cream and curd; handloom cloth, straw mats, baskets and brooms, etc. They used to carry baskets locally known as *Dokos* or *Kharpan* (suspended a thin, flexible wooden pole balanced across the shoulders) containing marketable goods with them. In this way, the local needs were met. This was due to the fact that in the historic time, there was no road in the valley, and therefore, regular visits by the city dwellers to the marketplaces, which were only in a few key places, were not possible. Even if they walked it took hours to reach the marketplaces. The study by the DHPP (1972) indicates that there has been a change in the valley's traditional marketing systems drastically when the valley was connected for the first time by road with the southern Tarai region (the main grains-producing area) in 1956. This thus replaced gradually both the manual conveyance means of goods and the traditional marketing system by the motor vehicles. Since then the valley has been connected by roads in all directions: north, east, south and west including India to the south and Tibet (China) to the north. As a result, the valley's urban population grew rapidly due mainly to the migrants from different parts of the country. These led to an increase not only in the number of marketplaces but also in to emergence of the modern marketing system. Along with these, the number of street vendors also grew remarkably since the early 1980s. The studies by Pradhan (1987) and Sharma (2018) show that the number of marketplaces containing 5 vendors and more grew considerably from 13 in 1987[3] to 87 (with 20 vendors and more) in 2018 and likewise, the number of petty vendors in the marketplaces grew from 317 to 13,935 during the same interval of years. The population censuses (CBS 2012) indicate that the valley's urban population increased by over 82% from 1981 to 1991, in which migration alone contributed 59%, which was the largest ever since the 1950s. The urban population in the valley grew by nearly 64% between 1991 and 2011. Table 16.1 shows that the net inflow of migrants in the valley in 2011 shared 36% of the valley's urban population, which is the largest net inflow of migrants among all urban areas in Nepal (CBS 2012). Now, the valley has become a main consumption and collection hub of local goods produced throughout the country.

(ii) *Classification of traded goods in informal marketing*

The valley's informal market system can be classified into different types in terms of form, size, and location. By form, the marketing system is either sedentary or mobile. In the sedentary form, the trading activity is carried out through marketplaces located at fixed sites or where a large number of individual traders have their own retail outlets. Mobile form refers to itinerant traders with no fixed locations or who move from locality to locality within the cities. However, both are petty traders but differ in employment structure. The retail outlets of the marketplaces employ two or more

[3] According the study by Pradhan (1987), the 13 marketplaces were associated with only Kathmandu city, the largest in the valley, but there was lack of information for other cities of the valley. However, the observation was that almost all major localities or *tols* of the valley's cities had petty vendors with few numbers to meet the daily marketing needs of local inhabitants. The study by Sharma (2018) indicates that the 87 marketplaces were related to the valley's cities. However, no comparison in the number of marketplaces between those two studies can be made due to variation in containing number of petty traders in the marketplaces.

Table 16.1 Population growth of five urban areas, Kathmandu Valley, Nepal

Municipalities	Population ('000)							Growth (%) 2001–11
	1952–54	1961	1971	1981	1991	2001	2011	
Kathmandu	106.6	121.0	150.4	235.2	421.3	671.8	1,003.3	4.93
Lalitpur	42.2	47.7	59.0	79.9	115.9	163.0	226.7	3.91
Bhaktapur	32.3	33.9	40.1	48.5	61.4	72.5	83.7	1.53
Madhyapurthimi	8.7	97.2	–	–	–	47.8	84.1	7.62
Kirtipur	7.0	5.8	–	–	–	40.8	67.2	6.45
Total/average	196.8	218.1	249.6	363.5	598.5	996.0	1,465.0	4.89

Source Pradhan et al. (2020)

family members involving both males and females while in the mobile form only a single family member including mostly males is engaged. The marketplaces across the cities are usually located at such sites, which are often crowded with pedestrians. However, mobile traders visit different localities or neighborhoods, which are some distance away from the main marketplaces, to seek potential customers for selling their goods. The mobile vendors in the valley are not only locally originated but also comprise migrants coming from the Tarai's different districts of the country, as well as from the adjoining border districts of India (Pradhan 1987). There is a huge difference in the variety and quantity of marketable goods between petty traders of marketplaces and mobile vendors. The marketplaces due to a large number of petty traders certainly have a wide variety of goods with large quantities such as vegetables, fruits, fresh flowers, and animal products (perishable), groceries, clothes, food and beverages, durable items (utensils, electrical, mobile telephone, etc.) and services such as hairdressing, tailoring and shoes repairing and polishing (Sharma 2018). In most cases, the marketplaces are characterised by a specialty in marketing goods. For instance, there a wide distribution of marketplaces containing fresh vegetables, fruits and animal products across the valley's cities and there is often crowded of customers in the morning and the evening, while the marketplaces of other than perishable products are located in few strategic places like public bus stops, downtown, etc. In contrast to the marketplaces' petty traders, the itinerant traders sell particularly the vegetables and fruits contained in the baskets and therefore the varieties and quantity are limited. They use bicycles carrying baskets containing marketable goods to visit or cover as many localities as they can, particularly during the morning and the evening.

(iii) *Spatial distribution pattern of the marketplaces*

The location of the marketplaces is randomly distributed across the Kathmandu Valley. The distance apart between the marketplaces is measured at approximately 340 m (Sharma 2018). However, two distinct patterns of spatial distribution of these marketplaces are discernible (Fig. 16.2). One is a concentrated pattern in the core area, and another is a dispersed pattern in the periphery area. These patterns are directly related to the degree of distribution or density of population between those two zones or areas in the valley.

This spatial distribution of the marketplaces can be related likely to satisfy the needs of low-income people. A large part of the basic needs of the poor, including the delivery of consumer goods, is provided through street vendors due mainly to two reasons. One is the number of small-sized marketplaces together with a comparably large number of small vendors being spread across the valley's entire area. The price of the basic goods and services of such marketplaces is generally cheaper. Another is that the street vendors carrying goods (comparably small volume and few varieties) by themselves tend to visit house to house to sell them. In so doing, the travel time or cost of the poor people is saved. It appears that the sites with transport facilities

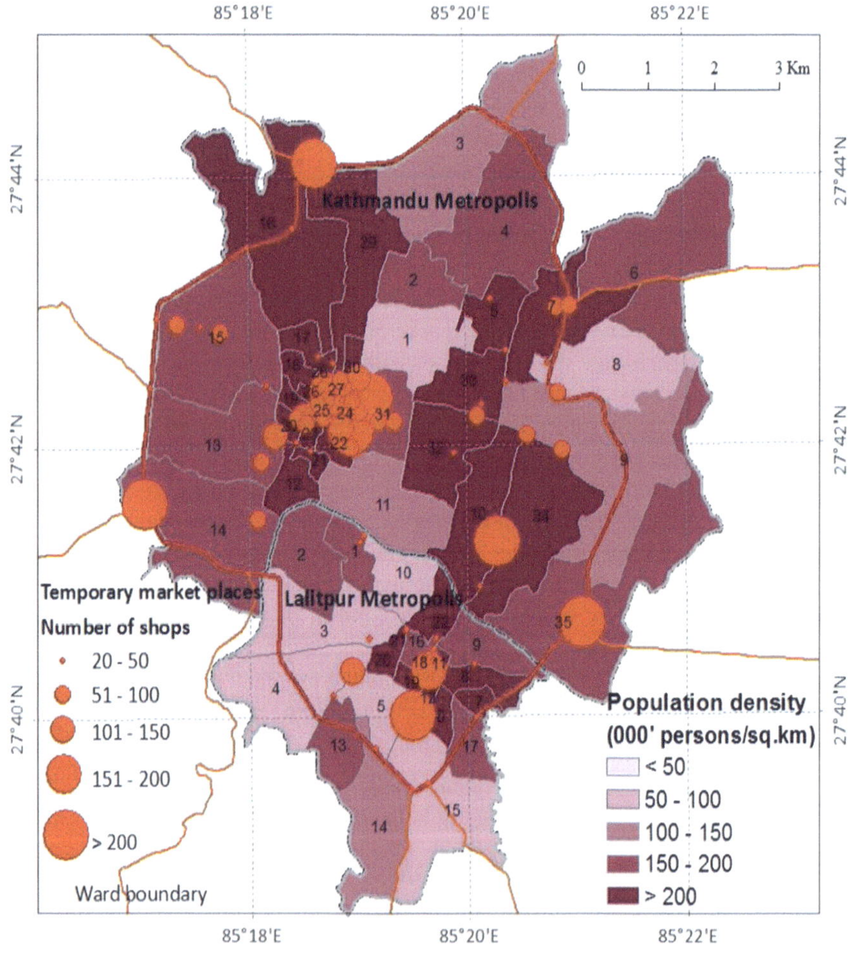

Fig. 16.2 Spatial distribution of marketplaces and population density (*Source* Sharma 2018)

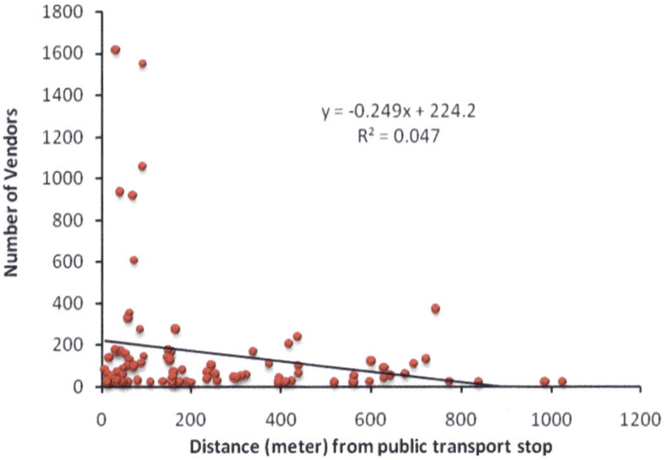

Fig. 16.3 Location of street vendors and transport facility (*Source* Sharma and Pradhan 2017)

are the most crucial factor for operating the vending business (Fig. 16.3). Two other favourable sites are public places and footpaths. This is because there are large pedestrians as potential customers, which means greater transactions and income. On top of them, the 'convenient' feature of the site to the potential customers is the most important factor (Sharma 2018).

(iv) *Employment and income*

The studies by Sharma (2018) and Pradhan (2010) state that informal marketing activities are essential in terms of providing employment and income, particularly to the poor and marginalized communities including women and migrants in the cities across the country. The activities seem to be important not only to increase access to basic goods and services but also to bring down the extent and intensity of poverty.

Available studies, for instance, by Dawadi (2008), Raut (2008), and Pradhan (2010) exhibit that the people living in the slums and squatter areas of the Kathmandu Valley adopt various types of informal activities as a means of livelihood. The studies indicated that informal activities have employed over 90% of the working population. Over 87% of the street vendors, mostly migrants were engaged in informal marketing activities (Sharma and Pradhan 2017)[4] and were mostly illiterate with semi-skilled or no formal education and training. It appears that they used to learn skills required for their job through the apprenticeship system from their elders. With the earnings of both men and women, families could acquire the most essential needs such as food, health, children's education and other amenities. However, during the lockdown

[4] Of the total sample 260 petty vendors from 36 sample marketplaces, 82% were employed in the sedentary marketplace and 18% in the footloose vendors with employed proportions of 51% and 49% of males and females respectively (Sharma 2018). The monthly average income of the petty vendor was $278 (NRs. 30,634).

period as a measure to control the spread of the current upsurge of the COVID-19 virus, most of them have been badly affected due to the loss of their marketing jobs.

The informal trading work is a place for employment and an alternative source of income for the women in the cities of Kathmandu Valley, who are usually engaged in the household chores. Regarding informal enterprising work, the women found to be engaged usually in producing goods based on home-cottage industries, sewing, food preparation, handicrafts, and so on. Also found is that the women workers outnumbered the male workers, particularly in the sedentary marketplaces (Sharma 2018). The main reason for the women being engaged in such activity is social security. Because in mobile or footloose vending, the vendors require special skills such as riding bicycles or driving push carts (*Thelas*) to move to different localities for trading until evening in the circumstance of lacking streetlights in most parts of the cities. In addition, the women also often need to look after their families or do household chores such as preparing food and other work at home in the morning and evening. In other cases, such as trading of goods during special occasions of festivals, cultural events around temples, or fairs in the cities, women were often found to be in larger number than males.

In the country, the informal sector has a predominant role in its economy, employing over 70% of the economically active population and contributing around 52% to the national GDP (CBS 2008). Its importance in employment generation and poverty alleviation has also been emphasized in the country's national plans (NPC 2002). Though the Nepal Labour Force Survey made the first attempt to gather some data on the informal sector across the country in 1998 (CBS 1999), there is yet lack of exhaustive data concerning the features and status of the informal sector available in the cities.

(v) *The environmental and sanitation problems of the marketplaces*

Despite the huge and significant roles and contributions, informal marketing or its street vendors often face different types of problems in operating business at the marketplaces. They include dust pollution during the dry months and muddy streets during the rainy months due to several dusty and unpaved roads and potholes across the valley, which are mainly due to snail-paced road expansion works and lack of regular maintenance of the broken-down streets or pavements. Next, the public toilets are extremely limited and their sanitation condition in terms of cleanliness and hygiene appear to be awfully poor. Further, most of the street traders are found encroaching upon public uses such as street sides, walkways, parks, alleys, traffic intersections or road junctions in the valley's cities. Furthermore, the problem they often face is harassment by the municipal police who either drive them away from the marketing sites confiscate their merchandises or charge them with heavy penalties. So, this has created a feeling of security lapse for them. According to the street traders, these problems should be managed and regulated by the municipal authorities by measures such as providing license, identity card, and credit facilities, regular monitoring, and other facilities (sanitation, water supply, waste containers, pavement of streets or walkways, etc.). As the informal marketing systems have been operated for a long and employ mostly the urban poor, the municipal authorities should think of the street vendors as opportunities rather than burdens and therefore they should consider the informal marketing sites to be included in the urban planning and management in a fair manner.

16.3.2 Utilization of the Urban Wastes

Waste management is a critical component of sustainable urban environment. Managing of ever-increasing volume of wastes, due to the rapid growth of migration, urbanization, improved living standards and changing consumption patterns, has been a great challenge and burden to the municipal authorities of the valley. Information is that over two-thirds of the total wastes are of organic origin, followed by recyclable non-organic materials (paper, plastic, glass, and metal), and other materials (ADB 2013). Of many, reduction of waste at source is one of the best ways as it involves cost while disposing it to the dumping sites. Recycling of wastes is a means of great potential for job creation and income generation, especially among the urban low-income groups (Muller 2002; Luitel and Khanal 2010). The practice of 3R (Reduce, Reuse, and Recycle) including the informal enterprising activity of scrap waste collection and composting of organic waste has been a good prospect not only for waste resource recovery and thereby reducing the volume of wastes at sources, but also to conserve urban environmental resources (ADB 2013).

16.3.2.1 Waste and Scrap Dealers

An important means of reducing the volume of waste at the source is to pick up a scrap of non-organic type for recycling. One of the virtues of the informal activity is to engage its participants in collecting recyclable waste materials generated in the cities. Studies show that quite many informal workers or scrap pickers are engaged in the collection of waste materials such as scrap metals, used bottles, waste paper, spent batteries, used tyres/tubes, wires and the like that come from household, commercial and institutional sources (CIUD 2006; CARE Nepal 2008). The scarp trading outlets scattered across the valley's cities sell such items to individuals for reuse or export them to the recycling plants.

The number of informal traders dealing with the scrap wastes, locally known as '*Kawādi*' has increased tremendously in the valley over the recent decades (Luitel and Khanal 2010). *Kawādi* has been home to more than 8,000 waste collectors and scrap dealers in the Kathmandu Valley (Practical Action 2014). There are around 250 scrap dealers in the valley who export around 116 tons of recyclable materials daily, largely within Nepal and some to India, and about 20% is recycled within the valley (MLD/JICA 2005; ICIMOD 2007). The waste pickers involve mostly women. The types of waste pickers include waste segregators, door-to-door collectors (with tricycles), cycle hawkers, scavengers and small scrap owners. It is estimated that each *Kawādi* dealer employs 5 persons (own families + collectors) on average. They often occupy the sites having wider space (at least 3,000 square feet surrounded by walls) for storing scrap commodities of different types, and road access for loading those materials in large vehicles such as trucks for export. The sites are usually surrounded by the localities (neighbourhoods) having construction works, commercial districts and high density of residences where there is potential for the availability of scrap wastes. The informal waste workers who live mostly in slums or squatter settlements are among the poorest, who earn their living from selling scrap materials to the *Kawādi* dealers. There is tax scrapped to the scrap dealers in Nepal.

Since most of these activities are carried out without the municipal government's intervention or support, explicit incentives such as access to social and health protection schemes and an increase in income to them can enhance their contribution to environmental conservation sustainably in the cities.

16.3.2.2 Composting of Organic Wastes

Composting of organic wastes including different methods such as vermi culture, aerobic degradation, etc. is the most important recycling measure to reduce the waste at sources significantly. Such practice has been carried out by several privates and communities in the valley over the past decades. They are located at sites of the cities' peripheries and collect vegetable and fruit wastes from the vegetables and fruits wholesale/retail marketplaces for producing organic manure. However, there is no information available on the number and size of such composting facility units. The daily total waste generation in the urban valley is estimated at 435 tons. In 1986, the valley initiated the first modern management of household wastes by collecting them from wastes containers located at sites serving surrounding neighbouhoods across the valley's cities, but the system could not run systematically due to the lack of seriousness of the municipal governments, irregular collection of wastes from the containers, avoiding the waste containers at the sites by the surrounding residents due to stink smell and political intervention. So, dispose of wastes in the open places and street corners in the cities began after some years. Then, the informal workers and local community organizations started to collect the waste (MLD/JICA 2005). The demand for manure produced from such organic wastes is estimated to be approximately 20,000–40,000 tons/year in the valley. The market price of such organic manure is far cheaper than the chemical fertilizer.[5]

To exemplify, one plant of composting of organic wastes called *Kalpabrikshya* is located at Chobhār to the south outskirt of the Kathmandu city. Set up about two and half decades ago with a processing capacity of 6 tons per day, the plant is known to have collected about 2 to 3 tons per day of fresh organic wastes from the vegetable and

[5] Based on the farm gate price of NRs. 15/kg for organic manure, against NRs. 20–45/kg for chemical fertilizer. Moreover, supply of essential chemical fertilizers is far below the total demand (MoAD 2014), and all are to be imported from other countries due to lack of own production plant. Even the supply of fertilizers is often not in time of need of farmers and quality is often poor, and so, farmers must get it at the black-market price, which is two to three times higher than the market price.

fruit wholesale market[6] with an annual production capacity of 200 tons of compost. The compost has a standard of international quality regarding nutrient content. The plant has employed a total of 36 local people (14 regular jobs + 22 daily wage workers). The operation of the plant is sustained through selling of the compost to local farmers, as well as to the retail outlets of agro-vet and nursery plants. This is an effort to meet the manure demand to some extent at the local level. Based on its own manure, the plant has also initiated commercial organic vegetable farming. It has a plan to expand its capacity by installing modern expanded technology.

To improve its capacities such as establishing a lab for making *Trichoderma* and others (plant pathology, etc.), the plant has also obtained financial support from the Public Private Partnerships for Urban Environment under its Public Private Partnership (PPP) programme, a joint effort of the Government of Nepal and the United Nations Development Programme (GoN/UNDP). It was to contribute to the creation of a healthy environment and the improvement of living conditions in urban areas of Nepal (Pradhan 2009). This was to promote partnerships between public and private sectors to assist small and micro enterprises in the sustainable provision of urban services. Given the scenario of the gap between waste generation and demand for manure as stated above, there is a huge scope to establish organic manure plants. More efforts are needed to attract entrepreneurs to set up more such organic manure plants, as well as other types of plants[7] not only for recycling of wastes generated, but also to drastically reduce the wastes at sources to lessen the cost burden of waste management in the cities.

16.4 Discussions and Conclusions

Urban informal sector is a great concern among urban planners for the development and management of economic, social, environmental, and spatial aspects in the cities of Nepal. They are important particularly for large cities, as they continue to grow due to the migration of the population for the availability of jobs, facilities and other opportunities (Sharma and Pradhan 2017). According to ILO (2000), poverty is caused by the lack of employment or inadequate incomes and so employment creation is perceived as the main goal of the urban government in developing countries. Small enterprises display a remarkable capacity to absorb labour, mostly poor adults, migrants, and women in full or part time and therefore equity distribution of income. So, it is reasonable for the urban government to promote small informal enterprises by

[6] The Kālimāti Vegetables and Fruits Wholesale Marketplace, the first and the largest marketplace in the Kathmandu Valley generates about 15 tons of organic wastes daily, which can be used for composting (Pradhan 2009).

[7] More plants like compost and plastic recycle, which has been operated since one and half decades ago based on waste plastic collection to produce 2–3 tons monthly of secondary product to be used as raw material for pipe factories located within the country under PPP programme, are essential for recycling of wastes generated by several vegetables and fruits markets and other non-organic products dumped onto public places.

providing the working environment, provision of credit at below-market interest rate, skill training, market access, sanitation and other facilities including streetlights, and to desist from destroying them (ILO 2005; CARE Nepal 2008). Of many, informal trading is the main and largest sector to employ petty traders, both men and women, as well as to contribute significantly to the functioning of the urban economy (Sanyal 1988).

One major concern of urban government in Nepal is the planning and management of the marketplaces and marketing systems, as they are mostly held at traditional sites. Their numbers are increasing due to the unprecedented growth of urban population. They are thus unavoidable, as well as an essential element of the city landscape. The Government of Nepal has made some efforts to address the issues in the urban informal marketing systems. For instance, it is stated in the national local governance policy that the municipal government can generate income by letting the traders to operate trading activities in the public places within the municipal boundary (MLD 1999). The provision allows the municipal authorities to regulate the operation of temporary marketplaces and vendors upon levying certain fees according to time shift, occupancy of site area, and others. The national plan and urban policy have incorporated efforts to include informal activities into the mainstream of urban development and to identify locations for operating informal trading activities by encouraging them by providing essential facilities like skill training, credit, and the like (DUDBC 2007; NPC 2007). The Kathmandu Metropolis has also made guidelines to regulate the informal marketplaces and street vendors (KMC 2007). In recent years, the private sector has also made efforts to provide marketplace facilities at crucial sites to the vendors to operate trading of vegetables and fruits and other services within the valley's cities, which are in fact praiseworthy, but such marketplaces lack maintenance of the pathway pavement, shop stall structure, storm drainage outlet, etc. Even the public marketplaces and marketing systems are disorganized. Both problems are related to the lack of serious responsibility, weak implementation of the regulations, often changing rules and regulations, and weak supervision and monitoring of the municipal governments (Pradhan 1987; Sthapit 1999; CIUD 2006; Dawadi 2008; Sharma and Pradhan 2017; Sharma 2018).

On the environmental planning and management front, there exists the Environmental Protection Act and Environmental Protection Rules to operate and regulate development projects like solid waste management facilities and activities in the cities. The National Plan during the years 2002–2007 highlighted the problem of finding disposal site for urban wastes as a major challenge mainly in the Kathmandu Valley. The Local Self Governance Act 1999 (MLD 1999) stipulated that large municipal governments like the valley's municipalities have the core operational responsibility of managing solid wastes within their boundaries. The acts described in the environmental protection do not allow operating any activities in open spaces containing historical and cultural monuments of national importance and likewise, the Public Road Act and Traffic and Transportation Management Act also do not allow operating any trading activities at the public places or footpaths.

There is a conflict in the complex nexus of the policies and efforts between the urban environment management and informal activities including marketplaces and marketing. The latter indeed have the perspective of employment and income generation to the poor. The consequence of this is that any efforts to improve the urban environment by reverting the nexus could affect the income and employment situation in the informal sector. So, there is a question of the implementation of all these acts and regulations together. Furthermore, from the critical geographic perspective, there is a danger with the urban poor, particularly the livelihood of street vendors in terms of economic, social, and cultural aspects possibly due to the adaptation of urban structural planning adjustments such as land use, road expansions and traffic congestion management by the urban government, even though the marketplaces are often located at the strategically accessible sites amidst of majority of the cities' residents.

Information on the location dynamics of the urban informal activities including marketplaces and waste management and their participants in the cities is important for understanding their distribution, growth, and transformation across the entire urban space. However, there is extremely lacking information on the location of the informal activities and their attributes such as number, size, employment structure, income, supply, and demand chains (forward and backward linkages), history of development and problems. Likewise, no design exists for long-term spatial and physical planning of the cities, including the Kathmandu Valley cities from economic, social and environmentally sustainable perspectives. These critical aspects can be for future research.

References

ADB (2010) Unleashing economic growth: region-based urban development strategy for Nepal. Asian Development Bank, Philippines

ADB (2013) Solid waste management in Nepal: current status and policy recommendations. Asian Development Bank, Philippines

Bromley R, Gerry C (1979) Who are the casual poor? In: Bromley R, Gerry C (eds) Casual work and poverty in third world cities. Wiley, New York, pp 3–23

CARE Nepal (2008) Study on status of unemployed slum youths and their possible linkages with job market in Greater Kathmandu. CARE Nepal, Kathmandu

CBS (1999) Nepal labour force survey 2008: statistical report. Central Bureau of Statistics, Kathmandu

CBS (2003) Population monograph of Nepal, vol II. Central Bureau of Statistics (CBS), National Planning Commission (NPC), Government of Nepal (GoN), Kathmandu

CBS (2008) Nepal labour force survey 2008: statistical report. Central Bureau of Statistics, Kathmandu

CBS (2012) National population and housing census 2011. Central Bureau of Statistics (CBS), NPC, Government of Nepal, Kathmandu

CIUD (2006) Study of street vendors in Greater Kathmandu. Centre for Integrated Urban Development (CIUD), Kathmandu

Dawadi GS (2008) Efficiency measures in informal street vendors of urban core area of Kathmandu Metropolitan City. Unpublished master's dissertation, Institute of Engineering, Tribhuvan University, Nepal

DHPP (1972) Inventory report of Kathmandu Valley. Department of Housing and Physical Planning, Ministry of Public Works, Transport and Communications, His Majesty's Government of Nepal, Kathmandu
DUDBC (2007) National urban policy. Department of Urban Development and Building Construction. Ministry of Physical Planning and Constructions, Government of Nepal, Kathmandu
Harper M, Soon TT (1979) Small enterprises in developing countries: case studies and conclusions. I.T. Publication, London
Harris B (1978) Quasi-formal employment structures and behaviour in the unorganized urban economy, and the reverse: some evidence from south India. World Dev 6(9/10):1077–1086
Hart JK (1973) Informal income opportunities and urban employment in Ghana. J Mod Afr Stud 11(1):61–89
HMG/UN/UNESCO (1975) Kathmandu valley: the preservation of physical environment and cultural heritage protective inventory, vol 1. Department of Housing and Physical Planning, His Majesty's Government of Nepal, Kathmandu
Hussmanns R (2004) Measuring the informal economy: from employment in the informal sector to informal employment, working paper no. 53 (December). International Labour Office, Geneva, pp 1–28
ICIMOD (2007) Kathmandu valley environment outlook. International Centre for Integrated Mountain Development, Kathmandu
ILO (2000) Current international recommendations on labour statistics. International Labour Office, Geneva
ILO (2002) Women and men in the informal economy. International Labour Organization. ISBN 978-92-2-113103-8. Accessed 18 Dec 2006
ILO (2005) A report on micro and small enterprise policy review in Nepal. Series 7. International Labour Organization, Kathmandu, Nepal
Kansakar VBS (1985) Indo-Nepal migration: problems and prospects. Contrib Nepal Stud 11(2):49–69
KMC (2007) Street and footpath business management guidelines. Kathmandu Metropolitan City, Kathmandu
Lewis WA (1955) The theory of economic growth. Allen and Unwin, London
Luitel KP, Khanal SN (2010) Study of scrap waste in Kathmandu Valley. Kathmandu University. J Sci Eng Technol 6(1):116–122
MLD (1999) Local self-governance act, 2055. Ministry of Local Development, HMG, Kathmandu
MLD/JICA (2005) The study on the solid waste management for the Kathmandu Valley. Final report: Volume I: Executive summary. Ministry of Local Development and Japan International Cooperation Agency, Kathmandu
MoAD (2014) Statistical information on Nepalese agriculture, 2013/14. Ministry of Agricultural Development, Agri-business Promotion and Statistics Division, Kathmandu, Nepal
Moser CN (1978) Informal sector or petty commodity production: dualism or independence in urban development. World Dev 6:1041–1064
MoUD (2017) National urban development strategy. Ministry of Urban Development, Government of Nepal, Kathmandu
Muller L (2002) Waste indicators in South Africa. J Sustain Waste Manag. Warmer Bulletin 82:86
NPC (2002) Tenth national plan (2002–2007). National Planning Commission, Government of Nepal, Kathmandu
NPC (2007) Three years interim plan (2008–2010). National Planning Commission, Government of Nepal, Kathmandu
Practical Action (2014) Pro-poor market facilitation approach to improve living condition of informal solid waste workers. Practical Action Aid, Kathmandu. http://practicalaction.org/
Pradhan B, Sharma P, Pradhan PK (2020) Urban growth and environment and health hazards in Kathmandu Valley, Nepal. In: Singh RB, Srinagesh B, Anand S (eds) Advances in geographical and environmental sciences: urban health risk and resilience in Asian cities. Springer Nature Singapore (e-book), pp 293–326

Pradhan PK (1987) Fresh vegetables marketing system in Kathmandu City. Tribhuvan Univ J XIII(2):69–76

Pradhan PK (2003) Population growth, migration and urbanisation, and environmental consequences in Kathmandu Valley, Nepal. In: Unruh J, Krol M, Kliot N (eds) Environmental change and its implications for population migrations. Kluwer Academic Publishers B.V. Dordrecht

Pradhan PK (2009) Report on PPPUE project evaluation. Public Private Partnerships for Urban Environment, NEP/04/001, the Ministry of Local Development/United Nations Development Programme, Kathmandu

Pradhan PK (2010) Living condition and livelihoods of the urban poor: case from greater Kathmandu. Mountain areas and globalization. IGU/UGI July 4–10, Switzerland, pp 1–16

Pradhan SM (1988) Kathmandu: metropolitan development and conservation issues. Reg Dev Dialog 9(3):203–221

Rabenau BV (1990) Issues in housing development. Urban Development through Local Efforts, Kathmandu

Raut UB (2008) Urban agriculture for sustainable urban development in Kathmandu. Unpublished master's dissertation, Institute of Engineering, Tribhuvan University, Kathmandu

Regmi MC (1978) A study of Nepali economic history (1768–1846). Manjushree Publications, New Delhi

Sanyal B (1988) The urban informal sector revisited: some notes on the relevance of the concept in the 1980s. Third World Plan Rev 10(1)

Sethuraman SV (1976) The urban informal sector: concept, measurement, and policy. Int Labour Rev 114(1):69–81

Sethuraman SV (ed) (1981) The urban informal sector in developing countries: employment, poverty, and environment. International Labour Organization, Geneva

Sharma P (2018) Urban informal sector: street trading and service activities in Greater Kathmandu. Unpublished PhD Dissertation, Central Department of Geography, Faculty of Humanities and Social Sciences of Tribhuvan University, Kathmandu

Sharma P (2021) Policy issues in informal marketplace and marketing system in Nepal: experiences from the cities of Kathmandu Valley. Geogr J Nepal (Central Department of Geography, Tribhuvan University) 14:113–130. https://doi.org/10.3126/gjn.v14i0.35555

Sharma P, Pradhan PK (2017) The petty street vendors and their livelihoods of the Kathmandu Valley Cities, Nepal. In: Li A, Deng W, Zhao W (eds) Land cover change and its eco-environmental responses in Nepal. Springer Geography 1:359–379. https://doi.org/10.1007/978-981-10-2890-8

Skeldon R (1985) Population pressure, mobility, and socio-economic change in mountainous environments: regions of refuge in comparative perspective. Mt Res Dev 5(3):233–250

Souza PR, Tokman VE (1976) The urban informal sector in Latin America. Int Labour Rev 355–65

Sthapit D (1999) Urban informal sector in Kathmandu City: a case study of commercial vendors. Unpublished master's dissertation, Department of Architecture and Urban Planning, Institute of Engineering, Tribhuvan University, Kathmandu

UNDP (2020) What are the sustainable development goals? United Nations Development Programme. Accessed 6 Sept

Chapter 17
COVID-19 Pandemic and Urban Informal Workforce: Challenges Faced by the Small Entrepreneurs in Silchar Town of Assam, India

Subhabrata Dutta

Abstract India is gifted with a large number of towns and cities. According to the 2011 census report, the country has 4041 cities and towns governed by local urban governing bodies (UCB). Governments at the central and state level consider the town and cities as the source of growth in various wage-earning programmes, including e-entrepreneurship development. The sudden outbreak of the COVID-19 pandemic has aggravated the situation seriously. The lockdown had jeopardised many workers—both in formal and informal sectors. The COVID-19 pandemic has seriously affected the global economy, and India is no exception to this. Silchar, the second largest city in the state of Assam, is the headquarters of the Cachar district. Various informal sources found that 2000 small and medium entrepreneurship are functioning in Silchar town. However, during COVID-19, due to the imposed lockdown (1st phase, which almost lasted for five months), most of them faced an acute financial crisis, and a few closed their enterprise.

Keywords COVID-19 · Urban workforce · Informal sector · Entrepreneurship

17.1 Introduction: Tackling Growing Urbanisation and Unemployment

Urbanisation is closely intertwined with economic development. Cities provide opportunities for increased growth and incomes, and most countries have achieved sustained economic development together with rapid urbanisation (Henderson 2010). However, the proliferation of urbanisation is unevenly linked to the development of a nation. This is one of the biggest problems that seriously impacted national development. Moreover, the problem has been terribly aggravated by the rising trend of

S. Dutta (✉)
Department of Social Work, Assam University, Silchar, India
e-mail: sduttasw@gmail.com

S. Anand et al. (eds.), *Sustainability in South Asian Cities*, Advances in Geographical and Environmental Sciences, https://doi.org/10.1007/978-981-97-7455-5_17

migration of the rural population to urban areas. Providing gainful employment opportunities to enable the population to lead a standard of living is necessary. However, the increasing number of migrated people has become a stumbling block to providing job opportunities. More recently, the slow growth of various urban economy sectors, particularly manufacturing and service sectors, has added fuel to the burning problem of unemployment in urban areas.

In a country like India, with surplus labour and a strong tendency for new technologies for mechanising industries, employment generation in both urban sectors faces a serious challenge and a huge influx from rural areas has made the challenge more severe. According to the Period Labour Force Survey (PLFS), a part of the National Sample Survey, the unemployment rate has neglected its highest peak (96%) in 45 years in 2017. This report also stated that unemployment in towns and cities had assumed a serious dimension. Due to surplus labour force and a sharp decline in wages and poverty, starvation and social unrest continue to rise. The urban economy has been gradually less labour-absorbent and is characterised by joblessness.

India is gifted with a large number of towns and cities. According to the 2011 census report, the country has 4041 cities and towns governed by local urban governing bodies (UCB) such as Municipal Corporations, Municipal Councils or Nagar Panchayat. The introduction of a new programme of urbanisation at the national level in the names of Smart Cities Mission and the Atal Mission for Rejuvenation and Urban Transformation (AMRUT) has blessed a handful of urban citizens. The majority of ULBs cannot provide basic minimum amenities due to the purity of funds. Moreover, rapid urbanisation has severely affected the ecological environment.

Governments, both at the central and state level, look upon the town and cities as the source of growth various programmes for wage-earning, such as Swarna Jayanti Shahari Rozgar Yojana (1997) are not all-inclusive. It only emphasises entrepreneurial skill. Encouraging entrepreneurial instincts will indeed produce sustainable results.

The pandemic has aggravated the situation seriously. The lockdown had jeopardised many workers—both in formal and informal sectors. The difficult situation is growing to be more different in the ensuing years. The demographic dividend is gradually becoming a demographic curse. According to the International Labour Organisation, workplace losses were significant; approximately 25 crore job losses globally in 2020.

The sudden outbreaks of the COVID-19 pandemic have produced various socioeconomic problems all over the world. It caused severe disruptive changes in all sectors of human life. The unanticipated changes created various serious problems in the health sector, business environments, industrial sector, and employment sector. In short, the COVID-19 pandemic has seriously affected the global economy, and India is no exception. The crisis had emerged in the form of trade disruptions, closure of industries, unemployment, rise and spread of hunger and poverty levels, ill health, lack of resources and depletion of foreign exchanges. The sudden increase in demand for adequate health services and life-saving drugs and products put a heavy strain on the nation's fiscal resources. The dismal socioeconomic scenario threatened peace and social unrest in many places. Different governments of developed countries struggled hard to address the problems and set the devastated economy on the right track. But India, a developing country facing a multiplicity of severe socioeconomic problems, suffered much more due to this pandemic. The struggle became more troublesome due to various regional and local diversities and variations in terms of development.

Various international organisations and agencies, such IMF, offered various data about the severe impact of the COVID-19 pandemic on the Indian economy. According to the IMF, the Indian economy suffered a sharp decline of 4.5% of GDP (Gross Domestic Product). The Reserve Bank of India had predicted a negative growth rate during the fiscal year 2020–21. The rate of industrial production recorded a 35.9% decline compared to pre-pandemic years due to the nationwide lockdown. The stability in the labour sector was severely affected. The Centre for Monitoring Indian Economy (CMIE) showed that the number of job-losing employees in July 2020 amounted to approximately 5 million, and the total job losses in both manufacturing and services sectors were about 18.9 million around. Another CMIE report recorded that about 17.7 million wage-earner lost their occupations during the month of April, and no sign of recovery was visible in the following months. The job losses continue in MSME especially.

The urban areas were severely affected by the COVID-19 pandemic the workmen of many countries were variously supported and protected in the form of wages and other supportive means. However, the limitation of the state capacity has created roadblocks for many countries. The states could not respond up to the expected level. Naturally, important issues of survival of the urban people came to the forefront (Khan 2020).

17.2 Entrepreneurship and Entrepreneurship Development

The economic development of a country somewhat depends on the scale and breath of entrepreneurship. This basic truth was realised many years ago (McClelland 1961; Guilder 1984; Drucker 1985; Mohanty 2005; Nandan 2013).

Various attempts have been made to understand the emergence of entrepreneurship, such as the need for achievement, creativity, and propensity to take risks, independence, leadership, and so on. However, the tradition-bound Indian society (Vasanthagopa 2008; Siwan 2012) is now undergoing a sea change, where backward communities and women are seeking gainful economic participation in several fields in increasing numbers (Swarajyalakshmi 1998). Thus, the opportunities for economic activities among poor and middle-income groups have increased drastically. It is necessary to increase the opportunity for employment for the unemployed through the development of entrepreneurship. However, with the changing scenario, they play the role of owners and decision-makers of non-traditional enterprises, where growth and employment generations are eminent. Therefore, there was a need to provide an appropriate environment to promote entrepreneurship among these groups and women (Vinze 1987; Mohanty 2005; Khanka 2007; Dutta and Chan

2013). Now there is a need for more outstanding commitment and involvement of financial and other supportive agencies to promote entrepreneurship among them.

Developed countries like the United States, Japan, and the UK have acknowledged that the 'spirit of enterprise' among their people has transformed their small agro-based industries into big industrial giants (Lalitha 1996). These countries further promoted entrepreneurship among the people looking towards the achievement of higher levels of economic growth. Therefore, entrepreneurs play a vital role among the factors of production and have enormous potential to galvanise the other components such as land, labour and capital towards generating increased output, income and employment, which directly hit the national development (Suresh 2012; Sharma 2017). Not only that, even the current Government of India is playing an active role of enabler, and facilitator by providing training and infrastructural facilities to prospective small and medium enterprises, including women entrepreneurs who have the potential and propensity to run a business.

It is well known that the Industrial Revolution in the West created a boost for entrepreneurship in the developed economies. However, after the partition of undivided India in 1947 into India and Pakistan (Das et al. 2022), the country found itself in poverty, unemployment, and chaos (Desai 2011). Only in the late 50s did interest in entrepreneurship development emerge mainly through small-scale sectors to achieve rapid economic growth in most parts of the globe, including India.

Uncertainty prevails due to poverty and unemployment and the effect of COVID-19 also falls on the small and unorganised business sectors in our country's rural and urban areas. At this juncture, India viewed the protection of small and medium entrepreneurship and cottage industries as an effective instrument to combat their many problems. India has already systematically developed entrepreneurship to enhance national production, promote balanced regional development, and create more employment opportunities (Lalngaihsaki and Lalnunthara 2021).

At present, Entrepreneurship has been recognised as a concept not only vital for starting industries and other enterprises but also in the development of livestock. The Government of India's planning on economic liberalisation evolved the transfer of economic and social activities to provide hands. Privatisation, in effect, means the orientation of the economy more and more to local, regional, national and international market forces though there are certain uncertainty prevails. The state needs to be more and more confine itself to public utilities and withdraw from competitive and commercial activities as it appears ambiguity.

17.3 COVID-19 and Micro, Small and Medium Enterprises in India

The problems faced by Small and Medium Enterprises (SMEs) have been an increasing area of service concern due to the onslaught of COVID-19. Losses of jobs and the closure of microenterprises and small businesses lead to poverty, unemployment, misemployment and underemployment. A depressed socioeconomic scenario is generated all around. Several studies have been conducted to assess the situation and its consequences (Lalngaihsaki and Lalnunthara 2021; MSMEs 2020; Soni 2021; Sowmya 2020; The Times of India 2021; Barik and Tripathi 2022). In fact, financial losses and joblessness experienced due to the pandemic by SMEs and small business owners can be somehow measured by statistical measures discussed so far.

A non-financial banking institution conducted a survey in the second half of May 2020 and charted the negative impact of COVID-19 on Micro, Small and Medium Enterprises (MSMEs) and made an excellent analysis of their attitude towards their loss and income. About 14,444 MSMEs were involved in the survey and about half of the respondents reported that their income sharply declined to 50%. Another study involving 360 enterprises indicated that 17% of annual sales had declined during the lockdown period, leading to a loss of two months of revenue. Again, there was a variation in losses (Shafi et al. 2020; WTO 2019). The smallest SMEs sustained the highest losses. A 24% reduction in annual sales occurred in the firm with less than eight employees. The situation differed in the case of the firms with more than 45 employees, and such firms experienced a much lower decline in annual sales. The survey also indicated that all kinds of MSMEs applied 75% of their operative capacity before the lockdown. During the lockdown period, they used 17% of their functioning capacities. It is evident from the survey that there was no transformity about the percentage of losses and application of operating capacity (The Times of India 2021).

Turning to the conditions of SMEs and small businesses, a quick survey was conducted on 1,416 micro enterprises all over the country. The survey fields covered different areas like sub-industries and manufacturing, services, and trading. Indian sub-industries with the minimum number of employees, ranging from 2 to 15 with a continuity of 12 years revealed a survey picture. The survey indicated that major sub-industries were closed during the lockdown period, and only the minimum number of smallest industries, i.e. 17%, remained functional either partly or fully 52% of the essential business remained fully closed. Moreover, the profits of the fully or partly functional business sharply declined by 72% compared to the pre-pandemic period. Again, the percentage of loss or profits was net equal in all sectors (Barik and Tripathi 2022). The manufacturing and service sectors incurred a heavy loss, varying from about 78%–81%, respectively, while the loss percentage was significantly lower in the trading sector. The functional microenterprises during the lockdown period reported that the loss of revenue was about 28%, compared to pre-lockdown income lives (Ramaswamy 2020).

Some observations of the pandemic's impact on the financial health of MSMEs and small businesses were available from various sources. Some observations need to be mentioned. An air freshener manufacturer located in Chennai, Tamilnadu, reported, "Our sales turnover has been badly hit a mid-impact for supply chain due to Covid".

The Retail Business survey data for the post-pandemic period shows signs of stability and recovery in the market. Retail business rose to 88% in August 2021 of the pre-pandemic levels, marginally higher than 72% in July 2021. Other retail business categories such as Quick Service Restaurants (QRS), food and groceries, salons, apparel footwear, and jewellery show signs of the rising trend in sales and witness rapid recovery in the post-pandemic period. The survey was done by the Retailer Association of India (RAI).

Under these circumstances, Covid-related stress has significantly affected MSMEs, particularly micro and small businesses across sectors in India. Various enterprises like supermarkets, grocery shops, beauty parlours, restaurants, pharmacies, manufacturers, and logistics service providers were seeking opportunities to sell their businesses (Soni 2021) due to various challenges since Covid struck last year (2021). The reasons were connected to credit, supply chain, competition, lack of customers, new business launches, etc. Under these circumstances, this study has been undertaken on Small and Medium entrepreneurship at Silchar Town of Assam, India.

17.4 Silchar Town

Silchar, the second largest city in the state of Assam, is the Cachar district's headquarters. About 90% of the population here comprises Bengalis who speak the local Sylhet language. The remaining 10% comprises people from Bihar, Manipur, Bishnupur, Rajasthan, and some tribal groups, including the Nagas. It is situated along the Barak River, popularly known as Barak Valley. The population of Silchar in 2020 was 164,353. Out of the total population, the number of literate people stands at 103,927 (55,263—males and 48,664—females). The total number of workers is 43,302 depending on multi-skill, 36,652 are men, and 6,650 are women. There is no official record of small and medium entrepreneurship in this town. However, from various informal sources, more than 2000 such entrepreneurs were working in 2018 (Dutta 2018). During COVID-19, due to imposed lockdown twice (in the 1st phase, which was almost five months duration 2020), most faced acute financial crisis, and few closed their enterprise.

17.5 Entrepreneurial Activities in Silchar Town

The Government has promoted entrepreneurial activities as income-generating activities through Microfinance and Self-help Groups (SHGs) with initiatives of local self-government (PRIs) in rural and through municipalities (by the National Urban Renewal Mission). In the case of women, it has been observed that women's participation in this entrepreneurial programme has gradually increased in the past few years. Many successful women entrepreneurs have come out through this region's SHGs programmes. However, it is quite challenging for all small entrepreneurs to establish themselves as entrepreneurs as the market and business environment is very uncertain after COVID-19, and socioeconomic instability also prevails.

Normally small entrepreneurs of this region produce food processing products like pickles, papads, masalas and candles, agarbattis and jute and bamboo-based handicrafts products. Local and state governments provide them with various facilities like financing, training, marketing of products, and so on. However, the growth of these entrepreneurship is not as high as expected. They encounter various problems, such as the delayed process of loan disbursement, marketing of their products, time availability of raw materials (especially for the Jute-based raw materials), etc. However, there is no way out of going reverse. To improve the socioeconomic conditions of this city and this district, livelihood opportunities will be increased for the unemployed and women through the development of entrepreneurship. So, whatever the volatile and uncertain business and socio-political environment prevailed after the pandemic, support and encouragement for the promotion of small and medium entrepreneurship must be provided. In this region, a good number of NGOs also engaged themselves in the promotion of small and women entrepreneurship.

Therefore, a study was conducted among 120 small and medium entrepreneurs, including women entrepreneurs who are engaged in making various products (food processing, manufacturing, etc.) and doing their business for more than 10 years in this town with the following objectives:

1. To know the socioeconomic status and educational attainment of the entrepreneurs;
2. To find out the types of products they are making or producing and their financial sources;
3. To study the income range and marketing opportunities of their products;
4. To know whether there is any level of significance exists between the education status and categories of work and income of the entrepreneurs;
5. To explore the challenges they face due to the COVID-19 pandemic situation.

Research Design

The present study is an effort to explore the socioeconomic condition of the entrepreneurs, marketing opportunities of their product, available government support, and challenges during COVID-19 lockdown period in the southern part of Assam based on exploratory research design.

Tools of Data Collection

Data was collected from the respondents through a structured interview schedule with 52 relevant questions.

Data Processing and Analysis

Data thus collected were processed through the SPSS programme in computer and analysed through the tables, charts and presented. A simple statistical method was used to calculate the percentage of different variables.

17.6 Results and Discussions

So far, the respondents' socio-economic background is a concern; most half of the members hailing from were from an economically low-income family (46.6% were found to be BPL category). However, as far as educational attainment is concerned, all are literate. More than 70% (85) are educated up to primary level, and 17.5% (21) are found to be qualified up to 10th standard, and the rest 11.7% (14), have completed their higher secondary education and significantly of them went for graduation, however, could not be completed. In this study, most members are from the OBC category (64.2%), followed by 17.5% are SCs, 7.5% STs, and the rest 10.8% belonged to the general caste (Table 17.1).

It has been found from the study that the entrepreneurs are engaged in producing five significant products such as food processing (like *Muri, Papad, Laddu,* dry fruits, etc. 43.4%), handloom products (23.4%), products of jute and bamboo (18.4%), candle and *agarbatti* making and pottery (10.8%). Very few are engaged in others like embroideries (4%) (Chart 17.1). As per as financial support is concerned, a good number of respondents (89, 74.1%) have received bank loan support, and out of them again, 42 (47% of the loan takers) have to get the benefit of subsidies, and all these members are belong to BPL category. 9 (7.5%) of the entrepreneurs have taken

Table 17.1 Socioeconomic status and education level of the entrepreneurs

Category of the respondents		Up to primary level				Up to 10 standard				Higher secondary qualified				
		OBC	SC	ST	Gen	OBC	SC	ST	Gen	OBC	SC	ST	Gen	Total
Below poverty line	56 (46.7%)	21	11	4	1	10	1	0	1	5	1	1	0	56 (100%)
Above poverty line	64 (53.3%)	36	7	2	3	4	1	1	3	1	0	1	5	64 (100%)
		57	18	6	4	14	2	1	4	6	1	2	5	
Total	120 (100%)	85 (70.8%)				21 (17.5%)					14 (11.7%)			120 (100%)

business loans from various commercial banks on their own (keeping mortgage of properties etc.).

According to the data, it has been observed that the majority (69, 57.5%) of entrepreneurs earn up to INR 6000/- per month, followed by 23.4% (28) earn between INR 6001/- to INR 8000/- range and rest 23 respondents (19.1%) have found the earnings are more than INR 8000/- per month (Chart 17.2). Therefore, the researcher has categorised them into lower, middle and higher income group and cross-checked (Table 17.2) with the education status of the entrepreneurs to find out whether any correlation exists. It has been found that its highly significant (Pearson Correlation: 0.365, Significant level is—0.000*-), which means that the higher level of education of entrepreneurs has the higher income.

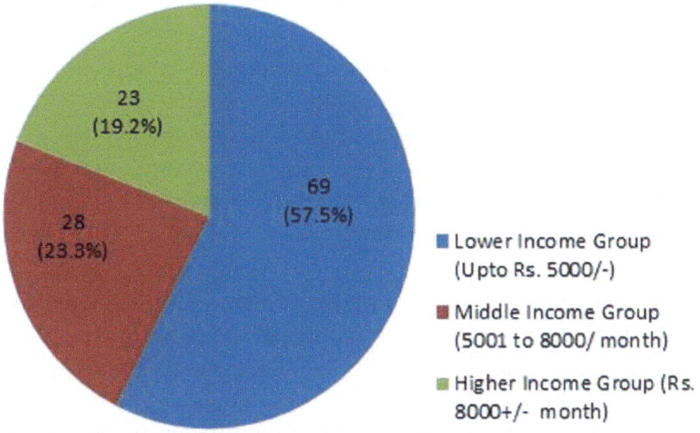

Chart 17.1 Distribution of monthly income (Rs.) status of the entrepreneur

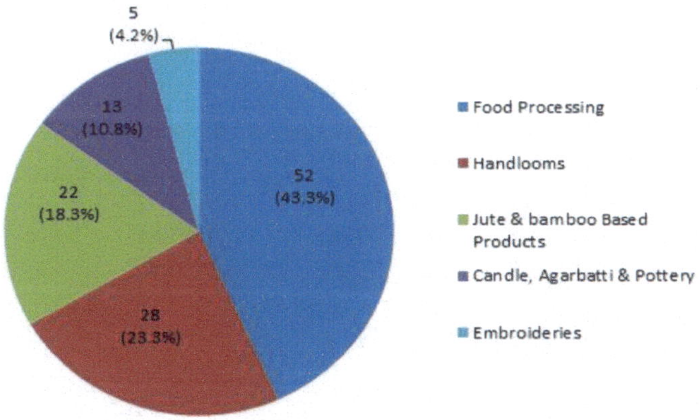

Chart 17.2 Distribution of the production of the entrepreneur

Table 17.2 Distribution of educational status and income status of the entrepreneurs

Educational status of the entrepreneurs	Monthly income status (Rs.) of the entrepreneurs			Total
	Lower inc. group (Up to 6000/-month)	Middle inc. group (Rs. 6001 to 8000/-month)	Higher income group (Rs. 8000⁺/-month)	
Up to primary level	57 (67.1%)	20 (23.5%)	8 (9.4%)	85 (100%)
Up to X standard	8 (38.1%)	4 (19.0%)	9 (42.9%)	21 (100%)
HS qualified and above	4 (28.6%)	4 (28.6%)	6 (42.9%)	14 (100%)
Grand total	69 (57.5%)	28 (23.3%)	23 (19.2%)	120 (100%)

Pearson correlation: 0.365, significant level is—0.000*-

Cross tabulation (Table 17.3) also made between the entrepreneurs' education status and work categories and found to be highly correlated (Pearson Correlation is 0.808 and level of significance is 0.000*), which means entrepreneurs' work categories are also depend on their levels of education. Even correlation was also found between the entrepreneurs' work categories and income status (Table 17.4). The level of Pearson correlation here was found to be 0.295, which is highly significant (0.000*).

It has been found from the study that entrepreneurs sell their products on their own arrangements, though in a few cases, municipalities of the towns and panchayats extended support in arranging space at different large and small *melas* in the region and in a few cases, some stalls provided at the shopping malls by the municipalities. Even sometimes, the exhibition is also arranged by different bodies and NGOs like Jute Corporation of India, KVIC, etc. However, the entrepreneurs had to bear the minimum cost of rent for those shops and exhibition registration fees.

Table 17.3 Cross tabulation of the educational status and work categories of entrepreneurs

Entrepreneurs work categories	Educational status of the entrepreneurs			Total
	Up to primary level	Up to X standard	HS qualified and above	
Food processing	39 (75.0%)	9 (17.3%)	4 (7.7%)	52 (100%)
Handlooms	22 (78.6%)	3 (10.7%)	3 (10.7%)	28 (100%)
Jute and bamboo based products	18 (81.8%)	2 (9.1%)	2 (9.1%)	22 (100%)
Candle, agarbatti and pottery	6 (46.2%)	5 (38.5%)	2 (15.4%)	13 (100%)
Embroideries	–	2 (40.0%)	3 (60.0%)	5 (100%)
Total	85 (70.8%)	21 (17.5%)	14 (11.7%)	120 (100%)

Pearson correlation: 0.808, significant level is—0.000*-

Table 17.4 Distribution of work categories and income status of the entrepreneurs

Work categories of entrepreneurs	Monthly income status (Rs.) of the entrepreneurs			Total
	Lower inc. group (Up to 5000/- month)	Middle inc. group (Rs. 5001 to 8000/ - month)	Higher income group (Rs. 8000+/- month)	
Food processing	40 (76.9%)	3 (5.8%)	9 (17.3%)	52 (100%)
Handlooms	15 (53.6%)	9 (32.1%)	4 (14.3%)	28 (100%)
Jute and bamboo based products	9 (40.9%)	8 (36.4%)	5 (22.7%)	22 (100%)
Candle, agarbatti and pottery	5 (38.5%)	5 (38.5%)	3 (23.1%)	13 (100%)
Embroideries	0	3 (60.0%)	2 (40.0%)	5 (100%)
Total	69 (57.5%)	28 (23.4%)	23 (19.1%)	120 (100%)

Pearson correlation: 0.295, significant level is—0.000*-

More than 50% of the entrepreneurs could not respond positively in connection with government support schemes; however, the rest, 50% of them narrated that the Government support schemes and the role of local self-government in making the process of bank loan, arrangements of raw materials, providing training, market space, conducting exhibition, facilities and collection of tax are found to be positive.

Regarding the challenges, especially at the post-pandemic stage, the following issues have been identified by the respondents.

- Entrepreneurs faced great difficulty in obtaining access to credit/financial support during and after the pandemic; it remains a significant challenge. Shortage of finance seems to be a problem for every entrepreneur. As most of them did not have movable or non-movable assets, they were denied the facilities even though they exhibited sufficient skills of entrepreneurship.
- They cannot go for technology-based products as lack of credit denies their access to technology, and demand for their products has not yet recovered from the impact of covid-19.
- Price fixation of the products is one of the significant challenges in post-pandemic situations, especially for those working to produce embroidery, jute and bamboo-based products.
- Demands for their products have not yet reached their pre-pandemic positions.
- There is no association among these entrepreneurs, limiting their bargaining power, which was essential in this post-pandemic situation.

17.7 Concluding Discussions

As per various economic reports, small entrepreneurs in this city have been significantly impacting all segments of the economy. However, due to uncertain socioeconomic conditions after COVID-19 and the bunnies' environment, which is not so stable and progressive, they are victims of regular finance, training, coordination, marketing opportunities, and so on, though as per their perception, they are ready to accept challenges and assume responsibilities in the fields of economic, social, political and more particularly in entrepreneurial stance. These enterprises are generally muscle-power technology based as they cannot buy proper tools and equipment to improve their products and services due to financial restraints. At the same time, their demand for training in business and technical skills is low because of their poor self-sphere system. Support systems were found to be less effective in reaching these small entrepreneurs with training services and technical assistance. The study reveals that education status, income level, and work categories are highly correlated (Table 17.5). It has been found that lower-educated entrepreneurs have lower incomes and make food-based products that need very low skill.

It is true that women's literacy has raised much. But their participation in different occupations remains minimal. Though the society has progressed much, women are still subservient to male authority. They are not yet capable of discovering their own identity and self-independence. But the COVID-19 pandemic provided women a chance to discover their potentiality outside their stereotyped roles.

Today, after the pandemic, the biggest challenge for these small and medium entrepreneurs is stiff competition in the market. It is either from big companies that provide branded goods to the quality-conscious high-income elastic group of customers or from a businessman who sells similar cheap products to the low-income elasticity consumers. For example, those making *muri* (puff rice) in the traditional system face threatening in cost cutting as few business people started making *muri* (puff rice) using machines. Again, if one business house agreed to buy embroidery and jute-based products, they are still turning down, claiming that business recessions are there though it was good during contact.

Under these circumstances, the author offers the following suggestions for the survival as well as improving the conditions of these entrepreneurs of this town.

Table 17.5 Correlation table of different variables

Categories	Table No.	Pearson correlation	Significant
Educational status versus income status	Table 17.2	0.365	0.000*
Educational status versus work categories of entrepreneurs	Table 17.3	0.808	0.000*
Work categories of the entrepreneurs versus income status	Table 17.4	0.295	0.000*

* = Significant

We cannot avoid the present volatile socioeconomic situation due to the pandemic, yet we have to move on with various strategies and changing strategies. First of all, there should be a thorough campaign by the Government involving local self-government, academic institutions, and reputed NGOs to inspire and promote small, rural as well as women entrepreneurship. There is no doubt that this should be enabled to improve the socio-economic scenario of the backward people of Silchar as well as the Cachar district of Assam. Second, government offices should be more flexible in making raw materials available, processing loans, subsidies, and training. Lastly, the Government must take initiatives to involve all the stake holders, that is, business houses, NGOs, academic institutions, and other government bodies, say Nehru Yava Kendra (NYKs) for the promotion of marketing of the products adding social values to it (by social marketing strategies).

References

Barik A, Tripathi S (2022) Sustainability of micro, small and medium enterprises in India during covid-19. Parikalpana KIIT J Manag 18(I):71–84

Das TK, Bhattacharyya R, Sarma PK (2022) Revisiting geographies of nationalism and national identity in Bangladesh. Geo J 87:1099–1120. https://doi.org/10.1007/s10708-020-10305-1

Desai V (2011) The dynamics of entrepreneurial development and management. Himalaya Publishing House, Mumbai

Drucker P (1985) Innovation and entrepreneurship. Harper and Row, New York

Dutta S (2018) Women entrepreneurship and marketing strategies of Indian rural products: a study on Barak Valley of Assam on jute diversified products. In: Shaw T, Malhotra AK, Chowdhury N (eds) Competitive strategies in emerging marketing. Mc Graw Hill, New Delhi, pp 215–227

Dutta S, Chan C (2013) Women empowerment through Entrepreneurship in Manipur. In: Chattopadhyay PK (eds) Inclusion and empowerment. N.D.Publishers, New Delhi, pp 51–60

Guilder G (1984) The spirit of enterprise. Simon and Schuster, New York

Henderson JV (2010) Cities and development. J Reg Sci https://doi.org/10.1111/j.1467-9787.2009.00636.x

Khan MA (2010) Effects of Human Resource Management practices on organizational performance: An empirical study of oil and gas industry in Pakistan. Eur J Econ Finance Adm Sci 24:157–175

Khanka SS (2007) Entrepreneurial development. S Chand & Company, New Delhi

Lalitha DR (1996) Women entrepreneur. A.P.H. Publication, New Delhi

Lalngaihsaki R, Lalnunthara R (2021) Impact of covid-19 crisis on micro-enterprises: a study of Lunglei Mizoram. Senhri J Multidiscip Stud 6(1):27–35

McClelland DC (1961) The achieving society. N.Y. Van Nostrand Co. New York

Mohanty SK (2005) Fundamental of entrepreneurship. Prentice Hall India Learning Private Limited, New Delhi

MSME (2020) Covid 19 Belief for MSME Sector (Report). https://msme.gov.in/sites/default/files/ActiononCOVID-19RelieforMSMESector.pdf

Nandan S (2013) Fundamental of entrepreneurship. Prentice Hall India Learning Private Limited, New Delhi

Ramaswamy KV (2020) Impact of COVID-19: micro, small and medium enterprises in India, pandemic shock of COVID-19 and policy response: a bird's eye view. IGIDR, Mumbai

Shafi M et al (2020) Impact of COVID-19 pandemic on micro, small and mediumsized enterprises operating in Pakistan. Res Glob 2(1):12–19

Sharma S (2017) Entrepreneurship development. PHI Learning Pvt. Ltd, New Delhi

Siwan M (2012) Women entrepreneurship. Lambert Academic Publishing, Germany
Soni S (2021) Covid horror: thousands of micro, small businesses in India are on sale right now. Financial Express (The Daily), September 28, 2021
Sowmya DN (2020) Repercussion of COVID-19 on MSMES—an Indian perspective. Seshadripuram J Soc Sci 2(2):19–25
Suresh J (2012) Entrepreneural development. Margham Publications, New Delhi
Swarajyalakshmi C (1998) Development of women entrepreneurship in India: problems and prospects. Discovery Publishing House, New Delhi
The Times of India (Daily) (2021) Pandemic has impacted 82% of small businesses: survey (report), PTI. April 21, 2022
Vasanthagopa R (2008) Women entrepreneurship in India. New Century Publications, New Delhi
Vinze MD (1987) Women entrepreneurs in India. Mittal Publications, New Delhi
World Trade Organization (2019) Natural disasters and trade: study I. World Trade Organization, Geneva
http://www.niti.gov.in/
https://gad.assam.gov.in/information-services/divisions
https://poonawallafincorp.com/blogs/impact-of-covid19-on-small-buisnesses-in-india.php

Chapter 18
Bolpur: A Historical Eco-friendly Townscape Promoting Sustainable City

Ankabehari Saha, Syeda Fahima Shahnaz Sultana, Payel Saha, and Madhushree Das

Abstract Bolpur, a historical town of Birbhum, West Bengal, existed during the mid-nineteenth century. The town quickly became prominent after establishing Visva Bharati, a university instituted by Rabindranath Tagore. Tagore was an environmentalist. As such, he stressed upon greenery around his beloved Santiniketan and Visva Bharati. Today's ideas of 'Smart City' and 'Sustainable City' also revolve around the basic premise that the environment within the cityscape should be preserved so that future generations can also enjoy it. Bolpur, since its inception, has changed according to human demands. Thus, the topic at hand tries to ascertain whether this historical town has retained its environment-friendly face. The idea of sustainability has also evolved with time. Thus, the topic also tries to determine whether Bolpur is considered sustainable in the modern idea of sustainability. The book chapter is divided into two sections. The first delves into the development of Bolpur since its inception. Though concrete data is not available in the early stages of its development, literature works of various renowned personalities can portray a solid account of the situation. It also discusses the evolution of the idea 'Sustainable City'. The second section of the chapter deals with the performance of Bolpur with various measures of the modern-day idea of a sustainable city. It examines how Bolpur stands up to the evolving idea of a 'Sustainable City'. Various indicators are calculated with the help of SPSS as well as mapped with ArcGIS. The study may help third-world cities in realizing the innate potential of achieving sustainability without dramatically altering their cityscape.

A. Saha (✉)
Department of Geography, Visva Bharati, Santiniketan, West Bengal, India
e-mail: ankabeharisaha@gmail.com

S. F. S. Sultana · P. Saha · M. Das
Department of Geography, Gauhati University, Guwahati, Assam, India
e-mail: shahnaz014@gauhati.ac.in

P. Saha
e-mail: payelsaha@gauhati.ac.in

M. Das
e-mail: madhushreedas@gauhati.ac.in

S. Anand et al. (eds.), *Sustainability in South Asian Cities*, Advances in Geographical and Environmental Sciences, https://doi.org/10.1007/978-981-97-7455-5_18

Keywords Sustainable city · Bolpur · Development · Smart city indicators · Environment

18.1 Introduction

A town emerged from the rural landscapes when on June 12th, 1950, the Bolpur municipality was established in the Birbhum district of the eastern state of West Bengal of independent India under the provision of the West Bengal Municipal Act. Local Self Government was already introduced in the Birbhum district during the Colonial rule. Priorly to being a small village under Supur Pargana, Bolpur was a region of less significance until a railway station was established by the East India Company in 1860, thereby enhancing the economy and attracting people. Gradually, court, police station, sub-registry office, etc. were established and Christian missionaries too came to the region and founded a church which was known as the "Mission compound" (Sen 2020). However, the significance of Bolpur increased manifold when Maharshi Debendranath Tagore established Shantiniketan, "the abode of peace". Captivated by the serene beauty of the landscape while taking a boat ride in the western-most corner of Bengal, Birbhum, in the year 1862, Maharshi Debendranath Tagore returned to open a spiritual center at the same landscape and established Shantiniketan in 1863 to connect people of diverse caste, creed and religion to spiritualism. Later, his son, the world-renowned literary face and the first Indian Nobel Laureate, Gurudev Rabindranath Tagore, founded the Brahmo Vidyalaya in 1888 and in 1901, the first open-air school, which was later expanded to form the famous Visva Bharati University in the year 1921. With Shantiniketan and Visva Bharati University in its vicinity, Bolpur grew as Shantiniketan grew (Pal 2016).

The etymology of Bolpur has its roots in the folklore of Raja Surath and his sacrificial offerings of one lakh goats to express his gratitude to the Goddess Bhavani, who helped him regain the capital of Supur. "Bolpur" thus is derived from the two words— "Boli" which in Bengali means "slaughter" and "Pur" representing "city" (Sen 2020). Famous for being an auxiliary town to Shantiniketan and Visva Bharati and being related to the Nobel Laureates, Rabindranath Tagore and Prof. Dr. Amartya Sen, Bolpur has grown as a town of eminence, attracting tourists from places far and wide (Garai et al. 2020). As an important tourist centre, urbanization in Bolpur municipality has been rapid in the past few decades. The idea that was core to Maharshi Debendranath Tagore in establishing the 'abode of peace' amidst the serene landscapes of greenery, and the changing Bolpur prompted by the surge in urbanization and the demands of time, are in sharp contrast with each other. According to the status of a 'City' in 2019, Bolpur is being developed sustainably by the state government (Ghosal 2016).

The concept of sustainable development of cities is not new. One of the earlier references of the term 'sustainable city' can be traced in Agenda 21 of the United Nations Conference on Environment and Development 1992, which amongst others, aimed at ensuring sustainable management of all urban settlements by including into the framework of constituting a sustainable city the factors—environmental, social, and economic (Koh et al. 2010). The Brundtland Report of 1987 highlighted the urban challenge and been the thrust of the sustainability concepts.

Cities drive a nation towards economic prosperity and are considered the 'engine of growth' (Lundqvist 2007). A city space is a place for the assimilation of people and cultures, the emergence of ideas and ideologies, innovation and creation, economic progress, human resource development and human advancement. A city thus reflects a nation. But not all glitter is gold. Cities or any urban space being the centres of economic drive, cultural and intellectual expression, and enhanced facilities in the various social sectors of growth pull migrants from regions within and across borders, thereby further accelerating urbanization. The process of rapid urbanization where on the one hand, facilitates the improvement of the quality of life of people and societies; on the other, too, deteriorates the same (Murali et al. 2018). The adverse impact of rapid and unplanned urbanization on the environment is immense, affecting the people and the growth process. These urbanization adversities induced the need for sustainable development and management of human settlements and the necessity of developing a 'sustainable city'. Goal 11 of the Sustainable Development Goals (SDGs), intended to be achieved by 2030, focuses on making cities and human settlements inclusive, safe, resilient and sustainable.

As Bolpur is an emerging city in the state of West Bengal, there arises the need to comprehend the status of Bolpur in making it a 'sustainable city' by considering the different components that are integral to the evolution of a city in a sustainable manner. This chapter is therefore focused on assessing the different dimensions of sustainability and the criteria which are requisite in aligning the development of Bolpur with the SDGs and in comprehending the present-day status of Bolpur as a 'sustainable city'.

18.2 Literature Review

18.2.1 Urbanization and the Need for Sustainable Development

Cities, towns and other urban spaces have grown in numbers and areas over the years. This increase is facilitated by the process of urbanization. McGranahan and Satterthwaite (2014) have loosely referred "urbanization" as the process of the increase in the proportion of people living in the urban spaces of a country which occurs mainly as a result of moving people to the same from the rural areas. This rural-to-urban transition involves population, land use, economic activity and culture. As of 2020, more than half of the world's population (56.2%) live in cities (Buchholz 2020). Asia has the highest percentage share of the urban population, accounting for 53.9%, which in 1960 was only 35.2% (Satterthwaite 2020). Our country, India, is one of the fastest-growing economies in the world, with its growth being propelled by its cities (Kapoor and Sinha 2020). The percentage of urban population to total population as per the Census of India 2011 was 31.6% which in 2001 was 27.81% and has increased as of now. Though the process of urbanization is strongly associated with economic success as it fosters economic growth and development as well as social development (Satterthwaite 2008), urbanization has adverse effects too. Associated with urbanization are environmental, social and human health threats. It affects not only the lives and environmental components within the urban regions but also the surroundings through continuous urban consumption and sprawl. The rapid urbanization across the world accompanied by urban growth and urban sprawl, and increasing urban consumption has further accelerated the pressure on resources and the environment, and this resulted in the inclusion of urban challenge in the Brundtland Commission Report of 1987 and the concept of urban settlement management in Agenda 21 for the sustainable development of urban areas.

18.2.2 Sustainable Development of Cities

18.2.2.1 Brundtland Commission's Report of 1987

The Brundtland Commission's report (1987), officially known as the UN Report *Our Common Future* by the World Commission on Environment and Development (WCED), which was headed by Gro Harlem Brundtland, the former Norwegian Prime Minister, enabled discussions and policy-making on sustainability and sustainable development at the global level with both, the developed and developing nations on board (Shaw 2018). The report made the first attempt to define sustainable development as meeting the needs of the present without compromising the ability of future generations to meet their own needs.

The Brundtland report under the 'Common Challenges' section discussed the urban challenge wherein the growth of cities, the crisis in the third world cities, the industrial cities, the factors getting affected and affecting urbanization such as services and facilities, resources, pollution, environment, society and health; the urban challenges in developing cities, the strategies, the policies and the internal and external cooperation requirements are delved into. The report also recognized the linkages between the environmental stresses and patterns of economic development on the one hand and the other, between environmental and economic problems and social and political factors. Thus economic, ecological and social factors are equally significant for improving the population condition and situation, and together they constitute sustainable development (Shaw 2018). Satterthwaite, in his presentation for the UN Population Division highlighted the strong development component which is to meet the people's needs and the strong environment component which is to not compromise the ability of the future generations to meet their needs in the definition of sustainable development by the Brundtland Commission.

Shaw (2018) has mentioned the critical outcome of the Brundtland Report's definition of sustainability to be the development of urban sustainability assessment based on economic, social and environmental dimensions. The urban sustainability assessment has influenced the choice of city development objectives and the criteria for evaluation of the same. For example, Newman (as cited in Shaw 2018) used eight criteria from social, economic and environmental areas in the city of Sydney's Plan of 2005.

The use of indicators or criteria for assessing the varied dimensions—social, economic, governance and environmental, varies from one city to another as each city differs from the other and so does the concept of sustainability with every city. Satterthwaite (as cited in Shaw 2018) has suggested

> prioritising on the basis of the most basic and foundational needs and then moving on to other goals. In his framework for assessing the environmental sustainability of cities, the provision of adequate water, sanitation, drainage and garbage collection come first followed by the goal to reduce chemical and physical hazards within the home, workplace and wider city. Thereafter the goal of achieving a high quality city environment for all city inhabitants becomes important followed by the goals to minimize the transfer of environment costs to inhabitants and ecosystems surrounding the city, and to minimize the transfer of environment costs to people and ecosystems beyond the city region and the future (Shaw 2018, 25).

Undoubtedly the concepts of sustainability and sustainable development as discussed by the Brundtland Report, led to the endorsement of the concepts of 'sustainable human settlement development' and 'sustainable city' in several global forums that followed (Shaw 2018).

18.2.2.2 Agenda 21

The alarmingly increasing share of the urban population and the attributed negative impacts of urbanization led to the incorporation of promoting sustainable human settlement development and the introduction of a 'sustainable city' under the Social and Economic Dimensions of Agenda 21, the non-binding action plan of the United Nations for sustainable development and the product of the Earth Summit (UN Conference on Environment and Development) held in Rio de Janeiro, Brazil, in 1992.

Chapter 7 of Agenda 21 is concerned with promoting sustainable human settlement development with program areas focusing on:

a. Providing adequate shelter for all;
b. Improving human settlement management;
c. Promoting sustainable land-use planning and management;
d. Promoting the integrated provision of environmental infrastructure: water, sanitation, drainage and solid-waste management;
e. Promoting sustainable energy and transport systems in human settlements;
f. Promoting human settlement planning and management in disaster-prone areas;
g. Promoting sustainable construction industry activities;
h. Promoting human resource development and capacity-building for human settlement development.

Each program area is defined by the basis for action and has an objective which is to be attained by activities and means of implementation. While the activities build up the framework for each objective, the means of implementation help accomplish the activities. Agenda 21 sought global cooperation and assistance in achieving sustainable human settlement development, and also human resource development, sound policies and the increased role of local authorities in pursuing sustainable development methods and ways.

18.2.2.3 UN-Habitat

The UN-Habitat, United Nations Human Settlements Program, is the United Nations program for human settlement and sustainable urban development. The program promotes towns and cities to be socially and environmentally sustainable and intends to provide adequate shelter for all. With the theme being "Innovation for Better Quality of Life in Cities and Communities" for the UN-Habitat Assembly, UN-Habitat works towards a better urban future with an integral and comprehensive approach to the sustainable development of human settlements and urban spaces (Garg n.d.)

18.2.2.4 2030 Agenda

The 2030 Agenda, officially known as *Transforming our world: the 2030 Agenda for Sustainable Development*, is an action plan for people, planet and prosperity, adopted by the United Nations General Assembly on the 25th of September, 2015, as a post-2015 development agenda at the global level. With a collection of 17 intertwined Sustainable Development Goals having 169 targets and numerous indicators, the 2030 Agenda intends to achieve sustainable development in its three dimensions—economic, environmental and social. The Sustainable Development Goals (SDGs) act as a blueprint to achieve a better and sustainable future for all (Table 18.1).

Of the 17 goals, Goal 11 is to "make cities and human settlements inclusive, safe, resilient and sustainable". Goal 11 is to be achieved through 10 targets, with each having one or more indicators. As these goals are not separate but related, and the urban issues have a converging nature, Goal 11 is also facilitated by the SDGs 1, 6, 7, 8, 9, 12, 15 and 17, among others (Sustainable Development Goals Knowledge Platform n.d.). The 2030 Agenda calls on for an inclusive "bottom-up" as well as horizontal approaches to address the issues through the goals and targets.

Pollock (2019) has discussed the lecturer's view to pedagogic methodology in learning to understand sustainability in cities, where while focusing on sustainability studies and the SDGs, the light was also thrown upon sustainable construction, sustainable transportation, water management and rest. The problems of the cities were to be discussed in brief. Pollock advocated the need for sustainability studies at the global level and pedagogic approaches.

Table 18.1 Sustainable development goals

Goal 1	End poverty in all its forms everywhere
Goal 2	End hunger, achieve food security and improved nutrition and promote sustainable agriculture
Goal 3	Ensure healthy lives and promote well-being for all at all ages
Goal 4	Ensure inclusive and equitable quality education and promote lifelong learning opportunities for all
Goal 5	Achieve gender equality and empower all women and girls
Goal 6	Ensure availability and sustainable management of water and sanitation for all
Goal 7	Ensure access to affordable, reliable, sustainable and modern energy for all
Goal 8	Promote sustained, inclusive and sustainable economic growth, full and productive employment and decent work for all
Goal 9	Build resilient infrastructure, promote inclusive and sustainable industrialisation and foster innovation
Goal 10	Reduce inequality within and among countries
Goal 11	Make cities and human settlements inclusive, safe, resilient and sustainable
Goal 12	Ensure sustainable consumption and production patterns
Goal 13	Take urgent action to combat climate change and its impacts
Goal 14	Conserve and sustainably use the oceans, seas and marine resources for sustainable development
Goal 15	Protect, restore and promote sustainable use of terrestrial ecosystems, sustainably manage forests, combat desertification, and halt and reverse land degradation and halt biodiversity loss
Goal 16	Promote peaceful and inclusive societies for sustainable development, provide access to justice for all and build effective, accountable and inclusive institutions at all levels
Goal 17	Strengthen the means of implementation and revitalise the Global Partnership for Sustainable Development

Source Transforming our world: the 2030 Agenda for Sustainable Development, 2015

Sustainable development of urban spaces is, thus, an integrated approach with the varied dimensions being addressed and the concept has emerged with the dawn of rapid urbanization.

18.3 Materials and Methods

18.3.1 Location

The present study is concentrated over the Bolpur municipality, which extends between latitudes 23.6431° N and 23.69870° N and longitudes 87.71323° E to 87.77294° E and covers an area of 13.13 km^2 in the south-east part of Birbhum district of West Bengal (Sen 2020). As per the 2011 Census of India, Bolpur being the home to a population of 80,210, has 40,468 (50%) males and 39,742 (50%) females, and with a density of around 6000 persons per km^2, the Bolpur municipality is the most populous municipality of the Birbhum district. The Bolpur municipality currently consists of 20 wards; the number of wards of the municipality increased from 9 to 14 in 1963, and in 2002 it increased to 18 (Sen 2020). The addition of two more wards, though it happened recently, the study considers only 18 wards of the Bolpur municipality for most of the analysis (Fig. 18.1).

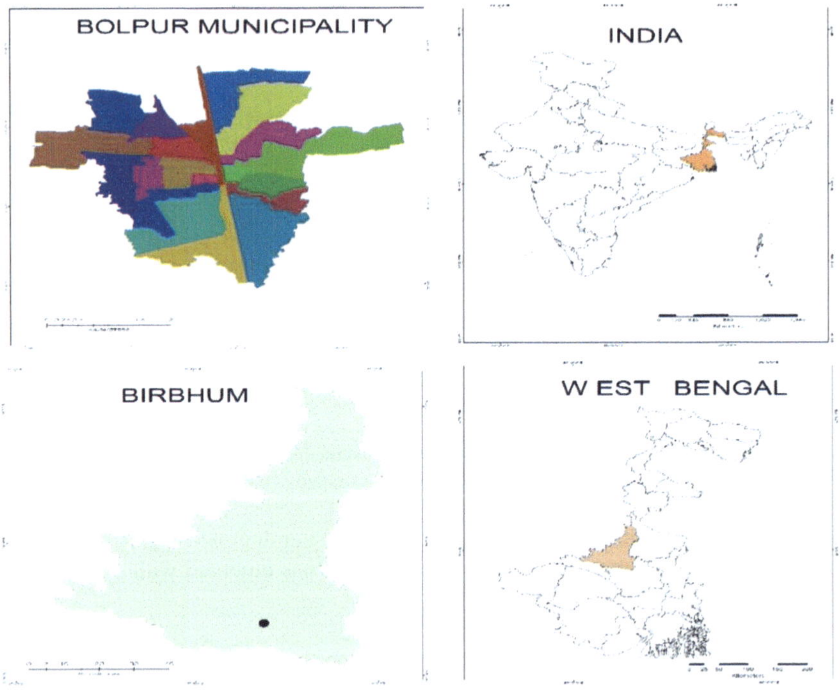

Fig. 18.1 Location map of the study area

Table 18.2 Data type and source

Data name	Type	Source
Field data	Primary	Personal interview
Bolpur municipality data	Secondary	S.E.S report 2008
Census data	Secondary	DCHB Part B, Census of India

18.3.2 Data Sources

The chapter used the Socio-Economic Survey and Analysis Report 2008 of Bolpur municipality for assessing the different indicators of sustainable development as no recent data on the socio-economic survey of Bolpur municipality has been made available yet. The District Census Handbook Part B from Census of India 2011 has also been consulted. The data available being more than a decade old, and the ward numbers as the administrative boundaries designated to each ward being modified several times; the authors have modified the numbers for the wards of the current map to that of the old 2008 ward numbers. This led to another problem: the lack of data regarding the latest two wards, Ward 19 and 20. However, a perception study has been carried out on the facilities and services rendered for all the wards including Ward 19 and 20, through the conduction of personal interviews (Table 18.2).

18.3.3 Methodology

To assess the sustainability of Bolpur municipality, the various indicators (whichever is applicable) have been chosen from the literature review. Various services and facilities provided such as education, number of taxpayers, streetlight availability, sanitation, waste management and so on are identified as indicators for assessing the sustainability of Bolpur. The indicators are grouped under three different dimensions of sustainable development—economic, environmental, and social (UN Economic and Social Commission for Asia and the Pacific 2015), and are also aligned with the Sustainable Development Goals whatsoever being applicable. The data collected from the S.E.S Report 2008 and Census of India 2011 have been analysed and plotted ward-wise, and maps and diagrams have been prepared. QGIS 3.14 has been used to prepare the maps, and Microsoft Excel 2019, the diagrams. In addition, the authors conducted personal interviews randomly in each of the wards to analyse the citizens' responses regarding the performance of various services rendered, the question of awareness regarding a sustainable city being focal (Fig. 18.2).

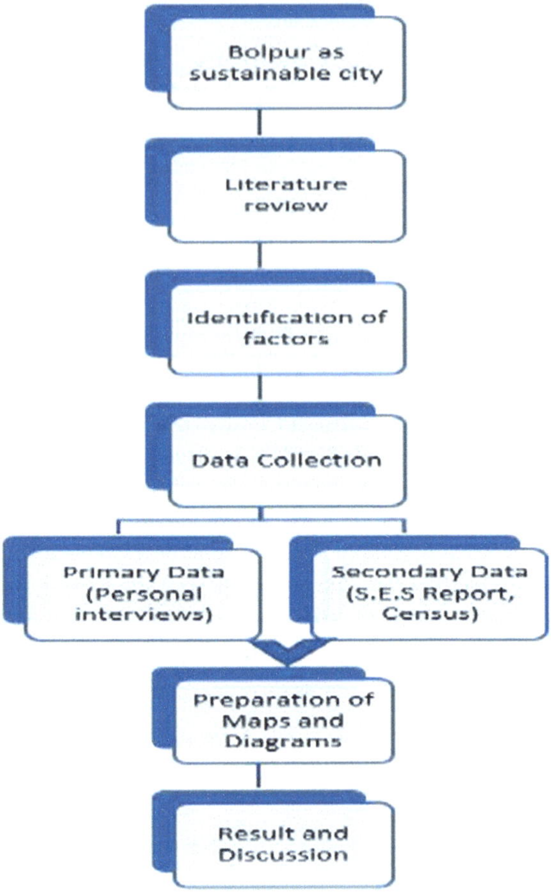

Fig. 18.2 Methodology of the study

18.4 Results and Discussion

As the concept of sustainability has become inseparable from the concept of development in recent decades, urban sustainability becomes core to the economic development of a country or state as the urban spaces push the economy towards progress and are also centres for human, social and environmental development (Satterthwaite 2008). Therefore, the urban sustainability of the city of Bolpur is assessed by taking into consideration the three different dimensions of sustainable development––economic, environmental and social, which are not separate but intertwined, with different indicators being evaluated to comprehend them. Moreover, the indicators of the three dimensions are also assessed with respect to the SDGs. This gives a better picture of Bolpur in pursuance of sustainability and development.

18.4.1 Economic Dimension

The economic dimension of sustainability focuses on achieving economic growth and economic development by taking into account the social and environmental consequences of economic activities and also in gauging the economic efficiency as well as technology (Team Prakati 2019; Globalization and Livelihood Options of People Living in Poverty [GLOPP] n.d.). For analyzing the economic dimension of sustainability for Bolpur, the indicators assessed are occupational status and the number of taxpayers. The engagement rate in the diverse occupations belonging to the various sectors of the economy and the number of taxpayers of the state showcase the efficacy of the economy.

18.4.1.1 Occupation Status

Occupational status gives an overview of the strength of the economy as the number of persons engaged in higher sectors of economic activities, the more advanced the economy is. As evaluated in Bolpur, occupational status is divided into six categories: business, service, professionals, daily rated, housework and others. Most workers are engaged as daily rated workers, followed by business and government service. House workers and other workers are meagre. Though the occupational status is assessed for the working age group, however for the age group 0–14 years in every ward, engagement is seen in occupational activities as daily rated, housework and others. This engagement of underaged children is mainly due to poverty and illiteracy (Fig. 18.3).

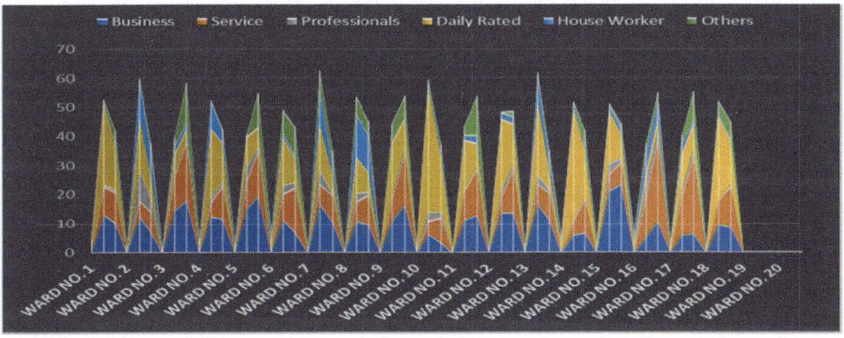

Fig. 18.3 Occupational status of the different wards

18.4.1.2 Tax Payers

Figure 18.4 indicates that more than 50% of people pay taxes in nearly every ward except for Ward no. 2, 10 and 15. The highest number of taxpayers, which accounts for 86.57% is found in ward no. 17, while the lowest (38.51%) is found in Ward no. 2. The more the tax-to-GDP ratio is in a state, the more government investments in the social development of the state (ET Online 2020).

18.4.2 Environmental Dimension

The environmental dimension of sustainable development addresses the concerns arising out of human needs and dependence on resources by optimizing the usage of the resources without undermining the ecosystem, the ecosystem services and the environmental components (GLOPP n.d.). For assessing the environmental dimension of Bolpur, the following indicators are taken into consideration.

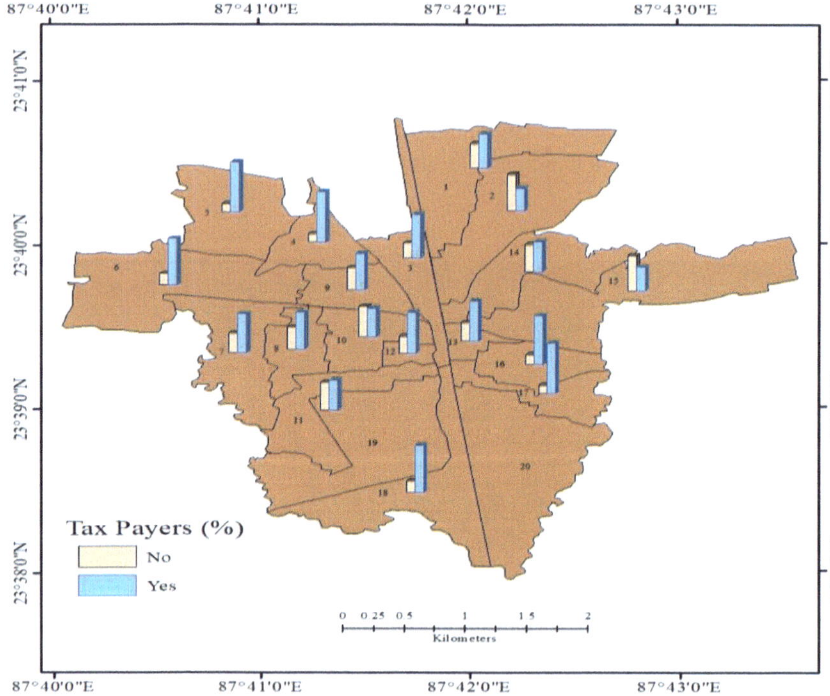

Fig. 18.4 Number of taxpayers in each ward

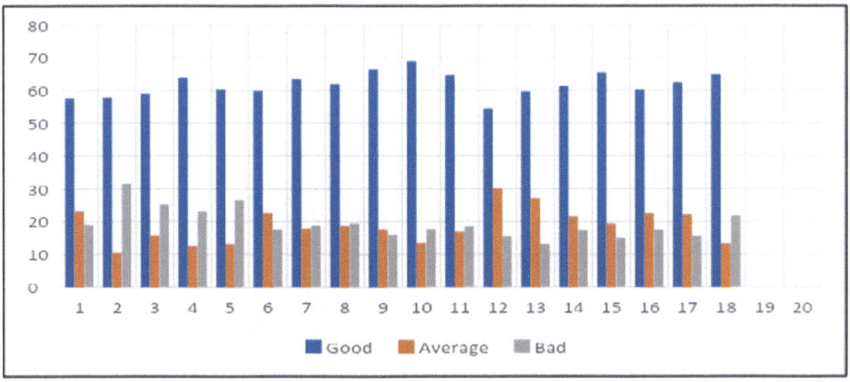

Fig. 18.5 Quality of water as marked by the Bolpur populace

18.4.2.1 Quality of Water

Access to safe water is a fundamental human right and is of utmost importance for maintaining quality of life, water being the most vital source for the existence of life on earth. 'Jalam Jeevamrutham' depicts the importance of water for survival. Water and health are interlinked. The quality of water determines and affects the health of the people, which proceeds to affect their well-being. In every ward, more than 50% of people are satisfied with the quality of accessible water. 65.52% of people in ward no. 15 marked the water quality as 'good', which is highest in the Bolpur municipality. 31.49% of ward no. 2 marked it bad while 30.09% of ward no. 12 marked it average (Figs. 18.5 and 18.6).

18.4.2.2 Waste Collection

Waste pollutes the components of the environment thus, making the proper collection and disposal of waste extremely significant to maintain the quality of the environmental components. However, waste collection is not regular in all the wards of Bolpur. 66.83% of people of Ward no. 16 agree about the regularity of waste collection while 61.18% of people of Ward no. 10 complained that waste materials are not collected from their homes. 84.08% of people in ward no. 17 disagree about the regularity of waste collection, which is quite high (Fig. 18.7).

18.4.2.3 Availability of Greenery

Greenery and maintenance of the same are integral to urban sustainability. Green spaces are considered one of the highly significant factors necessary for improving

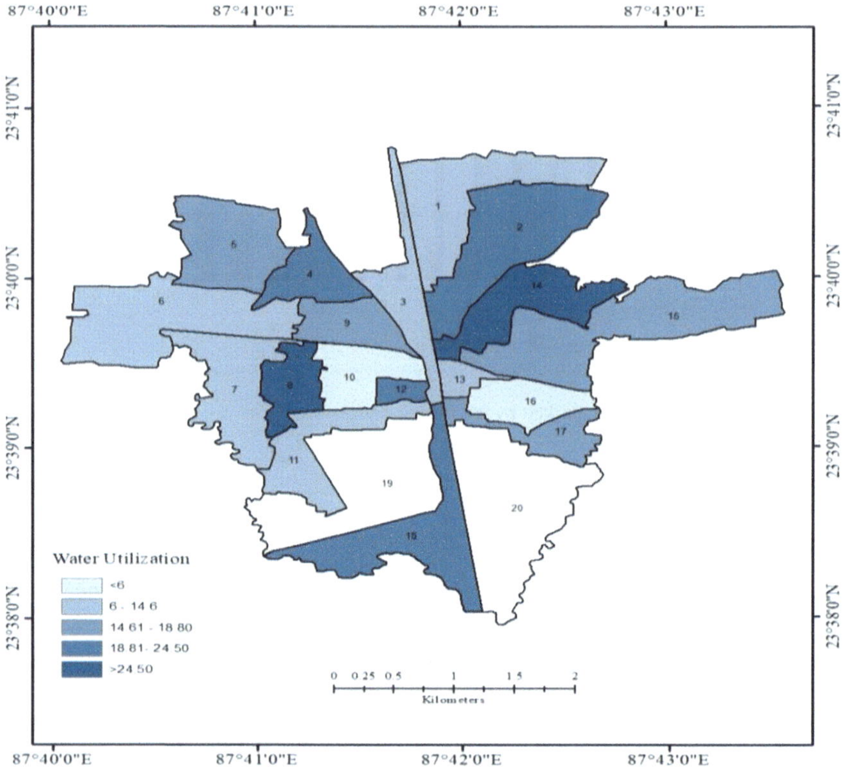

Fig. 18.6 Ward wise water utilisation in Bolpur municipality

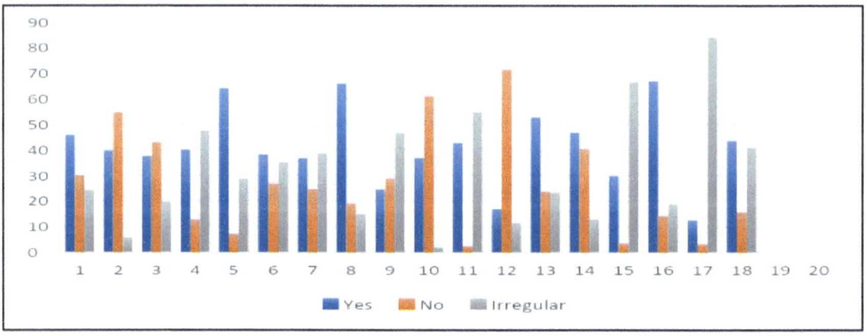

Fig. 18.7 Waste collection carried out in Bolpur municipality

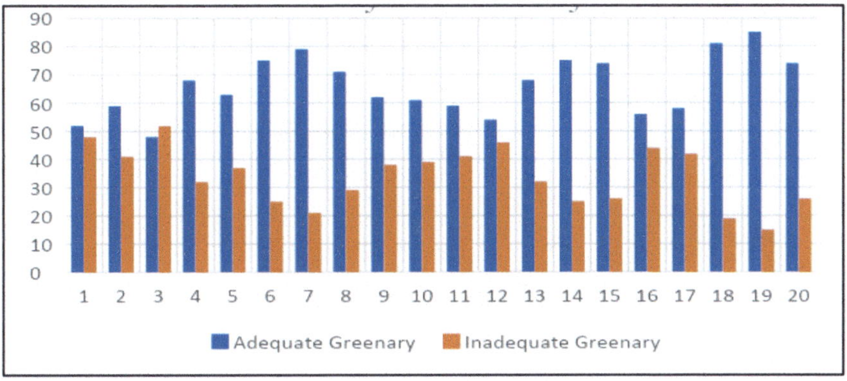

Fig. 18.8 Perception of availability of greenery amongst the Bolpur populace

sustainable urban development and are programmed, planned and managed for multi-dimensional purposes. They play ecological, recreational, cultural, aesthetical and other roles compatible with sustainable urban development (Karade et al. 2017). Wards 3, 4, 9, 14 and 18 each have one publicly managed green space, while ward no. 8 has two. Though official data is not available for overall greenery in the Bolpur municipality proper, residents were asked about their perception of greenery being adequate or inadequate. Figure 18.8 shows that the majority of the residents in each ward except ward 3 have given the impression of the availability of sufficient greenery in the municipal wards. Historically, Bolpur was flushed green after the inception of Visva Bharati, and clearly, the tradition continued.

18.4.3 Social Dimension

The social dimension is focused on achieving social sustainability in the process of urban development. A socially sustainable society, city or state is where the inhabitants have equal access to all facilities and amenities, equal rights, and equal say in decision-making. Social sustainability also involves improving the quality of life and fostering human resource development (GLOPP n.d.). Education, housing, sanitation, public amenities such as transportation, street light, water supply, government policies and satisfaction of services rendered are used as indicators to assess the social dimension of sustainable urban development.

18.4.3.1 Education

Education forms the base of human resource development, bringing about social sustainability by increasing accessibility to opportunities and upgrading the quality

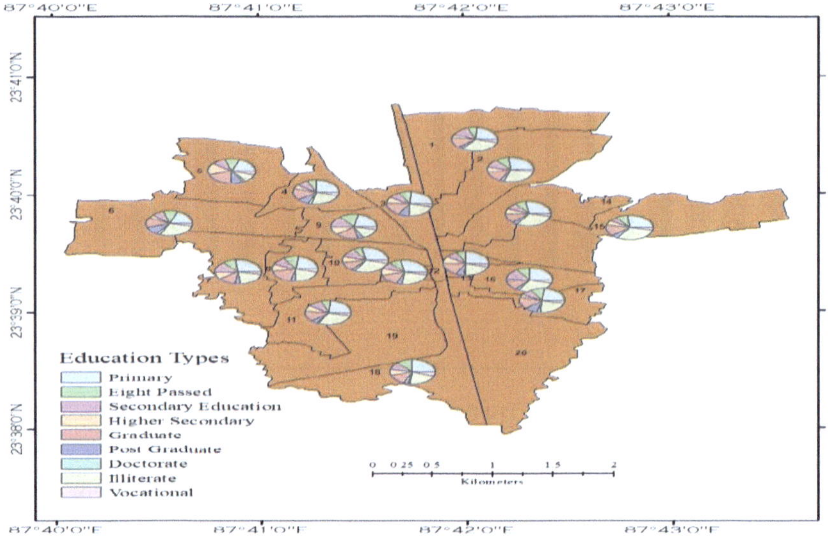

Fig. 18.9 Enrolment at various educational levels of the Bolpur municipality

of life. Figure 18.9 depicts the education scenario of the inhabitants of Bolpur municipality. To better understand the education system, the education levels have been classified into 9 primary categories, eight passed, secondary, higher secondary, graduate, postgraduate, doctorate, vocational and illiterate. The highest enrolment into primary education is seen in ward no. 12 whereas the lowest is seen in ward number 5. The number of doctorates is very less all over Bolpur. Maximum number of illiterate people is observed in ward no. 15, which accounts for 30.61% of the total population of this ward, and the reasons may be poor infrastructure, poverty, obsolete curriculum, superstitions, etc., while the minimum number of illiterates which is 10.02% of the total population of the ward is observed in ward no. 5. Maximum number of undergraduates and postgraduates are observed in ward no. 5 and 6, whereas, the minimum number of undergraduates and postgraduates are observed in ward no. 14 and 16, respectively.

18.4.3.2 School Dropouts

The number of school dropouts is inversely related to social sustainability. The more dropouts in an area, the lesser the human resource development and the lower the quality of life in that area. For Bolpur, the maximum number of dropouts is found in ward no. 2, and the minimum number of school dropouts is found in ward no. 9. Poverty, illiteracy and lack of immediate earning opportunities have spurned many families from formal education. This further plunged them into the depths of poverty, and the vicious cycle thus continues (Fig. 18.10).

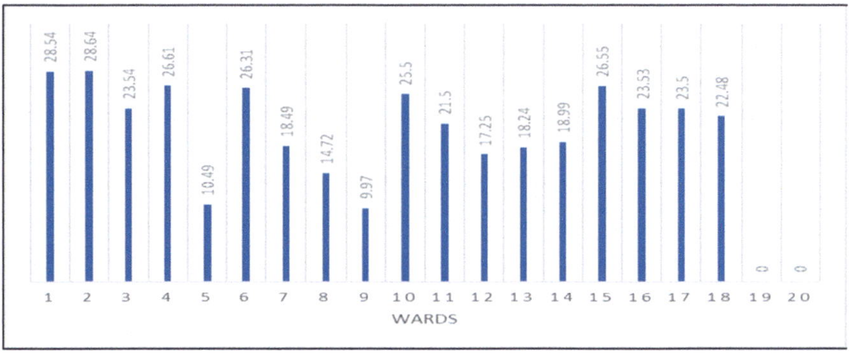

Fig. 18.10 School dropouts in the various wards of Bolpur municipality

18.4.3.3 Household Ownership and Slum Proportion

The household ownership type, whether rented or owned, reflects the socio-economic status of the area. It also depicts the population composition of the area, indicating the presence of migrants and residents in the area. The study found that 95.58% of people of ward no. 5 reside in self-owned houses, and 11.35% of people in ward no. 16 live in rented houses. 80.61% of ward no. 13 live in legally owned houses, while 2.69% of ward no. 17 reside in rented homes mainly due to occupation (Fig. 18.11).

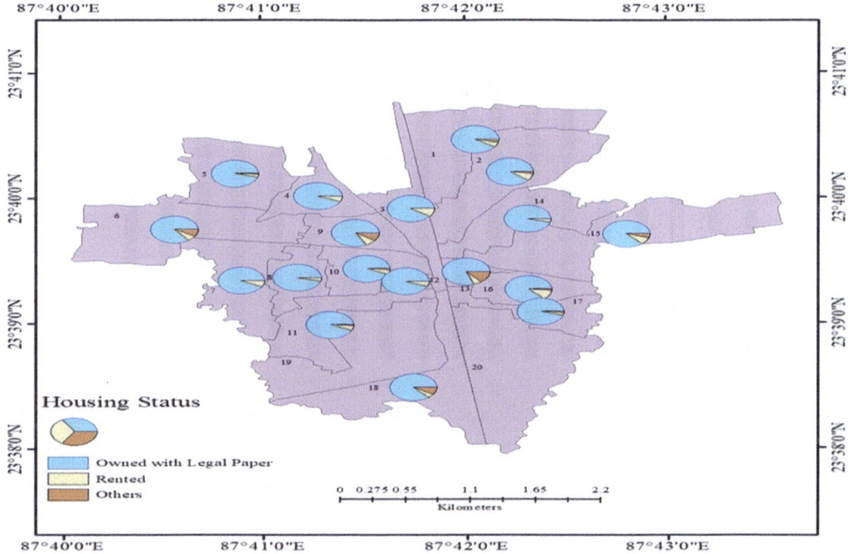

Fig. 18.11 House ownership type in Bolpur municipality

Moreover, the slum population is also seen in certain wards. The highest percentage of slum dwellers as against the total population for a ward is found in ward no. 1 where the percentage of slum population is 15%, which is followed by ward 20 (14%). Wards 2, 3 and 19 have 13% slum dwellers as against the total population whereas it is 125 for Wards 10 and 13 and 11% for Wards 11 and 14. In certain wards (4, 5, 6, 7, 8, 15 and 16), the slum population is not evident. The ward 12 has the lowest percentage of slum dwellers to the total population (6%).

18.4.3.4 Toilet Facility

The presence of a proper sanitation facility is a necessity. A toilet facility is one of the basic requirements to ensure a safe, hygienic and healthy environment for living, and its lack is a prominent threat to the health of the people and the environment. Open defecation, a noxious practice, makes people susceptible to diseases and life-threatening infections (Koner 2018). From Fig. 18.12, it is evident that more than 70% people of Bolpur have access to toilet facilities. The highest number of people, which accounts for 98.2% of the total population of ward no. 5 are accessing proper toilet facilities, the reasons being the presence of a more educated population, awareness among people about their health, availability of infrastructural facilities, etc. On the contrary, the lowest availability of toilet facilities is marked in ward no. 1 wherein only 70.27% of the total population gets access. This may be due to the prevalence of the age-old practice of open defecation, lack of awareness and infrastructure, etc.

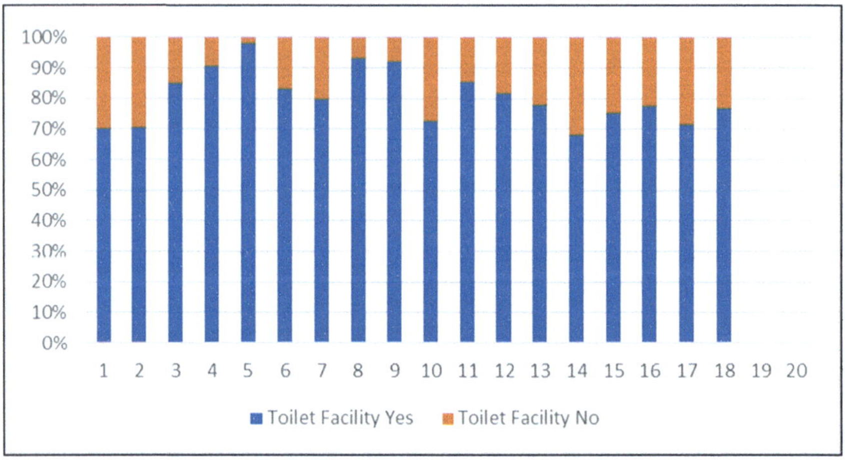

Fig. 18.12 Status of the presence of toilet facility per ward of Bolpur municipality

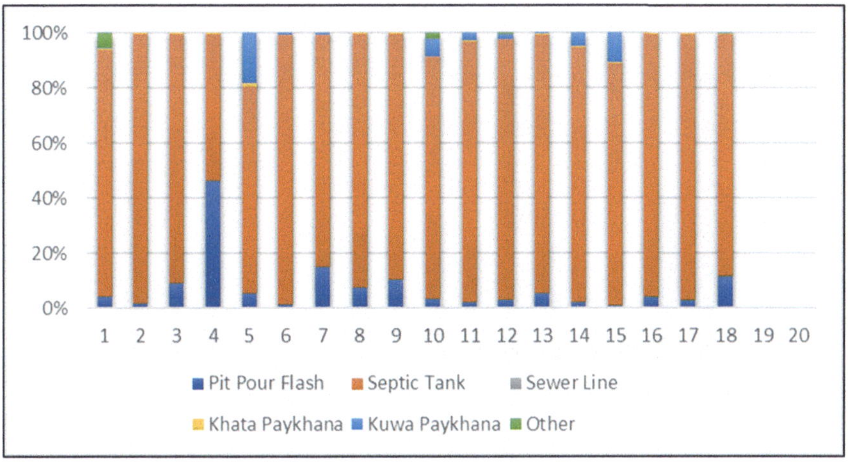

Fig. 18.13 Toilet types found in Bolpur municipality

18.4.3.5 Toilet Type

Not only is the availability of toilet facilities essential, but the type of toilet available is equally important. Toilet use is essential for survival. Open defecation causes several diseases like diarrhea, cholera, typhoid fever, etc. Therefore, from the public health perspective, proper toilet type and sewage facility must be used to flush down and dispose of wastewater contaminated with feces or urine, bacteria and germs. Otherwise, it affects human lives and the environment. In the study area, the toilet type is divided into six categories—Pit Pour Flash, Septic Tank, Sewer Line, Khata Paykhana, Kuwa Paykhana and others. No person is found to be using a sewer line here. Most people use septic tanks because they are environment-friendly, economical, low maintenance, and most importantly, they remove any bacteria before the water is released into the ground or soil, followed by pit pour flash because of its cleaning ease and less fear of falling into a pit. Very few people are using kuwa paykhana. The highest number of kuwa paykhana users are found in ward no. 5. Khata paykhana users are less all over Bolpur municipality and less than 1%. Maximum septic tank users are found in ward no. 2 (Fig. 18.13).

18.4.3.6 Community Toilet

The prevalence of community toilets or public toilets in an area indicates social health facilitation. The highest concentration of community toilets can be seen in Ward 3 followed by Wards 8, 17 and 18. On the other hand, the percentage is significantly lower in case of Wards 14, 15, 2, 5 and 6. Community toilets are prevalent in wards where markets, railway station, administrative buildings and sub-divisional court are located. Otherwise, the number is relatively low as Fig. 18.14 suggests.

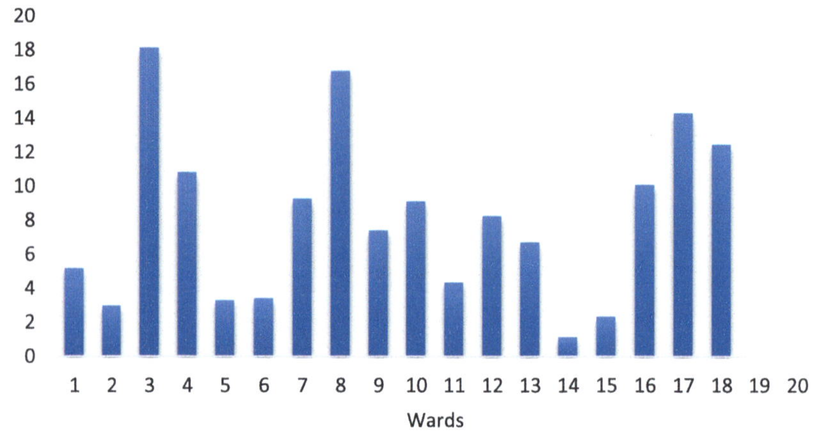

Fig. 18.14 Presence of community toilets in different wards of Bolpur municipality

18.4.3.7 Water Supply

The prevalence of a proper water supply system is a necessary civic amenity and important is too the source of supply. As the quality of water matters for sustainable development so does availability and accessibility. The water supply sources in the Bolpur municipality vary from municipal pipelines to municipal street taps to the municipal tube well, own tube well and other such sources. Water supply through the municipal pipeline is highest (74.86%) in Ward no. 5, covering only 18.34% of Ward 10. Through municipal street taps, the water supply covers 39.76% of the population for Ward 10, which is the highest and only 5.80% for Ward 5, which is the lowest. In the case of the municipal tube well, the water supply covers the maximum percentage of the population (39.06%) for Ward 13 while it is lowest (10.77%) for Ward 5. The water source as own setup tube wells are maximum for Ward 15 (3.64%) and minimum for Ward 8 (0%). The maximum percentage of people responding to the water supply to be sufficient occurs in Ward 5 and accounts for 86.88% of the total population of the ward, whereas the maximum percentage of people against the total population finding the water supply to be insufficient occurs in Ward 2 (43.73%) (Fig. 18.15).

18.4.3.8 Street Light

The availability of efficient street lighting facilities is another important civic amenity provided by the authorities. Figure 18.16 suggests that the maximum number of respondents have answered the availability of street lights as 'good'. Only in Ward no. 4 and 16 100% of street lights is available; in Ward number 14, street light availability is very weak, i.e., 28.86% due to poor maintenance.

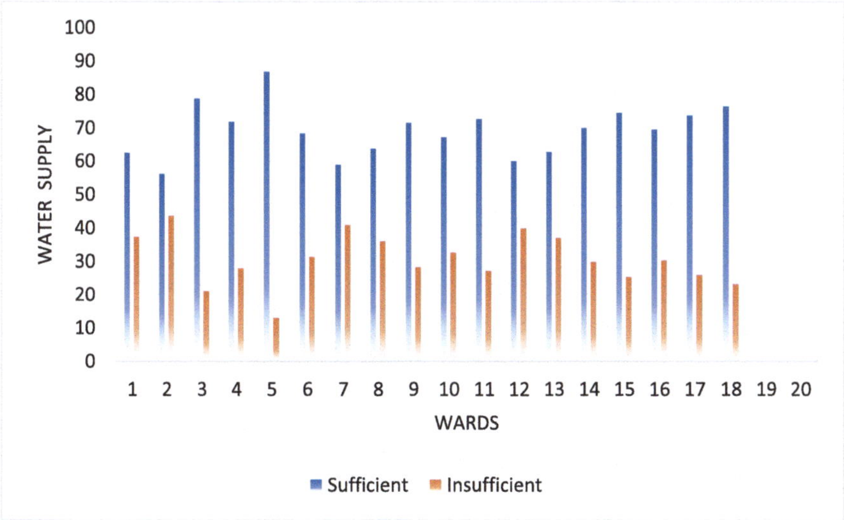

Fig. 18.15 Ward wise water supply in Bolpur municipality

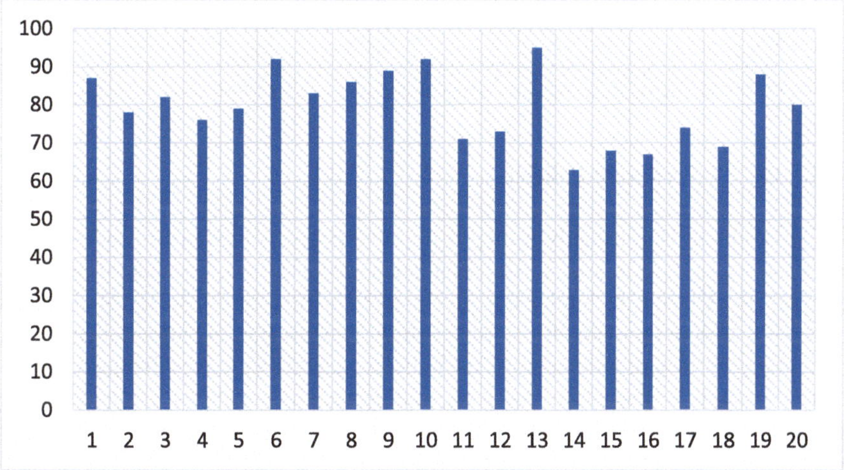

Fig. 18.16 Street light facilities in the different wards of Bolpur municipality

18.4.3.9 Public Transportation

Providing convenient accessibility to the public transportation system is a must on the part of the authorities. In the perception study carried out in all the 20 wards of Bolpur municipality, the perception rating of accessibility of public transportation in each ward is over 60%. For Ward 13, it is the highest among all wards (95%), while it is lowest in the case of Ward 14 (Fig. 18.17).

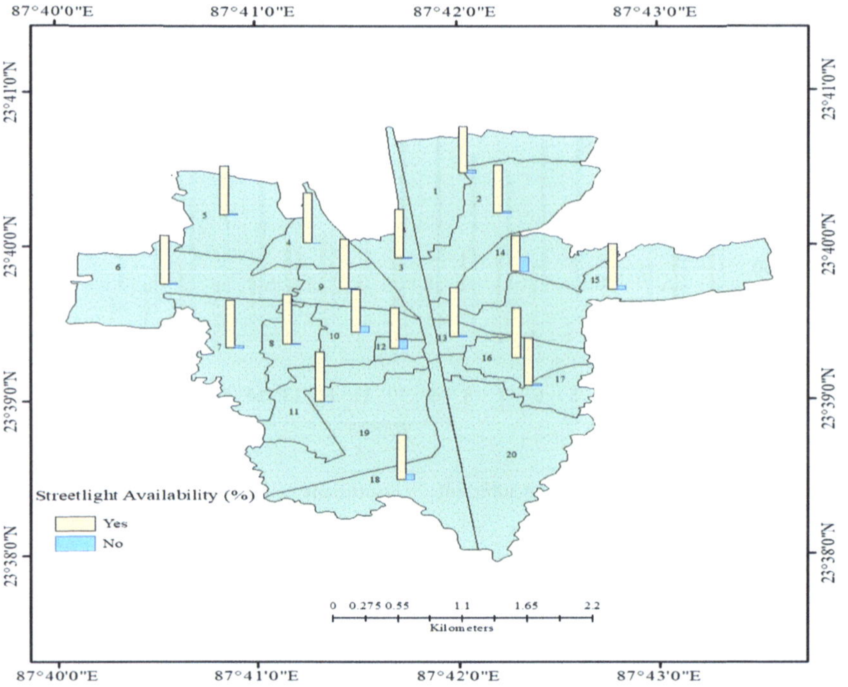

Fig. 18.17 Public transport accessibility rating in different wards of Bolpur municipality

18.4.3.10 Government Policies

The government is constantly trying to improve people's livelihoods and promote sustainable social, economic and environmental development. For that purpose, the government is trying to implement several policies to address society's concerns effectively. Bolpur municipality is no exception. Though there are no certain policies, particularly for Bolpur municipality, the policies framed by the West Bengal Government also apply to this place. Akansha, Gatidhara, Geetanjali and Amar Thikana, Jol Dharo Jol Bharo, Kanyashree, Karmashree, Karmatirtha, Lokprasar Prakalpo, Nirmal Bangla, Pathasathi, Pratyasha, Sabar Ghare Alo, Sabuj Sathi, Samajik Mukti, Shikshashree, Soofal Bangla, Shishu Sathi, Yuvashree and many more are some of the government schemes implemented in Bolpur municipality. A survey conducted by the authors and the responses from municipal officers has provided insight that these schemes cover the entire universe of the study area. Unfortunately, the researchers did not include any data table and maps and diagrams owing to the fact that 100% population of the study area are fully availing of and benefitted from these schemes.

18.4.3.11 Satisfaction of Municipal Citizens for the Services Rendered

Figure 18.18 suggests that the citizens of Bolpur municipality find the level of the education system more than adequate. 82% of the citizens feel that numerous schools and the solitary college within the vicinity of the municipality are more than adequate. The presence of the Visva Bharati University in Santiniketan is a boost to the education sector. Surveys showed that 68% of the citizens are satisfied with their jobs or earning opportunities. The streetlight's condition and availability have gotten a satisfaction rating of 71%. However, citizens have complained that streetlights often go missing. Taxpayers are highly satisfied with the amount of tax they are paying. 89% of the municipal citizens have corroborated that perception. The political climate and voting perception go hand in hand. 79% of the population vouched for political satisfaction, while 78% affirmed satisfactory voting participation. They also insinuated happiness with the current administration. Water availability is another criterion that has achieved a high satisfaction rating of 79% though water quality is relatively low according to the citizens. Waste management is another issue where the municipality service is lagging. It bagged a rating of 62% only. This is due to irregular cleaning, sweeping, and a lack of dumping ground. On the other hand, sanitation has been rated at 74%.

The different indicators help calibrate the three dimensions of sustainable development with the different Sustainable Development Goals. As the activity sphere and influence sphere of the different SDGs overlap, the indicators used in this chapter to show the levels of economic, environmental and social sustainability of the Bolpur municipality, the development of the municipality, whether in a sustainable manner or not, gets reflected.

The indicators of the social dimension of sustainability as household ownership, slum proportion, and public transportation, indicate adequate housing and informal settlements, and convenient access to public transport, which are the indicators of the targets of SDG 11. The indicators of water supply and toilet facility align with the indicators of SDG 6. Education and school dropout can be seen to be associated

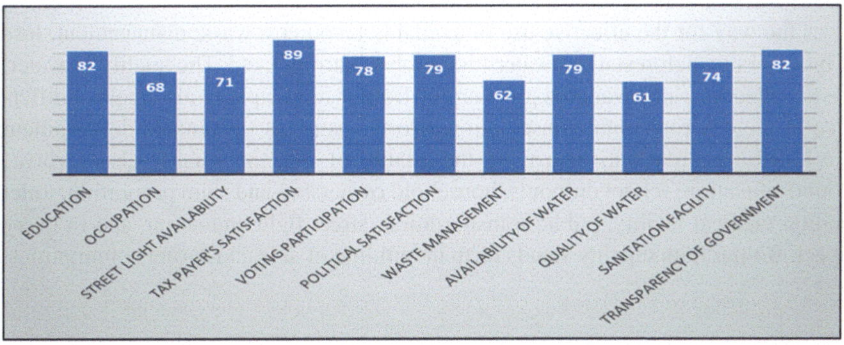

Fig. 18.18 Satisfaction assessment of the services rendered by the local authorities

with SDG 4. On the other hand, the environmental indicators used in this chapter, such as water quality, waste collection and availability of greenery, can be linked to SDGs 6 and 11. The occupation and taxpayer indicators of sustainability's economic dimension can also be connected to SDG 8.

18.5 Conclusion

Bolpur, though rapid growth of population, socio-cultural interaction, and economic growth is prevailing, still makes its way to retain its status as a 'sustainable city' with an objective of 'end poverty, protect the planet and ensure that all people enjoy peace and prosperity by 2030'. The history of Bolpur suggests that despite its rich cultural heritage, and tourist-attracting position of Bolpur, it still holds the status of an environment-friendly, green and healthy city from time immemorial. This chapter highlights how in this present era of urbanization and industrialisation, Bolpur is struggling to hold its position as a 'sustainable' almost in every field especially focusing on access to housing and basic services, sustainable transport system, sustainable urbanization, access to public spaces, per capita environmental impact of cities, and policies towards climate change, resource efficiency, disaster risk reduction, waste management, etc. But government intervention, policies, pollution control, water and waste management, and public awareness are required to make this status long-lasting.

The main theme of this book chapter is sustainable development goals and their implications on Bolpur. The evolution, growth, and changes of a small town named Bolpur and its journey towards a city sustaining natural resources and ecological balance are worth mentioning. Sustainable development goals address global challenges such as poverty, inequality, climate change, environmental degradation, peace, justice, etc. Here SDGs are successfully applied in this city through economic, environmental and social dimensions where sustainable economic growth is represented through occupational structure and tax payment that contributes to the country's GDP. Moreover, sustainable development, a tool for environmental sustainability, paves the way for the effective use of available resources, waste management, loss of biodiversity, deforestation, water loss, global warming, etc. The quality of water, waste collection, and available greenery of Bolpur municipality also make a difference in the environmental conservation effort. Lastly, as sustainable development is a holistic approach to improving the quality of life, this quality also revolves around education, school dropouts, household ownership and slum proportion, toilet facility, medical facility, public transportation, street light and so on, and in every aspect, Bolpur municipality stands forth in building of safe and caring communities.

References

Brundtland Commission (1987) Our common future. United Nations. Oxford University Press

Buchholz K (2020) UN: how has the world's urban population changed from 1950 to 2020? World economic forum. https://www.weforum.org/agenda/2020/11/global-continent-urban-population-urbanization-percent/

Census of India (2011a) District census handbook. Registrar General of India, Ministry of Home Affairs, Government of India

Census of India (2011b) Primary census abstract. Registrar General of India, Ministry of Home Affairs, Government of India

ET Online (2020) What is tax-to-Gdp ratio and where does India fare on this indicator? The Economic Times. https://economictimes.indiatimes.com/budget-faqs/what-is-tax-to-gdp-ratio-where-does-india-fare-on-this-indicator/articleshow/73222499.cms

Garai A, Ghosh B, Karmakar B, Mondal D, Mondal D, Mondal D, Mondal S (2020) Survey on drinking water supply management and mitigation of waterlogging problems in Bolpur Municipality of Birbhum District, West Bengal, Since Last Three Decades (1991 to 2011). Int J Res Eng Appl Manag 5(12):64–80

Garg AK (n.d.) UN Habitat and role of India. https://abhipedia.abhimanu.com/Article/IAS/MTIwMTQ4/UN-HABITAT-AND-ROLE-OF-INDIA-International-Affairs-IAS

Ghosal A (2016) West Bengal: Mamata dumps Centre's Smart City project, to develop cities on its own. https://indianexpress.com/article/india/india-news-india/west-bengal-mamata-dumps-centres-smart-city-project-to-develop-cities-on-its-own-2983994/

Globalization and Livelihood Options of People Living in Poverty (n.d.) The three dimensions of sustainable development. http://www.glopp.ch/A2/en/html/resear_area_present_1_3.html

Kapoor A, Sinha H (2020) View: India's urbanization challenges and the way forward. The Economic Times. https://economictimes.indiatimes.com/news/economy/policy/view-indias-urbanization-challenges-and-the-way-forward/articleshow/79443872.cms?from=mdr

Karade R, Kuchi V, Kabir J (2017) The role of green space for sustainable landscape development in urban areas. Acta Horticulturae 1181(September 2018):73–76. https://doi.org/10.17660/actahortic.2017.1181.9

Koh KL, Gunawansa A, Bhullar L (2010) "Eco-cities" and "sustainable cities"—Whither? Soc Space 84–92

Koner K (2018) Sanitation and hygiene of Darjeeling City: a crisis for women and adolescent girls. Space Cult India 5(3):89–105. https://doi.org/10.20896/saci.v5i3.292

Lundqvist M (2007) Sustainable cities in theory and practice

Mcgranahan G, Satterthwaite D (2014) Urbanization concepts and trends. International Institute for Environment and Development, June, pp 1–27

Murali M, Cummings C, Feyertag J, Gelb S, Hart T, Khan A, Langdown I, Lucci P (2018) 10 things to know about the impacts of urbanization. 10. https://www.odi.org/publications/11218-10-things-know-about-impacts-urbanization

Pal S (2016) #Travel tales: exploring Tagore's Santiniketan, an abode of learning unlike any in the world. https://www.thebetterindia.com/66627/santiniketan-rabindranath-tagore-bengal/

Pollock E (2019) Learning to understand sustainability in cities lecturer's view to pedagogic methodology. In: Pollock E (ed) Sustainable cities II. Essay, Metropolia University of Applied Sciences, pp 8–9

Satterthwaite D (2008) Urbanization and sustainable development [Presentation]. United Nations Population Division

Satterthwaite D (2020) An urbanising world. International Institute for Environment and Development. https://www.iied.org/urbanising-world

Sen M (2020) Urban expansion and population growth of Bolpur Town in Birbhum District, West Bengal. Int J Creat Res Thoughts 8(5):535–549

Shaw A (2018) Towards sustainable cities in India. In: Mukherjee J (ed) Sustainable urbanization in India, exploring urban change in South Asia. Springer Nature

Sustainable Development Goals Knowledge Platform. Sustainable cities and human settlements. United Nations. https://sustainabledevelopment.un.org/topics/sustainablecities

Team Prakati (2019) Dimensions of sustainability. Prakati. https://www.prakati.in/dimensions-of-sustainability/

United Nations Economic and Social Commission for Asia and the Pacific (2015) Integrating the three dimensions of sustainable development: a framework and tools. Greening of economic growth series. United Nations publication

Chapter 19
Pattern of Urbanisation, Its Sustainability and Spatial Set-up of Urban Centers in Uttar Pradesh, India

S. C. Bansal

Abstract Urbanisation is an index of economic development and social change. This is also true in the case of Uttar Pradesh state, the second largest urban population state of the country after Maharashtra, but as regards the proportion of the urban population in total population in the state it ranks 23rd position with a ratio of 22.88%. During the past three decades, its urban population has been increased by more than two times, but its ratio has noted an addition of 4.95% only. There are 71 districts, in 2011 (23 more than 1981), and each district has an average of 626,344 urban population. It has been increased nearly by 230,214 persons per district during the last three decades in spite of an addition of 23 more as new districts. The number of districts with high urbanisation is on the increase and also they have reported an increase in their share of total urban population. In 1981, the districts with more than 30% urban population were 6 in number with 30.53% population. In 2011, 13 districts with a ratio of more than 20% retained 50.94% urban population of the state. This change is an evidence of concentration of urban population in few selective areas. The increasing trend in urbanisation has encouraged the increase in number of urban centres. Nearly 304 urban centres have been added during the years 1991–2011. The state has noted 7 metropolitan centres with 28.07% urban population of the total. The state has 915 centres with an interval of 17.37 km in 2011. There are great variations in spacing at district level. The area and number both have their impact on this spacing. District Varanasi has the least spacing of these centres at an average distance of 6.71 km. On the other hand, Baharaich district has its 5 centres with an average space of 34.63 km. Districts with more number of urban centres have recorded low spacing. There are 17 districts with 334 centres that have the spacing of 14.10 km. It is also observed that more number of urban centres in a district means their close spacing, in that case, most of these are of small and medium size and thus do not have impressive their share in total urban population.

S. C. Bansal (✉)
J.V. Jain College, Saharanpur, Maa Shakumbhari University, Saharanpur, India
e-mail: scbansalgeog@gmail.com

S. Anand et al. (eds.), *Sustainability in South Asian Cities*, Advances in Geographical and Environmental Sciences, https://doi.org/10.1007/978-981-97-7455-5_19

Keywords Urbanisation · Urban centres · NCR-National Capital Region · Sustainable · Effective female literacy · Effective female workers

19.1 Introduction

Urbanisation is indicative of the growth of secondary, tertiary and quaternary activities. It means the per centage of people living in urban areas. The western part of U.P has the presence of 3 metros, i.e. Ghaziabad, Meerut and Agra. Meerut and Ghaziabad are now part of NCR. Centres with a population between 1 lakh and 9.99 lakhs are also on the increase, but their share in the total urban population has a slight decline in these years of study. In 2011, 57 cities noted 32.96% urban population. Small-medium size urban centres (10,000–19,999) are 301 in number, the largest one i.e. with 9.59% urban population. The state has noted 225 urban centres with population between 5000–9999. They give shelter only to 3.72% urban population. Urbanisation and its sustainability are interrelated. If urbanisation has been helpful in the socio-economic development of an area, it may be defined as sustainable urbanisation. In our country urbanisation and economic development have accelerated at a fast rate in the present century. There is a high increase in the number of urban centres. These are an indication of the change in demographic character of the urban population. So, here to determine the degree of sustainable urbanisation five indices have been taken into consideration: Per centage of urban population in total (22.3), Percentage of effective female literacy (69.2), Percentage of workers in total urban population (31.2), Percentage of other workers in total workers (79.8), and Percentage of female workers in female population above age 6 years (12.9). (Figures given in brackets show an average of the state). To determine the degree of sustainability, all the districts have been put into five categories. Districts that have scored the average values for all the above five indices have been categorized as very highly sustainable. In the same way, the other categories have been determined on the basis of number of score values such as high, moderate, poor and very poor sustainable urbanisation. It is observed that 17 districts of the state may be considered as sustainably urbanised, these have 51.57% urban population. Seventeen districts do not comply with the criterion adopted here to define sustainable urbanisation.

It is an indication of the transformation of society from rural to urban, village to towns, primary to non-primary activities. It is also an indication of developed agricultural economy, which by its stability and strong support gives sustenance to non-agricultural economy. It is taken to stand for relatively greater development and non-agricultural production and greater movement of urban-used goods and services to rural areas. Urbanisation is an instrument of change from rural development to other pursuits. It is an evidence of more independence, relative economic stability, decent standard of living, orderly and flexible social structure. Urbanisation is affected by two factors: **Firstly**, it has its natural growth by birth in urban areas, and **Secondly** the migration of countryside people to urban areas. This migration is related with the potentiality of livelihood in urban areas. It is observed that large size urban centres have more potentiality of job work than small size urban centres. Such practice makes the large centres more larger. It is also to be noted that the towns located in the fringe areas of big towns or cities also develop great potential to absorb the working population, to stay there, and thus develop them as dormitory town.

19.2 Objectives of the Study

The main objectives of the present research paper are to analyse the following aspects:

1. To study the trends in growth and distribution of the urban population according to the category-wise of urban centres during 1991–2011.
2. To analyse the spatial pattern of distribution of urban population among the districts in different census years.
3. To analyse the changes in urban population occupied by various categories in these census years.
4. To determine the degree of sustainable urbanisation, the average values of five indicators have been taken into consideration.
5. To examine the impact of distribution of urban population on the spatial set-up (areal spacing) of urban centres, district-wise and also to analyse its impact on the population size of urban centres.

19.3 Database and Methodology

The study is based on the secondary data, derived from census reports of the years in consideration. Uttar Pradesh is taken for the study. The study is based on district wise census data for the year 1991, 2001 and 2011, and also the data of urban centres is collected for these years. Urban centres have been put up into six categories as determined by our Census Department. I category of urban centres has been put into two groups-centres with more than 10 lakhs and centres with less than 10 lakh. The share in urban population of all centres category-wise is examined for all these census years. Category-wise change in number and their average size is also studied. District-wise spacing of urban centres is examined with the help of formula $1.07 \times \sqrt{\frac{A}{N}}$. All the districts have been categorized on two basis-**firstly**, Inter spacing and **secondly**, Number of urban centres therein. The impact of spacing is also examined on the average size of urban centres, share and percentage of urban population.

19.4 Study Area

Uttar Pradesh is extended between 23° 52′ N and 30° 25′ N latitudes and 77° 3′ and 84° 39′ E longitudes. This fourth largest state has an area of 240,928 km^2. It stands first with population 199,812,341 as per census 2011. It is divided into 71 districts (75 districts at the moment). The state is bounded by Uttarakhand and Himachal Pradesh in the North, Delhi and Haryana in the West, Rajasthan in South-west, Madhya Pradesh in South, Chhattisgarh and Jharkhand in south-east (Fig. 19.1). Bihar in east and also by Nepal in North. Most part of the state is a part of the vast homogeneous alluvial Plain of River Ganga and its tributaries, with an average elevation of 150 m above mean sea level. It has upland on its south margins. These uplands are the parts of the Bundelkhand and Baghelkahnd Plateaus. Districts lying along the border of Nepal are Terai districts and have a number of rivers, over flooded with water in the rainy season, thus leaving adverse effects on their economy.

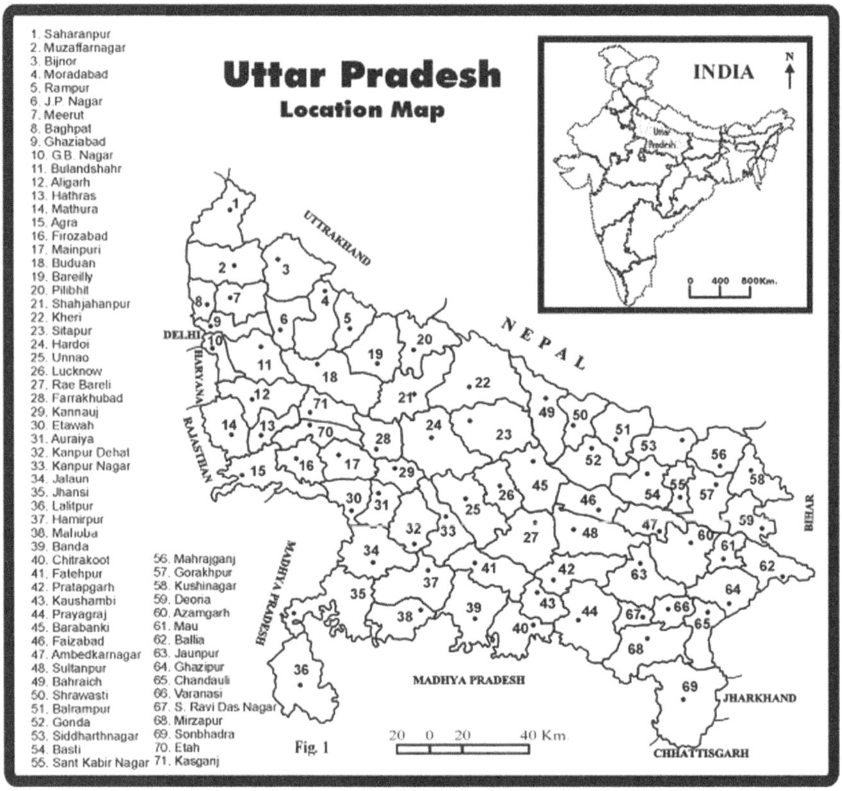

Fig. 19.1 Location map of study area. *Source* Prepared by Author, based on Census of India, 2011

19.5 Literature Review

Over the past six decades more and more geographers have turned their attention to the study of population. The first systematic study was made by G.S. Gosal, in which he studied all significant aspects of population in detail along with study of urbanisation. Gopal Krishnan, R.C. Chandana and S. Mehta of Panjab University also presented their valuable contribution on the various aspects of population in their doctoral research work. S.P. Chatterjee, examined the distribution of rural and urban population in seven macro-regions of India. A. Lal in a study revealed that accelerated migration of people to cities in search of employment has caused the consequent decrease of arable land. R.L. Singh presented a study on the growth of the population of Uttar Pradesh with particular reference to the influence of geographical phenomena on the growth of urban population etc.

Kinsley Davis firstly defined the urbanisation and discussed the trends of shifting people from rural to urban areas. A. Lal in another study made some comments on the urbanisation of India. A.S. Jauhri discussed the urbanisation in Punjab Plains. G.S. Gosal made an elaborate analysis of urban population in Punjab from 1881–1961. Omkar Singh made a detailed analysis of trends of urbanisation in Uttar Pradesh during 1901–1961. A.B. Mukherjee in a study examined the level of urbanisation in Uttar Pradesh. R.C. Chandana in his book on population geography in its various editions has also discussed such methods and techniques which are helpful in the study of various aspects of urban population.

Vandana Tyagi in a study on urbanisation in western Uttar Pradesh has also dealt with various aspects of the distribution of urban population. Archana Chauhan has presented a study on trends of urbanisation in Haryana state (1981–2011). R.B. Singh has made a comment in editorial address that the studies on urbanisation are the need of time to promote sustainable urban development. Henderson has remarked that an unhindered thirst for development has imbibed a dark side of human existence leading to possible failures for the cityscape itself.

A number of geographers have defined the dark side of urbanisation, though it is a symbol of development. As Chapman in a study has emphasized that urbanisation or to say the increase in the number of urban residents has serious health implications for them.

B.K. Chaudhary and A.K. Tripathi in a research paper entitled failing cityscape: urbanisation and urban climate have discussed its certain serious problems regarding human health.

19.6 Pattern of Urbanisation

Uttar Pradesh stands second just after Maharashtra as regards to urban population in the country. Besides this the state occupies 23rd place from the point of percentage of urban population in total population in 2011. It is far behind even from the smaller

Table 19.1 Pattern of growth of urban population (1981–2011)

Year	Total population	Urban population	% of urban population total	No. of districts	Growth of urban pop. in no.	Growth in % of urban pop.	Urban pop. per district
1981	106,026,298	19,014,255	17.93	48	–	–	396,130
1991	131,339,705	25,679,746	19.55	55	6,665,491	35.05	466,904
2001	166,197,921	34,539,582	20.78	70	8,859,836	34.50	493,423
2011	199,581,477	44,470,455	22.88	71	9,930,873	28.75	626,344
1981–11	+93,555,179	+25,456,200	+4.95	+23		+133.88	+230,214

Source Census of India, 2011

states as regards to urbanisation. The state includes 11.79% urban population of the country. The following table explains the percentage and growth in urban population of the state during the last three decades:-

It is evident here that the state is proceeding towards rapid urbanisation. Though It's ratio 22.88% of urban population is quite far below from the ratio of the country 31.17. Anyhow, this ratio has shown an increase of 4.95% during the past three decades. On the other hand, the rate of growth in urban population has gone downwards, in spite of that, it has shown an addition of 2.54 lakh urban persons during 1981–2011. There is an effective increase in the number of districts, in spite of that the average urban population per district has an addition of 230,214 persons (Table 19.1).

19.7 Urbanisation in 1981

This year the percentage of urban population is 17.93. There are found great variations in the number and percentage of urban population among the districts. Sixteen districts have this ratio above the average of the state, whereas it is less than the average in 32 districts. The least percentage of 3.30 is found in Sultanpur and the highest percentage of 46.32 is apparent in Kanpur district. On the basis of percentage, all the districts are categorized into four groups:-

(i) **Low Urbanisation**—The ratio of less than 10% is found in 14 districts. Their average ratio is 7.19%. They include 12.61% urban population of the state. The lowest ratio is recorded by Sultanpur 3.30 and the highest of the category 9.60 by Kheri. All these districts mainly occupy the eastern and terai region in continuity. Their economy is mainly based on agriculture. These are decorated with small and medium size towns and low accessibility (Table 19.2).

(ii) **Moderate Urbanisation**—Eighteen districts of the category have an average ratio of 13.86% and thus include 24.20% urban population of the state. Among these districts, Jalaun district has the highest ratio of 19.92%, while the lowest ratio of 10.01% is found in Bahraich. Most of these districts occupy the central

Table 19.2 Urban population and its percentage among districts (1981)

Sl. no.	Category	Percentage	No. of districts	Total population	Total urban pop.	% of urban pop. in total pop.	% of total urban pop.
i	Low	<10.00	14	33,340,291	2,397,642	7.19	12.61
ii	Moderate	10.01–20.00	18	33,207,084	4,601,093	13.86	24.20
iii	High	20.01–30.00	10	25,121,777	6,210,981	24.72	32.66
iv	Very high	30.01	06	14,357,146	5,804,539	40.43	30.53
		Total	48	106,026,298	19,014,255	17.93	100.00

Source Census of India, 2011

western part of the plains and plateau region. These are mainly found in medium accessible areas (Fig. 19.2).

(iii) **High Urbanisation**—All ten districts of the category have an average ratio of 24.72% with nearly 32.66% urban population of the state. Among these Bareilly district records the highest percentage 29.00 of the category, while Prayagraj district has the lowest of 20.37%. Such peculiarity is mainly of north-west terai districts of the state.

(iv) **Very High Urbanisation**—All six districts of the category have an average ratio of 40.43% with the lowest of 31.22 in Meerut and the highest 52.60% in Lucknow district. All these are highly accessible. They are also known for their large size towns. Ghaziabad, Agra, Jhansi, and Kanpur districts also, belong to this category.

19.8 Urbanisation in 1991

This census year recorded 19.55% population as urban showing not an impressive increase. Out of 55 districts, the ratio of urban population in 21 districts was more than the average of the state. Siddarth Nagar district recorded 3.48%, the least one, whereas Kanpur Nagar identified 84.24% persons as urban in the district. It shows a high variation in the urban percentage among the districts. These districts have been put into four groups:-

(i) **Low Urbanisation**—Below 10% urban population is the feature of 16 districts, and thus gives shelter to 11.51% urban population of the state. Most of these districts occupy east and north-east Terai Region and have the domination of the agricultural economy. Fatehpur district has a ratio of 9.99% urban population highest of the category (Fig. 19.3).

(ii) **Moderate Urbanisation**—Eighteen districts of the category have a ratio between 10 and 20%, with an average ratio of 14.75% of urban population. They give shelter to 20.90% urban population of the state. Among these districts,

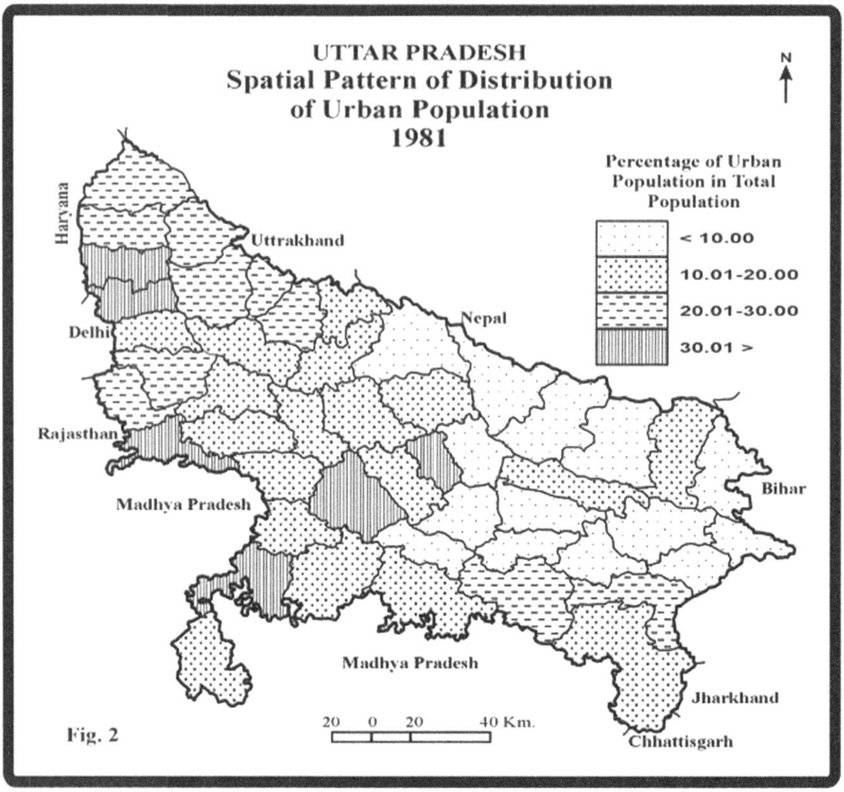

Fig. 19.2 Spatial pattern and distribution of urban population of Uttar Pradesh. *Source* Census of India, 1981

Kheri has 10.66%, the least ratio, while Gorakhpur with 18.76%, has the highest ratio of the category. Such characteristics are the feature of west-central plains and plateau regions (Table 19.3).

(iii) **High Urbanisation**—The average ratio of urban population is 24.41%, more than the state average, and is scored by 13 districts with 34.09% urban population. Among these, Moradabad district has 27.65%, the highest one, whereas the lowest percentage of the category is scored by Shahjahanpur district 20.76%. Most of these districts are found in the sugar belt of the north-west part of the state. In the eastern part, Prayagraj and Varanasi districts do have high urbanisation due to the presence of metropolitan cities.

(iv) **Very High Urbanisation**—Eight districts of the category record an average ratio of 48.84% and thus give shelter to 33.52% urban population of the state. Kanpur Nagar as a district has the highest ratio of 84.24%, whereas Haridwar district includes its 30.96% population in urban areas. Meerut, Ghaziabad, Agra, Bareilly, Lucknow and Jhansi districts belong to this category. Ghaziabad and Lucknow are in race for rapid urbanisation.

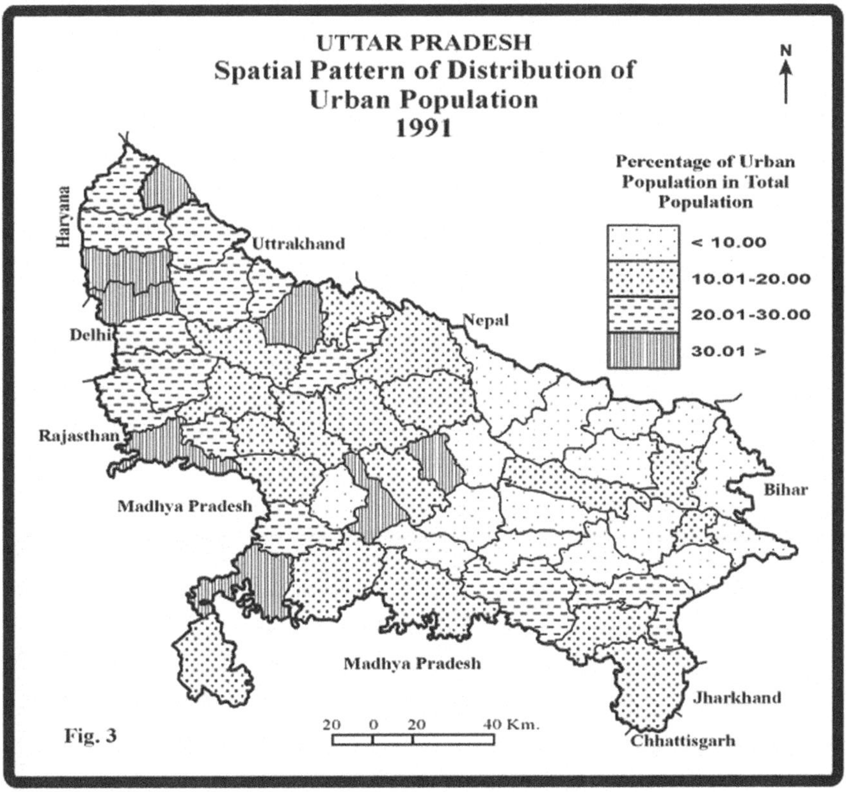

Fig. 19.3 Spatial pattern and distribution of urban population of Uttar Pradesh. *Source* Census of India, 1991

Table 19.3 Urban population and its percentage distribution—1991

Sl. no.	Category	Percentage	No. of districts	Total population	Total urban pop.	% of urban pop. In total	% of total urban pop
i	Low	<10.00	16	41,500,219	2,956,572	7.12	11.51
ii	Moderate	10.01–20.00	18	36,384,949	5,368,014	14.75	20.90
iii	High	20.01–30.00	13	35,828,017	8,746,967	24.41	34.06
iv	Very high	30.01>	8	17,626,520	8,608,193	48.84	33.52
		Total	55	131,339,705	25,679,746	19.55	100.00

Source Census of India, 1991

19.9 Urbanisation in 2001

The state has proceeded towards urbanisation at a rapid speed in the decade of 1991–2001. In 2001, the state recorded 34,539,582 population as urban, and their percentage in total was 20.78%. Twenty-four districts recorded the ratio above the average, whereas 46 districts were having the ratio less than the average. The least percentage was 1.35 in Mirzapur and the highest percentage 67.12% in Kanpur Nagar. This year Kanpur Nagar recorded its more growth in rural population as compared to urban population. All the 70 districts are put into four categories:-

(i) **Low Urbanisation**—This category includes 22 districts with an average ratio of 7.12% urban population. They include 11.07% urban population. Chitrakoot has highest ratio of 9.99% of the category. Most of these districts have occupied the eastern part of the state, which has the dominance in agricultural economy. Eastern Terai districts have mainly this feature (Fig. 19.4).

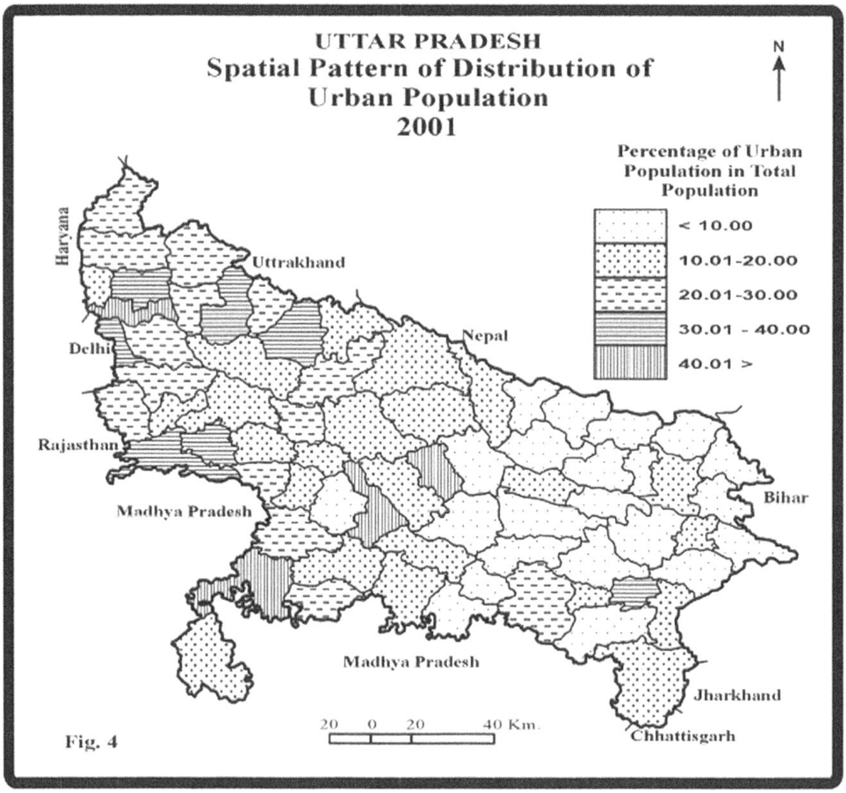

Fig. 19.4 Spatial pattern and distribution of urban population of Uttar Pradesh. *Source* Census of India, 2001

Table 19.4 Urban population and its percentage distribution—2001

Sl. no.	Category	Percentage	No. of districts	Total population	Total urban pop.	% of urban Pop. in total	% of total urban pop.
i	Low	<10.00	22	51,857,528	3,822,313	7.37	11.07
ii	Moderate	10.01–20.00	23	47,517,302	7,098,272	14.94	20.55
iii	High	20.01–30.00	14	33,530,713	8,264,978	24.65	23.93
iv	Very high	30.01	11	33,292,378	15,354,019	46.12	44.45
		Total	70	166,197,921	34,539,582	20.78	100.00

Source Census of India, 2001

(ii) **Moderate Urbanisation**—The average of 14.94% urban population is found in 23 districts with 20.55% urban population of the state. Among these Hathras district has the highest ratio of the category, 19.80 and Bahraich district has the lowest ratio of 10.01%. These are scatteredly located, but are mainly concentrated in central plains and plateau region. They form a continuous belt from Kheri in the north to Mainpuri in the south.

(iii) **High Urbanisation**—Fourteen districts have an average ratio of urban population of 24.65%, where 23.93% urban population resides. In this category, Aligarh district has the highest percentage of 28.90%, whereas Shahjahanpur has the lowest percentage 20.63 of the category. Most of these districts are mainly concentrated in north-west part of the state (Table 19.4).

(iv) **Very High Urbanisation**—Eleven districts have recorded an average of 46.12% urban population in their total. They give shelter to 44.45% urban population of the state. The percentage of 67.12 is recorded by Kanpur Nagar, the highest one, while Firozabad has 30.32%, the lowest one. Moradabad, Gautam Buddh Nagar, Bareilly, Meerut, Ghaziabad, Agra, Lucknow, Jhansi and Varanasi are the other districts of the category. The presence of metropolitans, dominance of industrial economy and high accessibility are responsible for their high urbanisation.

19.10 Urbanisation in 2011

During the decade of 2001–2011, the percentage of urban population has increased from 20.78 to 22.88%. The growth rate of urban population was 28.75%, which is rather below the growth rate (31.80) of the country. There is an addition of 9,930,873 urban population in the decade. The average urban population per district is 626,344 persons. The least percentage of 3.45 is in Shravasti and the highest percentage of 67.46 is found in Ghaziabad district. Ghaziabad has become the most urbanised district of the state for the first time. Out of 71 districts, 23 districts have the ratio above the average, giving an evidence that there is a concentration of urban population in few districts. Eight districts have invited the concentration of 36.88% urban population of the state. They have the average ratio more than 40%. The following table also explains this concentration.

It is apparent from the Table 19.5 that there is a heavy concentration of urban population among those districts which have high ratio of urban population. Thirteen districts, in all, include 50.94% urban population of the state, whereas the remaining 58 districts give shelter to only 49.06% urban population, giving evidence of low concentration.

i. **Low Urbanisation**—Nineteen districts have recorded a percentage of urban population less than 10%, with an average ratio of 7.12%. They include 8.61% urban population of the state. These are mainly concentrated in the Eastern Plains. Shrawasti has the lowest percentage of 3.45%, while the highest percentage of the category is found in Kanpur Dehat 9.66%. Pratapgarh (5.54), Basti (5.61), Jaunpur (7.45), Ghazipur (7.56), Sultanpur (5.28), Kushinagar (4.12) are the districts of the agro-based economy.
ii. **Moderate Urbanisation**—The ratio between 10 and 20% of urban population is scored by 24 districts. They have the distribution of 20.12% urban population of the state. The average ratio is 14.58%. The lowest percentage of 10.14 is recorded by Barabanki and the highest percentage of 19.70 by Shahjahanpur. The districts of the category are found mainly in Rohilkhand Plains, Plateau Region of Baghelkhand and nearby areas of River Ghaghra. Kheri, Pilibhit, Hardoi, Ambedkar Nagar, Deoria are the districts with 10 to 13% urban population.
iii. **High Urbanisation**—The ratio of urban population between 20 and 30 is the feature of 15 districts. Their average ratio is 24.76% Baghpat (21.05), Hathras (21.31), Farrukhabad (22.00), Mahoba (21.19) and Kashi Ram Nagar (20.05) districts of the category have a ratio less than the average of the state. The highest ratio of the category is 29.66% in Mathura. The other main districts are Muzaffarnagar (28.76), Bijnor (25.13), Jalaun (25.06), Amroha (24.83), Bulandshahr (24.80), Prayagraj (24.78) and Etawah (23.20). Most of these districts have occupied the western part of the state, which is quite prosperous in the production of commercial crops, thus encouraging urbanisation in the area. Districts of the category give shelter to 20.33% urban population of the state (Fig. 19.5).
iv. **Very High Urbanisation**—The ratio of urban population between 30 and 40% is found in five districts—Saharanpur (30.71), Moradabad (33.01), Aligarh (33.11), Firozabad (33.33) and Bareilly (37.43). All these districts include 14.06% urban population of the state. The average ratio is 33.12%. All these districts have their extension on the western fertile land of the state with the presence of large size cities.
v. **Extremely High Urbanisation**—Eight districts of the state have yielded ratio of urban population more than 40% and because of that, there is found 36.88% urban population of the state. Their average ratio is 56.54%. Among these Ghaziabad district has the highest ratio of 67.46%, showing a stage of rapid urbanisation. All the other districts of the category are Jhansi (41.78), Varanasi (43.43), Agra (45.87), Meerut (51.13), Gautam Buddh Nagar (59.56), Kanpur Nagar

Table 19.5 Urban population and its percentage distribution—2011

Sl. no.	Category	Percentage	No. of districts	Total population	Total urban pop.	% of urban pop. in total	% of total urban pop.
i	Low	<10.00	19	53,818,383	389,293	7.12	8.61
ii	Moderate	10.01–20.00	24	61,359,516	8,946,844	14.58	20.12
iii	High	20.01–30.00	15	36,521,539	9,041,334	24.76	20.33
iv	Very high	30.01–40.00	5	18,873,320	6,250,817	33.12	14.06
v	Highly excellent	40.01>	8	29,008,719	16,402,167	56.54	36.88
		Total	71	199,581,477	44,470,455	22.88	100.00

Source Census of India, 2011

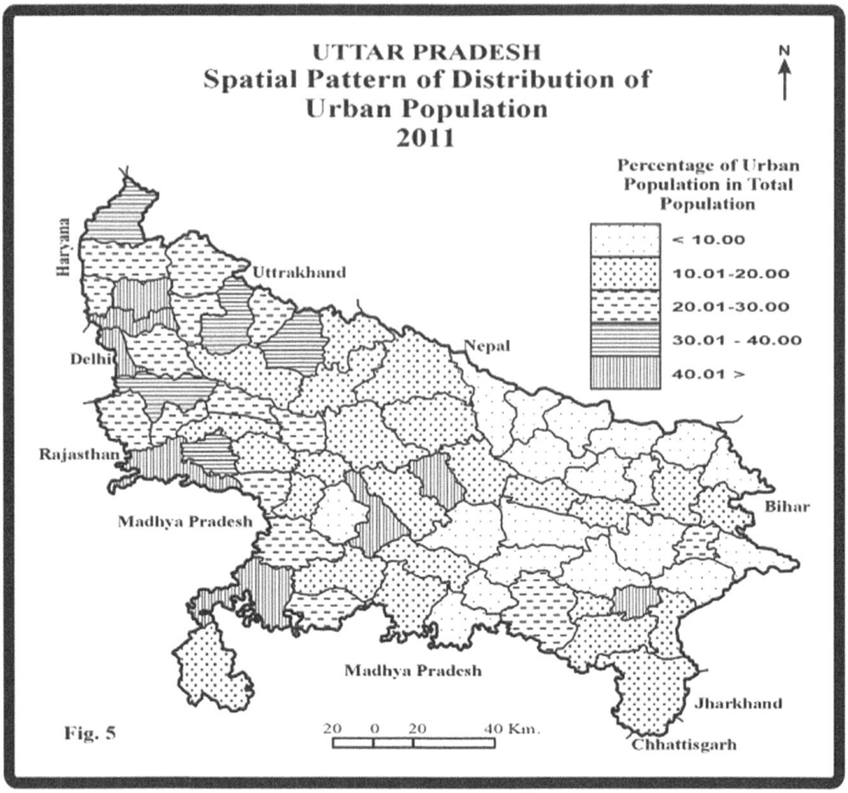

Fig. 19.5 Spatial pattern and distribution of urban population of Uttar Pradesh. *Source* Census of India, 2011

(65.93) and Lucknow (66.20). All our metro cities are located in these districts, which have accelerated the urban population. These districts mainly occupy the central and western parts, which are highly accessible. Jhansi and Varanasi districts have their isolated location.

19.11 Urban Centres and Their Size-Wise Role in Urbanisation

The increasing trends in the percentage of urban population have boosted the state by the high increase in the number of urban centres. These were 611 in 1991, and their number increased to 915 in 2011. It shows an addition of 304 towns during these two decades. The decade of 2001–2011 (211) has more additions than the decade of 1991–2001 (93). There is a specific change in the average size of these centres. Their addition in more number in the decade of 2001–2011, has reduced their average size.

On the other hand, the addition of 93 urban centres in the decade of 1991–2001 has enlarged their average size. There is an enormous increase in the number of upper category centres.

In 1991, the average size of an urban centre was 42,029 persons, which increased to 49,062 in 2001. It was slightly decreased to 48,629 persons in 2011, because of a heavy increase in their number. As a whole, there is an addition of 6600 persons after every centre during the last two decades with an addition of 304 urban centres. Such changes compel us to give our focus to discuss them on the basis of category-wise position of urban centres in the state in these census years (Table 19.6).

I. **Metropolitan Urban Centres (Population More than 10 Lakh)**—In 1991, Kanpur (18.7) and Lucknow (16.2) were the only metropolises of the state. These were having 13.60% urban population. In 2001, their number reached six with the inclusion of Agra (12.75), Meerut (10.68), Prayag Raj (10.18) and Varanasi (10.91). Kanpur (25.55) was still the first metropolis, followed by Lucknow (21.86). In all, these centres have sheltered 26.66% urban population (Table 19.7).

   It is apparent here that the average size of these centres decreased in 2001 due to the addition of four new centres. In 2011, Ghaziabad has joined this category with 16.48 lakh population. Its inclusion has helped in the increase of their average size. This city has emerged as the third-largest city in the state. Its near location to Delhi is responsible for the high increase in its population. All seven metropolises included 28.07% urban population of the state. Lucknow stood as the first metropolis with 28.17 lakh population. The other metropolis centres are Kanpur (27.68), Agra (15.85), Meerut (13.05), Varanasi (11.98), and Prayagraj (11.68). Meerut and Ghaziabad are the part of NCR. Among

Table 19.6 No. of urban centres and their average size (1991–2011)

Census year	No	Change in number	Average size	Change in size
1991	611	–	42,029	–
2001	704	93	49,062	+7033
2011	915	211	48,629	−433
		+304		**+6600**

Source Compiled by Author, based on Census of India

Table 19.7 No. of metropolitan centres and their average size (1991–2011)

Year	No. of metro cities	Population	% of total urban population	Average size
1991	2	3,493,524	13.60	1,746,762
2001	6	9,207,698	26.66	1,534,616
2011	7	12,491,814	28.07	1,784,545
1991–2011	+5	+8,998,290	+14.47	+37,783

Source Compiled by Author, based on Census of India

these Agra is situated in the south-west part of the state, whereas the other metros have their location in groups of two-such as Kanpur and Lucknow, Meerut and Ghaziabad, Prayagraj and Varanasi (Fig. 19.6).

II. **Very Large Urban Centres (1 Lakh to 999,999)**—These are defined as cities because of having more than one lakh population. Uttar Pradesh has always recorded the presence of these centres in an impressive number. In 1991, these were 36 comprising 41.13% population. In 2001 their number increased to 48, with an average size of 237,882 persons. They contained one-third urban population of the state (Table 19.8).

It is apparent here that in spite of the increase in their total population, their share and average size have declined. Though, their number is 21 more in 2011 than the year 1991. Their population has an addition of 41.04 lakh during the past two decades. It is observed that nearly all the district headquarters have emerged as cities. Among these Bareilly with 904,797 population was the largest one, while Kasganj was the smallest one with 101,277 people. The

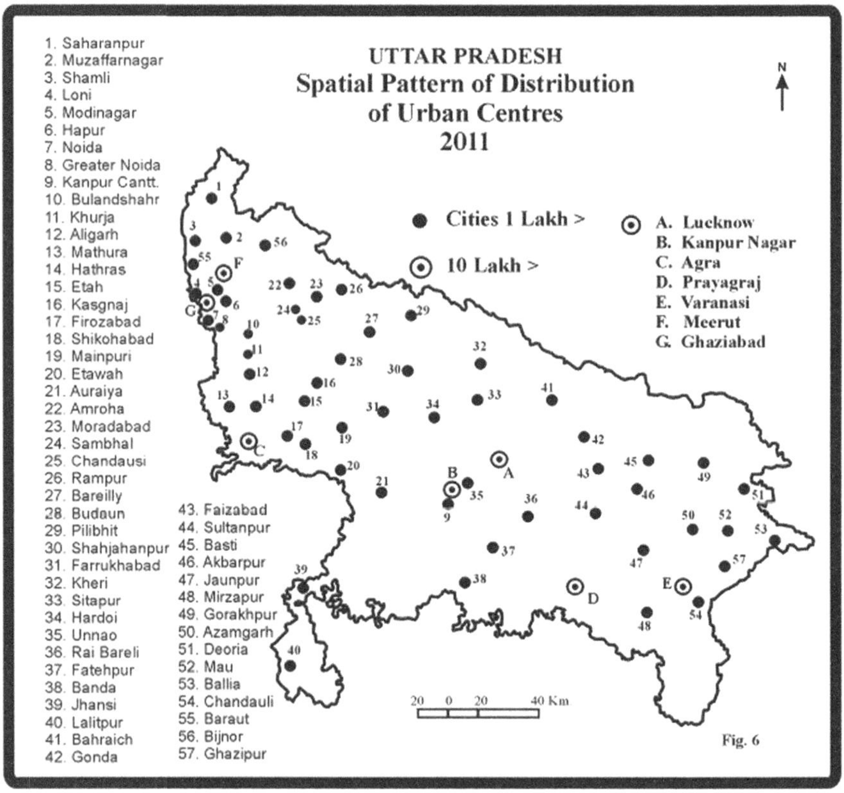

Fig. 19.6 Spatial pattern and distribution of urban centers, Uttar Pradesh. *Source* Prepared by the Author, Based on Census of India

Table 19.8 No. of very large size urban centres and their average size (1991–2011)

Year	No. of cities	Population	% of total	Average size
1991	36	10,561,538	41.13	293,376
2001	48	11,418,329	33.06	237,882
2011	57	14,665,834	32.96	257,295
	+21	+4,104,296	−8.17	−36,081

Source Compiled by the Author, based on Census of India

other cities with more than 5 lakh population are Firozabad (6.04), Gorakhpur (6.73), Jhansi (5.05), Loni (5.16), Moradabad (8.87), Noida (6.37, Saharanpur (7.05) and Aligarh (8.74). Among these except Gorakhpur, all the other above cities are situated in the western part of the state. It is observed that in 2021 they may join the list of metropolises. Loni a newly emerged city has developed near the border of Delhi, and its population increased by two times in the 2001–2011 decade. In 2001, Ghaziabad (9.68), Bareilly (7.20), Gorakhpur (6.22), Moradabad (6.41) and Aligarh (6.69) were the cities of having population more than 5 lakh (Population given in brackets is in lakh).

In 2011, Akbarpur, Basti, Ballia, Chandausi, Gonda, Kasganj, Gautam Buddh Nagar, Azamgarh, Mughal Sarai, Shamli, Shikohabad and Sultanpur were the new entrants.

It is evident from Fig. 19.6 that 31 cities are distributed only in western Uttar Pradesh. Here, these are rather very closely spaced, especially in the districts of Ghaziabad, Noida and Meerut. Their location close to Delhi is responsible for near spacing. In Terai districts they follow a linear pattern, and have distant location.

III. **Large Size Urban Centres (50,000–99,999)**—These urban centres have high fluctuations as most of these due to their favourable location, upgraded themselves to the first category with time. Their number was 44 in 1991. They included 12.44% urban population and their average size was 72377 persons. In 2001, their number reached 55 with an average size of 67368 persons. It is nearly 10,000 less than the figure of 1991. They contained 10.73% urban population of the state. The addition in number was not in favour of their share in total urban population (Table 19.9).

Table 19.9 No. of large size urban centres and their average size (1991–2011)

Year	No. of centres	Population	% of total	Average size
1991	44	3,184,577	12.40	72,377
2001	55	3,705,262	10.73	67,368
2011	62	4,310,577	9.69	69,525
	+18	+1,126,000	−2.71	−2852

Source Compiled by the Author, based on Census of India

It is apparent that in spite of the increase in their number, these large size towns have reduced their share in the total population. It has decreased to 9.69%. Anyhow there is a little increase in their average size of 2157 persons per center. This year Deoband was the largest one with 97,037 population and Baghpat was the smallest with 50,310 population. There are 8 towns having a population of more than 90,000. These are Bhadohi, Bijnor, Dadri, Mahoba, Meerut Cantt, Muradnagar, Nagina and Tanda. Seven of these towns are located in the western U.P. Twelve towns recorded a population between 80,001 and 90,000. It can be presumed that in the next census, these may get promotion for the first category. Twenty towns of the category have joined this category in this census year. Their population is less than 60,000 persons. Most of these towns are tahsil headquarters towns or have significant locations in their respective areas.

IV. **Medium Size Urban Centres (20,000–49,999)**—These medium-size towns are important large service centres feeding their countryside up to a considerable distant. Their number is always on the increase with every census year and they give shelter to nearly 15% urban population. In 1991, these were 125 in number with 14.94% urban population and their average size was 30,702 persons. In 2001, their number reached 186 with 15.58% population. Their average size was reduced and it reached to 28,940 nearly 1762 less than the previous census year. This decrease was caused due to the addition of 61 more new towns (Table 19.10).

It is apparent here, that the population of these towns has been nearly doubled during the last two decades. In 2011, there were 237 in number with 15.73% population of the total. Their average size was 29,518, a little more than the previous year. Padurauna with 49,723 was the largest one, while Jhusi Kohna of Prayagraj district with a 20,023 population was the smallest one. Thirty-one towns were having population between 40,000 and 49,999. On the other side, 148 towns noted a population between 20,000 and 30,000. These towns are really serving as rural service centres in their respective areas. It may also be assumed that they have occupied this category first time in this census year.

V. **Small-Medium Size Urban Centres (10,000–19,999)**—These are rather overgrown villages with urban infrastructure. They function as an agro-service centre. Their number is highest in all these census years. These were 211 in number in 1991 and increased to 262 in 2001 and 301 in 2011. These are

Table 19.10 No. of medium size urban centres and their average size (1991–2011)

Year	No. of towns	Population	% of total	Average size
1991	125	3,837,721	14.94	30,702
2001	186	5,382,837	15.58	28,940
2011	237	6,995,839	15.73	29,518
1991–2011	+112	3,158,118	+0.79	−1184

Source Compiled by the Author, based on Census of India

always one-third of the total number of towns in all the census years. In spite of the increase in their number, their share in the total urban population is on the decline. In 1991, they shared 12.12% population with an average size of 14747 persons. In 2001, 262 small-medium size towns contained 10.74% population with an average size of 14165 persons. The fall in average size denotes, that this category has entertained most of its towns from the lower category. The following table explains the changes in the urban centres of this category (Table 19.11).

It is evident here that these towns have added 11.55 lakh population with an addition of 90 new towns during the last two decades, but their share in total urban population aid also their average size have reduced. In 2011, Bahadurganj of Ghazipur with a 19,912 population is the largest town of the category while Ranipur of Kanpur Dehat with a 10,011 population was the smallest one. There are 104 towns which are having a population between 15,000 and 19,999. It shows that these were existing also in the last census year. The remaining 197 towns do have the population between 10,000 and 14,999. Most of these are new entrants. These are rather overgrown villages and function as central place for their rural countryside.

VI. **Small Size Urban Centres (5,000–9,999)**—These are newly developed small towns. These have a central location in their respective areas. Their number was 177 in 1991 and they included 5.54% urban population with an average size of 8031 persons. In 2001, their number is declined. These are 134 in number with 3.07% population. Their average size is 7,913 persons (Table 19.12).

It is evident here that there is an increase in their population, but due to heavy increase in their number, their average size is declined, giving an evidence that

Table 19.11 No. of small-medium size towns and their average size (1991–2011)

Year	No. of towns	Population	% of total	Average size
1991	211	3,111,597	12.12	14,747
2001	262	3,711,366	10.74	14,165
2011	301	4,266,862	9.59	14,176
1991–2011	+90	+1,155,265	−2.53	−571

Source Compiled by the Author, based on Census of India

Table 19.12 No. of small size towns with average size (1991–2011)

Year	No. of towns	Population	% of total	Average size
1991	177	1,421,505	5.54	8031
2001	134	1,060,388	3.07	7913
2011	225	1,656,544	3.72	7362
1991–2011	+48	+235,039	−1.82	−669

Source Compiled by the Author, based on Census of India

Table 19.13 No. of very small towns and their average size (1991–2011)

Year	No. of towns	Population	% of total	Average size
1991	16	69,284	0.27	4330
2001	13	53,702	0.16	4131
2011	26	108,048	0.24	4156
	+10	+38,764	−0.03	−174

Source Compiled by the Author, based on Census of India

it has an impressive entry of new entrants. In 2011, these towns are 225 in number and contain 3.72% urban population of the state. The average size is reduced because of a heavy increase in their number. Among these Lahsari (Gorakhpur) with 5051 population is the smallest one and Rampur Kharkera (Deoria) with 9943 population is the largest one of the category. All these towns according to their population are as follows—43 towns with less than 6,000 population, 57 between 6,000 and 6,999, 50 between 7,000 and 7,999, 36 between 8,000 and 8,999 and the remaining 39 towns have the population between 9,000 and 9,999. it is evident here that 100 towns in the category are very small in size. Most of these have qualified the definition of towns in this census year for the first time.

ViI. **Very Small Size Towns (Less than 5,000)**—These are defined towns for the first time in their own census years. These were 16 in number in 1991 with 0.27% population of the state. Their average size was 4330 persons. In 2001, their number was reduced to 16. They included 0.16% population only. Their average size was reduced to 4131 persons (Table 19.13).

In 2011, these very small towns were 26 in number with an average size of 4156 persons. They do give shelter to only 0.24% urban population. The smallest town of the category is Pavi Sarakhpur of district Ghaziabad with 757 population only. It is a census town and part of Ghaziabad urban agglomeration. Swamibagh (2039), and Dayalbagh (2830) census towns are part of Agra urban agglomeration. There are 20 towns having a population between 4,000 and 4,999. Among these Kataka of Mirzapur district is the largest one of the category with 4926 population. Most of these are census towns and are situated in or near the fringe areas of large urban centres.

19.12 Urbanisation and Its Sustainability

The interrelationship between urbanisation and sustainability has been of primary concern. It is related to such economic and social development which is in support of a large urban population. **Mills** and **Backer** have observed that urbanisation is closely related to a variety of indices of economic development. After Independence, especially after 1991, India's urbanisation and economic development have accelerated at a fast rate. Urban centres of different sizes have a rapid increase in their number and their regional distribution has left its impact on the development of economic and social activities. There are certain indices, i.e. percentage of urban population in total, percentage of effective female literacy, percentage of workers in total population, percentage of non-primary workers in total workers and percentage of female workers in female population above age 6 years, which denotes the change in status of urbanisation and its sustainability. It is assumed, that if the ratio in all these indices in an area is above the average of the region, may give an evidence of sustainable urbanisation. A district with a ratio for all these five selected indices above the average, will mean a district of highly sustainable urbanisation. On the other hand, if a district does not qualify for any one of these indicators it means that the urbanisation is not effective in the social and economic development of that district or area. All the districts have been put into six categories as shown in the Table 19.14.

It is apparent here that nearly 51.57% urban population corresponds to sustainable urbanisation as they have scored the average values for three to five indices under consideration. Such districts are 17 in number. Thirty-seven districts have comparatively poor sustainability as they have scored the value for only one or two indices. They give shelter to 36.18% urban population. On the other hand, 17 districts do not have sustainable urbanisation as per our definition adopted here. It is observed that those districts which have near location to NCR, or have the presence of metropolitan cities have developed sustainable urbanisation. The category-wise analysis reveals all these facts.

i. **Very High Sustainable Urbanisation**—Four districts Ghaziabad, Gautam Buddh Nagar, near NCR and Lucknow, Prayagraj have scored the value above the average for all the five indices. They give shelter to 19.43% urban population distributed among 76 urban centres. These are highly urbanised and seem to be prosperous for the provision of job opportunities (Fig. 19.7).
ii. **High Sustainable Urbanisation**—Six districts have scored the values for any of the four indices under consideration. They include 18.25% urban population

Table 19.14 Number of districts according to sustainability in urbanisation—2011

Sl. no.	No. of indices	Degree of sustainability	No. of districts	No. of urban centres	Urban population	% of total
i	5	Very high	4	76	8,648,725	19.43
ii	4	High	6	114	8,118,623	18.25
iii	3	Moderate	7	102	6,177,627	13.89
iv	2	Poor	22	304	11,574,533	26.01
v	1	Very poor	15	166	4,523,934	10.17
Vi	Nil	No	17	153	5,451,621	12.25
			71	915	44,495,063	100.00

Source Compiled by the Author, based on Census of India

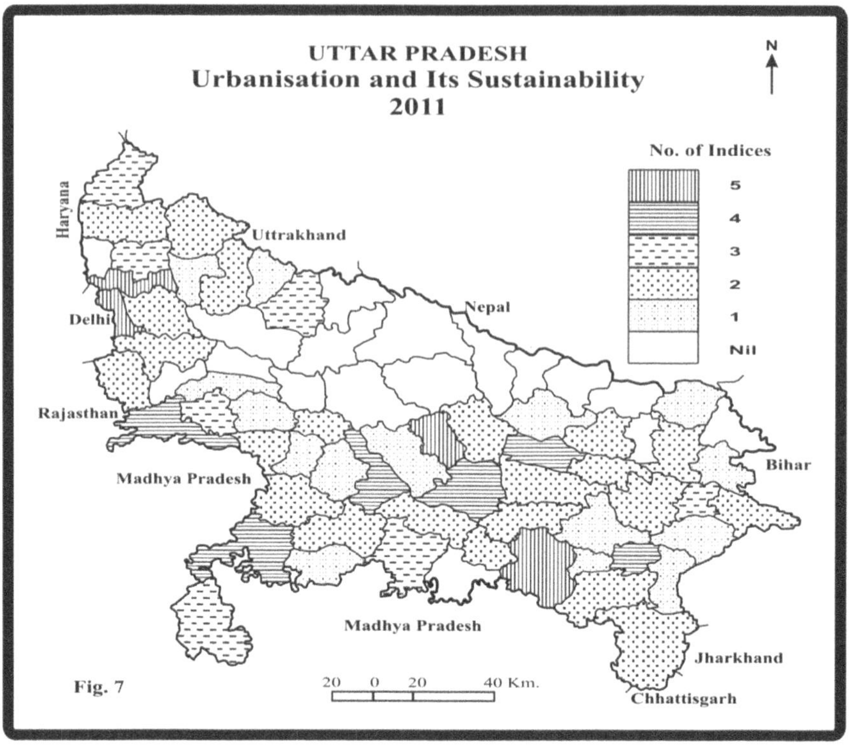

Fig. 19.7 Urbanisation and sustainability. *Source* Prepared by the Author, Based on Census of India data

with 114 urban centres. Agra, Kanpur Nagar and Varanasi districts of the category have metropolitan cities, whereas Rai Bareli, Faizabad and Jhansi districts have either near locations to urbanised districts or have their central location. Rae Bareli and Faizabad have not scored the value of per centage of urban population above the average. Agra district does not qualify for female literacy, while Kanpur Nagar district does have female workers less than the average. As a whole, the urbanisation in these districts is quite sustainable.

iii. **Moderate Sustainable Urbanisation**—Seven districts have occupied this category. These include 13.89% urban population distributed among 102 urban centres. These districts are found in scattered form. Saharanpur and Meerut districts have low participation rate of female workers. Bareilly, Firozabad, Mau have the dominance of female workers, because of their involvement in certain household industries like bangles making, carpet knitting. Lalitpur and Banda districts in the south are also known for effective female participation. These have low urbanisation ratio.

iv. **Poor Sustainable Urbanisation**—Twenty-two districts have scored the value above the average for any of the two indicators. They have 26.01% urban population distributed among 304 urban centres. These are spread mainly in central Doab, and Eastern Plains which are decorated by medium and small size towns. Most of the districts have the domination of agricultural economy. Among these is Moradabad. Aligarh, Gorakhpur districts have the presence of large size urban centres. Muzaffarnagar, Bulandshahr, Mathura, Etawah, Kannauj, Jalaun and Hamirpur districts have their location in the west and south-west parts of the state. Most of these have scored the values for the percentage of the urban population and work participation rate. Some of these have the upper hand for the index of female literacy. In the South-east Mirzapur and Sonbhadra districts of the category are known for mineral-based industries. Ballia, Azamgarh, Barabanki, Sultanpur, Pratapgarh, Ambedkarnagar districts are elongated along the River Ghaghra. Jalaun, Hamirpur, Fatehpur and Kaushambi districts also belong to the category. These are known for female literacy and work participation or female participation. Most of these have little percentage of the urban population, and thus are rural dominant districts.

v. **Very Poor Sustainable Urbanisation**—Fifteen districts of the state have scored only for one of the five indices. These include 10.17% urban population with 166 urban centres. All these districts have not scored the average value for work participation and other workers. Rampur and J.P. Nagar districts have qualified due to the above percentage of urban population. Siddharth Nagar, Mahoba, Sant Ravi Das Nagar and Etah have scored for the female workers' index. The remaining nine districts Mainpuri, Unnao, Auraiya, Kanpur Dehat, Gonda, Deoria, Jaunpur, Ghazipur and Chanduali have scored for female literacy index. Most of these districts are concentrated in Central Doab and the eastern most Part of the state.

vi. **Unsustainable Urbanisation**—The state has recorded 17 such districts that have not scored the average values for any of these five indices taken into consideration. Figure 19.7 shows, that most of these districts are distributed in eastern Rohilkhand Plains and Terai Region. In upper Doab, Baghpat and Hathras are such districts. In Rohilkhand Region Kasganj, Badaun, Shahjahanpur Farrukhabad, Pilibhit, Sitapur, Hardoi, Kheri are such districts. In Terai Region, Bahraich, Shrawasti, Balrampur, Sant Kabir Nagar Maharajganj, and Kushinagar districts have also such peculiarity. District Chitrakoot is the only plateau district.

It is revealed that sustainable urbanisation is the feature of high accessibility, high agricultural productivity, development of non-primary activities, social awareness, nearness to NCR and Metropolitan cities, developed social and economic infrastructure.

19.13 Spatial Set-up (Spacing) of Urban Centres and Urbanisation

The state has 915 urban centres at an average areal distance of 17.37 km in 2011. All of these centres have a great variation in spacing at district level. District Varanasi has 39 centres with an average spacing of 6.71 km the lowest one. On the other hand, District Bahraich has its 5 urban centres with an average spacing of 34.63 km, the highest one in the state. It is observed that the districts occupying the less fertile areas like plateau region and low accessibility like terai districts of border area do have the location of centres at high spacing. There are 26 districts, which have the spacing of urban centres more than the average of the state. It is also observed that those districts which have high spacing, are comparatively less urbanised, whereas, the districts having centres with low spacing are highly urbanised. These districts give shelter to nearly 84% urban population of the state. Urban centres of these districts are also of large size, as compared to the urban centres of districts with high spacing. It evidences that the areas with low spacing of urban centres, are more urbanised, and their centres are comparatively of large size. All the centres of the state have been put into six categories on the basis of areal spacing.

i. **Least Spacing**—Nearly 104 urban centres of five districts are interspaced at an average distance of less than 12 km. They are spaced at an average distance of 8.60 km. They have 14.52% urban population. Among these Ghaziabad and Gautam Buddh Nagar districts are situated in the NCR. and belong to highly accessible region. Ghaziabad district, alone has 26 urban centres at an average distance of 7.20 km. The remaining three districts Mau, Sant Ravi Dass Nagar and Varanasi are spread in the most eastern part. These districts are small in size and have a domination of rural growth towns.

ii. **Low Spacing Centres**—This spacing is observed among 334 urban centres of 17 districts. These centres are spaced at an average distance of 14.10 km. Most of these districts are highly urbanised. In all, they give shelter to 41.22% urban population. The average size of an urban centre is 54,910 persons. Agra, Prayagraj, Lucknow and Meerut districts of the category have the presence of metropolitan centres. Nine of these districts are located in upper and middle Ganga-Yaumuna Doab. As a whole, the western part of the state has been dominated by such spacing. In the eastern part, Azamgarh, Prayagraj, Kaushambi are such districts. The other districts are Bijnore Moradabad, Bareilly, Muzaffarnagar, Baghpat, Bulandshahr, Aligarh, Mathura, Hathras, Kasganj and Unnao.

iii. **Medium Spacing Centres**—Twenty-three centres have their 261 urban centres at a distance between 16 and 20 km. They have an average spacing of 17.32 km, and include 28.27% urban population. The average size of an urban centre is 48,199 persons, just near to the average of the state. These districts are found in all parts of the state, mainly in central Doab, and in the central-east part. Among these, Kanpur Nagar is an important one, 10 centres with a spacing of 19.00 km. All the other districts have domination of medium and high-medium size urban centres. Amroha, Auraiya, Badaun, Etah, Farrukhabad, Firozabad, Kannauj, Mainpuri, Pilibhit, Rampur and Saharanpur districts are spread in western part, whereas Ambedkar Nagar, Ballia, Barabanki, Chandauli, Deoria, Jaunpur, Sant Kabirnagar, Faizabad, and Gorakhpur occupy mainly the eastern part of the state. Jhansi and Kanpur Dehat do belong to this category (Fig. 19.8).

iv. **High-Medium Spacing Centres**—There are 140 urban centres scattered in 14 districts, located within a space of 20 and 24 km. Their average spacing is 21.89 km and gives shelter to 10.01% urban population of the state. They have an average size of 31,799 persons per centre. These districts are mainly dispersed in the eastern part. They have occupied mainly terai area, where Gonda, Kushinagar, Maharajganj and Siddarthnagar are known for such spacing of urban centres. The Baghelkhand Plateau Region of the state is occupied by Sonbhadra and Mirzapur districts also present such features. In the south-west, Etawah and Jalaun districts are known for such spacing. In Awadh Plain Shahjahanpur,

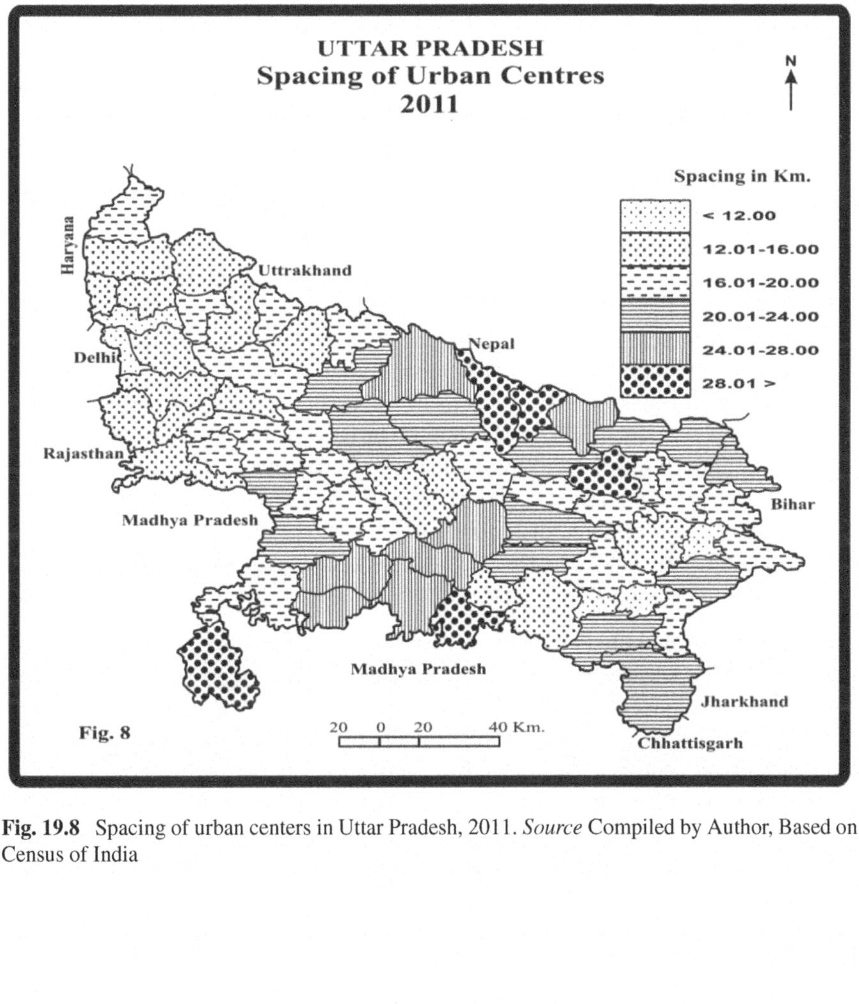

Fig. 19.8 Spacing of urban centers in Uttar Pradesh, 2011. *Source* Compiled by Author, Based on Census of India

Table 19.15 No. of urban centres with their spacing-2011

Sl. no.	Spacing (km)	No. of districts	Total urban centres	Area (km^2)	Average spacing (km)	Urban pop.	% of total	Average size
i	<12.00	5	104	6714	8.60	6,462,417	14.52	59,254
ii	12.01–16.00	17	334	58,026	14.10	18,339,943	41.22	54,910
iii	16.01–20.00	23	26	68,407	17.32	12,579,940	28.27	48,199
iv	20.01–24.00	14	140	58,588	21.89	4,451,865	10.01	31,799
v	24.01–28.00	7	56	31,363	25.32	1,928,398	4.33	34,436
vi	28.01>	5	20	17,820	31.94	732,500	1.65	36,635
	Total	71	915	240,928	17.37	44,495,063	100.00	48,682

Source Compiled by Author, based on Census of India

Hardoi and Sitapur districts do enjoy this spacing. Ghazipur, Sultanpur, Pratapgarh lying in Ghaghra-Ganga River tract also have high medium spacing of urban centres.

v. **High Spacing Centres**—This spacing of 24–28 km is observed in those areas which are either lying in the areas of poor accessibility or in between the developed areas like Kanpur and Prayagraj Region. Kheri and Balrampur districts are terai border districts and have low accessibility whereas Rai Bareli, Fatehpur, Mahoba, Hamirpur and Banda occupy the middle plateau area and transitional area. All these seven districts have a distribution of 56 urban centres at an average distance of 25.32 km. They give shelter to 4.33% urban population and have an average size of 34,436 persons (Table 19.15).

vi. **Very High Spacing Centres**—Only five districts with 20 urban centres present this characteristics. These centres are spaced at the farthest distance of 31.94 km. Their average size is 36,625 persons and do have only 1.65% urban population. The poor number of centres seems to be responsible for their high spacing. Lalitpur and Chitrakoot are plateau districts, with 4 and 6 urban centres, respectively. In Chitrakoot the average spacing is 30.34 km and in Lalitpur it is 31.01 km. The remaining three districts Bahraich (34.63 km), Basti (32.03 km) and Shrawasti (30.64 km) are rather Terai districts and are populated poorly. These are economically less developed.

It is concluded that the degree of accessibility, developed agricultural economy, presence of plain surface, free from the presence of number of unbridged rivers, good network of state and National Highways, high connectivity and nearness to the National Capital Region seem to be responsible for the close spacing and presence of big size urban centres in the state. The overgrown villages, which are located in agricultural productive area or are near the large size urban centres, and along the main Highways, attract the development of commercial and industrial activities, and thus with time these villages become part of urbanisation, urban centres of fifth and sixth categories are the best example of it.

References

Bansal SC (2017) Urban geography, vol 2. Meenakshi Prakashan, Meerut
Bansal SC (2018) Population geography (in Hindi). Books, New Delhi
Census of India (2001) U.P. census hand books—1991, 2001
Census of India (2011) Statistical abstract—Uttar Pradesh, Census of India, 2011
Chand M, Vishwakarma RK (1972) Urbanisation as a strategy of rural development. PTI, Library Press
Chandana RC (2016) A geography of population. Kalyani Publishers, Ludhiana
Chaudhary BK, Tripathi AK (2018) Failing cityscape urbanisation and urban climate. NGJI (Varanasi) 64:1–2
Chauhan A (2017) Urbanisation and environmental changes in India: problems and challenges. Jyoti Prakashan, Sonipat
Gosal GS (1956) A geographical analysis of India's population. PhD thesis (unpublished), Wisconsin University
Henderson A (2003) The urbanisation process sand economic growth: the so-what question. J Econ Growth 8:47–71
Jauhri AS. Trends of urbanisation in Punjab Plains during the twentieth century. NGJI 3:125–135
Kosambi M (1994) Urbanisation and urban development in India. ICSSR, New Delhi
Mills ES, Becker Charles M (1986) Studies in urban development. Oxford University Press, New York
Mukherjee AB (1961) Levels of urbanisation in Uttar Pradesh. GRI 35(1):31–42
Singh O (1967) Trends of urbanisation in Uttar Pradesh. NGJI 13:152–162

Chapter 20
Urban Sprawl and Future Land Resource Utilization Projections for Ghaziabad City, Uttar Pradesh, India, Using High-Resolution Remote Sensing Data and GIS Techniques

Virendra Kumar, Kamlesh Bhalavi, A. K. S. Rana, V. K. S. Rathore, and Harish Kumar

Abstract In almost all developing countries of the world, the extension of urban areas is a dynamic phenomenon over rural land resources. This chapter presents the results on the quantification of urban sprawl dynamics and land resource utilization rate and absorption coefficient for Ghaziabad city based on Survey of India topographical maps of 1971, IKONOS satellite data of 2009 and 2014 of Ghaziabad city. The results show that in 1971, the built-up area of the city based on the SOI topographical map was 2737.62 ha. Based on IKONOS satellite 1 m spatial resolution imageries of 2009 and 2014, it is 16,363.40 ha in 2009 and 18,411.80 ha in 2014, respectively. Urban spatial growth and per annum growth percentage between 1971 and 2009 is 13,625.58 ha (13.09%) and 2009–2014, the growth in built-up urban areas was observed as 2048.40 ha (2.50%), respectively. In time span of five years, from 2009 to 2014, a remarkable growth of 358.56 ha and 409.68 ha per annum has been observed during the period from 1971 to 2009 and 2009 to 2014 respectively. City extended toward the north-west direction along highways and major road corridors. The trend change between 2009 and 2014 has been observed as a remarkable decrease to − 18.82% in agricultural land and an increase in built-up urban 105.89%, respectively. The land consumption rate and absorption coefficient between the period from 1971 to 2009 and 2014 are 0.015 ha and 0.009 ha, and 0.005 ha, respectively, and the land

V. Kumar (✉) · K. Bhalavi · A. K. S. Rana · V. K. S. Rathore
Landuse & Urban Survey Division, Remote Sensing Applications Centre, Lucknow Sector-G, Jankipuram, Kursi Road, Lucknow (Department of Science & Technology, Uttar Pradesh), India,, Uttar Pradesh 226021, India
e-mail: drkvgis1989@gmail.com

V. K. S. Rathore
Department of Geography, Vardhman P.G. College, Bijnor, MJP Roh. University, 246701 Bareilly, India

H. Kumar
Department of Geography, Delhi School of Economics, Delhi University, Delhi, India

S. Anand et al. (eds.), *Sustainability in South Asian Cities*, Advances in Geographical and Environmental Sciences, https://doi.org/10.1007/978-981-97-7455-5_20

absorption coefficient is observed as 0.009 ha and 0.15 ha in the period of 1971–2009 and 2009–2014 respectively and for the period 2014–2021, it has estimated as 0.13 ha. The digital database generated for urban sprawl dynamics and land consumption rate for Ghaziabad city would be very useful for urban planners/decision-makers for better land use planning and management of land resources for cities.

Keywords Urban sprawl · Land resource utilization · IKONOS data · GIS techniques

20.1 Introduction

Almost all low and lower-middle-income group countries in the world, like India, are facing the problem of employment of people, and their ever-increasing basic needs in urban centres are ever putting pressure on our land. Developing countries like India are going through rapid urbanisation, reducing agricultural land dominancy and leading to urban sprawl. Approximately, there are about 7933 urban local bodies in the developing world. Uttar Pradesh, the largest populated state of India, is home to about 20 crores of population and presently has 763 urban local bodies (Census of India 2011). Hence, to analyse and monitor the present and past urbanization status and land-use land-cover (LULC) changes, multi-temporal satellite data integrated with GIS and GPS are the effective tools for urban growth analysis (Farooq and Ahmad 2008; Araya and Cabral 2008; Attua and Fisher 2011; Bhatta 2009; Cabral and Zamyatin 2006; Mohan et al. 2011; Gao et al. 2015; Sudhira et al. 2004; Gerten et al. 2019; Gregorio 2005; Himiyama 2004; Kumar et al. 2015, 2016; Lambin 1997; Lu and Weng 2007; NRSA 2003; Patkar 2003; Raziq et al. 2016) suggested that multi-temporal high-resolution satellite data in Geographical Information System environment and image classification techniques are effectively utilised and helpful to know the present and past status of urban land use/land cover dynamics, land transformation/land consumption, horizontal/vertical infrastructural development, change detection analysis in inner/outskirts of urban/rural areas. This chapter focuses on urban sprawl dynamics and future projections of land consumption for Ghaziabad city in the GIS domain. The study aims to prepare and analyze the detailed urban sprawl growth map, LULC change map, land transformation, trend change, per capita per person land absorption and land consumption of Ghaziabad city using IKONO'S (Google Earth) satellite 1 m spatial resolution data and GIS techniques.

20.2 Materials and Data Used

The district of Ghaziabad stretches from 28° 40′ North latitude and 77° 25′ East longitudes. Ghaziabad is fastest fastest-growing city in India. In the state of Uttar Pradesh, it is located in the northeastern part of the National Capital Region, at a distance of about 20 km from the National Capital, New Delhi.

The following dataset has been used to meet the set objectives

I. S.I.O. topographical map sheet No. 53H/1, 53H/5, 53H/6, 53H/9, 53H/10, 53H/13, 53H/14, 53L/1 and 53L/2 of Ghaziabad city on 1: 25,000/1: 50,000 scales surveyed in 1971.
II. Guide map of Ghaziabad city on 1:20,000 scale.
III. IKONOS 1m spatial resolution satellite data of study years i.e. 2009 and 2014.
IV. Ground truth data was collected using the Global Positioning System (GPS).
V. Census of India population data—1971, 2001, and 2011.

20.3 Methodology

The study is initiated by preparing and extracting the transport network using the survey of India (S.O.I) toposheet of the study area. The transportation network features like road (state/national highway), major/minor and canals are digitized as a polygon feature. Further, to update the road network the satellite data of IKONOS 1m spatial resolution is utilized of study years. S.O.I toposheet of 1971 and satellite data is used to prepare the urban growth and land-use maps for 2009 and 2014. Land consumption rate, ratio and the future land consumption have been predicted based on future population projection of the city and ground truth/field survey information was incorporated before the finalization of maps (Fig. 20.1).

20.4 Results and Discussions

The results and discussions are grouped under various sub-sections mentioned below for the present study, which covers urban spatial growth, LULC quantification, change and transformation.

20.5 Urban Spatial Growth, Rate and Direction

Urban sprawl is an extension of an urban built-up area beyond the city boundaries defined by municipal cooperation. The changing land use pattern is mostly directed by its influencing parameters such as population density, population growth rate, etc. The study and analysis of such parameters are crucial aspects for developing cities.

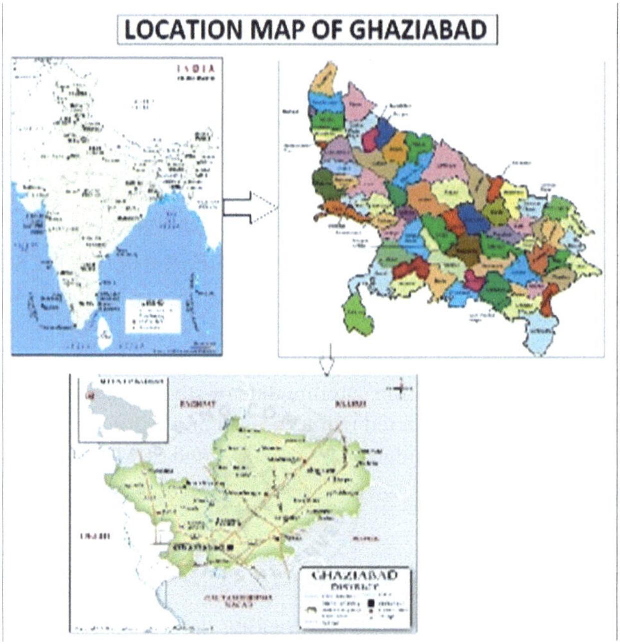

Fig. 20.1 Location map

It was observed from the results that the city grew concentrically. To analyze the urban sprawl of the city an overlay analysis was performed in which the vector layers are overlaid for the study years i.e. 1971, 2009 and 2014. It was observed that the 2737.82 ha area in 1971, which increased to 16,363.40 ha in 2009 and reached up to 18,411.80 ha in 2014 respectively as shown in Table 20.1. Further, the urban sprawl growth is revealed as 15,625.58 ha in time period of 1971–2009 and 2048.20 ha from 2009 to 2014 respectively. Table 20.2 asserts urban sprawls per annum growth and percentage per annum growth. Figure 20.6 and Table 20.1 narrate the urban expansion direction of the city. It is observed that the city mainly expanded towards the north to west and marginally in the east direction, where the prime agriculture and vegetated land is transformed to urban land with a remarkable growth of (13.09%) per annum, and it is compared to 2.50% per annum between 2009 to 2014 respectively.

Table 20.1 Urban spatial growth rate and percentage per annum

Year	Data	Area (ha)	Sprawl growth (ha)	Per annum growth (ha)	% per annum
1971	SOI toposheet	2737.82	–	–	
2009	IKONOS data	16,363.40	13,625.58	358.56	13.09
2014	IKONOS data	18,411.80	2048.40	409.68	2.50

Source Compiled by Researcher

Table 20.2 Land transformation statistics of study area between 2009 and 2014

S. no.	Class name	2009 (area in ha)	2014 (area in ha)	Change (area in ha)	Trend change (%)
1	Agriculture land	1068.4	682.8	−385.6	−18.82
2	Airport	923.4	1042.9	119.5	5.83
3	Built up	13,611.7	15,780.8	2169.1	105.89
4	Canal	131.6	175.9	44.3	2.16
5	Open space	487.7	553.5	65.8	3.22
6	Orchard/ Plantation	21.3	37.4	16.1	0.78
7	Park	38.6	41.4	2.8	0.14
8	Playground	18.9	24.4	5.5	0.27
9	Waterbody	61.8	72.7	10.9	0.53
	Total	16,363.40	18,411.80	2048.4	100

Land use/cover and Land Transformation of the study area are discussed as follows:

Land use/Land cover (1971, 2009 and 2014)

LULC map for the year 1971 is prepared using SOI topographical map, which is categorized in the four LULC classes. LULC has been depicted in Fig. 20.2. The four different categories are built-up, industrial, open space and orchard/plantation. These LULC classes occupy built-up 2624.25 ha (95.86%), open space 93.06 ha (3.39%), orchard/plantation 18.45 ha (0.67%) and waterbodies 2.06 ha (0.08%), respectively. Landuse categories based on 2009 imagery are agriculture Land 1068.4 ha (6.53%), built up 13,611.7 ha (83.18%), orchard/plantation 21.3 ha (0.13%), playground 18.9 ha (0.12%), park 38.6 ha (0.24%), open space 487.7 ha (2.98%), water bodies/pond 61.8 ha (0.37%), airport 1068.4 ha, open space 487.7 ha (2.98%) and canal 131.6 ha (0.81%), respectively (Figs. 20.3 and 20.4).

The nine LULC categories based on 2014 (Google data) are identified and mapped in the study area. Among these, the agriculture land occupies an area of 682.8 ha (3.71%), built up 15,780.8 ha (85.71%), orchard/plantation 37.4 ha (0.21%), playground 24.4 ha (0.13%), park 41.4 ha (0.23%), open space 553.5 ha (3.01%), water body 72.7 ha (0.39%), airport 1042.9 ha (5.66%) and canal 175.9 ha (0.95%), respectively (Figs. 20.5, 20.6 and 20.7).

20.6 Land Transformation Between 2009 and 2014

The present study shows that agricultural land has an aerial decrease of 385.6 ha from 2009 to 2014 and a change trend of −18.82%. Table 20.2 and Fig. 20.8 present the trend of changes of land-use categories where built-up (urban area) is observed to

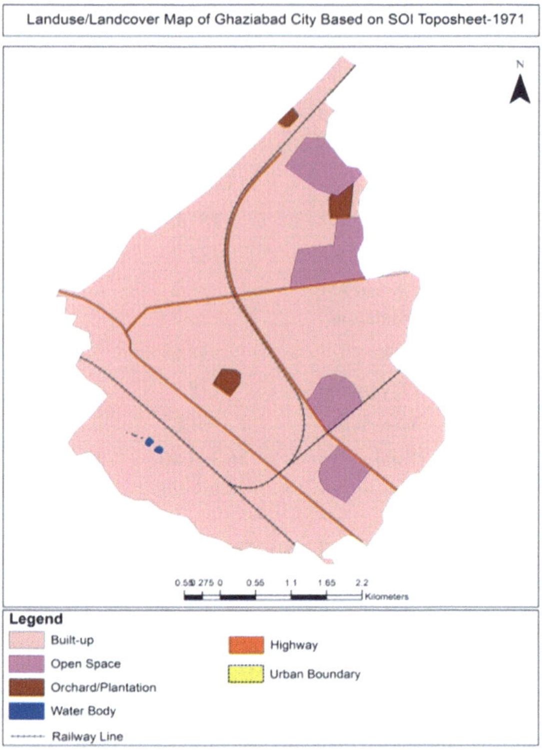

Fig. 20.2 Land use/cover map-1971

be increased 2169.1 ha (105.89%), canal 44.3 ha (2.16%), airport 119.5 ha (5.83%), parks 2.8 ha (0.14%), playground 5.5 ha (0.27%), Orchard/Plantation decreased to 16.1 ha (0.78), open space 65.8 ha (3.22%), and water bodies 10.9 ha (0.53), respectively. Some category's trend percentage values are negligible (Figs. 20.7 and 20.8).

20.7 Land Resource Consumption Rate and Future Projections

The rate of land consumption is observed to be 0.015, 0.009 and 0.005 for the years 1971, 2009 and 2014, respectively, as depicted in Table 20.3. Consequently, the land absorption coefficient is obtained as 0.009, 0.15 and 0.13 for the time period between 1971–2009, 2009–2014 and 2014–2021 projected, respectively.

Table 20.4 shows the population of study area from 1971, 2001 and 2011 and the estimated population for 2021.

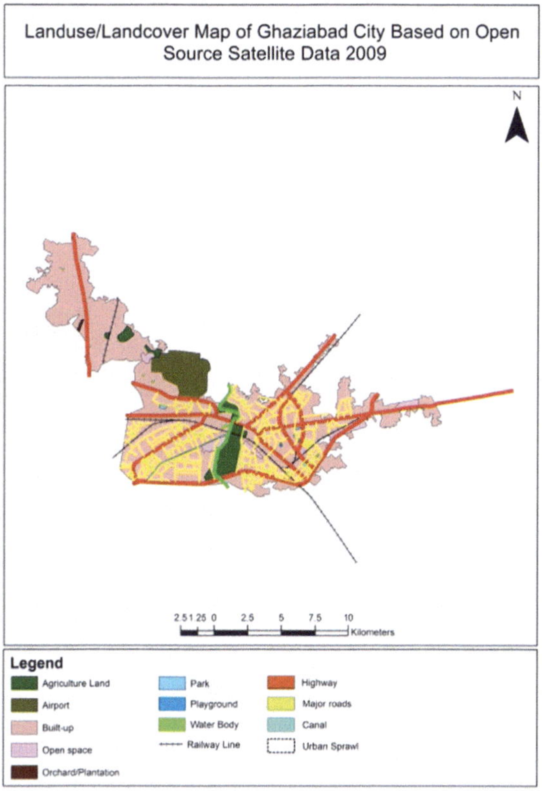

Fig. 20.3 Land use/cover map-2009

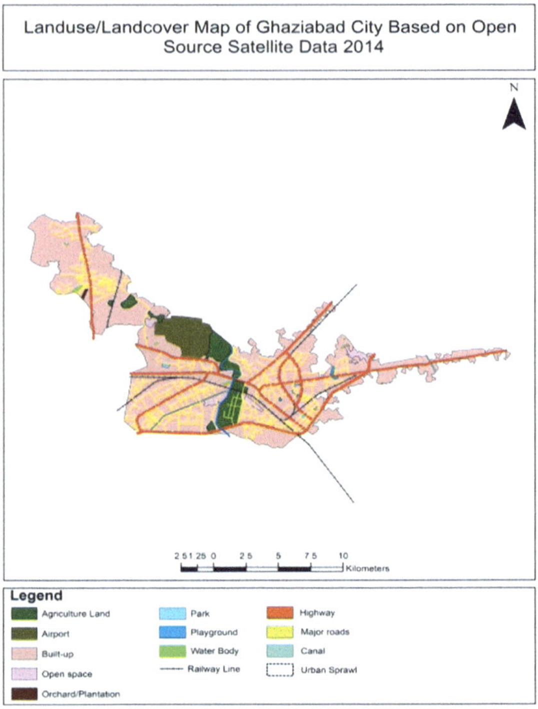

Fig. 20.4 Land use/cover-2009

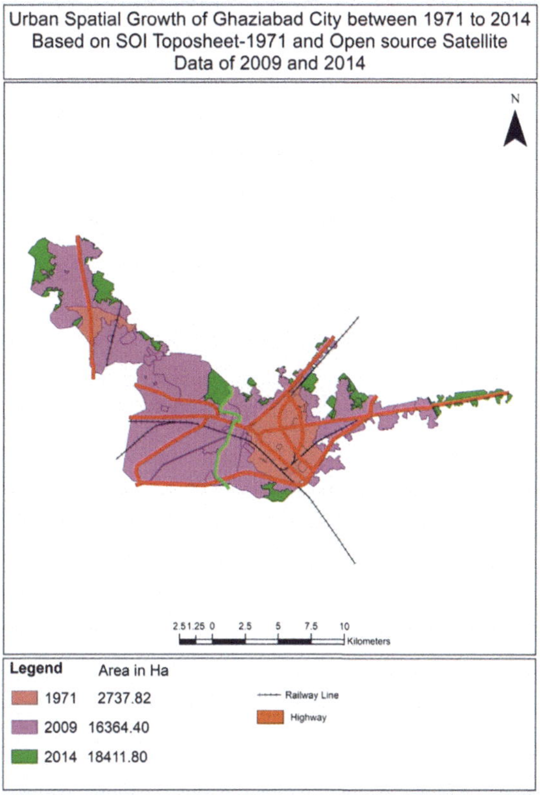

Fig. 20.5 Landuse/landcover-2014

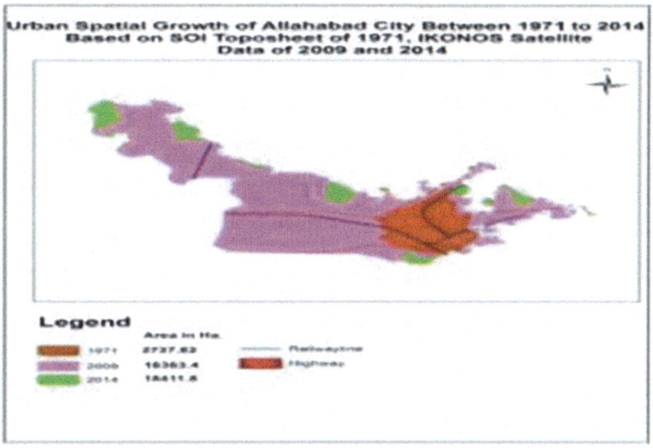

Fig. 20.6 Urban spatial growth—1971–2014

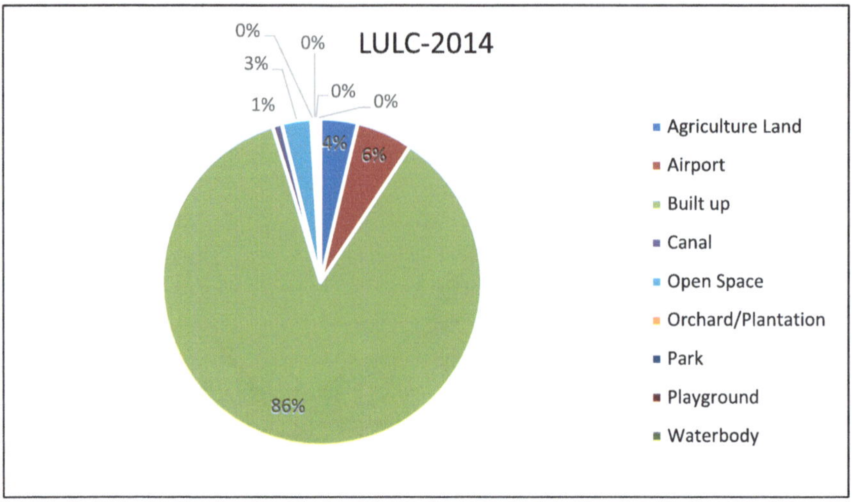

Fig. 20.7 Land use/cover-2009

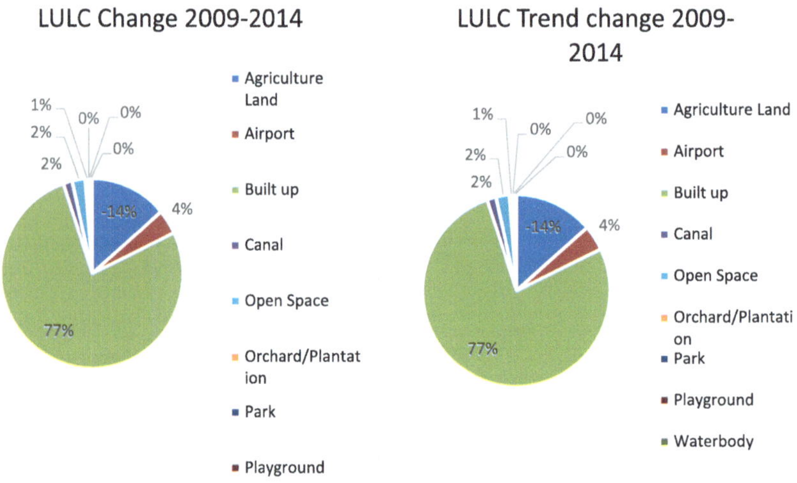

Fig. 20.8 Land use/cover change and trend change 2009–2014

Table 20.3 Land consumption projections

Year	Land resource consumption	Year	Land absorption coefficient
1971	0.015	1971–2009	0.009
2009	0.009	2009–2014	0.15
2014	0.005	2014–2021[a]	0.13[a]
2021[a]	0.13[a]		

[a] Estimated

Table 20.4 Population projections

Year	Population	Source
1971	1,81,832	Census of India
2001	13,67,575	Census of India
2009[a]	15,86,183	
2011	16,48,643	Census of India
2021[a]	23,44,700	

Source Compiled by the Author, based on Census of India Data
[a] Estimate

20.8 Conclusion

The study demonstrates the utility of high-resolution remote sensing data and GIS techniques for mapping and monitoring urban extension and proper management of land and other natural resources at local and global levels. From 1971 to 2014, various residential colonies in Ghaziabad city have been developed and permanently transformed into built-up urban.

References

Araya Y, Cabral P (2008) Urban land cover change detection analysis using GIS and remote sensing: a case study of Asmara, Eritrea

Attua EM, Fisher JB (2011) Historical and future land-cover change in a municipality of Ghana. Earth Interact 15(9):1–26. https://doi.org/10.1175/2010ei304.1

Bhatta B (2009) Analysis of urban growth pattern using remote sensing and GIS: a case study of Kolkata, India. Int J Remote Sens 30(18):4733–4746. https://doi.org/10.1080/01431160802651967

Cabral P, Zamyatin A (2006) Three land change models for urban dynamics analysis in Sintra-Cascais area. In: 1st EARSeL workshop of the SIG urban remote sensing Humboldt-Universität zu Berlin, 2–3 March 2006

Census of India (1971) Primary census abstract, Ministry of Home Affairs, Government of India, New Delhi

Census of India (2001) Primary census abstract, Ministry of Home Affairs, Government of India, New Delhi

Census of India (2011) Primary census abstract, Ministry of Home Affairs, Government of India, New Delhi
Farooq S, Ahmad S (2008) Urban sprawl development around Aligarh city: a study aided by satellite remote sensing and GIS. J Indian Soc Remote Sens 36(1):77–88. https://doi.org/10.1007/s12524-008-0008-0
Gao J, Wei Y, Chen W, Yenneti K (2015) Urban land expansion and structural change in the Yangtze River Delta, China. Sustainability 7(8):10281–10307. https://doi.org/10.3390/su70810281
Gerten C, Fina S, Rusche K (2019) The sprawling planet: simplifying the measurement of global urbanization trends. Front Environ Sci 7. https://doi.org/10.3389/fenvs.2019.00140
Gregorio A (2005) Land cover classification system: classification concepts and user manual. Food and Agricultural Organisation of the United Nations, Rome
Himiyama Y (2004) The trend, mechanism and impact of urban expansion in Japan. http://www.gisdevelopment.net/application/urban/overview/urban0042p1/htm, map Asia
Kumar V, Bhalvai K, Shukla A (2015) Quantification of land transformation using multi temporal satellite data and GIS techniques. In: National conference on recent advances in electronics computer engineering (RAECE). IEEE, pp 106–111 (2015)
Kumar V et al (2016) Urban spatial growth and land transformation analysis using high resolution remote sensing data and GIS techniques—a case study of Lucknow city, Uttar Pradesh, India. In: Proceedings of rapid urbanization and sustainable development in Asia, the 13th international urbanization conference, Badan Penarbit Fakultas Geographi (BPFG), Universitas of Gadjah Mada, Yogyakarta, Indonesia, pp 118–129. ISBN: 978-979-8786-58-7
Lambin E (1997) Modeling and monitoring land-cover change process in tropical regions
Lu D, Weng Q (2007) A survey of image classification methods and techniques for improving classification performance. Int J Remote Sens 28(5):823–870
Mohan M, Pathan et al (2011) Dynamics of urbanisation and its impact on landuse/landcover: a case study of Megacity Delhi. J Environ Protect 2
NRSA (2003) India Remote Sensing (IRS)-P6. National Remote Sensing Agency [online]. http://www.euromap.de/download/P6_data_user_handbook.pdf. Accessed Oct 2008
Patkar VN (2003) Directions for GIS in urban planning. GIS@development. Progr Phys Geogr 21
Raziq et al (2016) Monitoring of land use/land cover changes and urban sprawl in Peshawar city in Khyber Pakhtunkhwa: an application of geo-information techniques using of multi-temporal satellite data. J Remote Sens GIS5 174. https://doi.org/10.4172/2469-4134.10000174
Sudhira HS, Ramachandra TV, Jagadish KS (2004) Urban sprawl: metrics, dynamics and modelling using GIS. Int J Appl Earth Observ Geoinf 5(1):29–39

Part III
Environmental Challenges

Chapter 21
Analyzing and Mapping of Change in Water Surface Area Using Remote Sensing Data: Case Study of Band-E Amir National Park, Afghanistan

Dilek Eren Akyuz, Filiz Bektas Balcik, Ahmad Frahmand, and Fusun Balik Sanli

Abstract Surface water, such as rivers, lakes, swamps, and reservoirs, has an essential role in the Earth's water cycle, influencing the ecosystem and climate locally and globally. Because of their roles, the surface area change of the lakes is accepted as an important indicator for lake assessment. This chapter used remotely sensed data to determine spatial and temporal lake surface area changes in the selected four lakes (Band-e Zulfiqar, Band-e Haibat, Band-e Panir, Band-e Pudina) of Band-e Amir National Park, Afghanistan, with Google Earth Engine platform. To achieve the research objectives, the Modified Normalized Difference Water Index (MNDWI) was selected among the other indices to extract and quantify the lake water surface area variations between May 2000 and August 2019, using multi-temporal Landsat 7 Enhanced Thematic Mapper Plus (ETM +) and Landsat 8 Operational Land Image (OLI) data. A threshold value is determined to discriminate water surfaces from the other categories. Calculated accuracy assessment results showed that the selected method could provide the spatial distribution of lake water surface area changes with high accuracy. The results may help strengthen sustainable development goals (SDGs).

D. E. Akyuz (✉)
Department of Civil Engineering, Istanbul University-Cerrahpasa, 34320 Avcilar, Istanbul, Türkiye
e-mail: dilekeren.akyuz@iuc.edu.tr

F. B. Balcik
Department of Geomatics Engineering, Istanbul Technical University, Maslak, Istanbul, Türkiye
e-mail: bektasfi@itu.edu.tr

A. Frahmand · F. B. Sanli
Department of Geomatics Engineering, Yildiz Technical University, Esenler, Istanbul, Türkiye
e-mail: ahmad.shah@std.yildiz.edu.tr

F. B. Sanli
e-mail: fbalik@yildiz.edu.tr

S. Anand et al. (eds.), *Sustainability in South Asian Cities*, Advances in Geographical and Environmental Sciences, https://doi.org/10.1007/978-981-97-7455-5_21

Keywords Band-e Amir national park · Remote sensing · Modified normalized difference water index (MNDWI)

21.1 Introduction

Clean, fresh water has been a critical resource for promoting health in both developed and undeveloped countries since the beginning of humanity. For this reason, it has been used extensively and assumed to be an infinite resource. Moreover, the distribution of available water worldwide is unequal; now, 1 of 3 (2.2 billion) people do not have access to safe drinking water (Martínez-Santos 2017; WHO 2021). As a result, the Food and Agriculture Organization of the United Nations (FAO) places importance on a water-energy-food nexus sensitively linked (FAO 2023). Drought and flood may also lead to water-borne diseases such as tuberculosis and malaria (Kondo et al. 2002).

Given the significance of water, the sustainable development goals (SDGs) are directly or indirectly related to water and water issues, including sanitation, hygiene, water resources management, water quality, and water-related risks. From the global perspective, SDGs are directly related the nature, and the wanted future "*We recognize that planet Earth and its ecosystems are our home and that "Mother Earth" is a common expression in a number of countries and regions, and we note that some countries recognize the rights of nature in the context of the promotion of sustainable development*" (UN 2021). The SDGs envision an adequate quantity and high quality of water with holistic management, protection of ecosystems, reduced pollution, increased recycling, and finally, the world becoming *harmonized with nature*. Achieving this goal requires sustainable water withdrawal, efficient freshwater supply to minimize water scarcity, sustainable wastewater management, and reduced flood damage (Naddeo 2013).

The SDGs (UN SDGs 2021) and Paris Agreement (UNFCCC 2021) target climate change, while the Sendai Framework (UNDRR 2021) focuses on disaster risk reduction, protection, and restoration of ecosystems such as plants, animals, and other organisms, as well as weather and landscape, including mountains, forests, wetlands, rivers, aquifers, and lakes to ensure sustainable living for a healthy and wealthy future. Protecting and ensuring the sustainability of lakes is essential to achieve these targets (Ho and Goethals 2019).

The field of limnology, established by Forel, emphasizes the crucial role of lakes in nature with "*Every lake is an organ of earth*!" (Sioli 1975). Lakes are essential for primary cycles, such as those involving water, carbon, nitrogen, and phosphorous (Wetzel 2001). Despite their importance and essential roles, lakes are often used as wastewater sinks. The lakes help with the purification of water; the polluted water enters the lake and outflows, which has better water quality (Akyuz et al. 2014). Moreover, lakes are utilized as water supply and energy resources, such as biofuel (Sandar et al. 2019) and hydropower energy. Additionally, the fisheries (Lynch et al. 2019) that benefit human health and well-being (Sadeghi-Bazargani et al. 2019). Clearly, the importance of lakes in our lives increasing, and their protection is essential for sustainability.

Sustainable usage of lakes may help to achieve **water**, **energy,** and **food** security (Keskinen et al. 2015). Furthermore, some parts of sustainable development goals (SDGs) include no poverty(1), zero hunger (2), good health and well-being (3), clean water and sanitation (6), affordable and clean energy (7), decent work and economic growth (8), industry, innovation and infrastructure (9), sustainable cities and communities (11), responsible consumption and production (12), climate action (13), life below water (14), life on land(15), are directly or indirectly related to the lake (UN SDGs 2021).

Methods based on remotely sensed data have advantages over the traditional methods in surface water mapping due to the low cost, high frequency, and repeatable observations. In recent decades, freely available active and passive satellite data such as the moderate-resolution imaging spectroradiometer (MODIS), Landsat, and Sentinel images have been used to determine surface water areas at different scales, ranging from regional to continental, and global levels (Carroll et al. 2016; Xia et al. 2019). Additionally, several satellite-based approaches, such as classification methods and remote sensing indices, have been developed to detect surface water. To quickly and accurately extract temporal water surface areas at different scales, various water indices have been developed and applied (Albarqouni et al. 2022; Yagmur et al. 2021).

Many studies have been conducted using remotely sensed data and the Google Earth Engine (GEE) platform to determine changes in lake water surfaces. Mohsen et al. (2018) analyzed the spatiotemporal variation of Burullus Lake in Egypt using Landsat data sets from 1972 to 2015. Xia et al. (2019) examined the spatiotemporal dynamics of surface water areas in the Huai River Basin from 1989 to 2017 using the time series of Landsat TM/ETM + /OLI images on the GEE cloud computing platform. Wang et al. (2020) conducted a study to understand the impact of climate change on long-term changes in water surface areas in the Hetao Plain, using available Landsat images from 1989 to 2019 on the GEE cloud platform. Yagmur et al. (2021) used remotely sensed data and the GEE platform to determine long and short-term assessments of water surface area changes in 18 natural lakes in the Konya Closed Basin of Türkiye.

This study aims to calculate temporal changes in the surface area of lakes in Band-e Amir Protection Area, Afghanistan. Cloudless images were processed using the Modified Normalized Difference Water Index (MNDWI) among the available Landsat images in the GEE platform. Four lakes (Band-e Zulfiqar, Band-e Haibat, Band-e Panir, Band-e Pudina) were selected to calculate water surface area changes from 2000 to 2019 to evaluate the performance of the lake protection through comparison data from before (2000–2009) and after (2010–2019) the declaration of the protected area in 2009. The results show that the regulations for protection have helped to preserve the lakes as their water level did not change much despite Afghanistan experiencing a period of drought.

21.2 Methodology

This study aims to examine the spatiotemporal dynamics of surface water areas of Band-e Amir Lakes located within Band-e Amir National, Park from 2000 to 2019. This will be achieved by utilizing the time series of Landsat ETM + /OLI images, which will be accessed through the GEE cloud computing platform.

21.3 Study Site

The natural lakes of Band-e Amir are located at 34°50′ North latitude and 67°10′ East longitude (Fig. 21.1).

Band-e Amir National Park is one of the most beautiful landscapes of Afghanistan, and more importantly, it has the widest variety of biodiversity in the region. Despite the recent drought experienced in Afghanistan, as well as globally, the park's hydrological condition remains relatively stable. Given its natural treasure, it has the potential to be designated as a World Heritage Site, provided that appropriate management and community support are in place. If this is achieved, it could attract nature enthusiasts and become a significant nature tourism destination in the region.

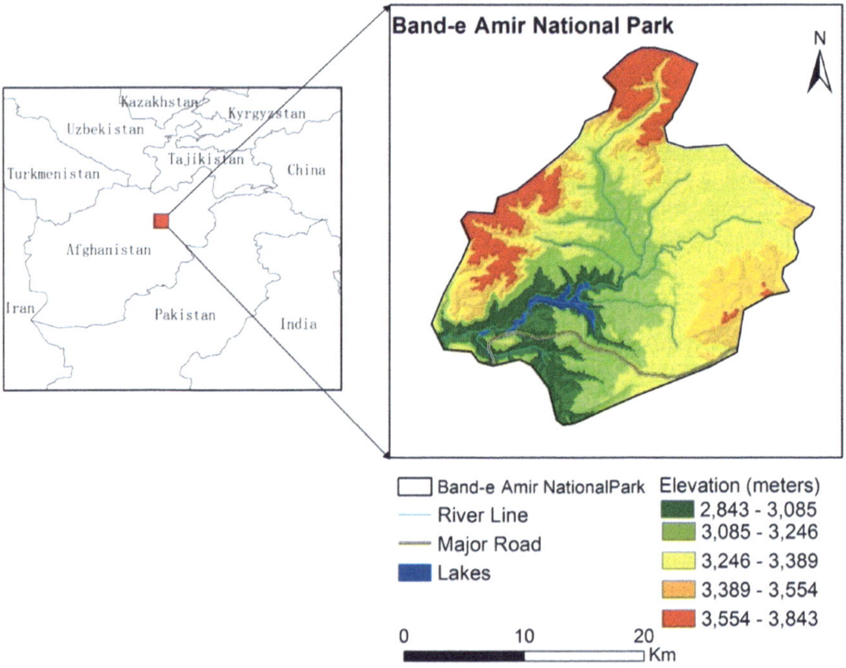

Fig. 21.1 Map of the study area. *Source* Prepared by the Authors

Fig. 21.2 Selected lakes: Band-e Zulfiqar, Band-e Haibat, Band-e Panir, and Band-e Pudina for the study area, in Afghanistan. *Source* Prepared by the Authors

The distance from Afghanistan's capital, Kabul, to Band-e Amir is 215 km. Band-e Amir is located north of Kohe Baba Mountain, which has a peak height of 5143 m. Band-e Amir consists of a chain of lapis lazuli lakes, which is east–west oriented, nestled between 300 m-high magenta rock walls in mountain ranges of the Hazaratjat massif.

These are namely Band-e Haibat (meanings of the names in English are given in parenthesis; Awe), Band-e Panir (Cheese), Band-e Pudina (The Mint), and Band-e Zulfiqar (the Sword of Ali), Band-e Gholaman (Slave), Band-e Qambar (The Groom). However, the last two were excluded in this study. Band-e Haibat and Band-e Zulfiqar are the two largest lakes, covering 490 ha and 90 ha, respectively (Fig. 21.2).

Band-e Amir lakes are not only known for their stunning beauty but also for their biological and geological features. The active encrusting waters of the lake allow for the growth of continental bioherms, making them an exceptional natural wonder (Bourrouilh-Le Jan et al. 2007).

The region's geomorphology comprises a vast plain consisting of slightly southwest-dipping upper Cretaceous marl sandstone marine sediments, which are over 700 m thick and covered by several-meter-thick Neogene marl conglomerates. The plain is situated at an elevation of 3300 to 3600 m and is bordered to the south by the Kohe-Baba massif. Additionally, a canyon-like valley 500 to 600 m cuts through a portion of the plain upstream of the Band-e Amir valley.

Jux and Kempf (1971) estimated that the investigated fragment of the Band-e Amir drainage basin covers an area of approximately 410 km^2 and receives an average annual precipitation of 450 mm, supplying approximately 185 million m^3 of water. However, the precipitation distribution varies greatly across seasons, with a significant increase in water supply observed in early summer due to the melting of snow and ice cover on the Kohe Baba massif (Pietruszka and Szarejko 1980). In terms of the water balance of the valley, evaporation plays a critical role, particularly given the desert steppe climate of the area. Snow cover is common from November until May and can reach up to 2 to 3 m thick, while the daily temperature during the months with clear skies can reach up to 30 °C, dropping below 0 °C at night.

The lakes' water is oligotrophic and calcareous, with a pH of 7.8. The lakes' stunning deep blue color is attributed to their high purity and high lime content with the surface water temperature reaching 14–17 °C during summer due to the precipitation of calcium carbonate (sinters and travertines). In winter, the lakes freeze over.

The four lakes have been naturally formed as a result of the valley being partitioned by travertine dams of the Haybat Series. This series is composed of calc-sinters and travertines, with inserts of limy sands, lakes marls, gyttja, and peat and Yakhak series (Lindner 1979). The flow in these Band-e Amir valleys occurs along gaps that have formed through the upper cretaceous sandstones.

Band-e Amir was recognized as a National Park in 2004 (BBC 2009), in the central high elevation of Bamyan Province, a national park in central Afghanistan. The national park was established on 22 April 2009, to promote and protect the natural beauty of a series of intensely blue lakes created by natural dams high approximately 3000 m of elevation in the Hindu Kush west of the famous Buddhas of Bamyan. It is a series of deep blue lakes separated by natural dams made of travertine (Alden 2019), a mineral deposit.

Band-e Amir is a unique natural wonder created by carbon dioxide-rich water leakage through the faults and fractures, resulting in the deposition of calcium carbonate precipitate in the form of travertine walls that store the water of these lakes. It is one of the few rare natural lakes in the world that is created by travertine systems.

The critical importance of Band-e Amir lies in its unique combination of features that meet the formal criteria for acceptance as a UNESCO World Heritage Natural Site. With proper management, it has the potential to become an essential eco-tourism destination (UNEPGRID 2021). Band-e Amir has been described as Afghanistan's Grand Canyon and attracts thousands of tourists annually (Ryan et al. 2009).

21.4 Data

Landsat satellite images are commonly used for long-term temporal change analyses of water surfaces and are widely used tools for environmental monitoring, as they provide free-of-charge optical images dating back to 1972. In this study, 158 Landsat images were downloaded but only 139 Landsat images were selected for further analysis based on a condition of 10% cloud cover (Table 21.1).

Table 21.1 Date of the images used for calculation

Date of total 139 images						
25/05/2000	05/06/2004	26/05/2006	19/08/2008	22/04/2011	01/08/2013	13/11/2016
12/07/2000	07/07/2004	11/06/2006	04/09/2008	09/06/2011	02/09/2013	31/12/2016
29/08/2000	23/07/2004	13/07/2006	20/09/2008	25/06/2011	18/09/2013	24/05/2017
01/11/2000	08/08/2004	29/07/2006	22/10/2008	11/07/2011	04/10/2013	25/06/2017
26/04/2001	24/08/2004	30/08/2006	23/11/2008	27/07/2011	20/10/2013	27/07/2017
29/06/2001	09/09/2004	15/09/2006	03/06/2009	13/09/2011	19/07/2014	28/08/2017
01/09/2001	11/10/2004	27/04/2007	05/07/2009	29/09/2011	20/08/2014	29/09/2017
04/11/2001	21/04/2005	13/05/2007	21/07/2009	15/10/2011	21/09/2014	15/10/2017
15/05/2002	08/06/2005	14/06/2007	06/08/2009	24/04/2012	10/12/2014	02/12/2017
18/07/2002	24/06/2005	30/06/2007	07/09/2009	10/05/2012	19/05/2015	27/05/2018
19/08/2002	10/07/2005	16/07/2007	23/09/2009	11/06/2012	22/07/2015	12/06/2018
20/09/2002	26/07/2005	01/08/2007	09/10/2009	27/06/2012	07/08/2015	30/07/2018
22/10/2002	11/08/2005	17/08/2007	19/04/2010	29/07/2012	23/08/2015	31/08/2018
18/05/2003	27/08/2005	18/09/2007	21/05/2010	14/08/2012	24/09/2015	16/09/2018
06/08/2003	12/09/2005	04/10/2007	06/06/2010	15/09/2012	10/10/2015	18/10/2018
23/09/2003	28/09/2005	20/10/2007	22/06/2010	01/10/2012	21/05/2016	14/05/2019
25/10/2003	14/10/2005	05/11/2007	08/07/2010	02/11/2012	22/06/2016	17/07/2019
10/11/2003	30/10/2005	29/04/2008	25/08/2010	29/05/2013	24/07/2016	18/08/2019
04/05/2004	24/04/2006	16/06/2008	26/09/2010	14/06/2013	25/08/2016	19/09/2019
20/05/2004	10/05/2006	18/07/2008	12/10/2010	16/07/2013	26/09/2016	

Source Prepared by the author

21.5 Extraction of Water Surface Area Using GEE

GEE is a spatial data processing service and is powered that uses the Google Cloud Platform to process satellite images with varying spectral, temporal, radiometric, and spatial resolution (GEE 2023). The purpose of the GEE is to provide an interactive platform that has had a significant impact on data-driven science in recent years. GEE is a cloud-based platform that includes a multi-petabyte catalog of a variety of remotely sensed data, together with historical earth images spanning more than 40 years. Ready-to-use data sets in GEE can be used to perform long-term analyses like forest and water surface studies from global to local scales. In this study, Landsat 7 and 8 images were selected from the GEE platform across the Band-e Amir National Park from 2000 to 2019. The collected images were masked using the Fmask function to remove low-quality pixels caused by clouds, snow, and shadows (Fig. 21.3).

Remote Sensing technology has a long-standing history in water resources assessment. Using bands (single band or multi-band) in remote sensing images, information about water surfaces are extracted. The single-band method involves selecting a band

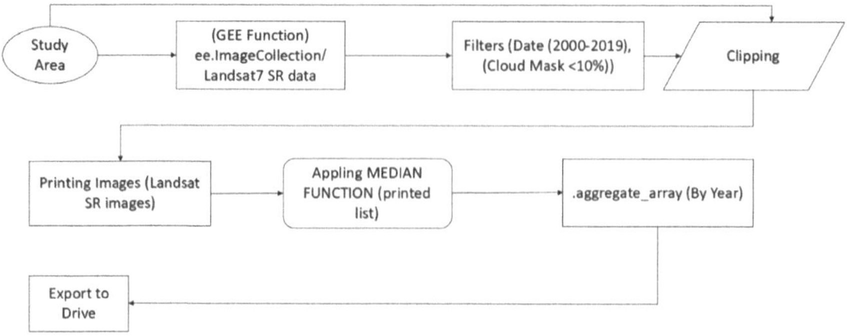

Fig. 21.3 Flowchart of the Google Earth Engine to extract lake shoreline. *Source* Prepared by the authors

from a multispectral image to extract water surface area, followed by the determination of a threshold to distinguish between water and land (Rundquist et al. 1987). However, selecting the threshold value may lead to an over- or underestimation of the water surface area.

The multi-band method utilizes the reflective differences of each involved band and is divided into two types. The first type involves analyzing the signature features of each ground target among different spectral bands and using an if-then-else logic to delineate land from water surfaces (Xu 2002). The second type of multiband approach uses two multispectral bands: one is taken from visible wavelengths and is divided by near-infrared (NIR) to suppress vegetation and land presence while enhancing water features. However, this method may not effectively remove non-water features. To address this issue, the Normalized Difference Water Index (NDWI) was proposed by McFeeters (1996). However, the NDWI may not efficiently suppress the signal from built-up land in most areas. To solve the mixed pixel problem, the Modified Normalized Difference Water Index (MNDWI; Xu 2006) was used in this study.

21.6 Modified Normalized Difference Water Index (MNDWI)

In the realm of remote sensing, various indices have been developed to identify water surfaces. In the present study, spatiotemporal changes in water were determined by using MNDWI. This index is widely recognized as an effective means of delineating open water areas, and is calculated using the green (band 2) and short-wave infrared (band 5) of Landsat TM (Han-Qiu 2005). MNDWI was calculated on a pixel basis, and extracted the surface water using the criteria of MNDWI value > 0.1. When the criteria were met, a pixel was classified as surface water which was then assigned to 1. The annual area of surface water area was calculated for selected lakes between 2000 and 2019 using this approach.

$$\text{MNDWI} = (\text{Green} - \text{SWIR}) / (\text{Green} + \text{SWIR}) \quad (21.1)$$

where Green and SWIR denote the reflectance of the green and shortwave infrared bands, respectively.

21.7 Results and Discussion

21.7.1 Lake Surfaces' Area Change

The basic statistical properties, such as minimum, maximum, average, and standard deviation, of the lakes' surface areas which are calculated from image processing are given in Table 21.2.

Three distinct time periods were chosen based on the declaration of the protected area in 2009. The first period spans from 2000 to 2009 and contains 72 lake water level values. The second interval covers the years 2010–2019 and includes 67 lake water level values. The last period selected from 2000–2019 includes 139 lake water level values. Comparison of the statistical properties for the selected lakes revealed that, for all periods, the minimum and maximum of the lake surface areas were generally larger in the 2010–2019 period except for Band-e Pudina. In addition, the

Table 21.2 Basic statistical parameters of the lakes' surface areas (km^2)

Parameter	Period	Band-e Zulfiqar	Band-e Haibat	Band-e Panir	Band-e Pudina
Minimum	2000–2009	4.173	0.531	0.008	0.008
	2010–2019	4.207	0.531	0.022	0.003
	2000–2019	4.173	0.531	0.008	0.003
Maximum	2000–2009	5.105	0.714	0.085	0.040
	2010–2019	5.278	0.733	0.092	0.036
	2000–2019	5.278	0.733	0.092	0.040
Max/Min	2000–2009	1.224	1.345	11.304	5.263
	2010–2019	1.255	1.381	4.261	12.421
	2000–2019	1.265	1.381	12.243	14.021
Average	2000–2009	4.632	0.637	0.055	0.017
	2010–2019	4.636	0.633	0.055	0.016
	2000–2019	4.634	0.635	0.055	0.016
Std Dev	2000–2009	0.156	0.040	0.015	0.009
	2010–2019	0.170	0.044	0.014	0.009
	2000–2019	0.162	0.042	0.015	0.009

Source Calculate by the author

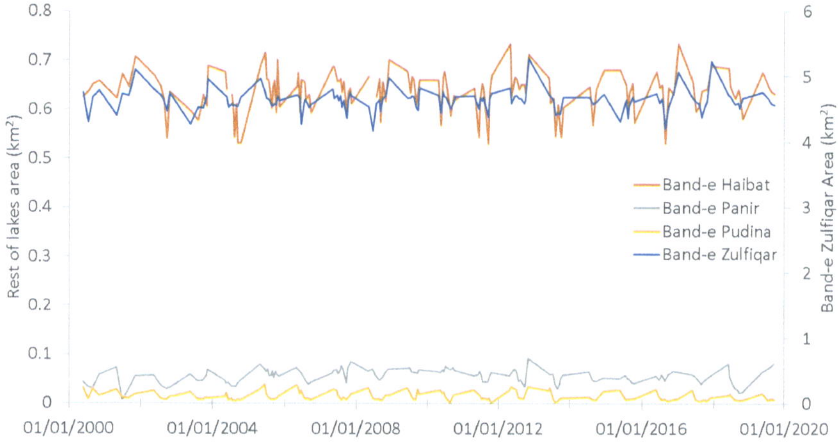

Fig. 21.4 The time series of the lakes' surface areas in km^2. *Source* Prepared by the authors

average surface area of Band-e Zulfiqar, the largest lake in the study area, was the largest in the 2010–2019 period.

The lakes' surface areas were calculated from image processing and their temporal changes are depicted in Fig. 21.4.

It was observed that the ratio between maximum and minimum values of the lakes' surface areas is inversely proportional to the size of the lakes' areas. This is due to the fact that smaller lakes are more susceptible to external forces and thus exhibit greater sensitivity to changes.

The correlation coefficient values of the lakes are also given in Table 21.3. The highest value of 0.616 is observed between Band-e Zulfiqar and Band-e Haibat, which are the two larger lakes. The low correlation coefficient values for the other lakes may be attributed to the fact that meteorological conditions may have a greater impact on the surface areas of the shallow and smaller lakes, making them less correlated with the larger lakes.

To compare the status of the lakes before (2000–2009) and after (2010–2019) the declaration of the protected area in 2009, Fig. 21.5 shows the lake surface area for the dry period (July and August) at the times of 2000, 2009, and 2019. Minimum values of lake surface areas were observed in 2000, except for Band-e Pudina.

Table 21.3 Correlation coefficient values of the lakes' surface areas

Band-e	Zulfiqar	Haibat	Panir	Pudina
Zulfiqar	1.000	0.616	0.299	0.289
Haibat	–	1.000	0.357	0.412
Panir	–	–	1.000	0.427
Pudina	–	–	–	1.000

Source Calculated by the author

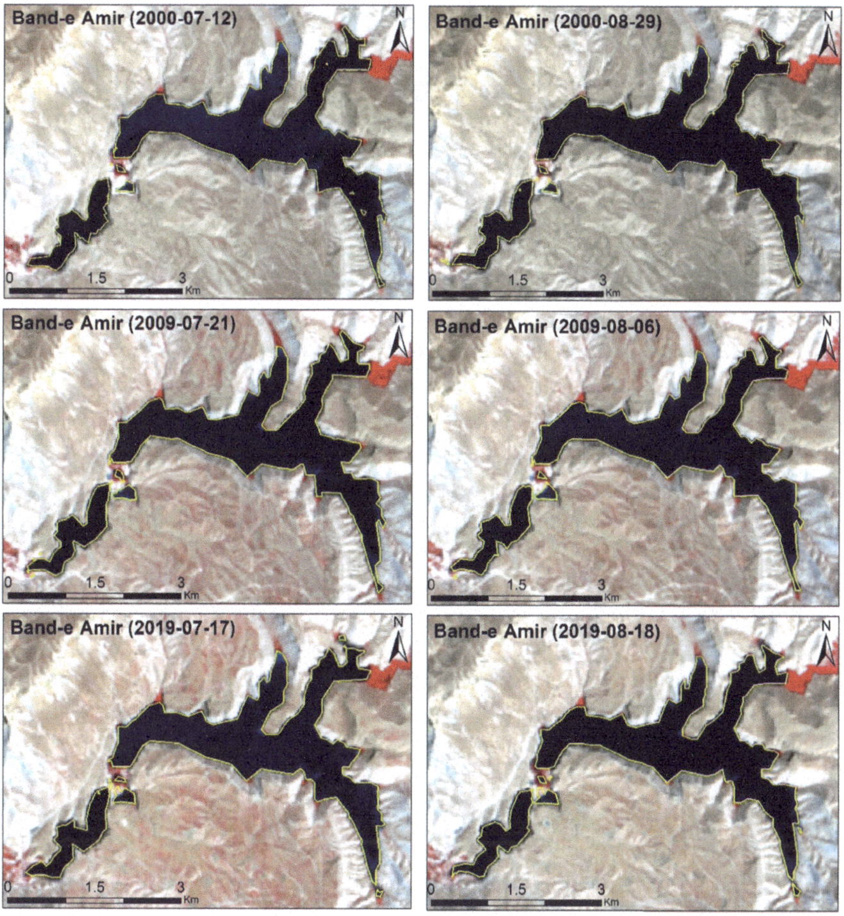

Fig. 21.5 The four lakes' surface area for dry period (July and August) in 2000, 2009, and 2019. *Source* Prepared by the authors

The highest change in the surface area of the four lakes occurred in 2008, as shown in Fig. 21.6. During the periods from 04 September 2008 to 20 September 2008 and from 22 October 2008 to 23 November 2008, there was a change of more than 5% of the total surface area of lakes. For example, the surface area of the Band-e Zulfiqar was changed from 4.70 km^2 to 4.97 km^2 within the time from 22 October 2008 to 23 November 2008, respectively.

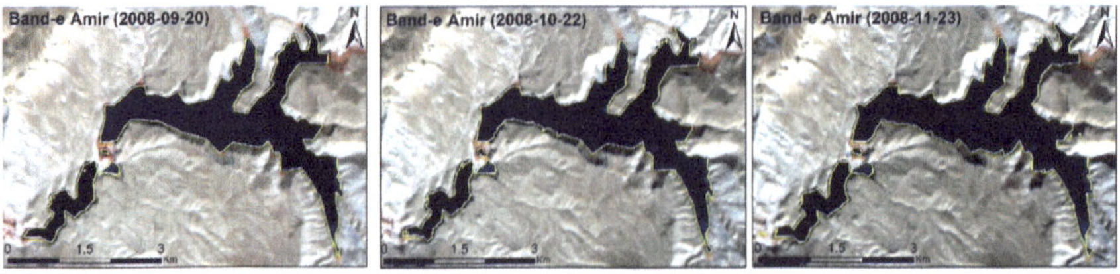

Fig. 21.6 The highest change of four lakes' surface area in the year 2008. *Source* Prepared by the authors

21.8 Relative Change of Lake Surface Area

To facilitate a more reliable comparison between the lakes, their surface areas were calculated by dividing the average surface area of each lake as shown in Fig. 21.7. The change ratio is highest in the smallest lakes, where it reaches up to 2.5 times. This is due to the fact that smaller lakes are more vulnerable to meteorological properties, which can vary greatly and affect their volume.

Figure 21.7 shows the relative surface areas of all the lakes, and Fig. 21.8 shows only Band-e Zulfiqar Lake's relative surface areas over time. Figure 21.8 indicates that the minimum water level occurred in 2008 before the protected area was declared in 2009, and maximum values is observed after 2009, surpassing those before the declaration.

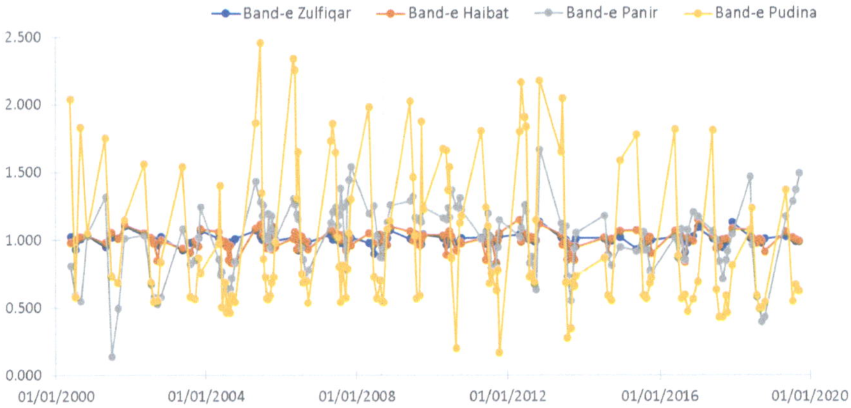

Fig. 21.7 Lake surface area divided by the overall average of each lake's surface area. *Source* Prepared by the authors

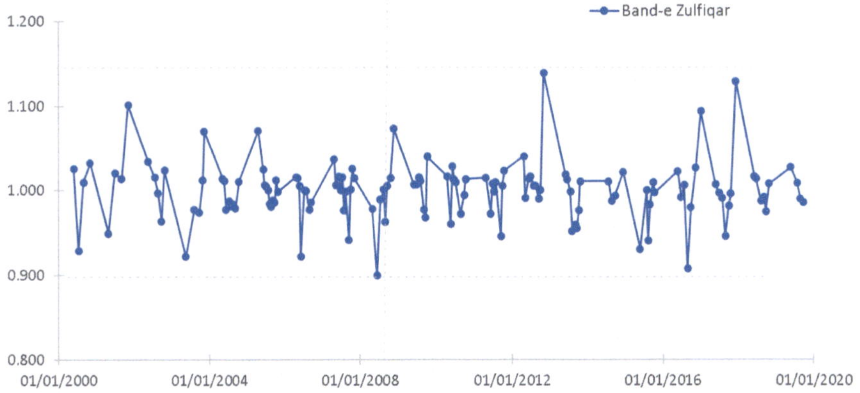

Fig. 21.8 Lake Band-e Zulfiqar surface area divided by the overall average of the lake's surface area. *Source* Prepared by the authors

21.9 Average of Monthly Lake Surface Area Change

Average monthly lake surface areas (km^2) were calculated with all 139 images and are presented in Fig. 21.9. However, the average monthly surface areas for January, February, and March could not be calculated due to the presence of clouds during this period.

First, the lake surface areas of all 139 images are separated according to the month they were taken, and then a monthly average of lake surface area was calculated for each lake. The results are presented in Table 21.4, where the relative average lake surface areas were calculated by dividing the all-over average (shown in bold characters). For example, the relative monthly lake surface areas of the biggest lake, Band-e Zulfiqar, varied from 1.07 to 0.97, with the lowest variation. On the other hand, the smallest lake, Band-e Pudina, had the highest variation.

To examine whether there is a change before and after the declaration of the lakes as a protected area, monthly calculations were repeated for two periods: 2000–2009 and 2010–2019, and are shown in Fig. 21.10. However, due to the cloudy weather

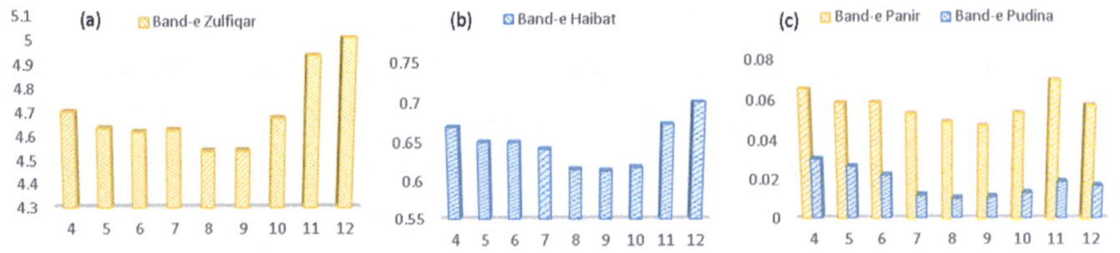

Fig. 21.9 The average of **a** Lake Band-e Zulfiqar, **b** Lake Band-e Haibat, **c** Lake Band-e Panir and Lake Band-e Pudina surface areas (km^2) in the months of all the periods: 2000–2019. *Source* Prepared by the author

Table 21.4 Average monthly lakes' surface areas in km^2 and relative (average monthly divided by the overall average) values of the lakes' surface areas in the 2009–2019 period

Parameter	Month	Band-e Zulfiqar	Band-e Haibat	Band-e Panir	Band-e Pudina
Average monthly	4	4.706	0.669	0.066	0.031
	5	4.637	0.649	0.059	0.027
	6	4.621	0.649	0.059	0.022
	7	4.629	0.640	0.054	0.012
	8	4.541	0.614	0.050	0.010
	9	4.542	0.613	0.048	0.011
	10	4.679	0.617	0.054	0.013
	11	4.938	0.673	0.071	0.019
	12	5.012	0.700	0.058	0.017
	Average	4.700	0.647	0.058	0.018
Relative average monthly	4	1.001	1.033	1.146	1.700
	5	0.986	1.004	1.027	1.494
	6	0.983	1.003	1.030	1.241
	7	0.985	0.989	0.935	0.659
	8	0.966	0.949	0.860	0.564
	9	0.966	0.947	0.828	0.609
	10	0.996	0.954	0.940	0.736
	11	1.051	1.039	1.226	1.059
	12	1.066	1.082	1.008	0.939

Source Calculated by the author

conditions, there is no data available for the month of December in the period of 2000–2009, which is excluded from the calculation.

Despite the increase in drought due to climate change in Afghanistan, the absence of any significant change in water surface areas of these lakes, with some months even having better averages, demonstrates that the declaration of the lakes as a protected area was a necessary step for their protection.

There is no statistically significant change in these lake surface areas, despite the fact that lakes around the world are experiencing deterioration and drying due to the shrinking of the world's lakes (Zhang et al. 2020). These results highlight the importance of continuing with the conservation plan in place for these lakes.

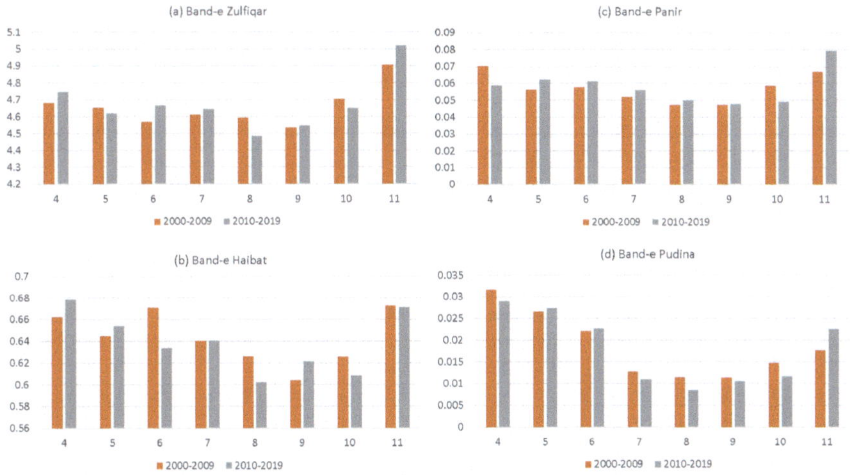

Fig. 21.10 The average of **a** Lake Band-e Zulfiqar, **b** Lake Band-e Haibat, **c** Lake Band-e Panir and **d** Lake Band-e Pudina surface areas (km^2) in the months of the periods: 2000–2009 and 2010–2019. *Source*: Prepared by the authors

21.10 Conclusions

This study utilized the water index to quantify the spatiotemporal variability of surface water bodies in Band-e Amir National Park, Afghanistan, from 2000 to 2019 using multi-temporal Landsat imagery through GEE. The results demonstrate the effectiveness of the selected method of monitoring the temporal and spatial variations of surface water bodies at a 30m spatial resolution. As anticipated, the smaller lakes exhibited the highest change due to the greater impact of climatic factors on small lakes. There was no drastic change observed in the maximum, minimum, and average variations of surface water areas in the four selected lakes in the past 19 years. The most significant change occurred in 2008, right before the area was declared as a protected zone in 2009. The comparison before (2000–2009) and after (2010–2019) the declaration of the protected area in 2009 indicates that protection helps the sustainability of the lakes. Protecting lakes plays an essential role in life sustainability and contributes to achieving SDGs worldwide. The results provide important references for water managers to continue to keep the area protected for biodiversity sustainability in the context of climate change.

References

Akyuz DE, Luo L, Hamilton DP (2014) Temporal and spatial trends in water quality of Lake Taihu, China: analysis from a north to mid-lake transect, 1991–2011. Environ Monit Assess 186(6):3891–3904. https://doi.org/10.1007/s10661-014-3666-0

Albarqouni MM, Yagmur N, Bektas Balcik F, Sekertekin A (2022) Assessment of spatio-temporal changes in water surface extents and lake surface temperatures using Google Earth engine for Lakes Region, Turkiye. ISPRS Intern J Geo-Inform 11(7):407. https://doi.org/10.3390/ijgi11070407

Alden A (2019) Thought co, science, tech, math. https://www.thoughtco.com/sedimentary-rock-types-4123132. Last access date: April 13, 2021

BBC (2009) Afghans get first national park. BBC News, 22 April 2009. http://news.bbc.co.uk/2/hi/south_asia/8013017.stm. Last access date: April 13, 2021

Bourrouilh-Le Jan FG, Akram B, Schvoerer M (2007) Band-e-amir lakes and Dragon Valley (Bamiyan): Myths and seismicity in Afghanistan. Geol Soc London, Spec Public 273(1):121–132. https://doi.org/10.1144/gsl.sp.2007.273.01.11

Carroll M, Wooten M, DiMiceli C, Sohlberg R, Kelly M (2016) Quantifying surface water dynamics at 30 meter spatial resolution in the North American high northern latitudes 1991–2011. Remote Sens 8(8):622. https://doi.org/10.3390/rs8080622

FAO (2023) Retrieved June 06, 2023 from https://www.fao.org/land-water/water/watergovernance/waterfoodenergynexus/en/

GEE (2023) Retrieved June 06, 2023 from https://earthengine.google.com/

Han-Qiu X (2005) A study on information extraction of water body with the modified normalised difference water index (MNDWI). J Remote Sens 5:589–595

Ho LT, Goethals PL (2019) Opportunities and challenges for the sustainability of lakes and reservoirs in relation to the sustainable development goals (sdgs). Water 11(7):1462. https://doi.org/10.3390/w11071462

Keskinen M, Someth P, Salmivaara A, Kummu M (2015) Water-energy-food nexus in a trans-boundary river Basin: the case of Tonle Sap Lake, Mekong River Basin. Water 7(10):5416–5436. https://www.mdpi.com/2073-4441/7/10/5416

Kondo H, Seo N, Yasuda T, Hasizume M, Koido Y, Ninomiya N, Yamamoto Y (2002) Post-flood-infectious diseases in Mozambique. Prehosp Disaster Med 17(3):126–133. https://doi.org/10.1017/S1049023X00000340

Lindner L (1979) Outline of stratigraphy of the Young Quaternary of Central Afghanistan in the light of studies on travertine dams of the Bande Amir Valley dammed lakes. Przegl Geol 27(5):286–291. Warszawa

Lynch AJ, Cooke SJ, Deines AM, Bower SD, Bunnell DB, Cowx IG et al (2016) The social, economic, and environmental importance of inland fish and fisheries. Environ Rev 24(2):115–121. https://doi.org/10.1139/er-2015-0064

Martínez-Santos P (2017) Does 91% of the world's population really have "sustainable access to safe drinking water?" Int J Water Resour Dev 33(4):514–533. https://doi.org/10.1080/07900627.2017.1298517

McFeeters SK (1996) The use of the Normalised Difference Water Index (NDWI) in the delineation of open water features. Remote Sens Lett 17(7):1425–1432

Mohsen A, Elshemy M, Zeidan BA (2018) Change detection for Lake Burullus, Egypt using remote sensing and GIS approaches. Environ Sci Pollut Res 25:30763–30771. https://doi.org/10.1007/s11356-016-8167-y

Naddeo V, Scannapieco D, Belgiorno V (2013) Enhanced drinking water supply through harvested rainwater treatment. J Hydrol 498:287–291. https://doi.org/10.1016/j.jhydrol.2013.06.012

Pietruszka J, Szarejko Z (1980) Comments on occurrence of penitents and their development in West Hindu Kush. Czasop Geograf 51(1):55–61

Rundquist DC, Lawson MP, Queen LP, Cerveny RS (1987) The relationship between summer-season rainfall events and lake-surface area JAWRA. J Am Water Res Assoc 1.23(3):493–508

Ryan J, Silvanto S (2009) The world heritage list: the making and management of a brand. Place Brand Public Dipl 5(4):290–300. https://doi.org/10.1057/pb.2009.21

Sadeghi-Bazargani H, Allahverdipour H, Asghari JM, Azami-Aghdash S (2019) Lakes drying and their adverse effects on human health: a systematic review. Iran J Public Health 48(2):227–237

Sandar S, Tibebu DT, Shromova OA, Kuittinen S, Turunen O, Pappinen A (2019) Lake bottom biomass as a potential source for the biorefining industry. Bioresou Technol Rep 7:100282. https://doi.org/10.1016/j.biteb.2019.100282

Sioli H (1975) Tropical rivers as expressions of their terrestrial environments. In: Golley FB, Medina E (eds) Tropical ecological systems. Ecological Studies (Analysis and Synthesis), vol 11. Springer, Berlin, Heidelberg. https://doi.org/10.1007/978-3-642-88533-4_19

UN (2021) Retrieved April 13, 2021 from http://www.harmonywithnatureun.org/

UNDRR (2021) Sendai framework. Retrieved April 13, 2021 from https://www.undrr.org/publication/sendai-framework-disaster-risk-reduction-2015-2030

UNEPGRID (2021) World heritage natural site. Retrieved April, 13 2021 from https://unepgrid.ch/en/resource/186C3B64#:~:text=Description,UNESCO%20World%20Heritage%20Natural%20Site

UNFCCC (2021) Paris agreement. Retrieved April 13, 2021 from https://unfccc.int/process-and-meetings/the-paris-agreement/the-paris-agreement

UN SDGs (2021) SDGs. Retrieved April 13, 2021 from https://sdgs.un.org/

Wang R, Xia H, Qin Y, Niu W, Pan L, Li R, Zhao X, Bian X, Fu P (2020) Dynamic monitoring of surface water area during 1989–2019 in the Hetao plain using Landsat data in google earth engine. Water 12:3010. https://doi.org/doi.org/10.3390/w12113010

Wetzel RG (2015) Limnology: Lake and river ecosystems. Academic Press, an imprint of Elsevier

WHO (2021) Health topics/Water, sanitation and hygiene (WASH). Retrieved April 13, 2021 from https://www.who.int/health-topics/water-sanitation-and-hygiene-wash#tab=tab_1

Xia H, Zhao J, Qin Y, Yang J, Cui Y, Song H, Ma L, Jin N, Meng Q (2019) Changes in water surface area during 1989–2017 in the Huai river basin using Landsat data and google earth engine. Remote Sens 11(15):1824. https://doi.org/10.3390/rs11151824

Xu HJI (2006) Modification of normalised difference water index (NDWI) to enhance open water features in remotely sensed imagery. J Remote Sens 27(14):3025–3033

Yagmur N, Bilgilioglu BB, Dervisoglu A, Musaoglu N, Tanik A (2021) Long and short-term assessment of surface area changes in saline and freshwater lakes via remote sensing. Water Environ J 35(1):107–122

Zhang Y, Luo P, Zhao S, Kang S, Wang P, Zhou M et al (2020) Control and remediation methods for eutrophic lakes in the past 30 years. Water Sci Technol 81(6):1099–1113. https://doi.org/10.2166/wst.2020.218

Chapter 22
Land Use/Land Cover Change Dynamics in Rapidly Growing City: A Potential Threat to Sustainability: A Case Study of the Peshawar District of Pakistan

Iffat Tabassum, Moazima Sultan, Ihsanullah, Janas Khan, and Syed NaeemUdDin

Abstract Land use is dynamic and is one of the main driving forces of global environmental changes. Urban expansion has increased the exploitation of natural resources and remarkably changed land use and land cover patterns. The land use/land cover change detection is of major concern in spatial planning, management and proper utilization of the land. This study aimed to detect the changes in the land use and land cover (LU/LC) of Peshawar district using remote sensing data. The study made use of the Landsat imageries of 1980, 2000 and 2020 taken from USGS Earth Explorer. Maximum likelihood classification method is used to classify the imageries. Four land use classes build-up area, water body, open area (barren land) and vegetation cover (agricultural land) were identified and mapped. The classification of images was analyzed and accuracy assessment was carried out using a confusion matrix which showed that accuracy for 1980 was 86.5%, for 2000 88.8% and for 2020 it was 91.85%. The results showed that in the past 40 years, there has been a significant change in the land use of the Peshawar District. The built-up area has increased to 24.82% over time, while there is a decrease in the vegetation cover, barren/open land and water body by 42.95, 31.42 and 0.81%, respectively. This trend is a significant threat to city sustainability and needs to be addressed urgently.

I. Tabassum (✉) · M. Sultan · Ihsanullah · J. Khan · Syed NaeemUdDin
Department of Geography, University of Peshawar, Peshawar, Pakistan
e-mail: iffattabassum@uop.edu.pk

M. Sultan
e-mail: moazimasultan786@gmail.com

Ihsanullah
e-mail: ihsanullah@uop.edu.pk

J. Khan
e-mail: janas_k@uop.edu.pk

Syed NaeemUdDin
e-mail: usyednaeem@gmail.com

S. Anand et al. (eds.), *Sustainability in South Asian Cities*, Advances in Geographical and Environmental Sciences, https://doi.org/10.1007/978-981-97-7455-5_22

Keywords City sustainability · Land use/land cover change detection · Geo-spatial tools · Peshawar · Pakistan

22.1 Introduction

Land use/ Land cover (LU/LC) for any area is the outcome of physical and natural features of the earth's surface which is used and transformed by humans with time (Rawat and Kumar 2015). Land use change is a historical process that explains how people use the land in the past. It alters the obtainability of different resources such as vegetation, soil and water (Ahmad 2014). These alterations can negatively affect the patterns of climate, natural hazards and socio-economic dynamics on a global and local scale (Chakilu and Moges 2017; Hegazy and Kaloop 2015Sewnet 2015). Change detection is valued in many applications related to land use and land cover (LU/LC), including cultivation dynamics, urban expansion and landscape changes (Hegazy and Kaloop 2015; Solaimani et al. 2010). Understanding landscape patterns, changes and interactions between human activities and natural phenomena is essential for proper land management and decision improvement (Rawat and Kumar 2015). Geographic Information systems (GIS) and remote sensing (RS) are identified as powerful and cost-effective tools for assessing the spatial and temporal changes of LU/LC (Herold et al. 2003; Serra et al. 2008; Yuan et al. 2005). This land use change detection has further become a useful and detailed way to improve the selection of areas considered for agricultural, urban and/or industrial purposes (Rawat and Kumar 2015). Remote sensing data is the most common source for the detection, quantification and mapping of LULC patterns due to its repetitive data acquisition, suitable for processing and accurate geo-referencing (Chen et al. 2005; Jensen 1996).

22.2 Literature Review

Urbanization can solve several issues of human society if controlled, coordinated and planned. However, unplanned urbanization is a curse. According to United Nations estimates, up to 2018, the world population living in cities was 55%, which will increase by about 70% by 2050 (United Nations 2018). Due to the acceleration of global urbanization in both intensity and area, there is a growing interest in understanding its implications with respect to a broad set of environmental factors including loss of arable land (Lopez et al. 2001), habitat destruction (Alphan 2003) decline in natural vegetation cover and climate at local, regional, and global scales (Grimm et al. 2000).

The conversion of rural areas into urban areas through development has occurred at an unprecedented rate in recent human history and is having a marked effect on the natural functioning of ecosystems (Song et al. 2018). Furthermore, the rapid increase in urbanization and accelerating anthropogenic activities (Al-Sharif and

Pradhan 2014) have strongly influenced the ecosystems in urban areas; therefore, more attention is currently being directed toward monitoring changes in urban land use/ land cover (LULC) (Stow and Chen 2002).

Land use and land cover is a critical component to understanding global land status; it shows the present as well as the past status of the earth's surface. Land use and land cover are two separate terminologies that are often used interchangeably (Dimyati et al. 1994). Land cover is a basic parameter that evaluates the earth surface's content as an important factor that affects the condition and functioning of the ecosystem. Land cover is a biophysical state of the Earth's surface, which is used to estimate the interaction of biodiversity with the surrounding environment (Prakasam 2010).). Land use refers to the evaluation of the land concerning various natural characteristics. It is the type of utilization to which man has put the land (Ndukwe 1997). It can also be said that Land use is a product of interactions between a society's cultural background, state and physical needs on the one hand and the natural potential of land on the other (Ram and Kolarkar 1993). Land use/land cover (LULC) changes are major issues of global environmental change. In developing countries, where urbanisation rates are high, urban sprawl significantly contributes to land use change. Sprawl commonly impacts the loss of agricultural land, open space and ecologically sensitive habitats in and around urban areas due to a lack of integrated and holistic approaches in regional planning (Sudhira and Ramachandra 2007).

With their repetitive nature, satellite remote sensing data have proved to be quite useful in mapping land use/land cover patterns and changes with time (Hudak and Wessman 1998). Such changes are quantifiable through GIS techniques even if the resultant spatial datasets are of different scales/resolutions (Garcia-Mora et al. 2012). Geo-informatics provides very efficient tools to collect and analyze the information required to detect changes in urban areas that conventional surveying technology cannot offer cost-effectively (Imhoff et al. 1997). The spatial pattern of urban sprawl over different periods can also be mapped systematically, monitored and accurately assessed from satellite data (remotely sensed data) along with conventional ground data (Lata et al. 2001).

Land use/cover change has become a central and vital component in current strategies for managing natural resources and monitoring environmental changes. Identifying patterns and trends of expansion in the built-up area helps appropriately plan land use and infrastructure facilities in the city (Sudhira 2004).

In Pakistan, the problem of urban expansion has affected the settlement pattern and land uses around the suburbs, increasing the area and population concurrently. Among the consequences of fast urban sprawl and population explosion, there exist many environmental and ecological challenges; some of them are the local climatic changes, eliminating and fragmenting negative habitats (Guru and Anubhooti 2015), energy crisis, water resources shortage, vegetation redistribution and agriculture high yield failure (Satterthwaite et al. 2010).

Therefore it is required to measure the existing situation and to foresee future developments accordingly (Giri et al. 2005). In Pakistan, the LU/LC changes are not updated on a regular basis and there is a lack of information regarding land use change and population. Presently it is at least needed to map the physical forms of areas to make available the indicators of their social and economic associations such as land use change and its impact on the population of the specific area. (Donnay and Unwin 2001).

Land use changes are taking place in all large cities of Pakistan as a result of economic growth and urban expansion (Lata 2005). Due to rapid urbanization built-up area is increasing in metropolitan cities like Peshawar. This resulted in the reduction of agricultural land within and around cities (Bhatt et al. 2005). Therefore it is essential to access and quantify the current and historic changes along with all types of population dynamics to promote sustainable growth of urban areas.

22.3 Materials and Methods

22.3.1 Description of the Study Area

Peshawar, the provincial capital of Khyber Pakhtunkhwa and the largest city in the province revels in tremendous historical, military, economic and political importance. It lies between 33° 44′ and 34° 15′ north latitude and 71° 22′ and 71° 42′ east longitude. The total area of this district is 1,257 square km. Peshawar is situated at an altitude of 347 m above sea level. The name Peshawar is derived from the word 'Pushpapura'. The word Pushpapura belongs to Sanskrit, meaning a city of flowers in many historical books.

The central part of the district consists of fine alluvial deposits. The cultivated tracts consist of rich, light and porous soil, composed of a pretty even mixture of clay and sand suitable for cultivating wheat, sugarcane and tobacco. It is approximately 1173 feet (358 m) above sea level. Peshawar bears a semi-arid climate with hot summers and mild winters. Winters in Peshawar start in mid of November and end in late March while summer months are from May to September. The mean maximum temperature in summer is over 40 °C (104 °F), and the mean minimum temperature is 25 °C (77 °F). The mean maximum temperature in winter is 18.35 °C (65.03°F), and the mean minimum temperature is 4 °C (39°F).

Peshawar City District is the main urban Centre of Peshawar. Its population was over 4.26 million (4,269,811) according to the 2017 census. The annual growth rate of Peshawar is increased from the density of the population of Peshawar is 3400 persons per square kilometer (8800 per square miles). According to the 2017 census, Peshawar's urban population is 1.97 million, while the rural population is 2.29 million. Since 2001, Peshawar District has been designated as a City District, consisting of four towns and a cantonment. There are 93 union councils in the District. According to the Khyber Pakhtunkhwa Local Government (Amendment) Act, 2019, the union councils are further divided into 216 village councils and 130 neighborhood councils.

Table 22.1 Detail of satellite imageries

S. no	Image acquisition	Satellite/sensor	Reference system/path/row
1	August 1979	Landsat 2	WRS2/151/036, 037
2	September 2000	Landsat 5 ETM +	WRS2/151/036, 037
3	August 2020	Landsat 8 OLI	WRS2/151/036

22.3.2 *Materials*

For the land use/land cover study of the Peshawar district, three Landsat imageries were acquired for the years 1979, 2000 and 2020. The image of 1979 was downloaded from the USGS (United States Geological Survey) Landsat Look Viewer website. The other image of 2000 was downloaded from the Landsat 5 ETM + (Enhanced Thematic Mapper Plus) website, while the Image of 2020 was downloaded from the Landsat OLI (Operational Land Imager) website. The data are preprocessed and projected to the Universal Transverse Mercator (UTM) projection system. The satellite data collected are shown in the Table 22.1.

22.3.3 *Methodology*

For the present study, the Peshawar district maps are extracted from the toposheets (1:50,000) of the survey of Pakistan. The toposheets are available in the Department of Geography, University of Peshawar. For the change analysis of the study area, three satellite images are downloaded. The processing of the imageries and image interpretation for developing Land use/ Land cover maps is done in the ESRI GIS 10.2 software. Other software used for the presentation are MapInfo 11.5, Microsoft Excel and Microsoft Word. The maps obtained after processing are then studied and analyzed to detect changes in Peshawar District's land use/ land cover. The methodology adopted in detail is shown in Fig. 22.2.

22.3.3.1 Image Processing

The downloaded images contained scenes for which the mosaicing was done to get the required area. Geometric rectification was done in the preprocessing. According to Mahendra et al. (2015), accurate geometric rectification or image registration of remotely sensed data is a prerequisite for combining different source data in a classification process. All the images contained different types of bands. There, stacking was performed to get the composite image. Other image enhancement techniques like panchromatic-sharpening techniques were also performed on each image to improve the quality of the image. The study area has been delineated with the help of Survey of Pakistan Toposheets of 1:50,000 and the city map obtained from the Department

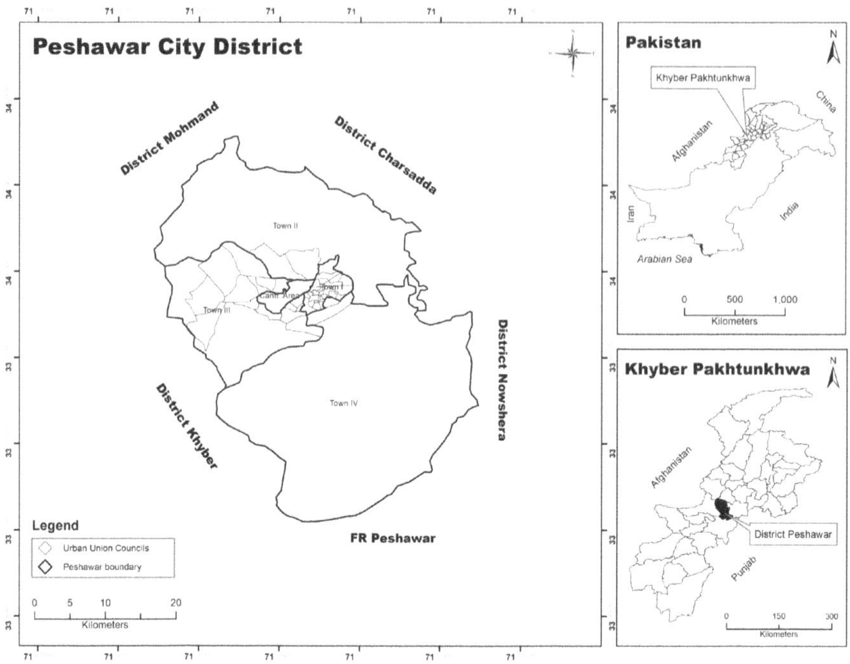

Fig. 22.1 Location of the Peshawar District (Modified from the map produced by the Planning and Development Department)

of Geography, University of Peshawar. The base layer so formed is used to subset the Landsat images.

22.3.3.2 Classification of Images

The preprocessed images were then classified by both supervised and unsupervised methods of classification. The former was done simply for reference and for the understanding of pixels with different digital numbers while the latter technique was the basic approach in which the spectral parameters are developed from the selected image which is given the name of the training site. The training sites are the representatives of land cover types on the basis of their pixel numbers. Supervised classification finally develops the land use/land cover image of the selected area in which four land cover categories are identified namely built-up area, vegetation cover, water bodies and open area/barren land.

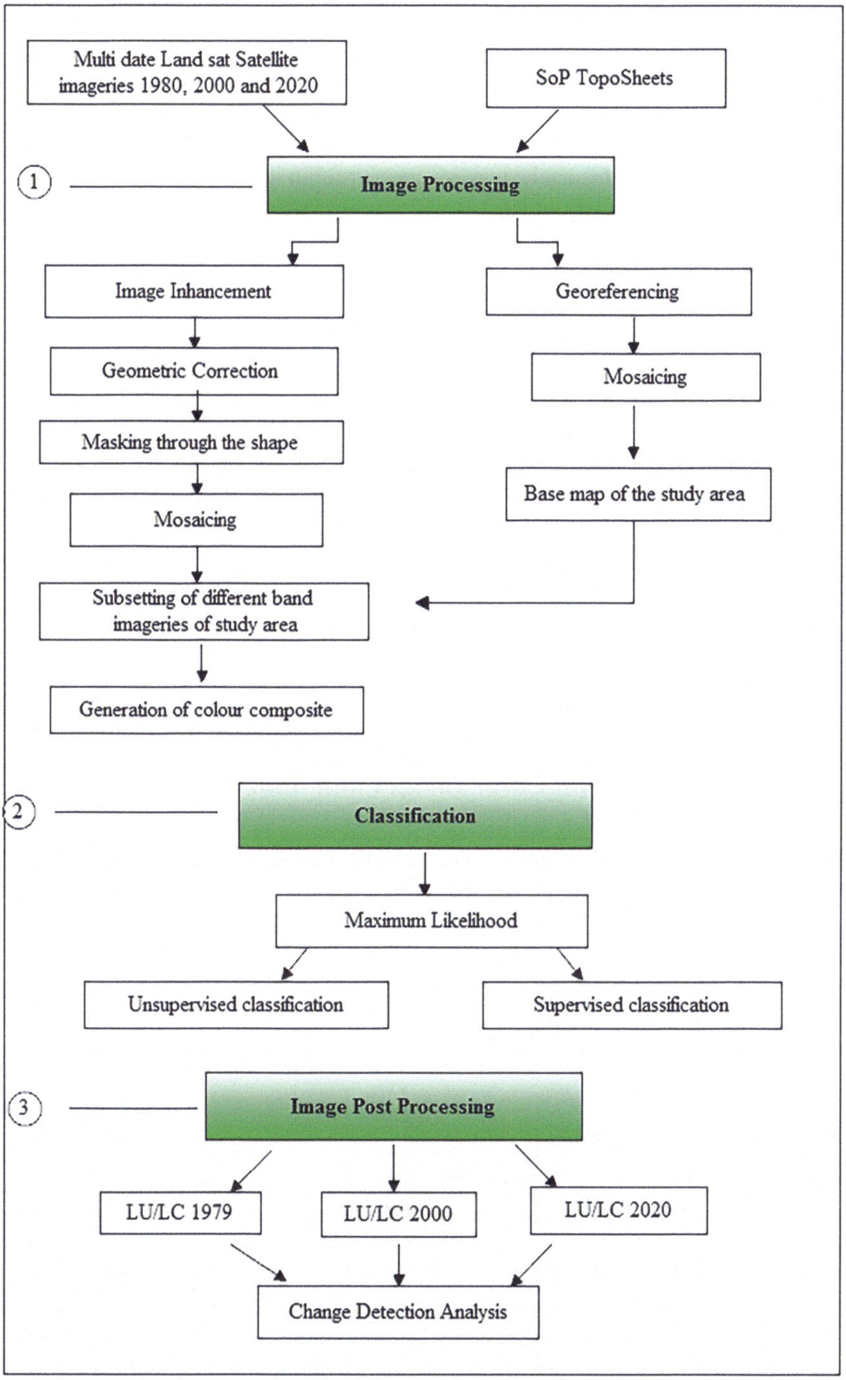

Fig. 22.2 Methodology adopted for the land use/land cover change detection

22.3.3.3 Change Detection Analysis

Change detection analysis describes and quantifies differences between images of the same scene at different times. The classified images of 1979, 2000 and 2020 were used to calculate the area of different land covers and observe the changes taking place in the data span. This analysis was very helpful in identifying various changes occurring in different classes of land use, like an increase in urban built-up area or a decrease in the vegetation cover.

22.3.3.4 Accuracy Assessment

The accuracy assessment, which was used for assessing data in the research, was a confusion matrix. In this process, the correspondence between the classification result and a reference image was done. First, the ground truth data was taken for the reference data then it was compared with the classified data. The tools used in the accuracy assessment were Extract value to point—frequency tools—pivot table. The formula used for the assessment of results collected through the classification of image was Accuracy assessment = (total true value / total sample value).

22.4 Results and Discussion

22.4.1 Land Use/Land Cover Images and Statistics

The classified images are obtained showing the Peshawar district's land use and land cover in the following Figs. 22.3, 22.4 and 22.5. These images provide information about the land use pattern of the study area. The built-up area, vegetation cover, barren land /open area and water body are the study area's major land use land cover classes. The statistics of the study area calculated from the satellite images after the supervised classification are shown in Table 22.2.

22.4.2 Change Detection Analysis

Land use and land cover change detection based on remote sensing images have been widely applied in research for LU/LC, natural resource management and environment monitoring and protection (Zhang et al., 2014). The satellite image classification has helped to gain information on speculative and extremely simplified visual images of the study area (Figs. 22.3, 22.4 and 22.5). These maps verify the dynamic population direction, urban extent and growth patterns. The rapid growth and development have resulted in a very congested inner city area and severely utilized vegetation cover

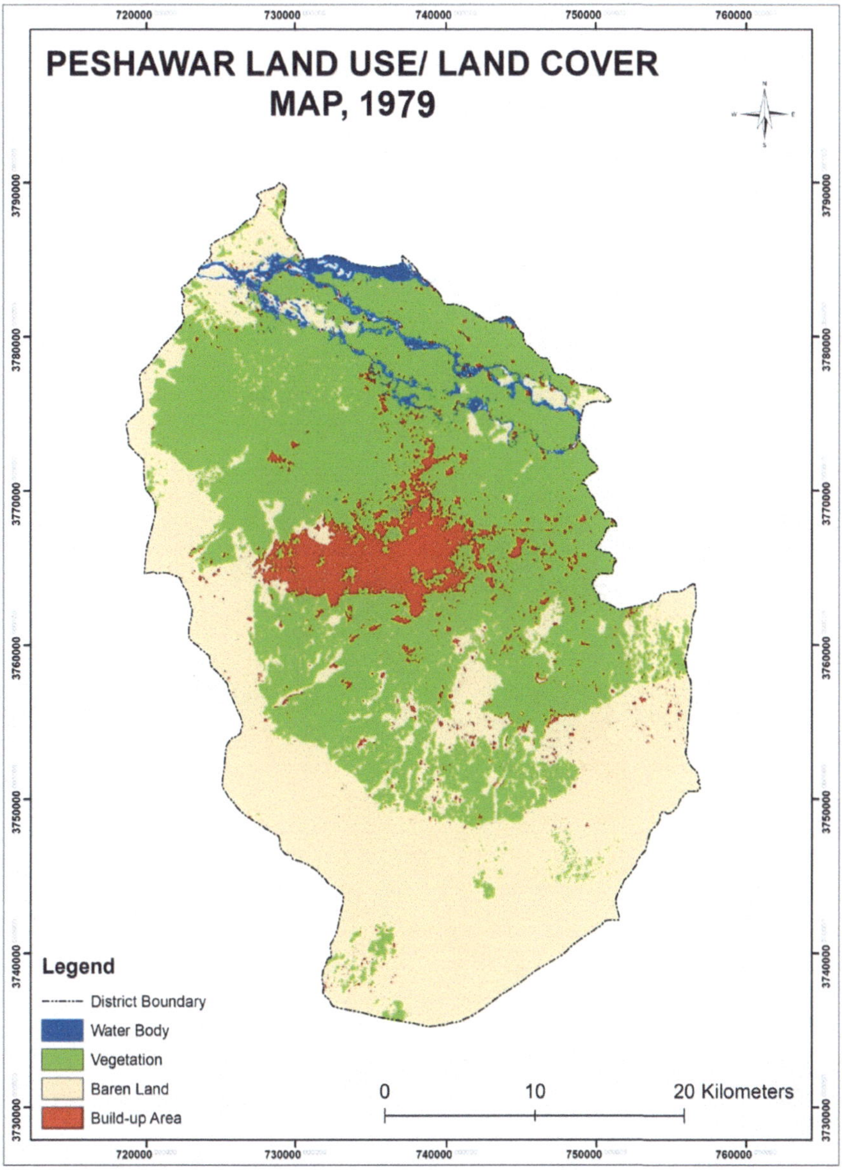

Fig. 22.3 LU/LC Map of Peshawar District 1979

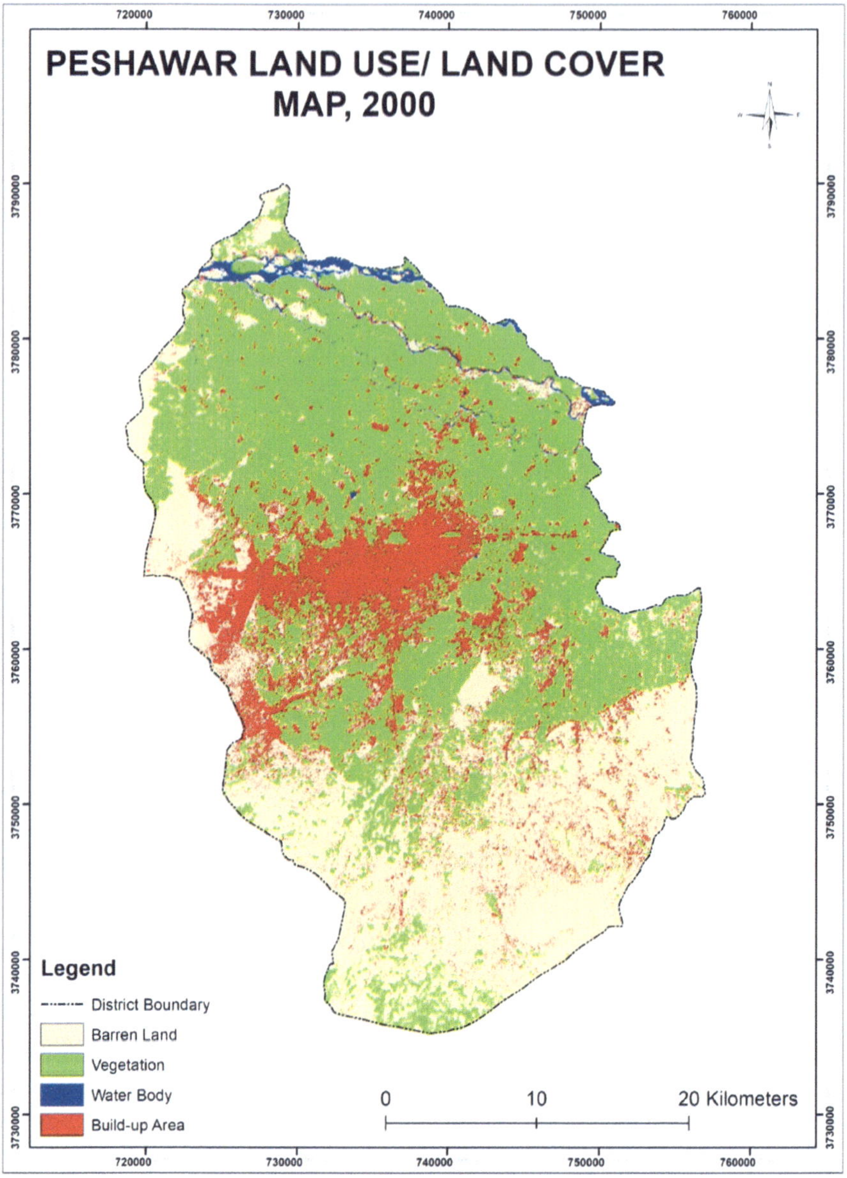

Fig. 22.4 LU/LC Map of Peshawar District 2000

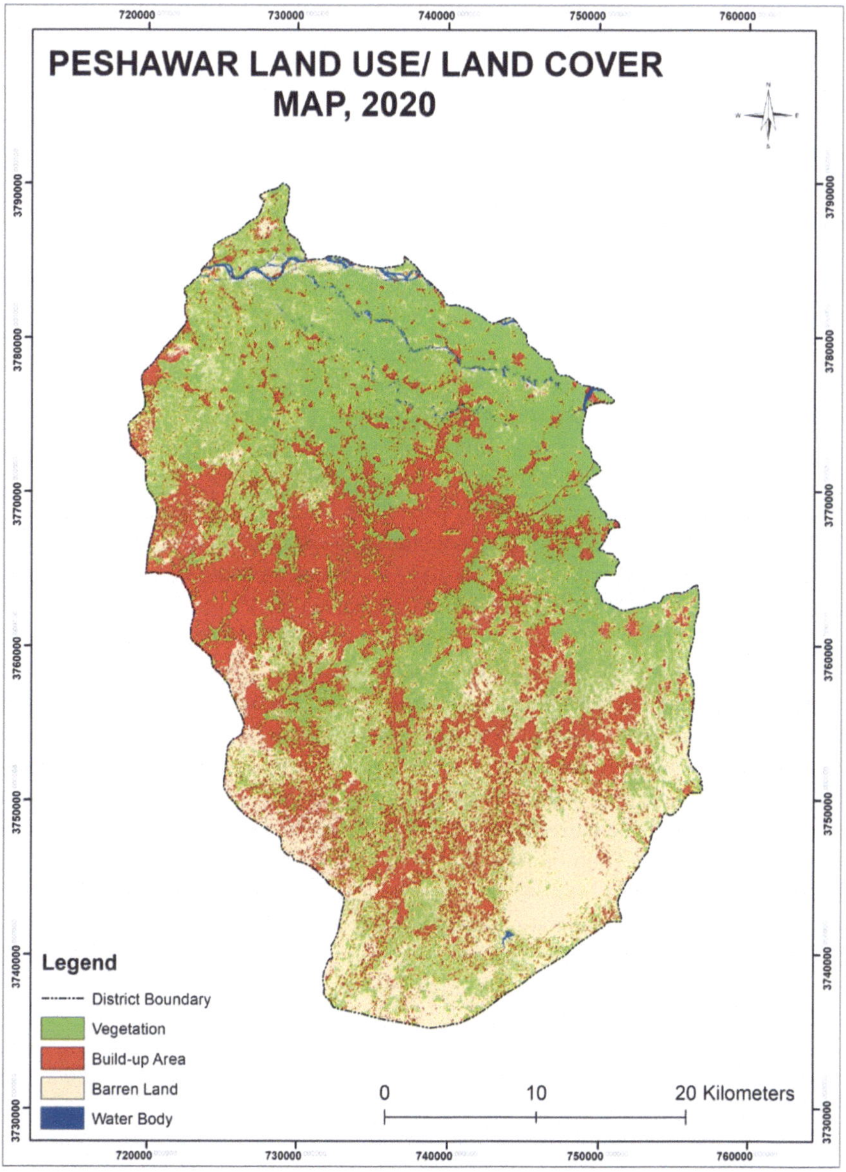

Fig. 22.5 LU/LC Map of Peshawar District 2020

Table 22.2 LU/LC of Peshawar District (1979, 2000, 2020)

S.No	Land use/land cover categories	August 1979		September 2000		August 2020	
		Area (Sq.km)	Area (%age)	Area (Sq.km)	Area (%age)	Area (Sq.km)	Area (%age)
1	Built-up area	79	6.28	156	12.41	312	24.82
2	Vegetative cover	605	48.13	570	45.34	540	42.95
3	Barren land/open area	546	43.43	515	40.97	395	31.42
4	Water bodies	27	2.16	16	1.28	10	0.81
	Total	1257	100	1257	100	1257	100

(mainly agricultural land) for residential purposes. The maps give the visual interpretation and the statistical information is also extracted from the maps of Peshawar District for the years 1979, 2000, and 2020 as described earlier. The changes in the land use and land cover during the study period can be clearly observed in (Fig. 22.6). The results of the image differencing concluded the increasing trend of the built-up area from 6.28% in 1979 to 24.82% in 2020 and the decreasing trend of vegetation cover from 48.13% in 1979 to 42.95% in 2020, barren land/ open area from 43.43% in 1979 to 31.42% in 2020 and water body from 2.16% in 1979 to 0.81% in 2020 (Table 22.3, 22.4 and 22.5). From the change analysis, it is clear that there is an inverse relationship between the change of built area and barren land. Both changes show the same pace over time. A decrease in the vegetation cover is also noticeable. The built-up area upsurge has many reasons. One of the main reasons is the access to basic necessities available in the core area. For this purpose, people started to convert the agricultural land and barren land into the built-up area surrounding the core area of Peshawar District. There is probably another reason, which is the migration of the people of tribal areas because of insecurity and natural disasters. These IDPs are being added to the local population occasionally and are playing a significant role in shaping Peshawar District. The gradual increase in the built-up area and decrease in the vegetation cover can be better visualized through (Fig. 22.7).

22.4.3 *Overall Accuracy*

The accuracy assessment was done which gives the overall accuracy of the confusion matrix. According to (Anderson, 1976), the minimum accuracy value for reliable land cover classification is 85%. The results showed that accuracy for 1980 was 86.5%, for 2000 88.8% and for 2020 it was 91.85%. The results not only fulfill the minimum (Anderson 1976) but also according to the classification scale given by (Moriasi et al. 2007) classification lies in an excellent range. Summary of the accuracy assessment from 1979–2020 is given in Table 22.6.

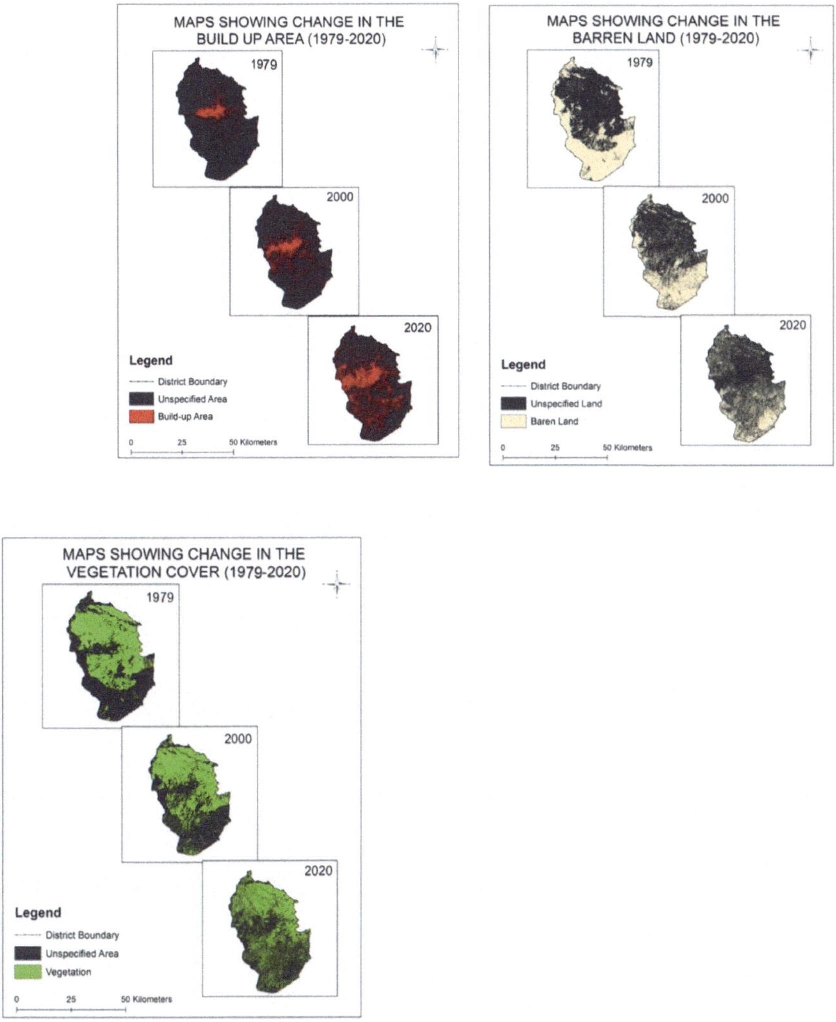

Fig. 22.6 Comparative study of different LU/LC categories from 1979–2020

22.5 Concluding Remarks

Land use plays a vital role in global environmental change and in the future availability of food and water. Modelling land use and land cover change presents new scientific challenges, particularly in integrating biophysical and socioeconomic data and processes (National Research Council, 2005). Considering the above importance of the land use land cover analysis, Peshawar District images are classified and analysed for the years 1979, 2000, and 2020.

Table 22.3 LU/LC change detection (1979–2000)

S.No	Land Use/land cover categories	August 1979		September 2000		Area change	
		Area (Sq. Km)	Area (%Age)	Area (Sq. Km)	Area (%Age)	Area (Sq. Km)	Area (%Age)
1	Built-up area	79	6.28	156	12.41	+77	+6.12
2	Vegetative cover	605	48.13	570	45.34	–35	–2.78
3	Barren land	546	43.43	515	40.97	–31	–2.46
4	Water bodies	27	2.16	16	1.28	–11	0.87
	Total	1257	100	1257	100	0	0

Table 22.4 LU/LC change detection (2000–2020)

S.No	Land use/land cover categories	September 2000		August 2020		Area change	
		Area (Sq. Km)	Area (%Age)	Area (Sq. Km)	Area (%Age)	Area (Sq. Km)	Area (%Age)
1	Built–up area	156	12.41	312	24.82	+ 156	+ 12.41
2	Vegetative cover	570	45.34	540	42.95	-30	-2.38
3	Barren land	515	40.97	395	31.42	-120	-9.54
4	Water bodies	16	1.28	10	0.81	-6	-0.47
	Total	1257	100	1257	100	0	0

Table 22.5 LU/LC change detection (1979–2020)

S.No	Land use/land cover categories	August 1979		August 2020		Area change	
		Area (Sq. Km)	Area (%Age)	Area (Sq. Km)	Area (%Age)	Area (Sq. Km)	Area (%Age)
1	Built-up area	79	6.28	312	24.82	+ 233	+ 18.53
2	Vegetative cover	605	48.13	540	42.95	-65	-5.17
3	Barren land	546	43.43	395	31.42	-151	-12.01
4	Water bodies	27	2.16	10	0.81	-17	-1.35
	Total	1257	100	1257	100	0	0

The methodology described in the study is feasible and useful for detecting rapid land use/land cover change. The study shows the increase in the built-up area from 6.28% to 24.82%in the last forty years. On the other hand decrease in the barren land and vegetation cover is found from 43.43 to 31.42% and 48.13 to 42.95%, respectively. Waterbody has also shown a decreasing trend from 2.16 to 0.81%.

This local-level study concluded that rapid increase in built-up areas without land use zoning can have disastrous consequences. Urban vegetation is essential for city

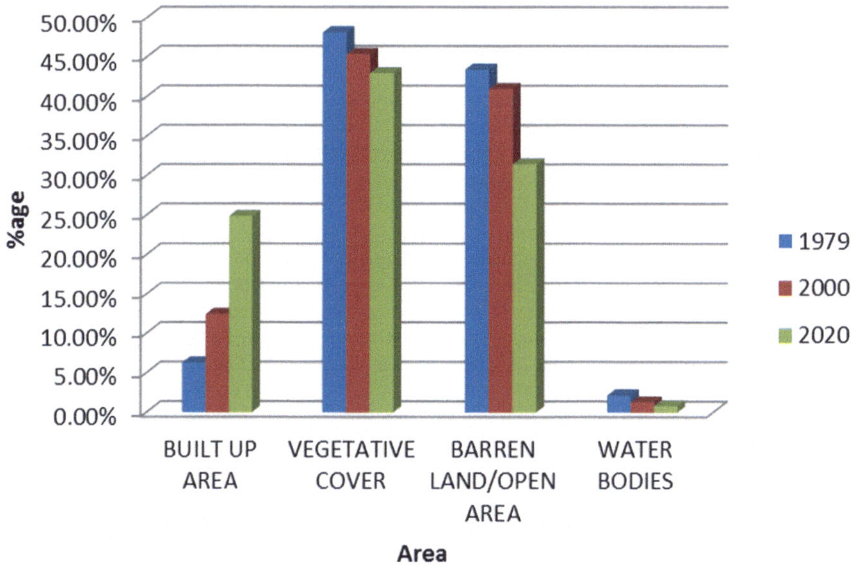

Fig. 22.7 Change detection analysis of Peshawar Land Use (1979, 2000 and 2020)

Table 22.6 Summary of the accuracy assessment from 1980–2020

Accuracy assessment for 1979	(%)age	Accuracy assessment for 2000	(%)age	Accuracy assessment for 2020	(%)age
Vegetation	85	Vegetation	83.5	Vegetation	87.2
Barren land	86.5	Barren land	86.2	Barren land	96
Settlement	88.7	Settlement	94.6	Settlement	95.1
Water bodies	81.1	Water bodies	90.8	Water body	89.1
Total	86.5	Total	88.8	Total	91.85

sustainability, which in this case, has decreased significantly. This will have impacts on urban climate, urban disasters and a number of social issues. More GIS and remote sensing techniques can assist decision-makers in preparing plans to find the appropriate solutions to urban encroachments.

References

Ahmadizadeh S (2014) Land use change detection using remote sensing and artificial neural network: application to Birjand, Iran. Comput Ecol Softw 4(4):276

Alphan H (2003) Land-use change and urbanization of Adana, Turkey. Land Degrad Dev 14(6):575–586

Al-sharif AA, Pradhan B (2014) Monitoring and predicting land use change in Tripoli Metropolitan City using an integrated Markov chain and cellular automata models in GIS. Arab J Geosci 7:4291–4301

Anderson JR (1976) A land use and land cover classification system for use with remote sensor data, US Government Printing Office

Balamurugan G (2015)Anubhooti Urban growth effects for land use changes leading to natural hazards in parts of Mumbai metro city, India. Disaster Advances 8(10):1-14.

Bhatt B et al. (2006) Application of remote sensing and GIS for detecting land use changes: a case study of Vadodara, GIS Development

Chakilu G, Moges M (2017) Assessing the land use/cover dynamics and its impact on the low flow of Gumara Watershed, Upper Blue Nile Basin, Ethiopia. Hydrol Current Res **7**(2)

Chen X et al (2005) A simple and effective radiometric correction method to improve landscape change detection across sensors and across time. Remote Sens Environ 98(1):63–79

Dimyati M et al (1996) An analysis of land use/cover change in Indonesia. Int J Remote Sens 17(5):931–944

Donnay J-P, Unwin D (2001) Modelling geographical distributions in urban areas. Rem Sens Urban Analy GISDATA 9:205–224

Foody GM (2002) Status of land cover classification accuracy assessment. Remote Sens Environ 80(1):185–201

García-Mora TJ et al (2012) Land cover mapping applications with MODIS: a literature review. Intern J Dig Earth 5(1):63–87

Geremew AA (2013) Assessing the impacts of land use and land cover change on hydrology of watershed: a case study on Gigel-Abbay Watershed, Lake Tana Basin, Ethiopia

Giri C et al (2005) A comparative analysis of the Global Land Cover 2000 and MODIS land cover data sets. Remote Sens Environ 94(1):123–132

Grimm NB et al (2000) Integrated approaches to long-term studies of urban ecological systems: Urban ecological systems present multiple challenges to ecologists—pervasive human impact and extreme heterogeneity of cities, and the need to integrate social and ecological approaches, concepts, and theory. Bioscience 50(7):571–584

Hegazy IR, Kaloop MR (2015) Monitoring urban growth and land use change detection with GIS and remote sensing techniques in Daqahlia governorate Egypt. Int J Sustain Built Environ 4(1):117–124

Herold M et al (2003) The spatiotemporal form of urban growth: measurement, analysis and modeling. Remote Sens Environ 86(3):286–302

Hudak AT, Wessman CA (1998) Textural analysis of historical aerial photography to characterize woody plant encroachment in South African savanna. Remote Sens Environ 66(3):317–330

Imhoff ML et al (1997) A technique for using composite DMSP/OLS "city lights" satellite data to map urban area. Remote Sens Environ 61(3):361–370

Jensen JR (1996) Introductory digital image processing: a remote sensing perspective, Prentice-Hall Inc

Lata K et al (2001) Measuring urban sprawl: a case study of Hyderabad. GIS Development 5(12). Reproduced with permission of the copyright owner. Further reproduction prohibited without permission, pp 8–13

López E et al (2001) Predicting land-cover and land-use change in the urban fringe: a case in Morelia city, Mexico. Landsc Urban Plan 55(4):271–285

Mahendra HN, Shivakumar BR, Praveen J (2015) Pixel-based Classification of Multispectral Remotely Sensed Data Using Support Vector Machine Classifier. National Conference on Advanced Innovation in Engineering and Technology (NCAIET-2015) Alva's Institute of Engineering and Technology, Moodbidri **3**(1) Special Issue 1, April 2015

Moriasi DN et al (2007) Model evaluation guidelines for systematic quantification of accuracy in watershed simulations. Trans ASABE 50(3):885–900

Ndukwe NK (1997) Principles of environmental remote sensing and photo Interpretation, New Concept Publishers

Prakasam C (2010) Land use and land cover change detection through remote sensing approach: a case study of Kodaikanal taluk, Tamil Nadu. Intern J Geom Geosci 1(2):150
Ram B, Kolarkar A (1993) Remote sensing application in monitoring land-use changes in arid Rajasthan. Int J Remote Sens 14(17):3191–3200
Satterthwaite D et al (2010) Urbanization and its implications for food and farming. Philos Trans Royal Soc B: Biol Sci 365(1554):2809–2820
Serra P et al (2008) Land-cover and land-use change in a Mediterranean landscape: a spatial analysis of driving forces integrating biophysical and human factors. Appl Geogr 28(3):189–209
Sewnet A (2016) Land use/cover change at Infraz watershed by using GIS and remote sensing techniques, northwestern Ethiopia. Intern J River Basin Manag 14(2):133–142
Solaimani K et al (2010) Land use/cover change detection based on remote sensing data (A case study; Neka Basin). Agric Biol J N Am 1(6):1148–1157
Song M et al (2018) How does the ecological well-being of urban and rural residents change with rural-urban land conversion? The case of Hubei, China. Sustainability 10(2):527
Stow DA, Chen DM (2002) Sensitivity of multitemporal NOAA AVHRR data of an urbanizing region to land-use/land-cover changes and misregistration. Remote Sens Environ 80(2):297–307
Sudhira H, Ramachandra T (2007) Characterising urban sprawl from remote sensing data and using landscape metrics. Proceedings of the 10th international conference on computers in urban planning and urban management, iguassu falls, Brazil, Citeseer
Sudhira H et al (2004) Urban sprawl: metrics, dynamics and modelling using GIS. Int J Appl Earth Obs Geoinf 5(1):29–39
United Nations (2018) Sustainable development changes. World Economic and Social Survey 2013. New York: United Nations Publications. https://www.un.org/en/development/desa/publications/world-economic-and-social-survey-2013-sustainable-development-challenges.html
Yuan F et al (2005) Land cover classification and change analysis of the Twin Cities (Minnesota) Metropolitan Area by multitemporal Landsat remote sensing. Remote Sens Environ 98(2–3):317–328
Zhang T et al (2014) An analysis of land use change dynamics and its impacts on hydrological processes in the Jialing River Basin. Water 6(12):3758–3782

Chapter 23
Impact of Bare Lands in the Cities of Kathmandu Valley, Nepal: Question for Urban Environmental Sustainability

Chhabi Lal Chidi

Abstract Over the recent decades, rapid urbanization in Kathmandu Valley, Nepal has rapidly increased the demand for land for constructing buildings and other infrastructure. During the past 15 years, the bare land area has increased rapidly in the surrounding hill area due to the haphazard extraction of construction materials and rapidly growing land plotting, resulting in heavy sediments in the rivers and rivulets. Heavy sediment loads in the river cause frequent flash floods during the rainy season. Uncontrolled mining of sand from the riverbed and deepening of riverbeds further damage the infrastructure and cause water pollution in the valley. This study assesses the bare land and sediment production and their impacts on the valley's cities. Land use/land cover data and Google Earth images were used to identify and measure the bare land areas. Fieldwork was conducted in different parts of the valley and an intensive field survey was conducted in the Bishnumati watershed area, one of the major river watersheds of the valley. It is argued that if the demands for land and construction materials are not met in a well-planned way, their negative impact will induce more vulnerability to the cities.

Keywords Bare land · Construction materials · Flood · Sediment · Infrastructure · Kathmandu Valley · Nepal

23.1 Introduction

Urbanization is essential for driving economic and social growth in its rural hinterland region. Environmental change is another concern of urbanization. Thus, urban areas cannot be considered separate entities from the environment as they, directly and indirectly impact the immediate and wider environment (Haase et al. 2014). Urban growth in the number and size and increasing demand for resources and energy pose a significant challenge to ensuring human welfare (Haase et al. 2013).

C. L. Chidi (✉)
Central Department of Geography, Tribhuvan University, Kathmandu, Nepal
e-mail: chidichhabilal@gmail.com

S. Anand et al. (eds.), *Sustainability in South Asian Cities*, Advances in Geographical and Environmental Sciences, https://doi.org/10.1007/978-981-97-7455-5_23

Urbanization creates an urban landscape with densification, expansion, and sprawling and shrinking patterns. This way of emerging patterns and their impact on land and the surrounding environment requires approaches that consider not only the complexity of the urbanization dynamic but also the interdependencies between drivers, impact, and responses to these dynamics. There are studies on the urban environment concerning mostly pollution, while the degrading state of the urban's surrounding environment due to haphazard urbanization is of little attention. Soil erosion and land degradation around the cities due to the high demand for land for the construction of buildings and infrastructure. Uncontrolled extraction of construction materials are emerging issues in the cities of developing countries (Sayami & Tamrakar, 2007). One issue is soil erosion from the hill slopes due to land use/ land cover change. Another issue is the rapid urbanization demanding large-scale construction materials such as sand, gravel, and stone causing soil erosion. While sand, gravel, and stone quarrying activities fulfil the growing demand of urbanizing areas in many developing countries of Asia, Africa, and South America. There are lacking regular monitoring and regulation systems for the uncontrolled mining of environmental resources (UNEP 2021). A fast-urbanizing country without proper land use planning creates haphazard construction of buildings. Particularly in the hill region, it highly aggravates soil erosion risk in its surrounding slope areas. Soil erosion in the upstream areas causes land degradation and flood hazard in the lower parts, where cities are often situated.

Nepal is one of the top ten fastest-urbanizing countries, and it is estimated to remain in the same position until 2050 (UN DESA 2014). Kathmandu Valley is the most populated and fastest-growing urbanization area in the South Asian region (MoUD 2015; Muzzini and Aparicio 2013). The valley's urban area accounts for 24% of the total urban population of Nepal (MoUD 2015). Rapid urban growth has resulted in high demand for land and construction materials. As a result, the urban expansion and extraction of construction materials are extending widely towards the valley's surrounding hill slopes. These activities have caused to increase in bare surface area in the hill slopes, resulting in huge soil erosion on the one hand and the deepening of riverbeds in the valley floor on the other hand. In addition, heavy sediment load in the river during the rainy season has often aggravated the floods on the valley floor (Sayami and Tamrakar 2007; Sada and Shrestha 2013). Unlike several studies concerning the urban environment, such as solid waste, air, and water pollution (Pradhan 1998; Pradhan and Sharma 2016; Saud and Paudel 2018; Pradhan et al. 2020), this paper assesses the impacts of bare land and extraction of construction materials on the environment in the cities of Kathmandu valley.

23.2 The Study Area

Kathmandu Valley lies in the middle mountain region of Nepal (Fig. 23.1). The valley [85°11′15"-85°31′30" E and 27°32′21"-27°49′01"N] covers approximately an area of 670 km^2. The valley's average altitude is 1,300 masl (meters above sea level), ranging from 1,095 m to 2,785 m. Geologically, the valley lies on the Kathmandu Nappe, located along the southern slopes of the Himalaya, which is composed of the Shivpuri Gneiss and marbles of the Bhimphedi Group. The early Paleozoic Tethyan rocks, the Phulchoki Group, overlie the Bhimphedi Group in the Kathmandu Valley area. The thickness of both these groups is 13 km (Paudyal et al. 2012). The valley's northern part comprises gneiss, schist, and granite, while its central part and other slopes consist of weakly metamorphosed Phulchauki Group. The basin is filled with clay, silt, sand, and gravel with more than 300 m depth (Sakai et al. 2001).

The valley's climate is characterized by mean air temperature ranging from 3°c in winter to 30°c in summer and average precipitation of 1,500 mm, 80% of which falls in the summer monsoon season (May to August). The Bagmati is the main river, and several tributaries flow down to the south through the valley floor. The valley region comprises surrounding hills with about 52% area and a valley floor with about 48%.

Administratively, Kathmandu Valley consists of three districts (Kathmandu, Lalitpur, and Bhaktapur), five major traditional cities, namely Kathmandu, Lalitpur, Bhaktapur, Madhyapurthimi and Kirtipur, and several towns. The annual growth rates of the population in the three districts were 4.88% during 1991–2001, 5.30% from 2001–2011 years (CBS 2015), and 1.9% from 2011 to 2021. These growth rates are relatively higher as compared to the national growth rates of 2.25%, 1.35% and 0.92% during the same three decades, respectively.

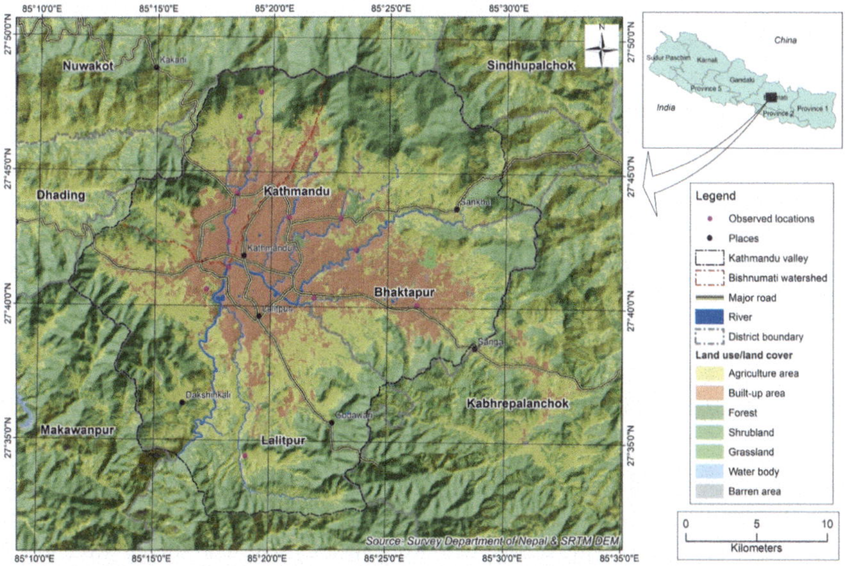

Fig. 23.1 The study area—Kathmandu Valley

23.3 Methods and Materials

Digital and ancillary imagery data were used. These were accompanied by field verification on the ground to assess the areas with soil erosion and land degradation in Kathmandu Valley. A detailed description of methods and materials is as follows.

23.3.1 Deriving Bare Land

A bare land area is defined as a bare surface without vegetation cover. Bare land areas of Kathmandu Valley for 1990, 2000, and 2010 were derived from the ICIMOD's land use maps (Bajracharya 2013; Bajracharya 2014a, 2014b) and verified and corrected with the Google Earth image. The bare land area for 2020 was derived from the Landsat8 image, which was freely provided by NASA. After the derivation of bare land using an object-based image analysis technique, it was then verified by a high-resolution Google Earth image.

23.3.2 Soil Erosion From Bare Land

The RUSLE model was used for soil erosion estimation, a suitable model for estimating the sheet and rill erosions in arable land (Chalise and Kumar 2020). The data on rainfall erosivity (R factor) was derived from the European Soil Data Centre (Panagos et al. 2015). Soil erodibility (K factor) was calculated using the soil properties data available from secondary sources. The topographic factor was calculated using the freely available 30 m resolution data of Aster DEM (Morgan and Davidson 1991). The cover management factor (C factor) was weighted 0.45 and the protection factor value was derived from previous studies in several parts of the world (Gwapedza et al. 2018; Jang et al. 2015; Tiruneh and Ayalew 2015).

23.3.3 Field Survey

The field survey was divided into two parts. The first one was an intensive field study of the Bishnumati watershed area (one of the tributaries of the Bagmati River), which is outside the city core and the second was inside the city. The field survey was carried out for three days in December 2020 in the Bishnumati watershed area and one day inside the valley's cities to acquire the ground reality. The Bishnumati watershed area covers approximately 102 km^2. There were altogether 17 observed locations; nine of them are in the Bishnumati watershed, and the remaining eight are in other areas of the valley. The locations being observed outside the Bishnumati watershed were to identify the frequency of inundation levels including their causes and consequences, which were identified by consultation with local people. The bare land and its ground reality from the headwater area of the Bishnumati River to the confluence point of the Bagmati River in the valley were carefully observed, which were identified from the land use/land cover maps and Google Earth images. Observation of construction activities and extraction of earth materials and their sediment production during the rainy season were made, which were assumed to be the possible causes of increasing bare land areas. In the valley floor, sediment flow was observed and deposited in the river during the rainy season. Their measurements at certain locations were done, while collection and extraction of sand from the riverbed at the foothill were also observed. Impacts of sand collected from the riverbeds were observed and recorded, and river deepening was measured at certain sites. Community-level discussions comprising five to nine local people were conducted at nine different locations. Informal discussions were also held with the local informants. During the field survey, the length, average width, and average depth of three tributaries of the Bishnumati River, namely the Mahadev Khola, the Koirala Khola, and the Sangla Khola in the foothill areas were measured, which were respectively 3408 m, 1.83 m, and 1.22 m. These were to estimate the total sediments carried by them in the rainy season.

23.4 Results

23.4.1 Distribution Pattern of Bare Land

The observation showed that the bare land areas comprised construction sites, land surface smoothing for building construction or plotting land by real estate, and extraction of earth materials such as stone, gravel, and sand. A few of them were also recent landslides.

Figure 23.2 shows the spatial distribution of bare land sites and their increment in Kathmandu Valley during four mapping years—1990, 2000, 2010, and 2020. In 1990, there were only 17 bare land sites, 10 of which were distributed in two major clusters—one around the middle of the valley and another in the northwest part from where the Bishnumati River flows down. Seven bare land sites were located at different locations; only three were over the hill slopes. From 2000 onward, the number of bare land sites has increased rapidly, particularly at the foothills on all sides of the valley, but more concentrated in the northern part. The spatial and temporal dynamics of the bare land sites are that their number has increased in the following decades, but their number has decreased gradually in the valley floor area and increased in the lower parts of the surrounding hills.

The geographical explanation of the uneven distribution of bare land sites in the valley is that the south-facing sunny side slope of the northern hills is preferable for housing due to longer sunny hours. This part gets sunshine in the early morning and remains for late hours in the evening during the cold winter than other parts of the

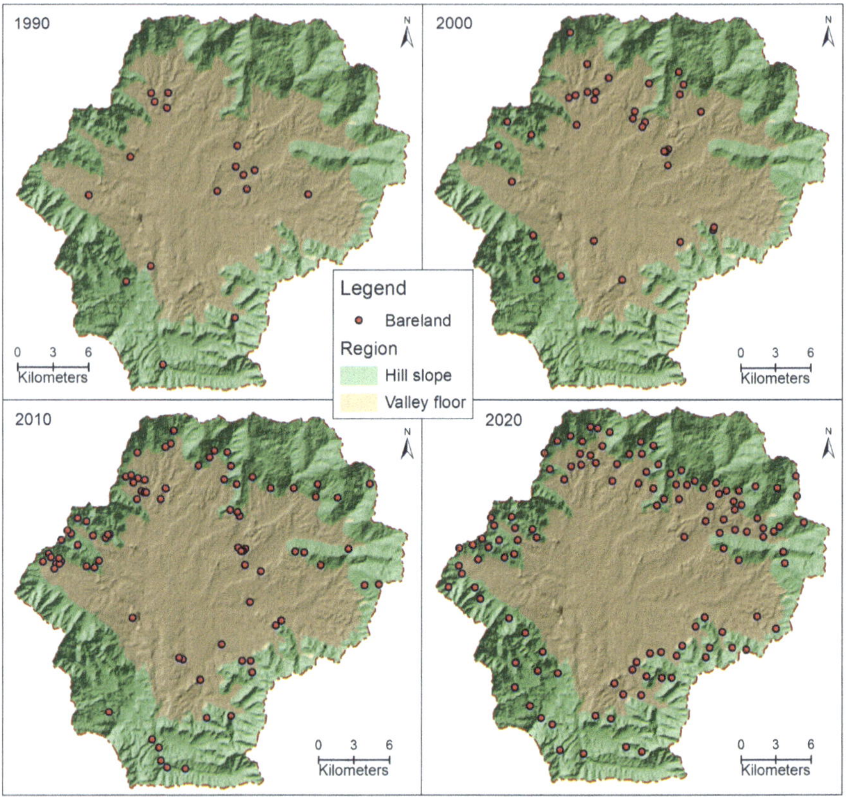

Fig. 23.2 Distribution of bare land sites in Kathmandu Valley. *Sources* Bajracharya (2014); Google Earth image and Landsat-8, NASA

valley. As little or no space was left due to the high demand for building construction in the northern foothill area, bare land sites began to appear in other parts of the valley. Secondly, some bare land sites in the far south have vanished because of vegetation regeneration in former landslide areas. Thirdly, bare land vanished around the central part and in most of the valley floor area due to densely built towering buildings and virtually no space left for further construction of buildings. Lastly, it is to be noted that the upper parts of the surrounding hills are the protected forest areas, which are the sources of rivers from where they flow down to the valley floor.

23.4.2 Bare Land Areas and Soil Erosion

Table 23.1 depicts the characteristic features of bare land sites, viz. number, area coverage, locations between hillslope and valley floor, slope gradient, descriptive statistics, and soil loss rate for Kathmandu valley in four mapping years (1990, 2000, 2010, and 2020). Firstly, the number of bare land sites has increased immensely, more than seven times, but at different rates during the last three decades. For instance, its number doubled from 1990 to 2000, more than double between 2000 and 2010, and less than 44% from 2010 to 2020. During the same decades, the area of the bare land sites has also increased constantly but with different ratios. Secondly, the mean area of bare land site units in 1990 measured 7.88 km^2 and decreased to 4.94 km^2 in 2020. Thirdly, the spatial distribution pattern of bare land sites is quite reversed between the hill slope and the valley floor. There is a constant increase (in %) in the area of the bare land sites in the hill slope, while its growth with a higher relative share is seen in the valley floor from 1990 to 2010 and then declined in 2020. Fourthly, bare land sites have shifted from lower slopes to higher slopes during the successive decades, signifying more vulnerability to the occurrence of soil erosion or landslides. The standard deviation values indicate the variation of upper to lower extreme values of the slope of bare land areas, but it remained constant after 2010, which indicates the variation of slope of bare land was already in the saturation stage in 2010. Slope steepness (degree) is one of the most important factors that has a straight relationship with the volume of soil erosion. There is a direct relationship between soil erosion rates of bare land sites with increasing slope gradients; the higher the slope gradient, the greater the volume of soil erosion. It is calculated that the annual average rate of soil loss from the bare land area was 64.6 tons/hectare which grew by and half-rate in 2020, which is a serious issue. This implies diminishing vegetative coverage or cultivated land over the hill slopes, causing flash floods and siltation in the valley floor.

Table 23.1 Characteristics of bare land area in Kathmandu Valley

Year	1990	2000	2010	2020
Number of bare land sites	17	34	85	122
Bare land area in km^2	134	201	402	603
Average area of bare land sites (in km^2)	7.88	5.74	4.73	4.94
Bare land area (%) over hill slope	0.1	0.1	0.8	0.9
Bare land area (%) in valley floor	0.3	0.5	0.9	0.6
Average slope (in degree) of bare land	5.5	8.7	13.2	14.2
STDEV of slope (in degree) of bare land site	5.4	6.8	7.1	7.1
Average rate of soil loss (t/h/y)	64.6	53.3	90.1	97.1

Source Field survey, 2020

23.4.3 The Ground Reality of Bare Land and Its Features—The Case of the Bishnumati River

The features of bare land sites, including the ground reality situation in the Bishnumati River watershed, are shown in Fig. 23.3 together with pictures taken during the field survey. The bare lands in the Bishnumati watershed and large parts of the cultivated land were found to be prepared for building constructions (Fig. 23.3; Photo a). The sandy parts of the hill slope were found to be excavated by using an excavator (Fig. 23.3; Photo b). The excavator was also used to dig out sand from the riverbed. Heavy sediment load in the Bishnumati River and its tributaries are often seen during the rainy season due to increasing bare soil patches over the hill slope.

Consequently, the river channels are filled with sediments and overflow of water from the riverbanks, damaging the cropland and flooding the built-ups in the valley floor. At the same time, this also offers local people employment and income opportunities. During the rainy season (May–September), local people collect sand and sell it to the sand contractor (Fig. 23.3; Photo c). This sand collection from the riverbed seems to be helpful in protecting the cropland and settlements from flash floods. According to the local people, an increase in sediment deposits in the rivers at the foothill has been seen over the last 15 years. The total length of stream channels of the Mahādev Kholā, the Koirala Kholā, and the Sānglā Kholā are measured at

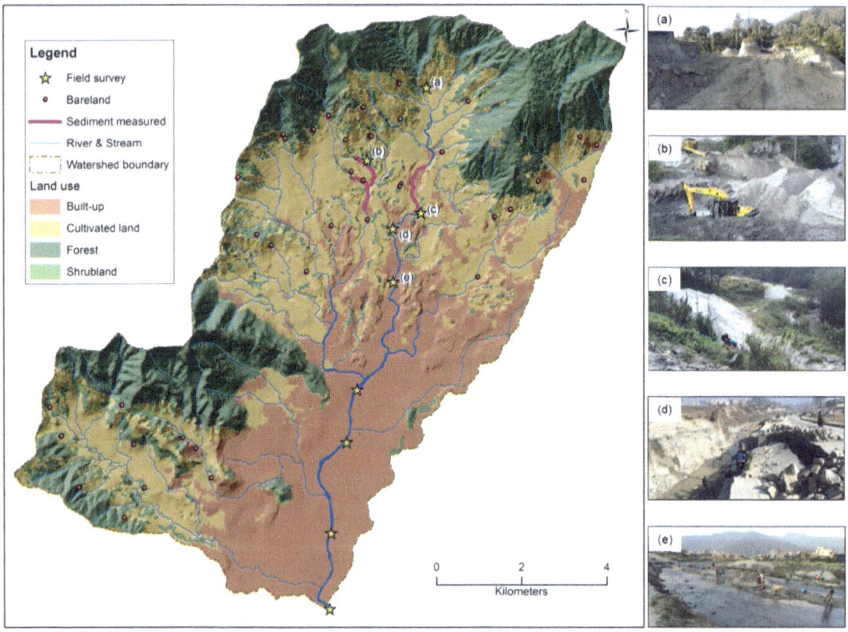

Fig. 23.3 Sand mining activity and road damage along the Bishnumati River (Field Survey 2020)

3408 m, and the annual volume of estimated sediment was 7609 m^3 in each rainy season (Fig. 23.3).

From Google Earth, it can be observed that most of the northern part of Kathmandu Valley has a similar situation of the sediment-loaded water flow in the streams and sand collection during the rainy season. It means this situation is common across the foothills of the valley. According to local people, this situation is dangerous. It needs to be controlled urgently at least to protect the ecology of the foothills, though they get the benefit of income from the sand collection, which is, however, only a temporary means. Further, sand mining from the riverbeds is seen even in nearby city neighbors of the valley floor (Fig. 23.3; Photo e), which has deepened the riverbed up to 7 m from the surrounding surface in some places (Figure; Photo d). This practice is dangerous to cause a slump in riverside and a risk to the road and other infrastructures such as bridges and buildings. Additionally, flash floods and water pollution in urban areas are often seen due to heavy sediment loads in the rivers. This activity calls for regulation urgently and thoughtfully from the concerned public agencies.

23.4.4 Flood Hazards in Kathmandu Valley

The information gathered from the field survey carried out in different parts of the valley exhibits that haphazard waste dumping and poor management of drainage channels are major causes of frequent floods in the rivers. Further, flooding in the rivers is also due to encroachment of their banks by construction works such as buildings, squatter settlements, bridges, and culverts. Sharp meandering of many stream channels in the valley floor is also a cause of frequent flash floods. There are several instances of these phenomena. In 2020, flash floods occurred in the Balkhu Kholā, the tributary of the Bagmati River, overflowing the retaining walls of both channel sides with about 2.5 m and affecting the streets and building structures in the Kalanki area near the central west of Kathmandu city badly. In 2018, large parts surrounding the Hanumante River, the tributary of the Bagmati River in Bhaktapur City towards the east of the valley, were inundated due to the flash flood in the Hanumante River. In the Jagati locality of Bhaktapur city, the flood level of the Hanumante River was 1.5 m above the riverbanks and a 6.2 m high water level from the riverbed. Three factors, viz. river channel encroachment, waste dumping in the riverbed, and lack of riverbank training, seem important to cause floods in the Hanumante River.

23.5 Discussions

The urbanization process increases the impervious surface, limiting the water infiltration capacity and thus raising surface runoff in urban areas (Konrad 2016; McGrane 2016). The rapid influx of people to the cities demands space for residential, industrial, services, and recreation activities. As a result, land use/land cover change takes place rapidly. Urban is not a single entity, which cannot be fully explained only concerning urban areas without interlinking it with total interconnected parts (Morris 2020). Rapid urbanization demands large volumes of construction materials such as stone, gravel, and sand to be supplied to the cities from the surrounding areas where they are available (Schiller et al. 2020). So, the rate of quarrying of such materials is directly related to the urbanization rate. If the implementation of the government policy towards controlling over mining of those resources is weak, haphazard mining of those resources will be increased. The nexus of rapid urbanization and uncontrolled exploitation of construction materials is a general scenario in developing countries, including the cities of Kathmandu Valley (Sayami and Tamrakar 2007). Rising unplanned land plotting over the hill slopes, thereby increasing bare land and building materials excavation, are the leading causes of supplying massive sediments into the rivers. These are harmful activities from the environmental sustainability and built-up structure protection perspectives.

Due to the rapid population increase, haphazard urbanization generates huge waste volumes. Uncontrolled waste and a poor sewerage management system aggravate flood and river pollution in the city region (Bhatta and Pandey 2020). Increasing extreme rainfall events due to climate change in recent decades have further aggravated the floods in the rivers affecting the cities and other settlements and properties across different parts of Nepal (Salike and Pokharel 2017). In Kathmandu Valley, frequent flash floods in the rivers during the rainy season are mainly due to heavy sediment load, particularly in the peri-urban areas, often neglected in the environment and sanitation conservation studies. Encroachment of the riverbanks is due to a lack of river conservation planning and management. Yet, the local and central governments are not serious about addressing these issues in Nepal.

23.6 Conclusion

Haphazard urbanization due to the lack of urban land use zoning plans and acts is a crucial issue in the cities across Nepal. Over the recent decades, the rapid increase in urban population has resulted from the high demand for land to construct buildings and other infrastructure and the uncontrolled extraction of construction materials in the surrounding hills of Kathmandu Valley. Poorly managed urban systems and weak implementation of urban plans and policies are responsible for the haphazard clearing of land for building construction on such environmentally vulnerable hill slopes. Moreover, heavy sediment loads in the rivers due to uncontrolled extractions of construction materials from rivers and hill slopes have aggravated the flood levels and frequencies and the water pollution in the valley's cities. Thus, for sustainable urban environmental planning, the valley's cities call for urgent and strict regulation by the concerned government agencies over increasing bare lands in the surrounding hills, haphazard and uncontrolled extractions of earth materials from the river channels, and river encroachments, it is argued.

References

Bhatta BP, Pandey RK (2020) Bhaktapur's flood-related disaster risk and strategy after 2018. J APF Comm Staff Coll 3(1):72–89

CBS (2015) Compendium of environment statistics. Central Bureau of Statistics, Kathmandu, Nepal

Chalise D, Kumar L (2020) Land use change affects water Erosion in the Nepal Himalayas. PLoS One (15). https://doi.org/10.1371/journal.pone.0231692

Gwapedza D, Slaughter A, Hughes D, Mantel S (2018) Regionalizing MUSLE factors for application to a data-scarce catchment. PIAHS 377:19–24

Haase D, Kabisch N, Haase A (2013) Endless urban growth? On the mismatch of population, household and urban land area growth and its effects on the urban debate. PLoS One, 8. https://doi.org/10.1371/journal.pone.006653

Haase D, Frantzeskaki N, Elmqvist T (2014) Ecosystem services in urban landscapes: practical applications and governance implications. Ambio 43:407–412. https://doi.org/10.1007/s13280-014-0503-1

Jang C, Shin Y, Kum D, Kim R, Yang JE, Kim SC, Jung Y (2015) Assessment of soil loss in South Korea based on land cover type. Stoch Environ Res Risk A 29:2127–2141

Konrad CP (2016) Effects of urban development on floods. U.S. Geological Survey (Fact Sheet 076–03). U.S. Department of the Interior, U.S. Geological Survey, Washington

McGrane SJ (2016) Impacts of urbanization on hydrological and water quality dynamics, and urban water management: a review. Hydrol Sci J 61(13):2295–2311. https://doi.org/doi.org/10.1080/02626667.2015.1128084

Morgan RP, Davidson DA (1991) Soil erosion and conservation. Longman Group, London

Morris GL (2020) Classification of management alternatives to combat reservoir sedimentation. Water 12:861. https://doi.org/10.3390/w12030861

MoUD (2015) National urban development strategy 2015. Ministry of Urban Development, Government of Nepal, Kathmandu

Muzzini E, Aparicio G (2013) Urban growth and spatial transition in Nepal. World Bank, Washington DC

Panagos P, Borrelli P, Poesen J, Ballabio C, Lugato E, Meusburger K, Montanarella L, Alewell C (2015) The new assessment of soil loss by water erosion in Europe. Environ Sci Pol 54:438–447

Paudyal YR, Yatabe R, Bhandary NP, Dahal RK (2012) Basement topography of the Kathmandu basin using microtremor observation. J Asian Earth Sci (62):627-637. doi.org/https://doi.org/10.1016/j.jseaes.2012.11.011

Pradhan B (1998) Water quality assessment of the Bagmati River and its tributaries, Kathmandu valley, Nepal. Unpublished Ph.D. Dissertation. University of Natural Resources and Life Science, Vienna, Austria

Pradhan PK, Sharma P (2016) Nepal: urban environment analysis. In RB Singh, P Prokop (eds.), Environmental geography of south Asia: advances in geographical and environmental sciences, Springer Japan, pp 305–334

Saud B, Paudel G (2018) The threat of ambient air pollution in Kathmandu, Nepal. J Environ Public Health 1:1–7

Pradhan B, Sharma P, Pradhan P (2020) Urban growth and environment and health hazards in Kathmandu valley, Nepal. In: R B Singh, et al. (eds) *Urban health risk and resilience in Asian cities*, Advances in Geographical and Environmental Sciences, Springer Nature Singapore, pp 293–324

Sakai H, Fujii R, Kuwahara Y, Upreti BN, Shrestha SD (2001) Core drilling of the basin-fill sediments in the Kathmandu valley for paleoclimatic study: preliminary results. J Nepal Geol Soc 25:9–18 (special issue)

Salike IP, Pokharel JR (2017) Impact of urbanization and climate change on urban flooding: a case of the Kathmandu Valley. J Nat Resourc Devel 7:56–66

Sayami M, Tamrakar NK (2007) Status of sand mining and quality in northern Kathmandu, Central Nepal. Bull Depart Geol, Tribhuvan University, Kathmandu, Nepal (10):89–98

Schiller G, Bimesmeier T, Pham ATV (2020) Method for quantifying supply and demand of construction minerals in urban regions—a case study of Hanoi and its hinterland. Sustainability 12:43–58. https://doi.org/10.3390/su12114358

Tiruneh G, Ayalew M (2015) Soil loss estimation using geographical information system in Enfraz watershed for soil conservation planning in highlands of Ethiopia. Intern J Agricul Res Innov Technol 5:21–30

Desa UN (2014) World urbanization prospects: 2014 revision. United Nations Department of Economic and Social Affairs, New York

Chapter 24
Urban Growth and Urban Land Use Dynamics of Pokhara City, Nepal: Challenge to Urban Sustainability?

Khagendra Raj Poudel, Rajesh Bahadur Thapa, Motilal Ghimire, and Ramesh Hamal

Abstract Over the recent decades, Pokhara has evolved immensely in its urbanization, urban functions, and administrative roles. With tremendous changes in land use pattern, environment, and socio-economic features, the city now is facing a challenge to its sustainable urban development. This paper analyses changes in land use pattern, demographic feature, and development activities vis-à-vis urbanization of Pokhara city. Data were gathered from satellite images of different years, field verification, and ancillary sources. The findings show that the city has evolved immensely from mere a trading and administrative centre to a tourism hub and a provincial capital. There has been a rapid increase in the urban population over the last three decades due mainly to boundary expansion and in-migration of the population. Availability of financial investment opportunity, boost in real estate business and change in socio-political phenomena were other factors of rapid urbanization. These changes have caused haphazard use of land, rising air pollution, and degrading environment, which thus call for immediate measures to be initiated regarding the environmental sustainability of Pokhara city.

Keywords Urbanization · Land use and land cover · Driving factors · Rural urban migration · Environmental degradation

K. R. Poudel (✉)
Department of Geography, Prithvi Narayan Campus, Tribhuvan University, Kirtipur, Nepal
e-mail: khagendra.poudel@prnc.tu.edu.np

R. B. Thapa
International Centre for Integrated Mountain Development, Kathmandu, Nepal

M. Ghimire
Central Department of Geography, Tribhuvan University, Kirtipur, Nepal

R. Hamal
Tribhuvan University, Kirtipur, Nepal

S. Anand et al. (eds.), *Sustainability in South Asian Cities*, Advances in Geographical and Environmental Sciences, https://doi.org/10.1007/978-981-97-7455-5_24

24.1 Introduction

Urbanization is a global phenomenon resulting primarily from changes in human behaviors such as changes in lifestyle from economic, political, and resource consumption perspectives. The term urbanization involves a multifaceted process that transforms the rural landscape to urban form, resulting in changes in the ecosystems and their functions, which ultimately impact the lifestyle of people (Thapa 2009). Urbanization also refers to the change and growth of human habitation, per capita energy consumption, resource utilization, and sociocultural and economic activities across the world (McDonnell and Pickett 1990; Mayhew 1997). Of all, rapid change in land use patterns including the farmland urban landscape over time is critical for urban development in developing countries (Li and Yeh 2000).

In Nepal, Pokhara is one of the fast-growing cities, with an urban growth rate of over 5%, followed by Kathmandu, the capital city of Nepal (MoUD 2017). This rapid growth of urbanization has resulted in various transitions such as spatial, demographic, and economic aspects in Pokhara like in many other large cities (Haack 2009; Tiwari and Joshi 2012; Rimal 2011; Rimal et al. 2017). Other visual impacts of rapid urbanization include irregular land use practices and environmental hazards in the city area (Pradhan and Sharma 2016). In this chapter, we aim to examine the change in the spatial pattern of urbanization in Pokhara city from 1990 to 2020 and explore their potential drivers.

24.2 Brief Introduction to Pokhara City

Pokhara, the metropolitan and capital city of Gandaki province, covers an area of 464.24 km^2 (83° 47’-84° 09’ East and 28°04’-28°20’ North) and is situated on the southern slope of the Annapurna Himalayan range (Fig. 24.1). The city extends over the entire Pokhara Valley and receives a mean annual precipitation of 4,234 mm, the highest in Nepal (CBS 2019). The valley often witnesses flash floods in the Seti River flowing through its middle part and landslides on the surrounding mountain slopes during the monsoon season. The city has a total population of 0.4 million, comprising different ethnic groups viz. Gurung, Newar, and Magar and other castes of Brahmin, Chhetri, Damai, Kāmi and Sārki (CBS 2012). It is a centre of the provincial administration, commerce, health and education, banking, transportation and communication, and tourism services.

Evolution of Pokhara as a town started about 250 years ago when the migration of 26 households of Newars (also known as ‘*chabbis-kuriya*’) from Bhaktapur, a city of Kathmandu Valley took place soon after King Prithvi Narayan Shah took over Kathmandu valley during the Nepal unification campaign (Gurung 1965). The Newar in-migrants with masonry skills, wood carving and rich cultural practices established settlements around the places currently named as Ganesh Tol, Bhimsen Tol, and Bhairav Tol (Fig. 24.2). According to local source, they not only initiated

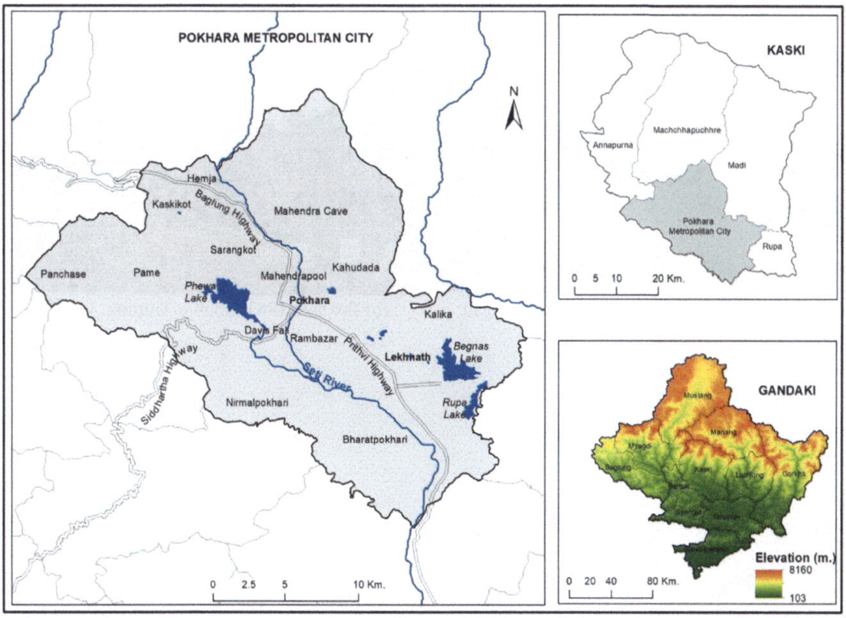

Fig. 24.1 Location of Pokhara City in Gandaki Province, Nepal

rice cultivation in the paddy fields but also produced handloom-based products such as woollen and cotton cloths and prepared Newari foods for selling through retail outlets and thus established a bazaar.

Historically, Pokhara had acted as a transit trade centre lying on the western Trans-Himalaya trade route between Tibet and India, and whatever trade existed tended to be transient, with goods such as musk, salt, woollen blankets, herbs, etc. from Tibet being carried through this town for sale or barter in Batauli (Butwal) lying in the Tarai, and goods including cotton made, kerosene, brass, and other goods manufactured in India being carried to the north for exchange in Tibet (Gurung 1965; Pradhan and Sharma 2022). The traders used to visit even the northern Indian cities such as Lucknow and Patna (Blaikie et al. 1983).

Over the past decades, Pokhara city has evolved tremendously. Historically, the city had got influx of outsiders from within Nepal in the mid-eighteenth century. It also became a refugee center for Tibetan refugees, followed by Muslim fugitives from Kashmir. Further, the town began to grow after the initiation of the malaria eradication program during the late 1950s. These have been a milestone to expand Pokhara town and so public investments and migration of population grew tremendously. The government declared Pokhara as an incorporated municipality in 1962, as the Western Development Regional headquarters in 1972, as a sub-metropolis in 1995, and as a metropolitan city as well as the capital of Gandaki Province in 2017. Construction of two national highways such as the Siddharth highway connecting Pokhara to Bhairahawā in the Tarai to the south bordering India in 1969 and the Prithvi highway

(a) Pokhara pre-1928, the first Newar settlement

(b) Pokhara 1963 (Source: Gurung, 1965)

(c) Field work, December 2020

Fig. 24.2 Changing building structure at Bhimsen *Tol* during different years

linking it to Kathmandu in 1971 accelerated the growth of the city by attracting investment in government offices and various economic activities, increased in migrants flows from different areas within Nepal, as well as from foreign countries, and flows of goods and services (Pradhan 1982; Adhikari 2000). It is one of the fastest growing cities of Nepal. In 1956, the city had only 4,000 inhabitants which increased to just over 5,000 in 1961 and to 20,611 in 1971. The population increased by about 49 times over the past five decades. During the 1981–1991 decade, the average annual growth of Pokhara city was 7.4%, which was the highest among all 58 urban areas of Nepal. Migration of the population to the city has been the main reason for the rapid urban growth. In 2001, the rural–urban migration constituted about 83% of the total migration in its Kaski district, the majority of which took place in Pokhara city. According to the 2011 census (CBS 2012), the metropolitan city had a total of population of 414,141. Likewise, investment in various infrastructure and facilities like administration, health, education, transportation, service sector, tourism, and hospitality industries also grew rapidly. Currently, it is widely known as the tourist city of Nepal (Fig. 24.3).

(a) Photo 1960 (b) Photo 2015

Fig. 24.3 Urbanization scenario on the northern part of Pokhara. *Source* Fieldwork (2020)

24.3 Materials and Methods

24.3.1 Data Sources

In this study, both primary and secondary data have been gathered (Table 24.1). Expert consultations and informal interviews with various concerned stakeholders were conducted during the field survey carried out in 2020. In addition, different types of ancillary data were collected from the public access sources such as the Central Bureau of Statistics (CBS) and other government agencies, which have been used to assess the population growth and urban expansion process. We used the UTM Zone 45N projection system and Global Positioning System (GPS) to collect cal/val data for land-cover and road network.

Table 24.1 List of primary and ancillary data sources

Data type	Year	Resolution	Sources
Landsat TM	1990-April 16	30 m	USGS (https://earthexplorer.usgs.gov/)
Landsat ETM +	2000-March 26	30 m	USGS (https://earthexplorer.usgs.gov/)
Landsat TM	2010-February 18	30 m	USGS (https://earthexplorer.usgs.gov/)
Sentinel -2	2020–11-12	15 m	Copernicus Open Access Hub (https://scihub.copernicus.eu/)
Aerial photographs	1978		Survey Department, Nepal 1998
Google Earth Image	Access 2020–12		Google Earth
SRTM – DEM data		30 m	USGS
Ancillary data	CBS 2001, 2011		CBS, Kathmandu
Local level boundary			Survey Department, Nepal
Field survey	2020		Pokhara area

24.3.2 Land Cover-Based Transition Analysis

The verification of Landsat TM and ETM, and Sentinel -2 images was carried out for geometric accuracy and the images were formed using the ENVI 5.3 software. Initially, knowledge-based visual depiction, texture, and association analysis were processed, followed by extracting information cautiously from high-resolution satellite images, field visit, aerial photographs, and urban sector documents for the preparation of land use classes. Three categories of urban land use, i.e., built-up area, non-built-up area, and water body were determined. Landsat image (30-m) was resampled with the help of nearest neighbour resampling method to balance the data resolution of Sentinel (15-m) for keeping spatial resolution consistency of the sources of data. Ground truths were verified with the help of GPS and Google Earth Imagery and loaded into view. With the help of the ground control point (GCP), training samples representing those three classes were created. The algorithm maximum likelihood was used to calculate the mean, variance, and covariance statistics. The accuracy of image classification processed by assessing the total classification accuracy along with geographically referenced vector datasets remained at 82% and overall accuracy at 87%. Figure 24.4 describes a sequence of processing methods for the transformation of those data into meaningful thematic layers.

The cross-function has been widely used to detect the transition matrix of LULC using the Modules for Land Use Change Evaluation (MOLUSCE) plug-in developed for QGIS software (Asia Air Survey and NextGIS—hub.qgis.org/projects/molusce). The plugin was designed to analyse model and simulated urban land use change including forestry. The transition matrix comprises rows and columns of landscape categories at times T1 and T2. In this work, QGIS software was used to perform the

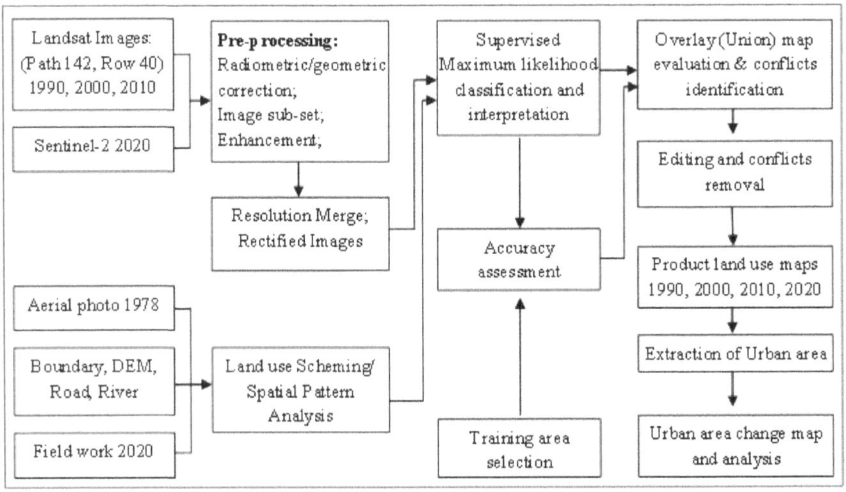

Fig. 24.4 Urban land use mapping design. *Modified* Thapa and Murayama (2009)

change on land cover layer for three pairs of time periods: 1990 and 2000, 2000 and 2010, 2010 and 2020.

24.3.3 Analysis of Spatial Patterns

Spatial metric is one of the useful measures for analysing spatial pattern of urban growth and sprawl, which helps in quantifying the urban structure, expansion, and trend (McGarigal and Marks 1995). To assess and compare the spatial patterns of urban development, five different spatial metrics, namely PLAND, PD, ENN, CIRCLE and SHAPE were used (Table 24.2). They have been used to calculate each of land use/cover layer at the class and landscape levels. The 8-cell neighbour rule was used to examine the association of every pixel to a patch, where the measurement of all four orthogonal and four diagonal neighbours of the focal cell was performed. In this approach, two cells of the same land cover class crosswise touching were taken as a portion of the same patch, in contrast to the 4-cell neighbor rule, as these were used as different patches (McGarigal et al. 2012). All these calculations were computed using the FRAGSTAT software (McGarigal 2015) as this software has been used by several authors in similar studies (Thapa and Murayama 2009; Estoque and Murayama 2017; Kamusoko 2017; Kamusoko and Chikati 2017). Image analysis and GIS data processing were carried out by using ENVI and ArcGIS software.

24.4 Results and Discussions

24.4.1 Urban Population Growth

Population growth is one of the crucial factors of urbanization in Pokhara city. The city has experienced high rate of population growth during the past decades. Having a mere 3,755 people in Pokhara in the early 1950s, its size grew by over five times to 20,611 in 1971 after it got municipal status in 1962. The urban population grew tremendously and reached 0.4 million in 2011 (CBS 2012). Table 24.3 depicts a massive jump in growth in the population size of Pokhara city between 2001 and 2011 (Fig. 24.5). This rapid population growth in the city is due to several factors, as described below.

24.4.1.1 Boundary Extension of Pokhara City

Firstly, the municipal boundary of Pokhara city has expanded constantly over the years. Table 24.4 shows that the municipal boundary extended from 27.4 km^2 in 1971 to 222.6 km^2 in 2014 and to 464.24 km^2 in 2017. The municipality then was

Table 24.2 Spatial metrics used for the built-up area (class-level)

Spatial metrics	Range	Unit	Description	Measure
PLAND area	0 < PLAND 100	Percent	Proportion of landscape; total proportion of built-up compared to the entire landscape (excluded water body)	Area
PD	PD > 0	Number/ 100 ha or / km^2	Patch density: refers to the total number of patches of built-up per unit area	Aggregation (subdivision/ fragmentation)
ENN (mean)	ENN > 0	Meter	Euclidean nearest neighbor distance: refers to the length of the nearest adjacent patch of the same type based on the least edge-to-edge distance	Aggregation (isolation/ dispersion)
SHAPE (mean)	1 SHAPE ∞	None	Shape index: calculates the complexity of patch shape as compared to a basic shape (square) of the same size; also provides the measurement of the irregularity of built-up patches. A low figure signifies low complexity	Shape complexity
CIRCLE (mean)	0 CIRCLE < 1	None	Related circumscribing circle; presents the measurement of overall patch elongation; computes the circularity of built-up patches. Low figures signify the circular patches	Shape (complexity)

Table 24.3 Population change in Pokhara city

Census year	Population	Growth rate (%)
1952/54	3755	-
1961	5413	5.52
1971	20,611	28.08
1981	46,642	12.63
1991	95,286	10.43
2001	156,312	6.40
2011	414,141	16.49

*Source*CBS (2012)

declared a sub-metropolis in 2014 and a metropolis in 2017. Accordingly, the number of wards, the smallest administrative units, has also been increased (Fig. 24.6). These extensions of municipal boundary have amalgamated surrounding rural areas into Pokhara city and consequently, their population has also been enumerated in the city

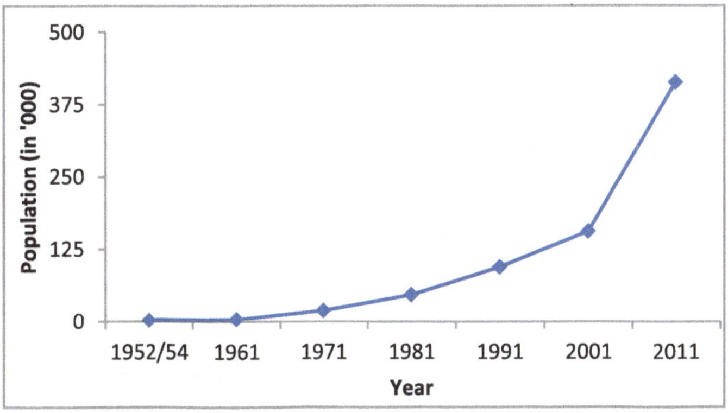

Fig. 24.5 Urban population growth trend, Pokhara City (CBS 2012)

Table 24.4 Municipal boundary expansion of Pokhara City

Year	Total wards	Area (km^2)
1971	13	27.38
1976	16	49.88
1981	18	49.88
1992	18	55.49
2014	29	222.60
2017	33	464.24

Source Pokhara Metropolitan City Office (2018)

population. Upgrading municipal status or declaring a rural town as a municipality due mainly to political decisions, but not to urban functions is usual practice in Nepal (Pradhan 2004).

24.4.1.2 *Change in Urban Land Use Patterns (1990–2020)*

There is a remarkable change in the urban land uses of Pokhara city. For instance, the city's built-up area has expanded almost by three times from 17.52 km^2 in 1990 to 50.54 km^2 in 2020 (Fig. 24.7 and Table 24.5). The annual rate of increment of the built area was 1.1 km^2/year over the past three decades. On the other hand, the non-built-up area decreased noticeably from 434.05 km^2 in 1990 to 399.98 km^2 in 2020. Interestingly, the waterbody area has increased from almost 13 km^2 to 14 km^2 during the same duration of years, which was due to an increase in area by constructing dam in five lakes, viz. Depang, Khaste, Gunde, Neureni, and Maidi.

Table 24.5 illustrates how the spatial metrics of the built-up area have changed in Pokhara City over the last three decades. The proportion of landscape (PLAND) metric refers to the measurement (in percent) of a particular class (built-up area) in a

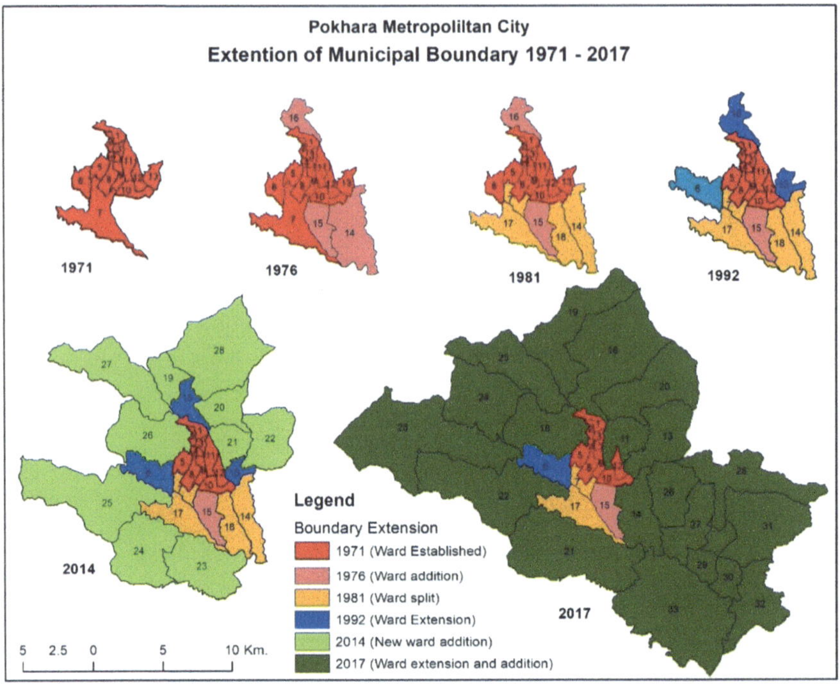

Fig. 24.6 Municipal boundary expansion. *Source* Toposheet, Survey Department 1998

particular year as compared to the entire landscape. In 1990, the built-up comprised a PLAND of 3.79% that rose to 4.93, 7.22, and 10.93% in 2000, 2010, and 2020 respectively. It means the built-up area has constantly expanded with a significant increment rate. Further, the patch density (PD) metric refers to the measurement of fragmented patches, which depends on how many numbers of patches are contained per unit area (1 km^2), where a patch has been derived from an 8-cell neighbor rule. The PD of the built-up area increased from 3.79/km^2 in 1990 to 10.93/km^2 in 2010, demonstrating that the patches of built-up got to be fragmented during the last 30 years. Moreover, it is interesting to note that the PD has more than doubled in 2020 over 2010, rising from 5.16 to 11.49. This indicates that the sprawl of newly built-up patches has been quite apparent in the past decade.

The measurement of dispersion on the ground in terms of the distance of a patch to the nearest adjacent patch of the same class is known as the Euclidean Nearest Neighbor distance (ENN) metric. The mean ENN of the built-up patches has dropped from 124.38 m in 1990 to 70.15 in 2020. However, there was a slight increment in the mean ENN from 112.21 in 2000 to 115.75 in 2010. Overall, the decrease in mean ENN during the past 30 years might be due to the increment of old built-up patches and the formation of new patches. This however is not essentially linked to the former patches, as shown by the rise in PD and PLAND during the same duration of years.

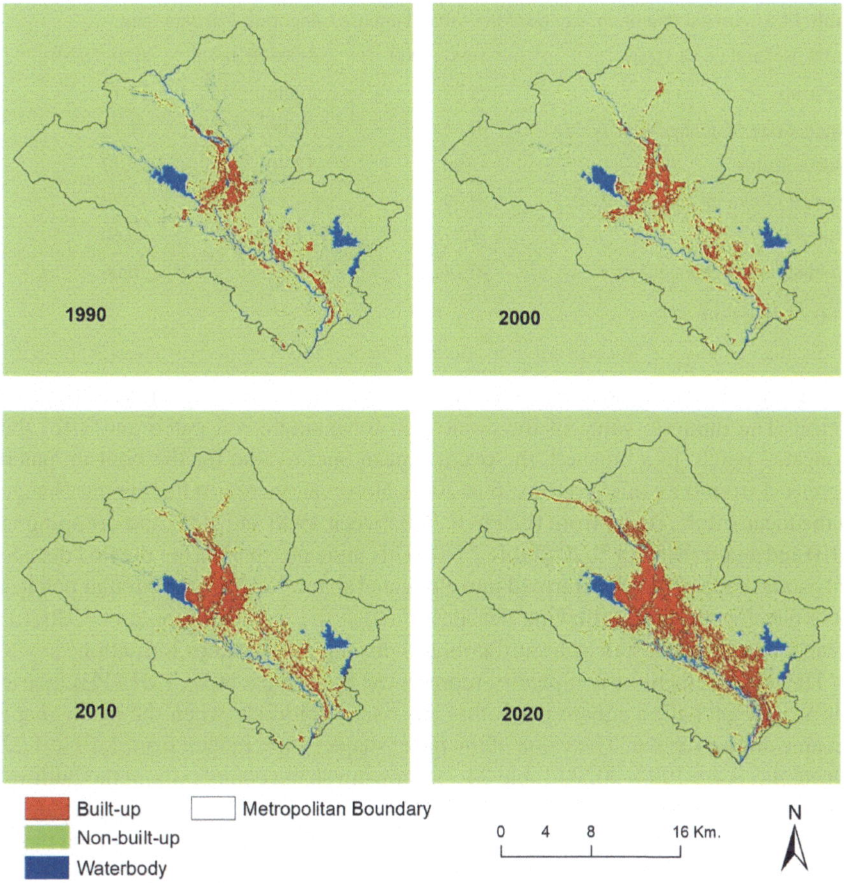

Fig. 24.7 Satellite imageries showing classes of urban land built-up, Pokhara City

Table 24.5 Change in area (km^2) of urban land use categories, Pokhara City

Types	1990	2000	2010	2020
Built-up	17.52	22.79	33.37	50.54
Non-built-up	434.05	430.15	417.75	399.98
Water body	12.9	11.53	13.35	13.95
Total	464.47	464.47	464.47	464.47

Source Data derived from Landsat Images (1990, 2000 and 2010) and Sentinel Image, 2020

Table 24.6 Annual rate of change (km^2) in urban land use categories, Pokhara City

Land use land cover types	1990–2000	2000–2010	2010–2020
Built-up	5.27	10.58	17.17
Annual rate of change (km^2/year)	0.53	1.06	1.72
Non-built-up	-3.90	-12.40	-17.77
Annual rate of change (km^2/year)	-0.39	-1.24	-1.78
Waterbody	-1.37	1.82	0.60
Annual rate of change (km^2/year)	-0.14	0.18	0.06

Source Landsat images

The circularity of patches is measured by the circumscribing circle (CIRCLE) metric. The numeric value of the circle is 0 for a single-cell patch and 1 for the elongated patch. In a nutshell, the overall mean circle value for the built-up patch increased from 0.29 in 1990 to 0.36 in 2020. However, there is a fluctuating change in the mean circle, rising from 0.29 to 0.36 between 1990 and 2000 and dropping in 2010 and again rising in 2020 (Table 24.7). This indicates that during the first decade only some built-up patches formed into elongated shapes while more circular patches were developed in the following decades. Finally, the increment trend of CIRCLE mean value for the last decade further highlights its linear pattern formation.

The measurement of complexity represented by a shape index (SHAPE) metric has a value of 1 for a square patch that has risen limitlessly when the patch shape became more irregular. The value of the mean shape metric ranged from 1.19 to 1.22 during the years 1990–2020 (Table 24.7), showing shape complexity in the built-up patches. Despite the variation in mean SHAPE values in different time periods, it decreased to 1.15 during the last decade, indicating that the complexity in the built-up patches was somewhat stable over the last 3 decades. Figure 24.8 shows how the built-up area has changed at every 10-year interval over the past three decades, while Fig. 24.9 depicts the contemporary status of a major urban area of Pokhara city.

Table 24.7 Change in urban land cover patterns, Pokhara City

Spatial matrix	1990	2000	2010	2020
PLAND (%)	3.79	4.93	7.22	10.93
PD (per sq.km)	3.74	4.12	5.16	11.49
ENN (mean in meter)	124.38	112.21	115.75	70.15
CIRCLE (mean ($0 \leq$ CIRCLE < 1)	0.29	0.33	0.29	0.36
SHAPE (mean $1 \leq$ SHAPE $\leq \infty$)	1.19	1.22	1.21	1.15

Source Calculated from Fragsat software

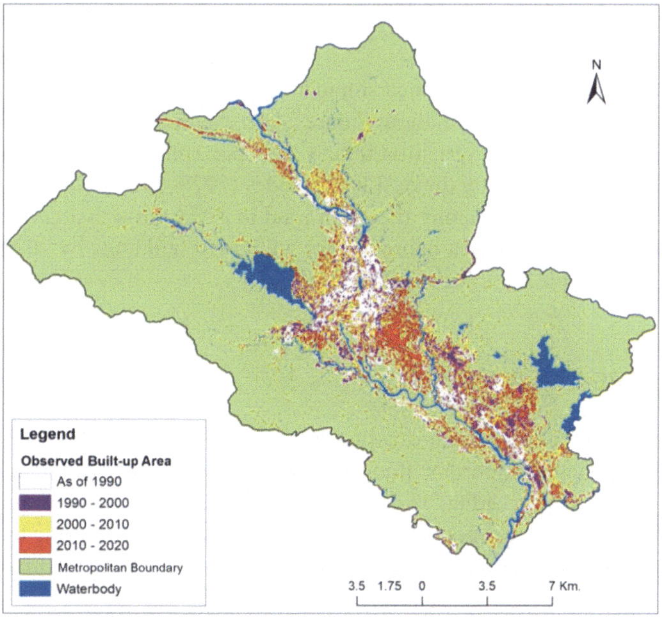

Fig. 24.8 Change in urban built-up area (1990–2020), Pokhara City

Fig. 24.9 An example of the urban core area, Pokhara City (Photo by Hamal, June 2020)

24.4.2 Driving Factors of Urbanization in Pokhara City

Undoubtedly, Pokhara has experienced rapid urban growth over the past few decades, showing a rapid urban transformation. Figure 24.10 exemplifies how the urban landscape has encroached upon agricultural land and open space areas of Chhorepātan, the city's southern part during the past five decades (1978–2020), where the locality evolved with few buildings along the main road in 1978 through a compact built-up area with virtually no space being left for additional buildings by 2020. Several factors are responsible for this transformation.

24.4.2.1 *Economic Opportunity*

Economic opportunity is one of the leading driving factors of urbanization in Pokhara. The city lying on the important trade route between the northern mountains of Nepal, adjoining Tibet (China) and the southern Tarai plain, neighbouring India is considered a booming economic centre of the country, with increasing the middle-class population (Gurung 1965; Rimal 2011). In historic times, the city acted as a crucial staging place between the western Trans-Himalayan and the east–west trade routes. The core factors consolidating the strength of Pokhara include the relatively flat geographical valley structure with a combination of magnificent landscapes of Annapurna and Machhapuchhre Himalayan ranges and several lakes, transport connectivity, economic opportunities, and conducive social environment (Muzzini and Aparicio 2013). The city represents an unprecedented degree of urban expansion in the Himalayan landscape and supports the tendency of urban expansion leading to South Asia (UN-Habitat 2012).

As shown in Fig. 24.11, a combination of naturally gorgeous landscapes, the ongoing construction of the international airport, well-equipped hospitals, Pokhara University, and many others such as tourism and business, health, and education advances, are one way to another facilitating economic progress.

Thus, the city offers diverse economic opportunities not only to the people from within the valley itself and other parts of the country but also to foreign investors. According to HAN (2021), Pokhara currently has enormous tourism and hospitality services including 2-five star, 2-four star and many 3-star hotels, banking, health, education, and diverse tourism activities (trekking, sight-seeing, religious, boating, cable car, bungee jumping, paragliding, ultralight flight, river rafting, homestay, diverse ethnic-cultural traditions). Road and air services connectivity to different parts of the country and an increase in foreign investment have proliferated the economic attractiveness of the city. This capital-centric development model, mostly popular in low and middle-income economies, can be considered a key factor for drastic land use land cover change in Pokhara.

(a) Aerial photograph, 1978

(b) Google Earth Image, 12-2004

(c) Google Earth Image, 21-12-2020

Fig. 24.10 Chhorepātan area—southern part of Pokhara

(a) Phewa lake side tourism area

(b) Pokhara international airport

(c) Manipāl teaching hospital

(d) Pokhara university

Fig. 24.11 Some of the important economic and social institutions. *Source* authors

24.4.3 Rural–Urban Migration

A huge number of urban-bound migrants has occurred in Pokhara, leading to its rapid urban transformation. Since the 1970s, there has been a significant rise in population in Pokhara, becoming one of the rapidly growing cities of Nepal (Ministry of Urban Development 2017). The economic backwardness, relatively low living standard, absence of basic amenities, and lack of other opportunities in rural areas act as push factors for the migration of people to Pokhara City (Muzzini and Aparicio 2013). Moreover, migration from rural areas to the city escalated because of civil unrest that took place between 1996 and 2006, resulting in the pouring of displaced people in Pokhara and other cities of Nepal (Williams and Pradhan 2009). Furthermore, after Pokhara became the capital of Gandaki province, according to the new constitution (Government of Nepal 2015), it attracted several provincial government facilities. So, Pokhara can represent one of the fastest-growing cities in developing countries, as the United Nations (2014) predicted that the number of people living in the urban areas was estimated to rise from 2.6 billion in 2010 to 5.2 billion in 2050. This rise in urban population is expected mainly due to migration from rural regions (World Bank 2011).

24.4.4 Real Estate Boom

The high rate of migration is a key to increase in the housing and land prices in urban areas (Lin, et al. 2018). In Pokhara like in many other large and rapidly growing cities of Nepal, the real estate business has boomed during the recent decades. The housing colonies with some modest amenities like private parking, advanced grocery stores, cafés and restaurants, road link, and other basic facilities are nowadays highly demanded by growing middle-class families. Though there is a lack of data on the exact number of real estate business agencies, our communication with the concerned local government officials indicated that there were currently more than a dozen agencies dealing with land procurements and housing in Pokhara, which was nowhere in a few decades ago. It appears that there is little or no government control over the real estate business in Pokhara due to a lack of land-acquisition rules and regulations.

24.5 Conclusion

Urban growth and expansion of Pokhara city will continue even in the future due mainly to migration of population and financial investment. Tourism is thriving rapidly and there is constantly growing investment in development activities, particularly the tourism-based infrastructure and services and remittance flow in the city. On the other hand, haphazard sprawl of urban built-ups has already occurred across the Pokhara city region, particularly in the environmentally sensitive areas such as riverbanks and hill slopes, causing pollution in lakes and rivers, and on land. Water hyacinth, for instance, is a recent phenomenon of aquatic plants being spread across the Phewa Lake and the Begnas Lake, which is due to nutrients formed by the organic wastes (Pradhan and Sharma 2016). The Phewa Lake is encroached upon by locals and invaded by weeds, making it extremely vulnerable and is now in real danger of extinction (Pradhan 2014). The downstream parts of the Phewa Lake and the Seti River are being polluted by dumping household wastes. The population living in the slum area, characterised by a deficiency in access to piped drinking water, education, and vulnerable houses with sharing toilets, has accounted for 3.6%, though it is far less than the national urban slum population of 7.1% (Gurung 2008; Lumanti 2010). Another environmental problem in the city region is river flash flood that occurs occasionally. For instance, in 2012 massive flash floods occurred in the Seti River due to broke down of the blockage of the river course by the avalanches upstream that swept away the Kharapani bazaar situated on the bank of the Sardi Khola, one of the tributaries of the Seti River. This was probably due to settlements on the bank and sand mining activity by people who didn't expect floods during the dry season (The Himalayan Times, May 13, 2013). According to the local inhabitants, a similar disaster had also occurred in 1959. In November 2013, several sinkholes happen to be developed in the northern part of the valley where the area consists of limestone calcium or carbonate rocks the underneath surface, creating uncertainty for further development of the area.

Indeed, there are several plans such as physical land use plan, tourism policy, tourism master plan, disaster plan, etc. to mention a few being devised for the development of Pokhara city and its urban region but most of those plans and policies either remained unimplemented or dropped. The haphazard urban growth-prone environmental problems as described above thus call for a modest land use zonation plan for the sustainability of the Pokhara urban region, based on suitability assessment of the land and other environmental resources.

References

Adhikari J (2000) Urbanisation, government policies and growing social and environmental problems in Pokhara, Nepal. Stud Nepali Hist Soc 5(1):43–87

Blaikie PM, Cameron J, Seddon D (1980) Nepal in crisis: growth and stagnation at the periphery. Clarendon Press, Oxford

CBS (2012) National population and housing census 2011(National Report). Government of Nepal, National Planning Commission Secretariat, Central Bureau of Statistics, Kathmandu

CBS (2019) Environment statistics of Nepal-2019. Central Bureau of Statistics, Kathmandu

Estoque RC, Murayama Y (2017) Trends and spatial patterns of urbanization in Asia and Africa: a comparative analysis. Urban development in Asia and Africa: pp 393–414

Government of Nepal (2015) The constitution of Nepal 2015. Nepal Gazette, Kathmandu

Gurung HB (1965) Study in regional geography of Pokhara valley. Unpublished Doctoral Dissertation, Faculty of Social Sciences of the University of Edinburgh, United Kingdom

Gurung L (2008) Status of slum dwelling women: a case study of Pokhara valley. Social Inclusion Research Fund (SIRF) Secretariat, Kathmandu

Haack B (2009) A history and analysis of mapping urban expansion in the Kathmandu Valley, Nepal. Cartogr J 46(3):233–241

HAN (2021) Hotel association Nepal. https://www.hotelassociationnepal.org.np/ Accessed 04 02 2021

Kamusoko C (2017) Dakar metropolitan area. Urban development in Asia and Africa. The urban book series, pp 257–273. https://doi.org/10.1007/978-981-10-3241-7_13

Kamusoko C, Chikati E (2017) Harare metropolitan area. *The urban book series*, pp 257–273. https://doi.org/10.1007/978-981-10-3241-7_13

Li X, Yeh A (2000) Modelling sustainable urban development by the integration of constrained cellular automata and GIS. Int J Geogr Inf Sci 14(2):131–152

Lin Y, Ma Z, Zhao K, Hu W, Wei J (2018) The impact of population migration on urban housing prices: evidence from China's major cities. Sustainability (Switzerland). https://doi.org/10.3390/su10093169

Lumanti (2010) Initiating people's process: ACCA programme. Bharatpur

Mayhew RJ (1997) Geography and literature in historical context: Samuel Johnson and eighteenth-century English conceptions of geography. University of Oxford School of Geography, Research Paper

McDonnell MJ, Pickett STA (1990) Ecosystem structure and function along urban-rural gradients: an unexploited opportunity for ecology. Ecology. https://doi.org/10.2307/1938259

McGarigal K (2015) FRAGSTATS HELP v4: spatial pattern analysis program for categorical and continuous maps. Computer software program produced by the authors at the University of Massachusetts, Amherst. LandEco Consulting

McGarigal K, Cushman SA, Ene E (2012) *FRAGSTATS v4: spatial pattern analysis program for categorial and continuous maps*. Fragstats

McGarigal K, Marks BJ (1995) FRAGSTATS: Spatial pattern analysis program for quantifying landscape structure. General Technical Report, US Department of Agriculture, Forest Service. https://doi.org/10.2737/PNW-GTR-351

Ministry of Urban Development (2017) *National Urban Development Strategy 2017*, Part B. Government of Nepal

Muzzini E, Aparicio G (2013). Urban Growth and Spatial Transition in Nepal. https://doi.org/10.1596/978-0-8213-9659-9

Parajuli BK, Paudel YR (2014) Impact of Nepal tourism year 2011 on tourist arrival in Pokhara. Himalayan J Sociol Anthrop. https://doi.org/10.3126/hjsa.v6i0.10687

Poudel KR (2008) Urban growth and land use change in the Himalayan region: a case study of Pokhara sub-metropolitan city, *Nepal*. GIS Ostrava, pp 110–122

Pradhan PK (1982) A study on traffic flows on Siddhartha and Prithvi highways. Himalayan Rev 14:1–8

Pradhan PK (2004) Rural urban relations with particular reference to Nepal. RUPP/UNDP, Kathmandu

Pradhan PK (2014) Pokhara city case study Nepal report. ADB (TA 7918–REG: Support to Urban Infrastructure Development and Financing), Manila

Pradhan PK, Sharma P (2016) Nepal: urban environment analysis. In: R.B. Singh and P. Prokop (Editors), Environmental geography, advances in geographical and environmental sciences. Springer Japan (e-book), pp 305–334

Rimal B (2011) Urban growth and land use/land cover change of Pokhara sub-metropolitan city, Nepal. J Theoret Appl Inform Technol 26(2):118–129

Rimal B, Zhang L, Keshtkar H, Wang LY (2017) Monitoring and modelling of spatiotemporal urban expansion and land-use/land-cover change using integrated Markov chain cellular automata model. ISPRS Int J Geo Inf. https://doi.org/10.3390/ijgi6090288

Runde D (2015) Urbanization will change the (developing) world. Forbes, pp 14–22

Thapa RB (2009) Spatial process of urbanization in Kathmandu valley, Nepal. Unpublished PhD Dissertation, University of Tsukuba. https://doi.org/10.13140/RG.2.1.2409.3207

Thapa RB, Murayama Y (2009) Examining spatiotemporal urbanization patterns in Kathmandu valley, Nepal: remote sensing and spatial metrics approaches. Rem Sens. https://doi.org/10.3390/rs1030534

Thapa RB, Murayama Y (2010) Drivers of urban growth in the Kathmandu valley, Nepal: examining the efficacy of the analytic hierarchy process. Appl Geogr 30(1):70–83. https://doi.org/10.1016/j.apgeog.2009.10.002

Tiwari PC, Joshi B (2012) Environmental changes and sustainable development of water resources in the Himalayan headwaters of India. Water Resour Manage. https://doi.org/10.1007/s11269-011-9825-y

UN-Habitat (2012) State of the world's cites 2012/2013. United Nations Human Settlements Programme. UN-Habitat, Nairobi

United Nations (2014) World urbanization prospects: the 2014 revision. United Nations Department of Economic and Social Affairs/Population Division, New York, USA

Williams N, Pradhan MS (2009) Political conflict and migration: how has violence and political instability affected migration patterns in Nepal? Nepal Study Centre. https://digitalrepository.unm.edu/nsc_research/28

World Bank (2011) *Rural-urban transition in developing countries*. http://documents1.worldbank.org/curatedar /547481468330918163/pdf/622590PUB0farm0476B0extop0id0186230.pdf

Chapter 25
Agriculture Practice in Kathmandu Valley, Nepal: Adaptive Strategy or Sustainable Livelihood?

Shobha Shrestha

Abstract Geographic perspective on questions like where the best places are to locate farming practices; are the farmer's decision-making and urban policy disjunctive; and how it influences sustainable urban development, etc. requires attentive analysis of urban land use and the dynamic interactions between humans and their environment as a function in space and of place. This study intends to explore the characteristics of urban agricultural practices and investigate strategies and decision-making of urban farmers in changing land use context in Kathmandu Valley. Agricultural practice is irrefutably one of the prominent activities, though with increasing urbanization, the most affected among all other resources is agricultural land. Increasing urbanization has changed the structure of the farming production system. Urban farmers are moving from traditional to modern and subsistence to mixed and commercial agriculture production systems. Spatial variability is notable in farming practice; commercial farming is dominant next to urban built-up in the valley floor followed by mixed farming over the foothills, while subsistence farming is located distant to built-up and over upper and middle hill slopes. Less influence of external structures such as national/regional market economy and more individual decision-making are evident in existing spatial agricultural structures. Thus, urbanization process has affected both the farm area and the farming operation. Urban farmers are adapting to the effects of urbanization through change and modification in farming operations. Urban agriculture in the study area at present is both demand-based and traditional. Risk perception and uncertainty is context-specific, and the decision is made within that specific context. To adjust to the pace and direction of the urbanization process, and to sustain their livelihood, urban farmers make tactical and strategic decisions.

Keywords Perception and decision making · Production system · Spatial structure · Urban farmer · Urban land use change

S. Shrestha (✉)
Central Department of Geography, Tribhuvan University, Kirtipur, Nepal
e-mail: shobha.shrestha@cdg.tu.edu.np

S. Anand et al. (eds.), *Sustainability in South Asian Cities*, Advances in Geographical and Environmental Sciences, https://doi.org/10.1007/978-981-97-7455-5_25

25.1 Introduction

An urban area is a spatial concentration of population working dominantly in non-agricultural activities and is largely dependent on peri-urban areas and hinterland for agricultural and other products. Kathmandu Valley with the country's capital is the largest urban area. Agriculture practice is nevertheless one of the prominent activities being carried out in the urban areas of Kathmandu Valley since the early times. In the valley, a large share of agricultural land was and is being converted into a built-up area that has gradually extended towards the rural periphery during different time periods (ICIMOD 2007; Thapa and Murayama 2010; Shrestha 2014). This process has led to heterogeneous and fragmented land use development in the valley. Main built-up growth patterns in agricultural land are attributed to factors like migration from rural areas, as well as from the inner urban core, institutional expansion, and so on. The urban expansion shows a linear pattern along the major roads radiating out from the urban core.

The 2011 census has indicated that urban-bound migration is a main contributing factor to rapid urban growth in Kathmandu Valley, with a 4.3% growth rate as compared to the national growth rate of 1.35% (CBS 2014). According to the preliminary census report of 2021, of the three large municipalities of the valley, Kathmandu has the largest population size, Lalitpur has the fourth highest population, and Bhaktapur has the highest growth rate (3.32%) in the country (CBS 2022). The population density of the valley's three districts is the highest among 77 districts in Nepal. Such an urban growth pattern increases the demand for space for urban investment and determines the form, scale, and location of urban land use. In 2002, it was projected that 3,600 hectares of land would be required for residential use every year in the Kathmandu Valley and if the existing rate of urbanization continues, four times more land would be required for residential use (KVTDC 2002). The biggest competition is thus between residential and existing agricultural land use.

Incompatible land uses, increasingly disorganized urban form, and declining neighborhood environment are major urban environment issues realized by the authorities and commons (MoUD 2017). Agriculture land use change and built-up area expansion due to urban growth implies a challenge to urban environmental sustainability in the Kathmandu Valley. The urban policy of growth around the peri-urban area versus the conservation of agricultural land has always been a central issue in urban planning (Marques-Perez and Segura 2018). In this context, Sustainable Development Goal 11 aims *to make cities and human settlements inclusive, safe, resilient, and sustainable* with the target (3) of *inclusive and sustainable urbanization* and indicator (11.3.1) to assess the ratio of land consumption rate to population growth. Within this framework, the current study tries to explore urban growth, land cover/land use change, and agricultural practice in the Kathmandu Valley.

Increased urbanization and urban growth in favorable regions are anticipatable, particularly in terms of opportunities. Spatial expansion of the urbanization process, however, is largely at the expense of a decrease in agricultural land. The importance of agricultural land, services, and products in an urban food supply system with a multi-functionality approach to land legitimizes sustainable agriculture and public intervention for conservation (Marques and Garcia 2015). In this context, the emphasis of the current study is on the changing spatial structure of agriculture land use and agricultural practices within the existing structure, characteristics of agriculture practices, and perception and decision-making towards sustainable agriculture practice.

25.2 Study Area

Kathmandu Valley is selected as the study area (Fig. 25.1) as it embraces dominant agricultural uses with heterogeneous land uses, though the valley is the most urbanized area and capital region of the country. Administratively, the valley comprises three districts (Kathmandu, Lalitpur and Bhaktapur), 2 metropolises, 16 municipalities and 3 rural municipalities. The latter three rural municipalities lying outside of the valley are not included in the study area.

The valley lying in the central hill region has an average elevation of 1,300 m while the elevation of the surrounding hills ranges from 1,500 to 2,800 m. The study area has approximately 700 km^2 calculated from the digital topographical data (Survey Department 1996). The Bagmati, Bishnumati, Manoharā, Dhobi kholā, and

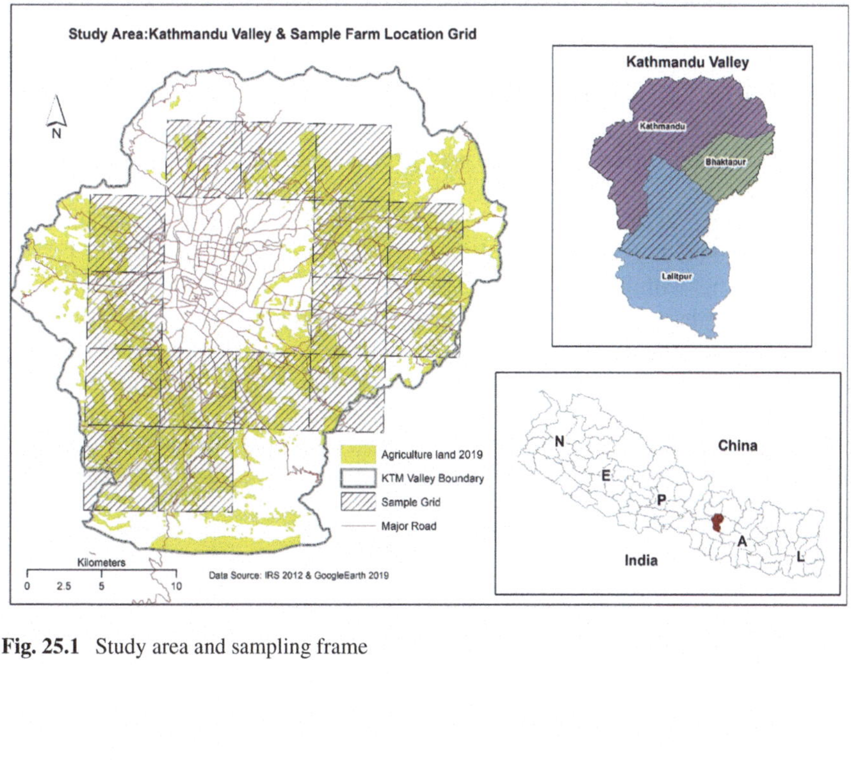

Fig. 25.1 Study area and sampling frame

Hanumante are major rivers flowing through the valley floor and eventually flow out of the valley to the south.

It is an intermontane basin surrounded by lower hills to the eastern and western parts and higher hills to the southern and northern parts. Geomorphology includes landforms like fertile flood plains, lower terraces, riverbeds, and hill slopes, and moderate to steep sloping hills, rocky outcrops. Based on location, the terraces in the valley can be divided into two types such as higher terraces along the fringe of hills ranging up to 160 m and lower terraces located in the central part ranging from 50 to 80 m, where intensive agriculture is practiced (DWIDP 2009).

The valley comprising the most fertile land is considered one of the earliest settlements in the country (Depuis 1962 as cited in Pradhan and Sharma 2016). Two primary landforms namely river floodplains and elevated river terraces are regarded as the most productive agricultural lands. It has been historically cited as an important trade center and a transit point for trade between India and Tibet (China). The settlements were largely concentrated within the compact clusters of three large traditional cities of Kathmandu, Lalitpur and Bhaktapur, and other smaller clusters with extensive agriculture fields. It was a food surplus region due to strong cultural practices to preserve agricultural land. Much of the floodplains and valley terraces were intensively cultivated and settlements were confined to less productive high terraces or upland plains (*Tarland*) and lower hill slopes with regular water supply through *Raj Kulos* (king irrigation channels). A large variety of food crops and vegetables are increasingly grown throughout the year providing fresh produce to the local population (Bohle and Adhikari 2002).

According to the preliminary census report of 2021, the total population of the valley is 29,96,341, comprising 787,725 households, 15,19,206 males and 14,77,135 females (CBS 2022). In 2011, the total population of the valley was 24,30,090 (comprising 6,08,252 households, 12,47,876 males and 11,82,214 females). Of the total population, the urban population was 59% and the rest 41% was in peri-urban area (CBS 2011).

Population density in the valley urban area has increased substantially in recent decades. In the urban built-up area, the density has increased from 8,494 to 11,736 persons per km^2 within a decade of 2001 and 2011. Likewise, the density of the whole Kathmandu valley has increased from 1,858 persons to 3,100 persons during the same decade. It indicates that there is an increase of more than 1,200 people per km^2. This has direct implications for the existing land use of the valley. Housing density is lower in the peri-urban area with 299 houses per km^2 as compared to 2,043 houses in the compact built-up area. The current population density of the valley districts is 3,209 persons/km^2, which is the highest population density in the country (CBS 2022).

25.3 Data and Methods

This study has used an integrated methodology. An exploratory and analytical framework is adopted for assessment. Qualitative and quantitative analysis is carried out for the information gathered from both primary and secondary sources.

The study has used three types of data, namely bio-physical, socio-economic, and spatial. Primary data was collected from the field in 2019 through a questionnaire survey, informal discussion with the key informants, observation using structured field protocol, and a checklist to record data. Secondary data was acquired from the reports, existing data and information of the public sources, existing base maps and digital data. Satellite images were archived from relevant public sources.

A multistage sampling process has been used for sample area selection through spatial stratification. The first step of spatial stratification was carried out to extract agricultural land (for sample farm location) after land use change detection between 2012 and 2019. Identification of the sample farm areas around the built-up area was made in the second step. A total of 15 grids (25 km^2) were selected as the sample sites for the agriculture household survey (see Fig. 25.1). Each sample spatial unit was tested for homogeneity and only those sample sites having homogeneity were included in the group. In the third stage, the purposive random sampling method was applied for the selection of farm households due to the unknown of the universe farm household. Four farm households were selected from each sample grid. Thus, a total of 60 sample farm households were selected for the questionnaire survey.

Urban land use/land cover change from 2012 to 2019 was carried out using high-resolution remote sensing images. Quick Bird images with a spatial resolution of less than 1 m were used for 2019, while Resource SAT (IRS) images with a 5-m spatial resolution were used for 2012 land cover land use, which was substantiated with imageries in the Google Earth platform. Ten training sample sites were selected for agriculture land use class and these sampling sites were field verified. Land use change analysis has been carried out in the GIS platform by overlaying land use data layers; change from one class to another was calculated; and spatial analysis was carried out based on the output map and table.

Characteristics of urban agriculture practice namely location of activities (where it is practiced), type of activities (e.g., food/non-food production, livestock or mixed), land tenure/ownership (private or public land), stage (production, processing, marketing, and distribution) and scale (farm area, coverage, level of intensity, purpose e.g., commercial or subsistence) as identified by Quon (1999) were explored and analyzed. The Chi-Square test was carried out to explore the association between agriculture practice and urbanization process.

The analytic Hierarchy Process (AHP method) for establishing priorities in multi-criteria decision-making introduced by Saaty (1987) was applied to analyzing urban farmers' perception and decision-making on agriculture resources and preference for alternatives for sustainable agriculture practice. The AHP consists of three stages: i) construction of the hierarchy of criteria and alternatives, ii) calculation of the priorities, and iii) aggregation of results for final decision. A relative comparison matrix (pair-wise comparisons with a weight range of 0–1) has been developed and applied to explore preference for alternatives. Priority vector (eigenvector)) was calculated for each criterion and decision alternative. The weighted average rating was calculated for each decision alternative. The one with the highest score was identified as the best decision alternative. The Consistency Ratio (CR) for a matrix was calculated from $CR = CI/RI$, where Consistency Index (CI): $(\lambda max\text{-}n)/(n\text{-}1)$ and RI was given a random index value.

25.4 Results and Analysis

25.4.1 Population growth and urbanization in Kathmandu Valley

Definition of the urban area and urban population in the Nepalese context is based on the government declaration of the municipality. Thus, the municipality area is regarded as an urban area and its population is urban population. This method does not consider non-primary function and degree of economic activities within the municipal boundary. It is argued that the spatial extension of boundaries of the existing municipalities and the addition of new municipalities from time to time are also key factors for urban growth in Nepal (Choe and Pradhan 2010). In Kathmandu Valley, there are a total of 18 municipalities that were established in different years (Table 25.1).

Kathmandu Valley having the merit of its location and both favorable internal and external conditions for urban growth, has the largest share of the country's urban population. The valley has witnessed unprecedented growth in the level of urbanization, and remained the most urbanized region in Nepal. It is illustrious that the valley has less than 0.5% of the total land area of the country, but the urban population has risen from only 47.4% in 1952/54 to 60.5% in 2001 (Sharma 2003). The urban population during the 1950–1990s increased by about 450% in the valley. The urban area that accounted for 26% in the 1970s rose to 46% in the 1990s. Between 1981 and 1991 it increased by 82%, in which migration of the population alone

Table 25.1 Urban (municipal) establishment in Kathmandu Valley

Number and name of municipalities	Date of establishment
Kathmandu, Lalitpur and Bhaktapur (3)	1953
Madhyapurthimi and Kirtipur (2)	1997
Gokarneshwor, Dakshinkali, Tarakeswor, Shankharapur, Chandragiri, Kageswori, Tokha, Nagarjun, Budhanilkantha, Godawwari, Changunarayan, Mahalaxmi, and Suryabinayak (13)	2014

Source MoFAGA (2016)

contributed 59%. The built-up areas together with slums and squatter settlements expanded remarkably encroaching upon agricultural land, forest area, riverbank, and open spaces (Pradhan 2004). The valley's urban growth rate of 4.23% (1991–2001) was nearly double compared to the country's urban growth rate of 2.25% (Gurung 2008). In 2011, the valley's share of urban population to the country's total urban population was 32.5%. The average annual population growth of the valley districts: Kathmandu (1.4%), Lalitpur (1.52%) and Bhaktapur (3.32%) remains higher than the national average annual population growth rate of 0.93% (CBS 2022).

The urbanization trend in the Kathmandu Valley is summarized in Table 25.2. The trend is that the valley's growth rate has increased by 5.21% whereas the urban growth rate declined to 4.32 from 5.05. Prominent is the increased growth rate of the surrounding area (New municipalities) from 4.42 to 5.82 indicating significant urban growth in the peri-urban area and an increased share of 39% urban population. The steady growth of the population in the peri-urban area has been apparent since 1991. As the data demonstrates, population increase and urban expansion towards the peri-urban area can be attributed to three major factors: (i) complete in-fill and no available land for built-up in older city areas, (ii) increasing land value in and near the core areas, and (iii) expansion of physical infrastructure particularly road network and transport service in the peri-urban/rural area.

However, other three municipalities, namely Bhaktapur, Kirtipur and Madhyapurthimi out of five municipalities established in 1997, still have less than one-third of the population engaged in a non-primary occupation, and more than 50% of land is still under agriculture (KVTDC 2008). Table 25.3 describes the projected population for the valley districts, based on the population and growth rate of 2011. Based on the urban growth rate of 4.63%, the projected population for the valley districts for 2021 is 3,298,118 (CBS 2014). It is apparent that more land is required for the growing demand for housing, putting immense pressure on existing agricultural land in the peri-urban area, as the core areas have virtually no space left.

The demand for residential land plots increased fourfold from 19.47 km^2 to 45.69 km^2 during the past 20 years (1981–2001) in the districts of Kathmandu and Lalitpur (UN-HABITAT 2010). With a growth rate of 4.63% and a projected population of

Table 25.2 Trend of urban population growth

		5 municipalities established in 1997			13 municipalities established after 1997		
Census Year	Valley Growth rate	Municipal Population	% of total population of the valley	Growth rate	Municipal Population	% of total population of the valley	Growth rate
2011	5.21	1,426,641	58.71	4.32	951,103	39.1	5.82
2001	4.80	995,966	62.36	5.05	601,244	37.6	4.42
1991	4.57	661,837	61.34	6.00	417,072	38.7	2.75
1981		413,608	55.85	–	327,011	44.2	

Source CBS, Population census 1981–2011

Table 25.3 Growth rate and projected population in Kathmandu Valley

Geographical unit	Growth rate	Projected population/growth rate 2021	Projected population/growth rate 2031
Kathmandu district	4.78	2,300,890	2,729,056
Lalitpur district	3.26	377,660	436,553
Bhaktapur district	3.01	585,982	680,157
Valley district total	4.63	3,298,118	3,845,766
Country	1.35	1.37	1.01

*Source*CBS (2014);UNFPA (2017)

3,845,766, land required for housing is estimated to be around 188 km^2 for 2031(CBS 2014; KVDA 2016). The valley has only 34.8% of constraint-free area for further planned development (KVDA 2016). This means there is no way to further encroach upon agricultural land and open spaces. Hence, to meet the target of sustainable urbanization of SDG by 2030, land use efficiency (i.e., the ratio of land consumption rate to population growth rate) needs to be maintained by slowing down the rate of urban sprawl through the compact built-up area or multistorey buildings or adopting satellite towns outside of the valley. Studies also show that compact cities use land more efficiently and are better placed to provide public goods and basic services at a lower cost (UN-HABITAT 2018).

25.4.2 Urban Land Use Change: Direction and Magnitude

Land cover and land use change analysis between 2012 and 2019 was carried out for mapping the pace and direction of land cover and land use change in the study area. The valley's land cover and land use are divided into six classes. The results show that, of the total area (699.94 km^2), land cover land use change with 238.94 km^2 area accounts for 34.14%. Within seven years (2012–2019), urban land use has shown a substantial change in the valley. Remarkable change in terms of pace and direction is evident in agriculture and built-up classes which signify a rapid urbanization trend in the valley. Agriculture land coverage has decreased to 22.2% from 39.3%, whereas built-up area has increased from 23.4% to 36.3% (Table 25.4; Fig. 25.2). Agriculture land has decreased by 17%, accounting for 119 km^2 area loss, whereas built-up area has increased by almost 13% with more than 90 km^2 gain. With the increasing urbanization, the most affected among all other resources is agricultural land.

Among six land use classes, five classes have positive change (area gain), whereas only agriculture land has negative change. Significant change is also visible in the forest cover, showing an increase of 3.59%.

Most of the agricultural lands next to the built-up core area have been converted to urban built-up during the 7-year period (2012–2019). Spatial pattern of urban growth is radial around the existing built-up and linear along the road network. Urban

Table 25.4 Land cover/land use change in Kathmandu Valley (2012–2019)

Land use class	Land use 2019		Land use 2012		Area change (km^2)	% change
	Area km^2	% Area	Area km^2	% Area		
Agriculture	155.6	22.23	275.07	39.3	–119.47	–17.07
Built up	254.28	36.33	163.97	23.43	90.31	12.9
Forest	278.51	39.79	253.41	36.21	25.1	3.59
Open space, barren	5.83	0.83	3.31	0.47	2.52	0.36
Shrub	2.18	0.31	0.72	0.1	1.46	0.21
Water body	3.54	0.51	3.45	0.49	0.09	0.01
Total	699.94	100.0	699.94	100.0		

Source Calculated using Quick Bird image (1 m) 2019 and ResourceSAT (IRS) image(5 m), 2012

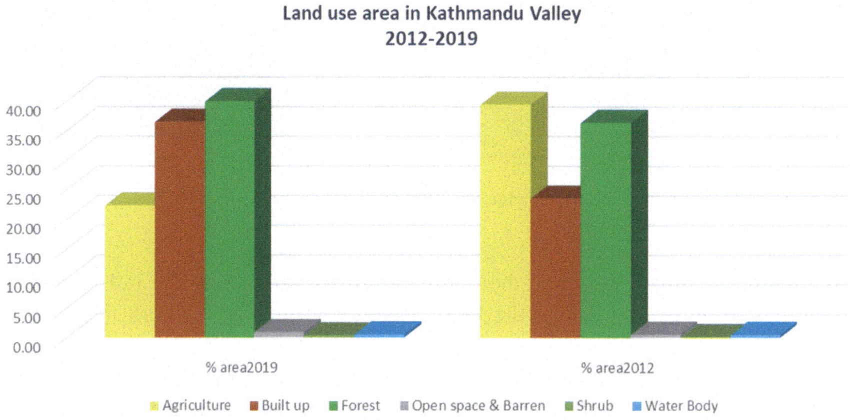

Fig. 25.2 Land use area in percentage

expansion is in all directions. However, obvious urban growth is visible in three different directions of the valley: central-east, north-west, and central-south. Earlier studies on the land use change between 2001 and 2009 showed that the metropolitan areas of Kathmandu and Lalitpur came under intense pressure to respond to the land use change due to the population increase and loss of prime agricultural lands to residential and commercial uses due to the uncontrolled urban growth (Hack 2009). During 1991–2000, the agricultural land transformation to urban built-up increased to 4.4% in the contest to a 3.9% increase during 1970–1990 (Thapa and Murayama 2012). Change in agricultural cropping patterns is also evident. High-value agriculture products (vegetables, fruits, cut flowers) have replaced traditional rice fields and terraces with easy access to water, road and market in the city fringe and nearby villages. The degrading environmental conditions due to rapid and haphazard urban built-up growth cause to reduction in urban efficiency. Land use planning with integrated transport and environmental planning is suggested with strict enforcement

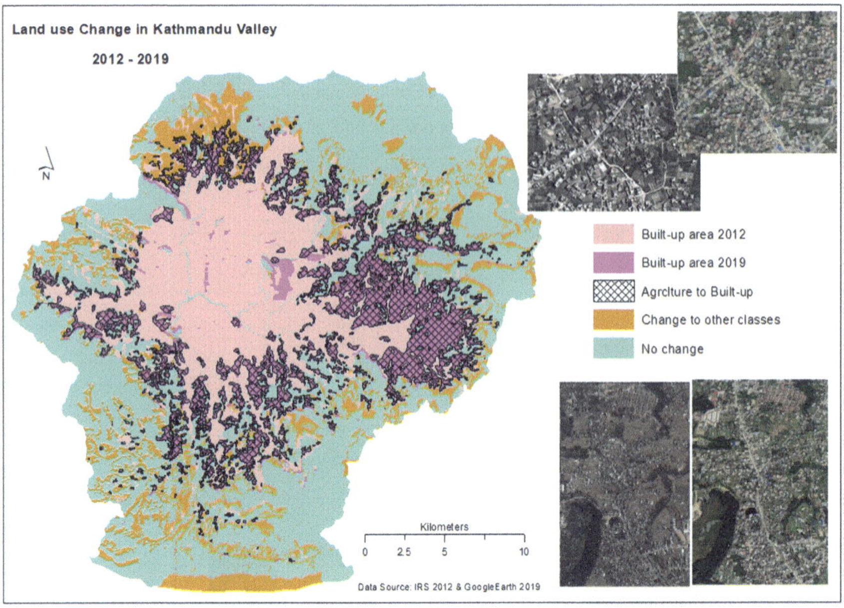

Fig. 25.3 Land cover and land use change in Kathmandu Valley (2012–2019)

and implementation of regulatory measures (Muzzini and Aparicio 2013). Haphazard urban expansion is also due to the lack of a long-term road network masterplan (Fig. 25.3).

Direction and magnitude of change from one land use class to another between 2012 and 2019 are provided in Table 25.5. Major land use conversion observed during the seven-year period is associated with the agriculture and built-up classes. Most of the agricultural lands next to the core area have been converted to urban built-up.

Among the land use classes, significant change is observed in the agriculture, built-up, and forest classes. More than 90 km^2 of agricultural land was converted into urban built-up during the period of 2012–2019. The built-up area has expanded even encroaching on the forest area to some extent, while noteworthy is, the conversion of agriculture and built-up area into forest. This transformation is attributed to the loss of building structures and agricultural land due to the major earthquakes of 2015. Besides, the conversion of agriculture and forests into open/barren and shrub is also visible. No significant land use transformation of other land use classes is observed during this period.

Table 25.5 Direction and magnitude of land use change by area (km^2)

		Land use 2012						
Land use 2019	Class	Agriculture	Built up	Forest	Open space, barren	Shrub	Water body	Total
	Agriculture	123.4	-	32.0		0.2		155.6
	Built up	90.2	155.8	5.0	3.0	0.1	0.1	254.3
	Forest	57.0	7.6	213.4	0.1	0.4		278.5
	Open space, barren	3.4	0.5	1.6	0.2			5.8
	Shrub	0.9		1.3				2.2
	Water body	0.2					3.3	3.5
	Total	275.07	163.97	253.41	3.31	0.72	3.45	699.94

Source Calculated using Quick Bird image(1 m) 2019 and ResourceSAT (IRS) image(5 m), 2012

25.4.3 Agriculture Practices and Production System

Agriculture activities in an urban area refer to urban agriculture, an activity that produces, processes and markets food, fuel, and other agricultural outputs (Mougeot 2005). Location, farm size, cropping pattern, production means, resource access and use, as well as household welfare is the criteria mostly used for the characterization of agriculture practice in the urban area (Mougeot 2005; Veenhuizen 2006; Adedeji and Ademiluyi 2009; Siti et al. 2019). Agriculture and urbanization are incompatible activities, competing for the access and use of limited land. Kathmandu valley from the very beginning has remained agriculturally the most fertile area. In the early periods, people followed judicious land use practices by using dry uplands for residential use and sparing fertile flood plains for cultivation. All the lowlands of the valley with the facilities of irrigation, extreme fertility (homemade manure) and productiveness were appropriated to the cultivation. Nearly all the inhabitants of the valley used to live on the high-level lands (Old-field 1980).

Urban agriculture in the study area has been practiced as a tradition since the early period. However, increasing urbanization has changed the structure of the farming production system from subsistence to demand-based adaptive strategy. Based on form or structure such as method of farming, use of inputs, and purpose of farming, the valley's agriculture production system can be grouped into commercial and subsistence, and the activities include cultivation, livestock, or floriculture. More than 50% of the urban farmers in the study area practice modern forms of urban agriculture. Being a capital city region, the valley has the advantages of technological development in the agricultural sector, road network and other service infrastructures. These advantages are utilized by most of the urban farmers. Agriculture land in the valley plain is dominated by commercial farming. Of the total sampled farm households, eighty-three percent practice commercial farming, whereas 38% practice mixed farming and only 17% practice subsistence farming. Among all these farmers, still, 44% are continuing with the traditional form of production system with basic agriculture inputs.

Another notable characteristic is that most farmers practice crop cultivation only as opposed to the traditional practice of mixed farming of livestock and crop. Dominant urban farmers are of mixed farming group constituting more than 40% followed by commercial farmers (32%). This shows that most urban farmers are happy with practicing farming mainly to address household food security with a surplus amount above consumption for selling to the market. Subsistence farmers who practice farming only for self-consumption constitute 22%.

Other characteristics of agriculture production systems are small land holding, combined with cropping patterns of vegetables and food crops, and family and neighbor or relatives constitute labor. The primary form of urban agriculture is agriculture land holding, which varies widely in size and scale due to limited availability of land and competition over other land uses. Small farm size is a characteristic feature of the study area. More than 70% of urban farmers have a farm size of 0.025 hectare whereas less than 1% farmers operate in 1 hectare farm size. Though the Gini coefficient value of 0.37 indicates a more equal farm size distribution, the situation of equal farm size distribution deviates from the type of production system. The largest number of commercial farmers (41.4%) operates in 0.025 ha or more farm area. On the contrary, 67% of subsistence farming is operated in 0.05 ha. The family-owned farm is a dominant land tenure type where 80% of farmers practice farming. Nearly all subsistence farmers operate in family-owned land whereas only half of the commercial farmers (50%) operate in family-owned farmland. This shows a diversity of land tenure and different production systems. The farmers operate in small farm sizes in most cases and framing is intensive and land is on average cultivated three times a year. No fallow farmland existed even during the winter (cold) season except land plots for selling or building construction. It is also apparent that diversity exists in urban agriculture in terms of production system, location characteristics, and land holding. Studies show that there is an inverse relationship between farm size and farm output with respect to productivity (Nugent 2000; Kekena 2006). Small farms are more efficient than large farms. Small farmers utilize farms effectively in terms of farm inputs as best as possible because it is the only source of their income. Labor is a major problem of the commercial and mixed farmers whereas irrigation is a major problem of subsistence farmers. Other problems include diseases in crops.

25.4.4 Structure and Change: Perspective on Urban Farming Practice

This section examines whether the changing spatial structure of agricultural land use influences the farming area and practices in the study area, causes of changes and the future perspective on agriculture activities.

Various characteristics are found among urban farmers in terms of farm area and farm operation. In the case of farm area, most of the farmers continue farming with the same farmland area where they started farming before and few of them have got their farm areas increased or decreased. Most of the farmers, who have been.

practicing farming for more than ten years, have changed their farming operations either in terms of production system or agricultural inputs. Conversely, more than 33% of them appear to have been involved in the past five years. This indicates the involvement of various age groups of farmers with varied experience in urban agriculture. Majority of the farmers with 61% have been operating farming in the same farm area without any change and the rest (39%) have changed their farm area (Table 25.6). Those farmers practicing farming for more than 20 years constitute 10% but those practicing farming for five years are the most volatile in case of dynamics of farm area. Of all, nearly 37% of farmers of 1 to 5 years practicing changing in farming area constitute the highest. Construction followed by subdivision of land parcels among the farmer family members and selling of land are major causes of the decrease in farm size. This implies the increasing of agricultural land fragmentation.

Parallel to change in area, change in farm operation is a characteristic feature either in the form of a production system or agricultural input. Less than 40% of the farmers reported area change whereas 65% of farmers reported change in farm operation since they started farming. Of the total farmers who changed the farming operation, 31% constituted switching to modernization, and among which 71% were commercial farmers. Thirty-five percent of farmers have continued the same traditional operation, among which, subsistence farming constituted 72%. However, just opposite this practice, farmers believed that people continue extending farming to marginal land to grow more food due to the increasing demand of the urban population.

It is hypothesized that the urbanization process causes a change in the existing spatial structure of urban land use. Farming practice in the Kathmandu Valley is independent of changes in the spatial structure of urban land use change. Number of farming years and state of farm area (change or no change) have been taken as associated variables. With a calculated Chi-square value of 10.44 with a degree of

Table 25.6 Direction of farm area and operation

Farm operation			Farm area		
Change in operation	f	Percent	Factors of change	f	Percent
Farming system/modernization	19	31.7	Construction	9	15.0
Increase in investment	7	11.7	Erosion/landslides	2	3.3
Increased varieties	13	21.7	Sub-division of land	6	10.0
Change	39	65.0	Selling of land	6	10.0
No change	21	35.0	Total	23	38.3
Total	60	100.0	No change	37	61.7
			All total	60	100.0

Source Field survey, 2019

freedom value of 4 at 0.05 significance level (value 9.488), the null hypothesis has been rejected and so, it is inferred that the farming practice in the valley is independent of change in spatial structure of urban land use caused by the urbanization process. However, regarding the perspective of the farming practice in the coming five years, 65% of the urban farmers reported that they would be in favor of changing the farming practice among which 22% were subsistence farmers, 25% were commercial farmers and 43% were mixed farmers, whereas only 35% reported to continue with the farming practice of which 71% were commercial farmers.

25.4.5 Perception and Behavior of Farmers

It is argued that, within an agricultural environment each farmer sets goals, which influence his/her choice and decisions on the type of farming system and practices. Equally influential is the institutional attitude and behavior. The goals are dynamic so are perception and knowledge of the agricultural environment and the factors that influence decision-making also change. These account for the spatial variations in agriculture land-use patterns (Dessart, et. al. 2019).

Perception and decision were examined among the commercial, mixed and subsistence farmers group, regarding extension of agricultural land to grow more food for the increasing urban population. Just opposite to this practice, farmers view that they believe in extending agricultural land to grow more food to meet increased urban demand, but the majority of the farmers (61%) had not expanded the farm area (see Table 25.6). This entails that there is a difference between perception and practice. Likewise, there is an inconsistency between behavior and decision-making. More than 55% of farmers perceived that more land would be used to increase production due to increasing urban demand, but they had not expanded the farm size, whereas 43% thought that farm area remains to be the same, but more food would be produced. According to them, agricultural labor and irrigation water were the major problems as compared to availability of agriculture land because modernization in agriculture could increase the production. Again, the Chi-square test was carried out to examine whether farming practice is perceived as a long-term adaptive strategy to the process of urbanization in terms of change or no change of the farming practice. It is assumed that the urbanization process and government policy create uncertainty in farming practices and farming decisions. Farming practice in the urban area is, hence, short-term adaptive strategy for survival. In the use of the Chi-square test ($\chi = 9.40$ α at 0.05 with a degree of freedom 4), null hypothesis is accepted. It implies that urban agriculture is not adopted as a short-term survival strategy.

The urban farmers' perception and decision was also examined on what agriculture resource and what alternatives they prefer for improving urban agriculture and sustaining their agriculture practice. An Analytic Hierarchy Process (AHP) was applied to assess whether the importance as perceived by farmers would match between enablers (criteria) and preferences (alternatives) and which decision would prove better. In the first step, a single pair-wise comparison matrix was developed for

Table 25.7 Comparison matrix of enablers and alternatives

Enablers	Bank credit	Local cooperatives	Government subsidy
Bank Credit (BC)	1	7	3
Local Cooperatives (LC)	0.5	1	0.4
Government Subsidy (GS)	0.14	0.5	1
Alternatives	Increase production	New agricultural inputs	Increase area
Increase production (IP)	1	3	7
New Agricultural Inputs (NI)	0.333	1	3
Increase Area (IA)	0.2	0.333	1

criteria and decision alternatives. The third root of the product was calculated for three criteria (bank credit, local cooperatives, and government subsidy) and three alternatives. The comparison matrix is shown in Table 25.7. The third step was to normalize the root product and weight calculation (priority vector). Finally, the consistency ratio (CR) was calculated to assess against the judgments made completely at random.

Finally, a weighted average rating table was obtained to get the final score. The result as given in Table 25.8 shows that new agriculture input (NI) has the highest value of 0.5699. It indicates that the decision of the farmers who chose to get new agriculture input through local cooperatives seemed to be the best decision, among all others. Bank credit seemed to be the most important enabler for increasing agricultural benefits as perceived by many of the commercial and mixed farmers and it is likely that they would choose such, whereas the calculated index proves that it would be better to choose local cooperative for new agriculture input which was preferred by few farmers. This entails that the choice which was favored by many might not be the right one.

25.4.6 Urban Agriculture Practice: Problems, Synergies and Conflict

Major problems associated with the urban agriculture practice in the study area are related to agriculture inputs, bio-physical and economic aspects. Labor shortage is identified as a major problem. Labor shortage as an agriculture input, remains an obvious problem in the urban context as agriculture labor remains the least preferred activity among all other operating informal activities in the urban area. It is followed by bio-physical problems such as limited irrigation/water shortage, and weeds/diseases. Economic factors are perceived as less problematic compared to the other two categories. Distinctively visible is the perception of a shortage of land for agricultural practices though more than 60% of farmers perceived that there is no shortage of land for agricultural practice.

Table 25.8 Weighted average rating score for enablers and decision alternatives

Criteria	BC	LC	GS	Weighted rating score (S)
	Priority vector values for Criteria			
	0.7617	0.1765	0.0618	1.0
Alternatives	Priority vector values for alternatives			
IP	0.4534	0.0523	0.0458	0.3574
NI	0.5247	0.9090	0.1590	0.5699
IA	0.0219	0.0386	0.7952	0.0726
	1.00	0.9999	1.00	0.9999

Note Bank Credit (BC), Local cooperative (LC), Government Subsidy (GS), Increase product (IP), New Agriculture Input (NI), Increase Area (IA)
Priority vector PV = $\sqrt[3]{Pi} / \Sigma \sqrt[3]{Pij}$
(PVC for criteria, PVA for alternative)
Weighted Rating Score, S = Σ (PVCi * PVAi) +... (PVCj * PVAj
Noteworthy differences and some similarities are seen in the perception and resultant decisions regarding resources for farming and future perspectives. Farmers' decisions are influenced by bio-physical, socio-economic, production and institutional factors. The decision-making in a spatial context is therefore a more complex process. Attitudes and behavior of the farmers in respect to the environment in the agricultural sector demonstrate a strong influence of practical knowledge and personal experience. Relationship derived between attitude and behavior shows the links among the socioeconomic, cognitive, and other contextual factors that form and influence the farmers' environmental attitudes, behavior, and decisions (Vogel 1996). An important part of this analysis is the way in which commercial, mixed and subsistence farmers make sense of enablers through the processes of perception and how they behave. These perceptions and behavior possibly will have a major implication on urban agriculture, the method they adopt, use of the land, products produced and the resultant spatial pattern. It is therefore important to understand the cultural and social factors as well as the government efforts in place that facilitate or hinder the implementation or adaptation of urban agriculture in a sustainable way.

Assessment of attitudes towards urban agriculture practice reveals some interesting differences between the two groups, such as urban farmers and urban development officials. Household food security and supplementary income are the prime benefits of urban agriculture, according to 72% of urban farmers. They consider it as a means of self-employment and adopt it as a cultural value. In contrast, 60% of institutional personnel viewed that it is not a significant employment generator or a secure food source. Moreover, farmers acknowledge the environmental role of urban agriculture though equally fear associated with environmental problems such as pollution. On the other hand, risks of urban agriculture identified by the government authorities include environmental and health risks from pesticides and insecticides, water contamination and scarcity, and occupancy of economically productive land. The synergies and conflict regarding the factors of urban agriculture from an urban planning perspective as perceived by the urban farmers and the urban planning authorities are summarized in Table 25.9.

However, some common synergies between urban farmers and urban planning authorities are linked to implications of urban agriculture practice, which include

Table 25.9 Agriculture practice: synergies and conflicts

	Urban Farmers	Urban Planning Authorities
Synergies	- Urban agriculture is grass-roots efforts that inspire change on a larger scale through a bottom-up approach - Most of the urban farmers have their own land. So, in general, secure land tenure for urban agriculture in Kathmandu Valley	- Top-down efforts at the institutional level improve the coordination of urban agriculture activities and maximize the multifunctional benefits - Attainable strategies for improving a complex urban ecosystem and effective regulation and facilitation
Conflicts	- Provide cheap, fresh, and nutritious food in larger varieties. Less need for packaging, storage, and transportation so reduced environmental pollution - Agricultural production is the primary goal of urban farmers as household food security and cultural value	- Measures to reduce the health and environmental risks associated with urban agriculture. The control needs to be more stringent because farming is near dense human activities - Agricultural production is the subordinate goal of the urban farmers, and it is not appropriate urban function

land fragmentation, conflict over the use of land, and no implementation of land use zoning by the related planning authorities.

25.5 Conclusion

Despite all spatial dynamics and decreases in agricultural land use area due to urban expansion, the overall spatial extent of urban agricultural land use and agriculture practices remain a significant urban landscape. A higher extent of change in farm operation as opposed to change in farm area holds that the urbanization process has a larger effect on the farming practice than on the farm area. Urban farmers are adapting to the effects of urbanization through change and modification in farming operations for sustaining their livelihood. However, the prospects and constraints faced by the farmers are different as they respond to their needs and priorities differently. An area of further research calls for evaluating the impacts of agricultural systems on the planning and sustainability of the urban environment. Studies are also suggested to analyze the contribution of urban agriculture to urban sustainability in terms of food security, use of land and space, employment, and income generation. Assessing sustainability on a larger scale would make an important contribution to the literature.

References

Adedeji OH, Ademiluyi IA (2009) Urban agriculture and urban land use planning: need for a synthesis in metropolitan, Lagos, Nigeria. J Geogr Reg Plan 2(3):43–50

Bohle HG, Adhikari J (2002) The metropolitan food system of Kathmandu conceptual considerations and empirical evidence. Erde 133:401–421

CBS (2022) Population census 2021 preliminary report (Rastriya Ganana 2078: Prarambhik Natija). Central Bureau of Statistics, National Planning Commission, GoN, Kathmandu, Nepal

CBS (2014) Population monograph of Nepal, Vol 1: population dynamics. National Planning Commission, Kathmandu, Nepal

CBS (2011) Population census 2011 preliminary report. National Planning Commission, Kathmandu, Nepal

Dessart FJ, Hurle JB, Bavel RV (2019) Behavioral factors affecting the adoption of sustainable farming practices: a policy-oriented review. Eur Rev Agric Econ 46(3):417–471. https://doi.org/10.1093/erae/jbz019

DWIDP (2009) The preparation of flood risk map of Kathmandu valley. Department of Water Induced Disaster Prevention, DWIDP Final Report, Kathmandu

Gurung H (2008) *Nepal regional strategy for development*. NRM Working paper series 3, ADB, Nepal Resident Office

Haack B (2009) A history and analysis of mapping urban expansion in the Kathmandu Valley, Nepal. Cartogr J 46(3):233–241

ICIMOD (2007) Kathmandu valley environment outlook. International Center for Integrated Mountain Development, ICIMOD, Kathmandu Nepal

Kekena DS (2006) A socio-economic analysis of urban agriculture: the Soshanguve project case study. Unpublished Master's Degree Thesis, University of Pretoria, South Africa

KVDA (2016) Vision 2035 and beyond: 20 years' strategic development master plan (2015 - 2035) for Kathmandu valley. Kathmandu Valley Development Authority (KVDA), Ministry of Urban Development, Nepal

KVTDC (2008) Final project completion report preparation of detail digital land use map of municipalities of Kathmandu valley. Kathmandu Valley Town Development Committee. Kathmandu, Nepal

KVTDC (2002) Kathmandu valley long term development concept 2020: Kathmandu 2020. Kathmandu Valley Town Development Committee, Kathmandu, Nepal

Li A, Lei G, Cao X, Zhao W, Deng W, Koirala HL (2017) Land cover change and its driving forces in Nepal since 1990. In Li A, Deng W, Zhao W (Eds) *Land cover change and its eco-environmental responses in Nepal*, pp 41–66. Springer Nature, Singapore. https://doi.org/10.1007/978-981-10-2890-8

Marques-Perez I, Segura B (2018) Integrating social preferences analysis for multifunctional peri-urban farming in planning an application by multi-criteria analysis techniques and stakeholders. Agroecol Sust Food Syst 42(9):1029–1057. https://doi.org/10.1080/21683565.2018.1468379

Marques-Perez I, García del Río BS (2015) Identifying functionality of peri-urban agricultural systems: a case study. Chap 5. In: M. Samer (Ed) Urban agriculture. IntechOpen. https://doi.org/10.5772/61603

Mougeot LJA (2005) AGROPOLIS: the social, political and environmental dimensions of urban agriculture. International Development Research Center, IDRC, EarthScan, London

MoUD (2017) National urban development strategy (NUDS), 2017: part A main document. Government of Nepal, Ministry of Urban Development, Urban Development and Physical Planning Division, Kathmandu, Nepal

Muzzini E, Aparicio G (2013) Urban growth and spatial transition in Nepal: an initial assessment. The World Bank, Washington DC

Nugent R (2000) The impact of urban agriculture on the household and local economies. In: N Bakker, M Dubbeling, S Guendel, USH Koschella, de Zeeuw (Eds) Growing cities, growing food, urban agriculture on the policy agenda, DSE, Feldafing

Old-Field HA (1980) Sketches from Nepal. W.H. Allen and Co., London

Pradhan PK, Sharma P (2016) Nepal: urban environment analysis. In RB Singh, P Prokop (Eds) *Environmental geography: advances in geographical and environmental sciences*, Springer Japan, pp 305–334

Pradhan PK (2004) Population growth, migration and urbanization: environmental consequences in Kathmandu valley. In JM Unruh, S Krol, N Kliot (Eds), *Environmental change and its implications for population migration,* Kluwer Academic Publication, London

Quon S (1999) Planning for urban agriculture: a review of tools and strategies for urban planners. IDRC CFP Report series, Report 28, Ottawa

Saaty RW (1987) The analytic hierarchy process-what it is and how it is used. Mathem Model 9(3–5):161–176

Siti AS, Musaab A, Zawiyah MH, Laili FMI, Siti HS, Hashim M (2019) Urban farming activities in southeast Asia: a review and future research direction. MATEC Web Conf 2010. https://doi.org/10.1051/matecconf/2019I201826626602010ConBEE

Sharma P (2003) Urbanization and development. In: *Population monograph of Nepal, Vol. I.* Central Bureau of Statistics, CBS, National Planning Commission, Kathmandu, Nepal, pp 375–412

Shrestha S (2014) Urban agriculture: structure and dynamics of urban landscape in Kathmandu valley. Unpublished doctoral dissertation, Faculty of Humanities and Social Sciences, Tribhuvan University, Nepal

Thapa RB, Murayama Y (2012) Scenario based urban growth allocation in Kathmandu valley, Nepal. Lands Urban Plan 105(1–2):140–148

UNFPA (2017) Population growth rate (2001–2011) (Population situation analysis of Nepal: with respect to sustainable development). United Nations Population Fund, UNFPA, Nepal

UN HABITAT (2018) Metadata on SDGs indicator 11.3.1 indicator category: tier II. United Nations Human Settlements Program (UN HABITAT), Nairobi, Kenya

UN-HABITAT (2010) Nepal urban housing sector profile. United Nations Human Settlements Program, UN-HABITAT, Nairobi, Kenya

Veenhuizen R (2006) Cities farming for the future - urban agriculture for green and productive cities. RUAF Foundation, IDRC and IIRR

Vogel S (1996) Farmers' environmental attitudes and behavior: a case study for Austria. Environ Behav 28(5):591–613

Index

A

B

C

S. Anand et al. (eds.), *Sustainability in South Asian Cities*, Advances in Geographical and Environmental Sciences, https://doi.org/10.1007/978-981-97-7455-5

J

L

M

T

The manufacturer's authorised representative in the EU is Springer Nature Customer Service Centre GmbH, Europaplatz 3, 69115 Heidelberg, Germany. If you have any concerns regarding our products, please contact ProductSafety@springernature.com

Printed and bound by CPI Group (UK) Ltd, Croydon, CR0 4YY
01/10/2026
02225941-0016